OXIDATIVE STRESS AND SIGNAL TRANSDUCTION

OXIDATIVE
STRESS AND
SIGNAL
TRANSDUCTION

EDITED BY

HENRY JAY FORMAN
ENRIQUE CADENAS

UNIVERSITY OF SOUTHERN
CALIFORNIA

CHAPMAN & HALL

I(T)P® International Thomson Publishing
New York • Albany • Bonn • Boston • Cincinnati • Detroit • London • Madrid • Melbourne
Mexico City • Pacific Grove • Paris • San Francisco • Singapore • Tokyo • Toronto • Washington

Cover design: Saïd Sayrafiezadeh, *Emdash Inc.*

Library of Congress Cataloging-in-Publication Data

Oxidative stress and signal transduction / edited by Henry Jay Forman and Enrique Cadenas.
    p.   cm.
  Includes bibliographical references and index.
  ISBN 0-412-07681-0 (alk. paper)
    1. Active oxygen in the body. 2. Stress (Physiology). 3. Cellular signal transduction. 4. Second messengers (Biochemistry)
5. Genetic regulation. I. Forman, Henry Jay. II. Cadenas, Enrique.
  RB170.0959   1997
  616.9'8–dc21   96-47089
                                                                CIP

British Library Cataloguing in Publication Data available

To order this or any other Chapman & Hall book, please contact **International Thomson Publishing, 7625 Empire Drive, Florence, KY 41042.** Phone: (606) 525-6600 or 1-800-842-3636. Fax: (606) 525-7778. e-mail: order@chaphall.com.

For a complete listing of Chapman & Hall titles, send your request to **Chapman & Hall, Dept. BC, 115 Fifth Avenue, New York, NY 10003.**

Contents

Section II: The Connection of Oxidative Induced Changes in Second Messengers and Transcription Factors with Activation of Gene Expression

Section III: Gene Expression in Response to Oxidative Stress

Introduction

Henry Jay Forman and Enrique Cadenas

In the past few years there has been increasing recognition that the effects of oxidative stress are not all mediated through damage of cellular constituents. Indeed, evidence suggests that reactive oxygen species can alter cell function by acting as, mimicking, or affecting the intermediates (second messengers) in signal transduction. As examples, oxidative stress can signal the induction of antioxidant enzymes, mimic the effects of insulin, and activate protein phosphorylation. The effect of oxidants on signaling mechanisms probably account for the role of oxidative stress in inflammation, aging, and cancer among other biological effects and, not insignificantly, adaptation to the stresses that underlie those pathologies. The underlying mechanisms whereby oxidative stress causes effects on signal transduction are the subject of intense investigation that links two major areas of current research.

We have asked several of the most prominent investigators in this emerging field to report on the state of knowledge in their respective areas. The work is designed to include critical aspects of signal transduction related to oxidative stress and is organized in three levels, reflecting somehow the cellular sequence of events: (1) immediate cellular responses to oxidative stress and the production of second messengers, (2) the connection of oxidation induced changes in second messengers and transcription factors with activation of gene expression, and finally (3) gene expression in response to oxidative stress. This chapter serves as a general introduction to oxidative stress. We then raise a number of questions concerning the participation of oxidants in signaling, many of which are addressed in the individual chapters in detail.

Oxidative challenge to the cell comes from both endogenous generation of free radicals, such as superoxide (O_2^-) and hydrogen peroxide, and exogenous agents. These agents include environmental pollutants, such as NO_2 and O_3; redox cycling xenobiotic compounds, such as quinone-derived drugs; and O_2 itself when used therapeutically at high concentration. How O_2^- and H_2O_2 are generated from the reduction of a redox cycling compound is illustrated by the following where PQ^{2+} is the herbicide paraquat

$$PQ^{2+} + NADPH \xrightarrow[\text{reductase}]{\text{NADPH–P-450}} NADP^+ + PQ^+ \tag{1}$$

$$PQ^{\cdot +} + O_2 \longrightarrow O_2^{\cdot -} + PQ^{\cdot 2+} \tag{2}$$

$$2\ O_2^{\cdot -} \longrightarrow H_2O_2 \tag{3}$$

The last reaction is rapid at physiological pH but can be markedly accelerated by superoxide dismutase. In the presence of reduced iron, H_2O_2 can initiate lipid peroxidation, resulting in the production of lipid hydroperoxides (LOOH) and reactive α,β-unsaturated aldehydes. The endogenous production of $O_2^{\cdot -}$ and H_2O_2 is due primarily to nonenzymatic autooxidation of ubisemiquinone in the mito- chondrial electron transport chain analogous reaction to reaction 2, followed by dismutation (reaction 3).

While much research has naturally focused on the damaging effects of these oxidants, a realization has come about that H_2O_2 and products of nonenzymatic oxidation of cellular constituents, such as the F2-isoprostanes, may act as or mimic physiologic second messengers and thereby affect function. In the past few years, since the discovery that the essential hormone, endothelial-derived relaxing factor, is actually the free radical, nitric oxide, an even greater realization has occurred that free radicals are not necessarily "bad."

Nonetheless, one must consider that the range of concentration within which these reactive species exert specific effects on signaling is often narrow. Slightly higher concentration than that which may cause a maximal effect on signaling can enter the range in which cellular damage occurs. Thus, the antioxidant defenses of the cell can modulate the balance between effects on signaling and injury. Cellular defense against xenobiotic redox cycling compounds largely depends on conjugation of the agent and/or its metabolites with glutathione (GSH) or reduction or scavenging of the resultant free radicals and H_2O_2. While catalase can eliminate H_2O_2, most H_2O_2 and LOOH is metabolized by glutathione perox- idase:

$$2\ GSH + LOOH \xrightarrow[\text{peroxidase}]{\text{Glutathione}} LOH + GSSG \text{ (oxidized glutathione)} \tag{4}$$

GSSG is reduced by NADPH through the action of GSSG reductase:

$$NADPH + GSSG + H^+ \xrightarrow[\text{reductase}]{\text{Glutathione}} NADPH^+ + 2\ GSH \tag{5}$$

NADPH can be restored through pentose shunt activity.

GSSG may also interact with protein sulfhydryls to produce mixed disulfides:

$$\text{Protein-SH} + \text{GSSG} \Leftrightarrow \text{GSH} + \text{Protein-SSG} \qquad (6)$$

The reactive aldehydes can react with GSH to form conjugates by Michael addition or proteins by both Michael addition and Schiff base formation. The consequence of these reactions is that components of cells may be altered, while the cell must expend its reductive capacity to prevent cell injury. Faced with chronic oxidative stress, cells can respond period by increasing their production of antioxidant enzymes and the rate of de novo GSH synthesis. One of the most interesting aspects of the relationship between oxidative stress and signaling is in the regulation of the transcription and mRNA stabilization of the antioxidant enzymes and GSH-synthesizing enzymes by oxidative stress.

Some of the questions that next arise are How do oxidants act as signals? Do products of nonenzymatic oxidation, such as α,β-unsaturated aldehydes, or products of antioxidant defense, such as GSSG, act as second messengers? Do oxidants act by altering physiological signaling, such as inositol-trisphosphate-mediated Ca^{2+} release? Does H_2O_2 or NO· mediated release of Ca^{2+} come from the same source as in receptor mediated signaling? Are adenylate and guanylate cyclase, heterotrimeric G proteins, small G proteins, protein kinases, protein phosphatases, and phospholipases directly or indirectly altered by oxidants? Which genes are directly responsive to oxidative stress? How are the oxidant responsive genes in eukaryotes regulated? How are oxidant-responsive transcription factors, such as NF-κB, the antioxidant response element (ARE), and the STAT regulated? What are the consequences of oxidative signaling on cellular function? These questions are addressed in the chapters that follow.

The reader of this volume may find some inconsistencies among the chapters in the answers to these questions. While systems differences account for some actual variation in pathway of signaling, differences of opinion necessitate speculation at a time when there is incomplete understanding of the complexities of cell signaling and even more limited knowledge of how oxidants interact in these systems. Nonetheless, significant advances have been made in the past few years and the contributors have been among the pioneers. Through their efforts we have been able to provide a state-of-the-art view of this emerging area of research. Undoubtedly, there are other questions that have not yet been asked or even imagined. What we hope to accomplish in this book is to stimulate pursuit of the unimagined answers concerning the interrelationships between oxidative stress and signal transduction and the consequences of those interactions.

List of Contributors

Angelo Azzi*
Institut für Biochemie and Molekular
  Biologie
Univertät Bern
Bühlstasse 28
3000 Bern 9
Switzerland

Patrick A. Baeuerle*
Tularik Inc.
270 East Grand Avenue
South San Francisco, CA 94080

Manuel Bauer*
Medical Clinics
Department of Internal Medicine
Otfried-Muller-Str. 10
D-72076, Tubiugen
Germany

Joseph S. Beckman*
Department of Anesthesiology
University of Alabama at Birmingham
933 THT, 619 19th Street
Birmingham, AL 35233

Daniel Boscoboinik
Institut für Biochemie and Molekular
  Biologie
Univertät Bern
Bühlstasse 28
3000 Bern 9
Switzerland

Regina Brigelius-Flohé
Gesellschaft für Biotechnologische
  Forshung mbH
Mascheroder Weg 1
D-38124 Braunschweig
Germany

Roy H. Burdon*
Department of Bioscience
Biotechnology University of Strathclyde
31 Taylor Street
Glasgow G4 ONR
United Kingdom

Orazio Cantoni
Institut für Biochemie and Molekular
  Biologie
Univertät Bern
Bühlstasse 28
3000 Bern 9
Switzerland

* Senior author.

Brent H. Cochran
Department of Medicine
Tufts University School of Medicine
NEMC #257
750 Washington Street
Boston, MA 02111

Zhen-Hai Chen
Department of Cell and Neurobiology
University of Southern California
2025 Zonal Avenue
Los Angeles, CA 90033-4526

Gregory P. Downey*
Department of Medicine
University of Toronto
Respiratory Medicine
The Toronto Hospital
Eaton Wing North 10, Room 212
200 Elizabeth Street
Toronto, Ontario
Canada M5G 2C4

Alvaro G. Estévez
Department of Anesthesiology
University of Alabama at Birmingham
933 THT, 619 19th Street
Birmingham, AL 35233

Barry L. Fanburg*
Department of Medicine
Tufts University School of Medicine
NEMC #257
750 Washington Street
Boston, MA 02111

Leonard V. Favreau
Schering-Plough Research Institute
2015 Galloping Hill Road
Kenilworth, NJ 07033-0539

Agata Fazzio
Institut für Biochemie and Molekular
   Biologie
Univertät Bern
Bühlstasse 28
3000 Bern 9
Switzerland

Léa Fialkow
Department of Medicine
University of Toronto
Respiratory Medicine
The Toronto Hospital
Eaton Wing North 10, Room 212
200 Elizabeth Street
Toronto, Ontario
Canada M5G 2C4

Leopold Flohé*
Gesellschaft für Biotechnologische
   Forshung mbH
Mascheroder Weg 1
D-38124 Braunschweig
Germany

Rayadu Gopalakrishna*
Department of Cell and Neurobiology
University of Southern California
2025 Zonal Avenue
Los Angeles, CA 90033-4526

Usha Gundimeda
Department of Cell and Neurobiology
University of Southern California
2025 Zonal Avenue
Los Angeles, CA 90033-4526

Louis J. Ignarro*
Department of Molecular Pharmacology
University of California at Los Angeles
School of Medicine
Los Angeles, CA 90024

Anil K. Jaiswal*
Department of Pharmacology
Fox Chase Cancer Center
7701 Burholme Avenue
Philadelphia, PA 19111

Yvonne M. W. Janssen*
Department of Pathology
University of Vermont
Medical Alumni Building
Burlington, VT 05405-0068

L. Albert Jiminez
Department of Pathology
University of Vermont
Medical Alumni Building
Burlington, VT 05405-0068

Pius Joseph
Department of Pharmacology
Fox Chase Cancer Center
7701 Burholme Avenue
Philadelphia, PA 19111

Richard J. Kulmacz
Department of Internal Medicine
University of Texas Health Science
  Center
Houston, TX 77030

Renato Laffranchi
Division of Endocrinology and
  Metabolism
Department of Internal Medicine
University Hospital Zurich
Ramistr. 100
CH-8091 Zurich
Switzerland

William E. M. Lands*
Division of Basic Research
NIAAA, NIH
6000 Executive Boulevard, MSC 7003
Bethesda, MD 20892-7003

Dominique Marilley
Institut für Biochemie and Molekular
  Biologie
Univertät Bern
Bühlstasse 28
3000 Bern 9
Switzerland

Brooke T. Mossman
Department of Pathology
University of Vermont
Medical Alumni Building
Burlington, VT 05405-0068

Viswanathan Natarajan*
Department of Medicine
Indiana University School of Medicine
OPW425 Wishard Hospital
1001 West 10th Street
Indianapolis, IN 46202

Valerie O'Donnell
Department of Anesthesiology
University of Alabama at Birmingham
619 19th St. South, 946 THT
Birmingham, AL 35233

Nesrin Kartal Özer
Department of Biochemistry
Faculty of Medicine
Marmara University
81326 Haydarpasa Istanbul
Turkey

Cecil B. Pickett*
Scherling-Plough Research Institute
2015 Galloping Hill Road
Kenilworth, NJ 07033-0539

Rafael Radi*
Department of Biochemistry
University of Uruguay
Facultad de Medicina Universidad de la
  Republica
Montevideo
Uruguay

Venugopal Radjendirane
Department of Pharmacology
Fox Chase Cancer Center
7701 Burholme Avenue
Philadelphia, PA 19111

Julia Rashba-Step
Department of Molecular Pharmacology
  and Toxicology
University of Southern California
1985 Zonal Avenue
Los Angeles, CA 90033

Christoph Richter*
Laboratorium für Biuochemie
ETH Zentrum
Universitatstrasse 16
Zurich CH-8092
Switzerland

Stefan W. Ryter
Physical Carcinogenesis Unit
Swiss Institute for Experimental Cancer
  Research CH-1066
Epalinges
Switzerland

Gary L. Schieven*
Bristol-Myers Squibb Co.
3005 First Avenue
Seattle, WA 98121-1035

Klaus Schulze-Osthoff
Institute of Biochemistry
University of Frieburg
Hermann-Herber Str. 7
D79104, Frieburg
Germany

William M. Scribner
Department of Medicine
Indiana University School of Medicine
OPW425 Wishard Hospital
1001 West 10th Street
Indianapolis, IN 46202

Alex Sevantian*
Department of Molecular Pharmacology
  and Toxicology
University of Southern California
1985 Zonal Avenue
Los Angeles, CA 90033

Amy R. Simon
Department of Medicine
Tufts University School of Medicine
NEMC #257
750 Washington Street
Boston, MA 02111

Nathan Spear
Department of Anesthesiology
University of Alabama at Birmingham
933 THT, 619 19th Street
Birmingham, AL 35233

Stefan Spycher
Institut für Biochemie and Molekular
  Biologie
Univertät Bern
Bühlstasse 28
3000 Bern 9
Switzerland

Shirin Tabataba-Vakili
Institut für Biochemie and Molekular
  Biologie
Univertät Bern
Bühlstasse 28
3000 Bern 9
Switzerland

Andrea Tasinato
Institut für Biochemie and Molekular
  Biologie
Univertät Bern
Bühlstasse 28
3000 Bern 9
Switzerland

Cynthia R. Timblin
Department of Pathology
University of Vermont
Medical Alumni Building
Burlington, VT 05405-0068

Rex M. Tyrrell*
Physical Carcinogenesis Unit
Swiss Institute for Experimental Cancer
  Research CH-1066
Epalinges
Switzerland

Suryanarayana Vepa
Department of Medicine
Indiana University School of Medicine
OPW425 Wishard Hospital
1001 West 10th Street
Indianapolis, IN 46202

Markus Vogt
Institute of Biochemistry
University of Frieburg
Hermann-Herber Str. 7
D79104, Frieburg
Germany

Sebastian Wesselborg
Institute of Biochemistry
University of Frieburg
Hermann-Herber Str. 7
D79104, Frieburg
Germany

Edgar Wingender
Gesellschaft für Biotechnologische
    Forshung mbH
Mascheroder Weg 1
D-38124 Braunschweig
Germany

Christine L. Zanella
Department of Pathology
University of Vermont
Medical Alumni Building
Burlington, VT 05405-0068

SECTION I

Immediate Cellular Responses to Oxidative Stress and the Production of Second Messengers

1

Activation and Regulation of the Nitric Oxide–Cyclic GMP Signal Transduction Pathway by Oxidative Stress

Louis J. Ignarro

1.1 Introduction

Nitric oxide (NO) has come a long way in a relatively short period of time, with our knowledge rapidly progressing from the finding that NO is a major component in polluted air to the discovery that NO accounts for the pharmacological actions of nitroglycerin and to the discovery that NO is a widespread biological mediator in the mammalian species.[1,2] The initial observations that NO is a potent vascular smooth-muscle relaxant and inhibitor of platelet aggregation actually predates the discovery in 1980 of endothelium-dependent vasorelaxation and endothelium-derived relaxing factor (EDRF).[3] In 1979, together with the finding that NO is a potent vasorelaxant came the observation that cyclic GMP is responsible for mediating the intracellular action of NO.[4] Cyclic GMP was also found to mediate the antiplatelet effects of NO and chemically related NO donor agents.[5] The principal receptor for NO turned out to be a macromolecular complex with an extremely high binding affinity for NO, namely, heme-iron.[6] NO binds to the heme-iron prosthetic group of the cytosolic enzyme, guanylate cyclase, causing a conformational change at the porphyrin binding site leading to a reduced Km for Mg^{2+}-GTP and a marked increase in the reaction V_{max}, thereby resulting in a greatly increased rate of cyclic GMP production (Fig. 1.1).[7,8]

The exquisite potency of nitroglycerin as a vasorelaxant both in vitro and in vivo, together with the findings that NO itself is a potent vasorelaxant,[4] suggested that nitroglycerin or some closely related chemical agent might occur naturally in mammalian cells. On learning that NO is responsible for the vasodilator actions of nitroglycerin, other organic nitrate esters, organic nitrite esters, nitroprusside, and a series of N-nitrosoguanidines,[9] it became clear that NO itself might represent an endogenous nitrovasodilator. In 1986, this view was substantiated by the observations that the biochemical and pharmacological properties of NO were indistinguishable from those of EDRF, and in 1987, chemical evidence was

3

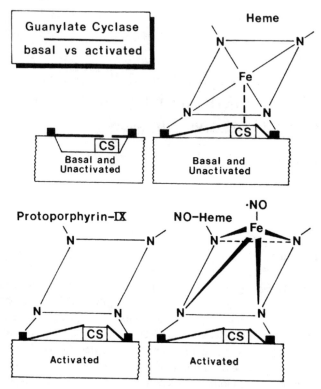

Figure 1.1. Schematic illustration of the mechanism of heme-dependent activation of cytosolic guanylate cyclase by NO. Cytosolic guanylate cyclase is a hemoprotein in which the heme, which can be detached by anionic exchange chromatography, functions as a prosthetic group to enable NO to bind to and activate the enzyme. The catalytic site (CS), which binds the enzyme substrate (Mg^{2+}-GTP), is envisioned to be buried within a specific region such that the substrate has only little access to the CS, and product formation is minimal. Heme binds to the porphyrin binding site adjacent to the CS as a five-coordinate complex and causes a conformational change, resulting in exposure of the CS, but the axial ligand between the heme-iron and a histidine residue in the protein creates sufficient steric hindrance to block access of substrate to the CS. NO binds to the heme-iron resulting in the formation of a five-coordinate complex, thereby implying disruption of the axial ligand with consequent elimination of the steric hindrance to substrate binding at CS. This results in increased binding affinity of Mg^{2+}-GTP and 100- to 200-fold increase in the V_{max} of the reaction.[6-8] Protoporphyrin-IX, which may be analogous to the nitrosyl-heme complex (no axial ligand), causes guanylate cyclase activation by mechanisms that are kinetically indistinguishable from that of the nitrosyl-heme complex.[6-8] Recent studies have confirmed and extended this original hypothesis.[134-136]

presented for the identification of EDRF as NO.[1,10] This pivotal discovery that NO must be generated by mammalian cells led to the elucidation of the relatively complex process by which the basic amino acid arginine is converted to NO.[11-15]

The realm of biological actions of NO soon expanded well beyond the original two actions discovered on vascular smooth muscle and platelets. NO plays additional important physiological roles as a neurotransmitter, autacoid, modulator of cell proliferation, and modulator of transcriptional expression; pathophysiological roles as a mediator of the inflammatory process, cytostasis against invading tumor cells, bacteria, and protozoa; and cytocidal actions under certain environmental conditions.[10,16-18] These varied functions of NO are dependent on and dictated by the sources of NO in different cell types. Three distinct isoforms of NO synthase (NOS) are responsible for the endogenous formation of NO, and each isoform has a special cellular distribution. NOS present in vascular endothelial cells (eNOS) is continuously present (constitutively present) in that no special induction or transcriptional event is required to signal its appearance, although various factors can cause up- or down-regulation. Basal catalytic activity is relatively low but is increased rapidly and markedly in a calmodulin-dependent manner by small increases in intracellular calcium concentration (discussed below). Neuronal NOS (nNOS) is present constitutively in neurons within the central and peripheral nervous systems, including the autonomic nervous system, and is responsible for biosynthesis of the NO that functions as a neurotransmitter. Like the constitutive eNOS, the nNOS isoform is calmodulin- and calcium-dependent. The third isoform of NOS has been termed immunologic NOS (iNOS) because its major role appears to be a pathophysiologic one in mediating inflammation and host-defense reactions. iNOS is catalytically active on synthesis of enzyme protein and is not dependent on added calmodulin or calcium.

The constitutive isoforms of NOS (eNOS and nNOS) generally are responsible for the calmodulin-dependent and calcium-stimulated production of relatively small quantities of NO that elicit biological actions immediately at its sites of synthesis. In view of the very short half-life of NO in biological fluids (less than a few seconds),[2] NO can function only locally as an autacoid or neurotransmitter. The inducible or immunologic isoform of NOS is responsible for the calcium and calmodulin-independent, high-output production of NO. Although the catalytic activity of all three NOS isoforms is similar, the quantity of NOS protein that can be induced in various cell types far exceeds, by several orders of magnitude, the quantity of eNOS or nNOS that is present constitutively. The high-output production of NO is largely responsible for the pathophysiological effects of NO, including cytotoxic actions on invading and host cells as well as profound influences on cell function such as pronounced and nearly irreversible vasodilation leading to states of hypotensive shock.[10]

The major mechanism for the termination of biological action of NO is its reaction with O_2 to form NO_2, which in aqueous solution results in the formation first of N_2O_3 and then NO_2^-.[19] NO_2^- is then further oxidized by a variety of endogenous oxidants to NO_3^-, which is relatively innocuous. In aqueous solution, the biological generation of NO does not result in any appreciable accumulation of NO_2. In a reaction that is faster than the superoxide dismutase (SOD)-catalyzed dismutation of O_2^- to H_2O_2, NO reacts with O_2^- to form $ONOO^-$,[20,21] which represents a much less biologically active metabolite of NO (discussed below). NO also reacts rapidly with oxyhemoproteins to undergo oxidation to NO_3^-. Other reactions also occur that would tend to diminish the biological actions of NO, such as interactions with protein-bound metals of low oxidation states and –SH groups under certain conditions. All of these reactions account for the very short biologic half-life of NO in mammalian tissues.

1.2 Biosynthesis and Termination of Action of Nitric Oxide

NO is synthesized from the semiessential basic amino acid, arginine, by a five-electron oxidative reaction catalyzed by NOS and yielding citrulline as the second product.[22-28] NOS requires NADPH and O_2 as cosubstrates and flavin adenine dinucleotide (FAD), flavin mononucleotide (FMN), tetrahydrobiopterin, and calmodulin as cofactors. NOS isoforms that are present more or less constitutively, such as eNOS (type III) and nNOS (type I), also require calcium, whereas iNOS (type II) does not. nNOS is a soluble or cytosolic protein (homodimer; $M_r = 160$ kDa), iNOS is also a cytosolic protein (homodimer; $M_r = 130$ kDa), whereas eNOS is a membrane-bound protein (homodimer; $M_r = 135$ kDa). There is a great deal of amino acid homology between certain regions in the NOS isoforms and NADPH cytochrome P-450 reductase. These observations are consistent with the closely similar catalytic mechanisms among all of these enzymes. The sequence of eNOS contains a myristoylation consensus sequence at the N terminus, which is absent in nNOS and eNOS.[29] Myristoylation may be responsible for membrane attachment of eNOS as no regions corresponding to classical transmembrane domains have been observed.

Calcium activates eNOS and nNOS by promoting the binding of calmodulin, which in turn facilitates electron transfer from the reductase domain to the oxygenase or heme-containing domain, thereby completing the electron flow cycle from NADPH to heme-iron and resulting in the reduction of heme-iron to the ferrous or Fe^{2+} state so that O_2 binding can occur.[30] iNOS already has calmodulin bound as a subunit and, therefore, no calcium is required to activate iNOS.[31] The enzyme-bound FAD and FMN allow NOS isoforms to be termed flavoproteins and, as such, function to channel or direct electron flow from NADPH, which binds to a site adjacent to the flavins, to heme-iron (Fig. 1.2). The catalytic role of heme-iron is supported by the observations that heme-iron ligands such as CO, NO, and CN^- can interfere with the oxidation of arginine to NO plus citrulline.[22,32] The role of tetrahydrobiopterin in the NOS catalytic mechanism is as yet unresolved. Studies indicate that this cofactor may not function stoichiomet-

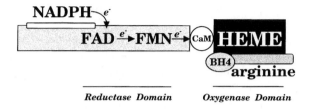

Structure-Function of NO Synthase

Figure 1.2. Schematic illustration of electron transfer in NO synthase. NO synthase can be visualized as a homodimer of the monomer depicted in the schematic. NO synthase contains a reductase domain and an oxygenase domain separated by a calmodulin (CaM) binding region. NADPH binds to a specific region in the reductase domain and transfers electrons (e^-) to the flavins (FAD and FMN), as shown. CaM functions to facilitate electron transfer from the reductase domain to the heme-iron contained in the oxygenase domain, thereby resulting in the reduction of Fe^{3+}-heme to Fe^{2+}-heme to enable the heme-iron to bind dioxygen. The arginine substrate binding site is in the oxygenase domain adjacent to the heme-iron. Heme-bound oxygen serves as a cosubstrate and is incorporated into both products (NO and citrulline) of the catalytic reaction. Tetrahydrobiopterin (BH4) is believed to bind at a site close to the heme-iron and arginine binding sites, although its precise function is still unclear.

rically in the reaction but rather may act either allosterically or to stabilize NOS during catalysis.[33] We have proposed that tetrahydrobiopterin functions to stabilize the catalytic activity of NOS by preventing the direct negative feedback action of NO on NOS.[32]

The only chemically well identified intermediate in the NOS reaction is N-hydroxyarginine, which also requires the same cosubstrates and cofactors for subsequent oxidation to NO plus citrulline.[15] Details pertaining to the conversion of N-hydroxyarginine to NO plus citrulline are less clear. We have proposed that HNO (nitroxyl [NO^-] in the nonprotonated state) is another intermediate that immediately precedes generation of NO.[34] As HNO is chemically labile and undergoes rapid oxidation to NO in the presence of numerous biological oxidants including O_2 and hemoproteins,[35] the formation of NO from HNO is not dependent on NOS. Therefore, NOS may catalyze a four-electron oxidation of arginine to HNO, with citrulline formed as the second product. HNO would then be rapidly oxidized to NO under physiological conditions.[34] Evidence for this hypothesis was provided by the observation that SOD catalyzed the oxidation of HNO to NO and markedly increased NO but not citrulline formation from arginine in the presence of catalytically active iNOS or nNOS.[34] Preliminary data indicate that HNO (trapped as the N_2O dimerization product) is generated from arginine by NOS under oxygen-limiting conditions. (J. M. Fukuto, A. J. Hobbs, and L. J. Ignarro, unpublished observations).

Regardless of whether HNO or NO is the terminal oxidation product of the

NOS reaction, it appears likely that NO is the biologically important species resulting from the NOS-catalyzed oxidation of arginine. HNO is chemically labile and rapidly oxidized to NO. Clearly, the biosynthesis of NO involves a relatively complex redox enzymatic pathway that is subject to modulation by factors that can alter the redox function of the cosubstrates and cofactors required for normal NOS catalytic activity. Depletion of O_2 (hypoxia or anoxia) or NADPH-derived reducing equivalents leads to diminished NO production. Calmodulin depletion or antagonism and chelation of calcium leads to diminished NO production by nNOS and eNOS. Interference with the electron transfer function of the flavins causes an interruption of NO production. Heme ligands or oxidants also interfere with NO production by virtue of their capacity to interfere with O_2 binding to the heme-Fe^{2+} prosthetic group. A good example is NO itself, which feeds back to inhibit NOS catalytic activity by directly binding to the heme-Fe^{2+} in NOS.[32,36] The binding affinity of the NOS heme-Fe^{2+} for NO is greater than that for O_2 or CO, and enzymatically generated NO can tie up and inhibit as much as 80 to 90% of the NOS during catalytic oxidation of arginine under certain conditions.[36] The inhibition of NO production by NO itself represents a classical example of negative feedback modulation of enzyme product formation.

1.3 Regulation of Nitric Oxide Production

The three known isoforms of NOS are regulated or have the potential to be regulated by similar as well as dissimilar mechanisms. For example, NO can act as a negative feedback modulator of all three isoforms.[32,37,38] The two isoforms of NOS that are generally present constitutively (nNOS and eNOS) are regulated by calcium in a calmodulin-dependent manner, as discussed above. iNOS must be newly synthesized to be present, and its catalytic activity is independent of calcium or fluctuations in calmodulin levels. Although nNOS is considered to be constitutively present, there are some indications for expressional regulation. For example, estrogen may up-regulate nNOS mRNA in various tissues[39] and axotomy may promote long-term up-regulation of nNOS gene expression in brain.[40] There are other examples as well.[41] Like nNOS, eNOS is considered to be constitutively present primarily in vascular endothelial cells but also in certain other cell types.[41] Nevertheless, the eNOS gene can undergo expressional regula-tion. Shear stress caused by blood flow through blood vessels causes an up-regulation of eNOS expression[42] via interaction with a shear stress-response element in the promoter of the eNOS gene.[43] Cytokines such as tumor necrosis factor alpha (TNF-α) down-regulate eNOS mRNA expression.[29,42,44] Estrogen appears also to increase eNOS mRNA expression and enzymatic activity.[41] Trans-forming growth factor $\beta 1$ (TGF-$\beta 1$) has been reported to produce an up-regulation of eNOS mRNA and protein.[45] Changes in oxygen tension have been shown to variably affect eNOS mRNA expression and protein synthesis.[41]

iNOS is regulated by induction, primarily at the level of transcription. Lipopolysaccharide (LPS) and a variety of cytokines (gamma interferon [IFN-γ], TNF-α, interleukin-1β [IL-1β]) can cause iNOS induction either alone or in various combinations and in a diversity of cell types and species.[41] In certain cell types, cyclic AMP-elevating agents, protein kinase C activating agents, and certain growth factors can promote iNOS induction. Inhibitory cytokines (IL-4, IL-10) and growth factors (TGF-β) interfere with the induction of iNOS in certain cell types. Tyrosine kinase inhibitors (genistein, herbimycin A, tyrphostin), inhibitors of NF-κB activation (dithiocarbamates) and glucocorticosteroids also prevent iNOS induction in many cell types. Finally, NO itself can interfere with the transcriptional expression of iNOS[46-49], as discussed below. Details pertaining to possible mechanisms of action of stimulators and inhibitors of iNOS induction have been recently reviewed.[41]

1.4 High-Output Production of Nitric Oxide and Role in Pathophysiology

The biosynthesis of relatively small quantities of NO by the two isoforms of NOS that are present constitutively in certain cell types (eNOS and nNOS) appears to be important for the maintenance of normal physiological function in the cardiovascular, nervous, pulmonary, immune, and inflammatory systems.[10,14] There is little or no experimental evidence that the small quantities of NO generated by eNOS or nNOS under normal physiologic conditions are cytotoxic either against host cells or invading cells. In special cases, nNOS can generate sufficient amounts of NO leading to target cell damage in the central nervous system (CNS).[14] On the other hand, the biosynthesis of relatively large quantities of NO by the inducible immunologic isoform of NOS appears to be associated universally with widespread cytotoxicity not only of invading cells and viral particles but also of certain host target cells. iNOS can be induced in a wide variety of cell types.[10,16-18] There is a great deal of experimental evidence that high-output production of NO results in cytostatic and/or cytocidal actions against numerous cell types including invading host tumor cells, bacteria, viruses, protozoa, other parasites as well as normal host cells such as neurons, oligodendrocytes, and other cell types.[16-18,50-53] The high-output production of NO can also lead to the development of pathophysiological conditions that are not necessarily attributed to cytotoxicity per se. A prime example is the profound and life-threatening hypotension that results from extensive iNOS induction in vascular tissue of patients exposed to endotoxin (LPS). So much NO is generated in or near the vascular smooth muscle to cause marked relaxation that the administration of vasoconstrictor drugs generally cannot overcome the marked NO-elicited vasodilation.[54-56]

The mechanism(s) by which NO exerts its antimicrobial and antiviral actions or cytotoxicity in general has been a hotly debated topic and is discussed in greater detail elsewhere in this volume. NO may interact directly with critical cellular components such as protein-bound iron or iron-sulfur clusters and cause their destruction, thereby leading to cytostatic or cytocidal actions.[57] A second mechanism that is popular today is the interaction between NO and O_2^- to generate ONOO$^-$, which has been proposed to be the principal cytotoxic species involving NO.[58] Under aerobic and slightly acidic conditions, NO or N_2O_3 could nitrosylate free −SH groups of proteins and thereby interfere with their function.[59,60] In addition, NO appears to be capable of causing release of zinc from zinc-sulfur cluster proteins, presumably via S-nitrosylation.[61,62] These and other less clearly defined mechanisms can lead to a diversity of cytotoxic effects including inhibition of cell proliferation, viral replication, protein synthesis, enzyme protein function, DNA repair, and binding of certain transcription factors to DNA.[53] NO can induce mutagenicity by mechanisms associated with nitrosative deamination.[63,64] Higher oxides of nitrogen such as N_2O_3 rather than NO itself may be primarily responsible for reactions involving deamination and cross-linking,[65] and such reactions can lead to apoptosis or programmed cell death.[53]

The issue of whether or not the very rapid reaction between NO and O_2^- to yield ONOO$^-$ actually represents a cytotoxic manifestation of NO is open to debate. Very high concentrations, and perhaps unattainable concentrations *in vivo,* of ONOO$^-$ appear to be cytotoxic. Lower concentrations of ONOO$^-$ do not appear to be cytotoxic and actually elicit some of the biological actions of NO, albeit with much less potency.[66-69] The possible pathological significance of the reaction between NO and O_2^- to yield OONOO$^-$ was first realized in 1990 by Beckman and coworkers.[70] Subsequent studies[71] revealed the biological oxidizing capacity of ONOO$^-$, although these reactions are relatively slow. Biological target molecules for ONOO$^-$ have been identified by monitoring formation of nitrotyrosines,[72] but alternate pathways leading to nitrotyrosine formation are possible, and there is no experimental evidence that nitrosation of tyrosine residues on proteins is cytotoxic. Moreover, the generation of appreciable quantities of ONOO$^-$ would demand the formation of relatively similar quantities of NO and O_2^- not only simultaneously but also exactly in the same location.[73] Thus, it is unlikely that ONOO$^-$ accounts for the cytotoxic actions of high-output production of NO in all or even most situations.

Perhaps the reaction between NO and O_2^- is merely a means of rapidly terminating the biological actions of NO. If ONOO$^-$ is only a weak or partial agonist of NO receptors, such as the heme-iron in guanylate cyclase or the iron in other proteins, then the localized generation of O_2^- may serve to terminate the action of NO. The observations that the rate of this reaction is very rapid and that ONOO$^-$ undergoes rapid spontaneous degradation to the biologically inactive NO_3^- support this view.

There may very well be alternative explanations for some of the cytotoxic actions of NO, particularly on cell proliferation. Although there is experimental evidence that relatively high concentrations of NO can promote both cytostatic and cytocidal effects, it is possible that iNOS induction causes cytostasis by mechanisms that are not directly related to any actions of NO. Recent studies in this laboratory on arginase induction and the regulation of intracellular levels of arginine in vascular cells and macrophages have led to the observation that when such cells are activated by LPS and certain cytokines, the result is marked iNOS induction with consequent formation of very large quantities not only of NO and citrulline but also NOHA (Fig. 1.3). Rat aortic endothelial cells challenged with LPS + IFN-γ + TNF-α + IL-1β causes the accumulation of approximately 500-μM NO (as NO$_2^-$ + NO$_3^-$) and citrulline as well as 170-μM NOHA in the cell culture medium. These findings indicate clearly that NOHA serves, not only as an intermediate in the arginine-NO pathway, but also as a final stable end product of the iNOS reaction pathway. The presence of such unexpectedly high concentrations of NOHA suggests that NOHA is a biological effector molecule in its own right.

If NOHA is a distinct, enzymatically generated effector molecule, what is its biological action? One likely function of NOHA is to inhibit arginase activity

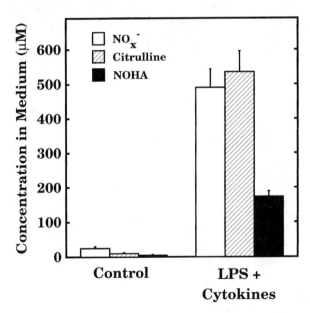

Figure 1.3. Synthesis and accumulation of *N*-hydroxyarginine by rat aortic endothelial cells treated with LPS and cytokines. Rat aortic endothelial cells (3 $\times$ 10^5/ml) were incubated in DMEN-HEPES medium containing 1% fetal bovine serum and 3 mM arginine in the absence (control) or presence of LPS (100 (μg/ml), IFN-γ (100 U/ml), TNF-α (1000 U/ml), and IL-1β (400 U/ml) for 24 h. Culture media were separated from cells and assayed for NO$_x^-$ (NO$_2^-$ + NO$_3^-$; chemiluminescence detection), citrulline (spectrophotometric analysis) and NOHA (HPLC) as described previously.[137] Data represent means ±SE of duplicate determinations from three to four separate experiments.

in cells that possess this enzyme. NOHA has been reported to be a potent inhibitor of hepatic and macrophage arginase activity,[74,75] possessing a K_i of about 35 μM.[74] A major function of arginase in cells that do not have a complete urea cycle is to generate ornithine from arginine, thereby providing the principal precursor for the biosynthesis of the polyamines required for cell proliferation. Interference with arginase activity and ornithine formation would presumably lead to inhibition of cell proliferation, just as inhibition of conversion of ornithine to putrescine by ornithine decarboxylase inhibitors causes termination of cell proliferation.[76] Thus, NOHA itself could elicit cytostasis and, in fact, there is one report to this effect.[77] Studies in this laboratory indicate that the K_i for NOHA as an inhibitor of arginase under assay conditions that more closely mimic intracellular conditions (pH 7.4 rather than pH 9.6; 0.5 to 1 mM rather than 20 mM arginine as substrate; use of non-heat/Mn^{2+}-activated arginase) is 10 to 12 μM, about threefold lower than the reported value of 35 μM.[74]

Based on these observations, we forward the hypothesis that NOHA generated from arginine by iNOS represents a distinct, biological effector-signaling molecule that functions as an endogenous arginase inhibitor to suppress cell proliferation. NOHA may turn out to be more important than NO as a cytostatic agent under physiological or pathophysiological conditions as low concentrations of NOHA sufficient to inhibit arginase (10 to 50 μM) can be readily achieved as a consequence of iNOS induction. In contrast, the concentrations of chemically labile NO required to cause cytostasis are much higher and may not be attainable in cells as a product of the arginine-NO pathway.

1.5 Oxidative-Stress-Mediated NF-κB Activation and Consequent Induction of Nitric Oxide Synthase

Nuclear factor-κB (NF-κB) is a higher eukaryotic transcription factor that plays a key role in the inducible expression of many genes that encode proteins involved in host defense and inflammation.[78,79] Soon after the realization that relatively high concentrations of NO serve as a mediator of the inflammatory process, we supposed that iNOS might be one of many proteins involved in promoting inflammation and host defense and that the induction of iNOS might therefore involve the actions of NF-κB. To test this hypothesis, we conducted experiments to ascertain whether an established inhibitor of NF-κB activation could interfere with high-output NO production by LPS-activated macrophages. Pyrrolidine dithiocarbamate (PDTC), a compound possessing antioxidant and metal-chelating properties, which was shown to inhibit NF-κB activation,[80] was shown to prevent NO production by rat alveolar macrophages activated by LPS + IFN-γ.[81] Inhibition of NO production by PDTC was attributed to interference with expression of the iNOS gene, as indicated by Northern blot analysis of mRNA (Fig. 1.4). Soon thereafter, similar results were reported for diethyldithiocarbamate,[82] and the

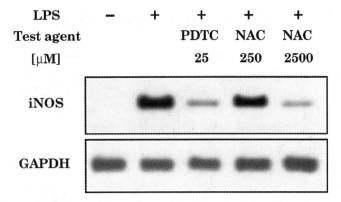

Figure 1.4. Inhibitory action of pyrrolidine dithiocarbamate and *N*-acetylcysteine on LPS-induced iNOS mRNA expression in murine macrophages. Raw (264.7) cell macrophages (10^6 cells/ml; 2×10^7 cells total) were incubated in MEM containing 10% fetal bovine serum and 1.2-mM arginine in the absence or presence of test agents, as indicated, for 6 h. Cells were harvested, total RNA was extracted, and Northern blot analysis was performed as described previously,[123,125] using GAPDH as the house keeping gene. PDTC (pyrrolidine dithiocarbamate) and NAC (*N*-acetylcysteine) were added to cell cultures 30 min prior to addition of 75 ng/ml LPS. Data illustrated are from a single experiment and are representative of a total of three separate experiments.

promoter region of the iNOS gene was shown to possess a consensus site for NF-κB.[83–85]

The findings that PDTC and other antioxidants could interfere with NF-κB activation suggested that oxidants were involved in bringing about NF-κB activation. At first, it was unclear whether O_2^- or H_2O_2 was the principal reactive oxygen species responsible for NF-κB activation, but more recent studies indicate that H_2O_2 is the primary species involved.[86] To better understand the influence of oxidants on NF-κB activation, it is necessary to briefly review the steps involved in NF-κB-mediated gene expression (Fig. 1.5). NF-κB is a heterodimer composed mostly frequently of the DNA-binding subunits p50 and p65 (also termed RelA), and resides in the cytoplasm as a latent form stabilized by an inhibitory protein termed IκB.[87] As a consequence of cell stimulation, the IκB dissociates from the NF-κB, thereby allowing the liberated NF-κB to translocate into the nucleus and bind to the appropriate regulatory elements in enhancer and promoter regions on target genes. This process results in gene activation. The rapid post-translational activation of NF-κB can be initiated by a plethora of pathophysiological stimuli including inflammatory cytokines, bacterial and viral infections, radiation, and certain oxidants.[88–90] Many, if not all, of these cell stimuli cause oxidative stress, as manifested by increased production of reactive oxygen species such as O_2^-, H_2O_2, ·OH, and related chemical species.[91,92] Experimental evidence from numerous sources suggests that there is a convergence of

NF-κB Activation & iNOS Induction

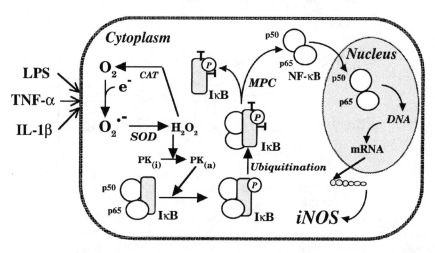

Figure 1.5. Schematic illustration of the activation of NF-κB by LPS and cytokines. LPS, TNF-α, and IL-1β interact with distinct receptors on the extracellular cell surface resulting in the intracellular generation of O_2^-, which undergoes dismutation to H_2O_2 catalyzed by SOD. By unknown mechanisms, H_2O_2 triggers a protein kinase to phosphorylate IκB-α (IκB), thereby tagging IκB-α as a substrate for subsequent ubiquitination. The ubiquitinated IκB-α then serves as a substrate for the proteasome (multicatalytic protease complex [MPC]), which degrades IκB-α and thereby promotes its detachment from NF-κB. Thus, the heterotrimeric complex of p50, p65, and IκB is split into degraded IκB plus NF-κB, a heterodimer of p50 and p65. NF-κB then translocates into the nucleus, binds to a consensus site on the promoter or enhancer region of the iNOS gene, and initiates transcription to generate iNOS mRNA. The mRNA translocates into the cytosol and binds to ribosomes, resulting in synthesis of iNOS protein.

distinct stimuli at the level of the cell membrane to generate a common reactive oxygen species, probably H_2O_2, that serves as the intracellular second messenger in mediating NF-κB activation.

The experimental evidence in support of oxidative stress as a key factor resulting in NF-κB activation has been reviewed fairly extensively.[80,86,92-95] Briefly, many chemically unrelated antioxidants interfere with NF-κB activation in response to diverse stimuli. Dithiocarbamates such as PDTC have been used frequently to inhibit NF-κB activation. Chemical agents that increase O_2^- and/ or H_2O_2 production in certain cells rapidly trigger NF-κB activation. Stable overexpression of catalase impairs, whereas overexpression of Cu/Zn SOD enhances NF-κB activation in response to certain cell stimuli. Whether or not oxidative stress represents the final common pathway for NF-κB activation remains to be determined. Although evidence exists that certain activators of NF-κB such as okadaic acid and calyculin A work by mechanisms that are seemingly independent of reactive oxygen species,[96] other investigators have shown that such mechanisms are in fact dependent on reactive oxygen species.[97]

How is the generation of reactive oxygen species like H_2O_2 coupled to the activation of NF-κB? Some recent evidence exists that an early phosphorylation step, which is requisite for NF-κB activation, is redox-controlled.[86] One of the earliest steps in NF-κB activation is phosphorylation of the IκB subunit of the heterotrimeric complex residing in the cytoplasm (Fig. 1.5). A protein kinase that has yet to be identified brings about the rapid phosphorylation of the IκB-α region of the IκB protein on serine 32 and serine 36 in response to diverse stimuli that also generate reactive oxygen species.[98,99] This phosphorylation of NF-κB-bound IκB-α tags the IκB-α subunit as a high-affinity substrate for a ubiquitin conjugating enzyme system.[100,101] Ubiquitination of IκB-α, in turn, results in covalently modified IκB-α that serves as a substrate for the proteasome, a calcium-independent and nonlysosomal multicatalytic proteinase complex.[97,100] Thus, phosphorylation-controlled ubiquitination signals a rapid and extensive degradation of IκB-α, resulting in the liberation of the NF-κB heterodimer. This process is transient and is terminated through delayed induction of IκB-α by NF-κB.[102,103] Proteasome inhibitors block NF-κB activation and lead to the intracellular accumulation of phosphorylated IκB-α still bound to NF-κB. Antioxidants block NF-κB activation with no resulting accumulation of phosphorylated IκB-α because the phosphorylation step itself is prevented. Exactly where antioxidants act to block NF-κB activation remains elusive. The answer to this question awaits more information on both the precise mechanisms by which reactive oxygen species are generated and how they initiate protein phosphorylation.

Despite our lack of understanding of the mechanisms associated with oxidative-stress-induced NF-κB activation, it is clear that the activation of this widespread transcription factor and the consequent expression of many genes encoding proteins involved in inflammation and host defense is signaled by reactive oxygen species. One of the many efferent pathways that is stimulated by NF-κB activation is the induction of iNOS, with its consequent high-output production of NO. Thus, together with the expression of other genes and induction of corresponding proteins, all initiated by NF-κB in response to oxidative stress, iNOS may play a key role in mammalian host defense and inflammation.

1.6 Regulation of High-Output Production of Nitric Oxide by Factors Affecting NF-κB Activation

In view of the convincing experimental evidence that NF-κB activation is the principal mechanism for the initiation of iNOS gene expression and high-output production of NO, it follows that inhibition of any rate-limiting step that results in interference with NF-κB activation will likewise result in interference with iNOS induction and NO production. Contrariwise, factors leading to enhanced expression of the iNOS gene would in turn lead to further high-output production of NO, provided none of the substrates and cofactors for iNOS become limiting in the cell. Even before it was recognized that NF-κB activation was requisite for iNOS induction, we found that certain antioxidants such as PDTC could markedly interfere with the induction of iNOS and production of NO by alveolar macrophages stimulated with LPS and IFN-γ.[81] Indeed, these observations first suggested that NF-κB might be closely linked to iNOS induction, as was clearly shown later by others.[83-85]

Despite the knowledge that the dithiocarbamates possess antioxidant properties, we have found that not all antioxidants are capable of interfering with iNOS gene expression. For example, up to 0.25-mM N-acetylcysteine elicited no appreciable influence on NF-κB activation or iNOS induction in rat or murine macrophages stimulated with LPS, whereas 10 to 25 μM PDTC produced 70 to 90% inhibition (Fig. 1.4). Much higher concentrations (2.5 mM) of N-acetylcysteine, however, did inhibit iNOS gene expression. The failure of N-acetylcysteine to elicit an effect at 10-fold higher concentrations than PDTC may be attributed to its relatively weak antioxidant properties, but PDTC is not generally considered to be a more potent or effective antioxidant than N-acetylcysteine. The inhibitory actions of PDTC on NF-κB activation and iNOS induction could be attributed to mechanisms other than its antioxidant effect. For example, PDTC has the capacity to chelate metals such as copper and iron.

Several distinct classes of chemical agents inhibit the high-output production of NO by interfering with the induction of iNOS. In each case, there is evidence that these chemical agents prevent either the activation of NF-κB or its binding to DNA in the nucleus of cells. Antioxidants were discussed above. Other agents include glucocorticosteroids, nonsteroidal anti-inflammatory agents, nonselective protease inhibitors, and proteasome inhibitors. Dexamethasone and other glucocorticosteroids inhibit iNOS induction in various cell types[104-107] and also interfere with the binding of NF-κB to specific promoter or enhancer regions in DNA of certain genes.[108-110] Dexamethasone may inhibit the transcriptional action of NF-κB also by promoting the induction of IκB-α inhibitory protein, thereby preventing the liberation and translocation of NF-κB to the nucleus.[111] Several chemically distinct nonsteroidal anti-inflammatory agents including salicylates, indomethacin, and ibuprofen have been shown to inhibit the expression of the iNOS gene in rat alveolar macrophages.[112] Aspirin and sodium salicylate were reported to interfere with NF-κB activation.[113] Unpublished observations from this laboratory indicate that this diverse class of nonsteroidal anti-inflammatory agents including diclofenac and naproxen also interfere with NF-κB activation, and this may account for their capacity to inhibit iNOS induction.[112] The mechanism by which these agents interfere with NF-κB activation is unknown, but one explanation may be an antioxidant effect at relatively high concentrations.

The observations that both steroidal and nonsteroidal anti-inflammatory agents interfere with high-output production of NO lend further support to the growing evidence that NO is an important mediator of the inflammatory process.[114-118] We have also obtained *in vivo* experimental evidence that nonsteroidal anti-inflammatory agents interfere with the high-output production of NO. Rats injected with LPS show a 15- to 20-fold increase in plasma levels of NO_3^- at 5 h, and this effect is attenuated by pretreatment of rats with nonsteroidal antiinflammatory agents (Fig. 1.6). Thus, like the glucocorticosteroids, the nonsteroidal anti-inflammatory agents appear to elicit their pharmacological effects by multiple mechanisms of action, including inhibition of high-output production of NO.

Various nonselective protease inhibitors such as $N\alpha$-p-tosyl-L-phenylalanine chloromethylketone (TPCK), $N\alpha$-p-tosyl-L-lysine chloromethylketone (TLCK),

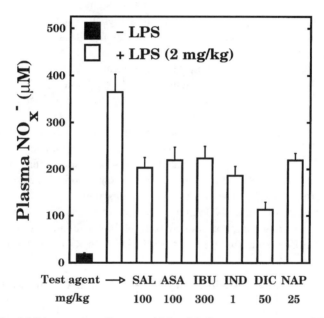

Figure 1.6. Inhibitory action of nonsteroidal anti-inflammatory agents on LPS-induced increases in plasma nitrate levels in rats. Rats (Sprague-Dawley, males, 250 g) were given one of the nonsteroidal anti-inflammatory drugs at the dose indicated by intraperitoneal injection 1 h prior to the intravenous injection of LPS. One group of rats was given saline intraperitoneally and intravenously (-LPS) and another group of rats was given saline intraperitoneally and LPS intravenously as shown. SAL (sodium salicylate), ASA (acetyl-salicylic acid), IBU (ibuprofen), IND (indomethacin), DIC (diclofenac), and NAP (naproxen) were administered in saline at a constant dosing volume of 10 ml/kg. Blood was drawn by cardiac puncture from anesthetized rats at 5 h after LPS administration. Plasma was analyzed for NO_x^- (NO_2^- + NO_3^-) by chemiluminescence detection. Data represent means ± SE of duplicate samples from six rats per group. Each of the drugs tested elicited a significant decrease ($p < 0.01$) in plasma NO_x^- as compared with the control (LPS without drug).

and phenylmethylsulfonyl fluoride were shown to inhibit NF-κB activation, probably by multiple mechanisms.[102,119] These protease inhibitors were also shown to inhibit iNOS-catalyzed NO production by macrophages activated with LPS and cytokines, although the mechanism was not understood.[120–122] More recently, we and others conducted studies with a series of nonselective protease inhibitors and revealed that these agents interfere with NO production by preventing the induction of iNOS.[123,124] In addition to the nonselective protease inhibitors, more selective calpain inhibitors and proteasome inhibitors were found to inhibit the induction of iNOS in macrophages.[123] These agents included calpain inhibitor 1 (α-N-acetyl-Leu-Leu-norleucinal), calpain inhibitor 2 (α-N-acetyl-Leu-Leu-methioninal), N-benzyloxycarbonyl-Ile-Glu-(O-t-Bu)-Ala-Leucinal (Z-IE[O-t-Bu]A-Leucinal), and N-benzyloxycarbonyl-Leu-Leucinal. We then provided evidence using electrophoretic mobility gel shift assays that these selective protease inhibitors markedly interfere with NF-κB activation.[125] This study revealed that the critical protease inhibited by these agents was most likely the multicatalytic proteinase complex known as the proteasome because the most potent inhibitor was Z-IE(O-t-Bu)A-Leucinal, a potent proteasome inhibitor.[126] The latter compound produced over 75% inhibition of NF-κB activation and iNOS induction at concentrations as low as 10 to 100 nM.

In view of the knowledge that NF-κB activation plays such a vital role in the induction of the iNOS gene, it stands to reason that any compound that interferes with NF-κB activation by any mechanism will likewise interfere with iNOS induction and high-output production of NO. As such, chemical agents are discovered to interrupt NF-κB activation by selective mechanisms, it is likely that they will also be found to inhibit iNOS induction. Such chemical agents might include specific protein kinase inhibitors, potent SOD inhibitors, ubiquitination inhibitors, inducers of IκB-α, and inhibitors of NF-κB binding to DNA in the nucleus.

1.7 Negative Feedback Modulation of NF-κB-Mediated Transcriptional Expression by Nitric Oxide

In as much as oxidative stress appears to be the major extracellular signal for NF-κB activation and the consequent initiation of numerous complementary pathways involved in host-defense and inflammation, chemical agents that prevent or attenuate oxidative stress should likewise interfere with NF-κB activation and host-defense processes. Certain antioxidants such as α-tocopherol (vitamin E) and perhaps other naturally occurring or endogenous antioxidants may function in this capacity. Since H_2O_2 may be the final reactive oxygen species that triggers the first step in NF-κB activation (perhaps phosphorylation of IκB-α) and because the immediate source of intracellular H_2O_2 is likely to be O_2^-, chemical agents that interfere with the production or dismutation of O_2^- are likely to interfere with NF-κB activation. One attractive candidate for an intracellular molecule that could interfere with the dismutation of O_2^- to H_2O_2 is NO. The chemical reaction between NO and O_2^- to generate $ONOO^-$ is extremely rapid, even more rapid than the SOD-catalyzed dismutation of O_2^- to H_2O_2.[21] This means that the concomitant production of NO and O_2^- in close vicinity could result in diminished H_2O_2 production as a consequence of increased $ONOO^-$ production.[73]

The above interaction between NO and O_2^- could serve to control NF-κB activation via a negative feedback mechanism involving multiple pathways. That is, an increase in oxidative stress (increased O_2^- production) would lead sequentially to NF-κB activation, iNOS induction, high-output production of NO, increased $ONOO^-$ and decreased H_2O_2 production, and ultimately decreased NF-κB activation. Such a potential negative feedback mechanism by NO could serve to regulate the production not only of NO itself but also all other proteins that rely on transcription factor NF-κB for expression.

Experimental evidence in support of the above hypothesis derives from both unpublished and published observations. We observed that the addition of NO in the form of NO donors such as *S*-nitroso-*N*-acetylpenicillamine (SNAP), DETA/NO, and nitroprusside to murine macrophages (RAW 264.7) at concentrations of 50 to 100 μM causes marked inhibition of LPS-induced iNOS gene expression, whereas addition of NOS inhibitors such as N^G-methylarginine (1 mM) and *S*-ethylthiourea (0.1 mM) causes upregulation of iNOS by enhancing gene expression (Fig. 1.7). Although the precise mechanisms involved need to be elucidated, one attractive possibility is that NO interferes with the formation of H_2O_2 from O_2^-, as discussed above. These observations are consistent with other unpublished findings from this laboratory that the inhibitory effect of NO on iNOS gene expression is mimicked and potentiated by catalase but diminished by SOD, agents that mimic SOD, catalase inhibitors, pyrogallol (generates O_2^- in aqueous solution containing oxygen), and H_2O_2 (0.2 to 1 mM). Preliminary experiments utilizing electrophoretic mobility gel shift assays to monitor NF-κB activation indicate that NO inhibits, whereas NOS inhibitors, pyrogallol, and H_2O_2 enhance NF-κB activation. Figure 1.8 illustrates data for DEA/NO, DEA/NO plus pyrogallol, and SNAP.

Several recently published studies are entirely consistent with the view that NO can interfere with NF-κB activation and thereby interfere with the expression of certain genes involved in host defense. One study revealed that an iNOS inhibitor caused an increase in NF-κB activation and iNOS expression in murine macrophages challenged with LPS or silica.[127] Another study showed that low concentrations (10 μM) of the NO donor agent DEA/NO enhanced, whereas higher concentrations (200 μM) diminished iNOS expression in murine macrophages challenged with LPS plus IFN-γ.[47] NO donor agents were found to suppress NF-κB activation and iNOS expression in human microglial cells treated with LPS plus TNF-α[48] and also in human vein endothelial cells challenged with TNF-

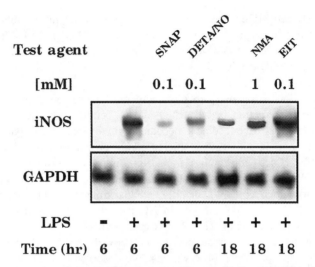

Figure 1.7. Differential actions of NO and NO synthase inhibitors on LPS-induced iNOS mRNA expression in murine macrophages. Raw (264.7) cell macrophages (10^6 cells/ml; 2×10^7 cells total) were incubated in MEM containing 10% fetal bovine serum and 1.2-mM arginine in the absence or presence of test agents as indicated for 6 h or 18 h, as shown. Cells were harvested, total RNA was extracted, and Northern blot analysis was performed, as described previously,[123,125] using GAPDH as the house keeping gene. SNAP (*S*-nitroso-*N*-acetylpenicillamine), DETA/NO,[138] NMA (*N*G-methylarginine), and EIT (*S*-ethylisothiourea) were added to cell cultures 5 min prior to addition of 75 ng/ml LPS. Data illustrated are from a single experiment and are representative of a total of three separate experiments.

α.[49] In the latter study, experimental evidence was provided showing that NO causes a cyclic-GMP-independent induction of IκB-α mRNA expression and stabilizes the IκB-α against rapid degradation, thereby leading to diminished NF-κB activation. The authors proposed this as a mechanism by which NO attenuates atherogenesis. Two reports by another group using human and Jurkat cell lines indicated that NO produced the complete opposite effect, namely, NF-κB activation,[128,129] but no explanation for this discrepancy is readily apparent.

The majority of the above observations from this and other laboratories suggest convincingly that NO plays a potentially important role in regulating transcriptional expression of numerous genes involved in host defense and inflammation by interfering with the post-translational activation of NF-κB. In as much as oxidative stress is the principal signal for NF-κB activation and NO possesses unique antioxidant properties as a scavenger of O_2^-, it is easy to understand why and how NO can function to attenuate NF-κB activation. The fact that NF-κB-mediated expression of many genes that encode proteins involved in host defense also results in iNOS gene expression and high-output production of NO implies

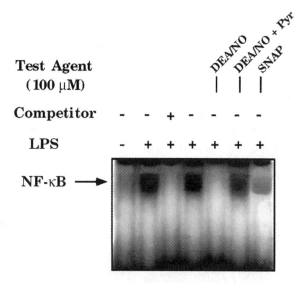

Test Agent
(100 μM)

Competitor

LPS

NF-κB ⟶

Figure 1.8. Inhibitory action of NO on NF-κB activation in murine macrophages treated with LPS. Raw (264.7) cell macrophages (10^6 cells/ml; 2×10^7 cells total) were incubated in MEM containing 10% fetal bovine serum and 1.2-mM arginine in the absence or presence of test agents as indicated for 1 h, as shown. Cells were harvested, nuclei were isolated and extracted, and electrophoretic mobility shift assays were conducted, as described previously.[125] Binding assays were performed using 20-μg nuclear extract protein and 4-μg competitive oligonucleotide, as indicated. DEA/NO,[138] DEA/NO + Pyr (pyrogallol), and SNAP (*S*-nitroso-*N*-acetylpenicillamine) were added to cell cultures 5 min prior to addition of 75 ng/ml LPS. Data illustrated are from a single experiment and are representative of a total of three separate experiments.

that NO plays a role as a negative feedback modulator of NF-κB-mediated host defense. Moreover, the widespread findings that NO is a mediator of acute and chronic inflammation enables one to draw the conclusion that NO can function as a negative feedback modulator of its own biosynthesis by attenuating iNOS gene expression. This is a distinct mechanism from the direct inhibitory influence of NO on the heme prosthetic group of NOS.[32,36–38,130–133]

Acknowledgments

The author wishes to thank all members of his laboratory for the inspiration and motivation to prepare this research article. The work reported here was supported in part by National Institutes of Health grants HL 40922 and HL 35014, the Laubisch Cardiovascular Fund, the Tobacco-Related Disease Research Program, and the American Heart Association.

References

1. Ignarro, L. J. 1989. Biological actions and properties of endothelium-derived nitric oxide formed and released from artery and vein. *Circ. Res.* **65**:1–21.

2. Ignarro, L. J. 1990. Biosynthesis and metabolism of endothelium-derived nitric oxide. *Annu. Rev. Pharmacol. Toxicol.* **30**:535–560.

3. Furchgott, R. F., and J. V. Zawadzki. 1980. The obligatory role of endothelial cells in the relaxation of arterial smooth muscle by acetylcholine. *Nature* **288**:373–376.

4. Gruetter, C. A., B. K. Barry, D. B. McNamara, D. Y. Gruetter, P. J. Kadowitz, and L. J. Ignarro. 1979. Relaxation of bovine coronary artery and activation of coronary arterial guanylate cyclase by nitric oxide, nitroprusside and a carcinogenic nitrosoamine. *J. Cyclic Nucl. Res.* **5**:211–224.

5. Mellion, B. T., L. J. Ignarro, E. H. Ohlstein, E. G. Pontecorvo, A. L. Hyman, and P. J. Kadowitz. 1981. Evidence for the inhibitory role of cyclic GMP in ADP-induced human platelet aggregation. *Blood* **57**:946–955.

6. Ignarro, L. J., K. S. Wood, and M. S. Wolin. 1984. Regulation of purified soluble guanylate cyclase by porphyrins and metalloporphyrins: a unifying concept. *Adv. Cyclic Nucl. Res.* **17**:267–274.

7. Wolin, M. S., K. S. Wood, and L. J. Ignarro. 1982. Guanylate cyclase from bovine lung: a kinetic analysis of the regulation of the purified soluble enzyme by protoporphyrin IX, heme and nitrosyl-heme. *J. Biol. Chem.* **257**:13312–13320.

8. Ignarro, L. J., B. Ballot, and K. S. Wood. 1984. Regulation of guanylate cyclase activity by porphyrins and metalloporphyrins. *J. Biol. Chem.* **259**:6201–6207.

9. Ignarro, L. J., H. Lippton, J. C. Edwards, W. H. Baricos, A. L. Hyman, P. J. Kadowitz, and C. A. Gruetter. 1981. Mechanism of vascular smooth muscle relaxation by organic nitrates, nitrites, nitroprusside and nitric oxide: evidence for the involvement of S-nitrosothiols as active intermediates. *J. Pharmacol. Exp. Ther.* **218**:739–749.

10. Moncada, S., R. M. J. Palmer, and E. A. Higgs. 1991. Nitric oxide: physiology, pathophysiology and pharmacology. *Pharmacol. Rev.* **43**:109–142.

11. Förstermann, U., H. W. Schmidt, J. S. Pollock, H. Sheng, J. A. Mitchell, T. D. Warner, M. Nakane, and F. Murad. 1991. Isoforms of nitric oxide synthase. Characterization and purification from different cell types. *Biochem. Pharmacol.* **42**:1849–1857.

12. Förstermann, U., E. I. Closs, J. S. Pollock, M. Nakane, P. Schwarz, I. Gath, and H. Kleinert. 1994. Nitric oxide synthase isozymes. Characterization, purification, molecular cloning and functions. *Hypertension* **23**:1121–1131.

13. Stuehr, D. J., and O. W. Griffith. 1992. Mammalian nitric oxide synthases. *Methods Enzymol. Rel. Areas Mol. Biol.* **65**:287–346.

14. Bredt, D. S., and S. H. Snyder. 1994. Nitric oxide: a physiologic messenger molecule. *Annu. Rev. Biochem.* **63**:175–195.

15. Marletta, M. A. 1993. Nitric oxide synthase structure and mechanism. *J. Biol. Chem.* **268**:12231–12234.

16. Xie, Q.-W., and C. Nathan. 1994. The high-output nitric oxide pathway: role and regulation. *J. Leukocyte Biol.* **56**:576–582.

17. Kröncke, K.-D., Fehsel, and V. Kolb-Bachofen. 1995. Inducible nitric oxide synthase and its product nitric oxide, a small molecule with complex biological activities. *Biol. Chem. Hoppe-Seyler* **376**:327–343.

18. Nussler, A. K., and T. R. Billiar. 1993. Inflammation, immunoregulation and inducible nitric oxide synthase. *J. Leukocyte Biol.* **54**:171–178.

19. Ignarro, L. J., J. M. Fukuto, J. M. Griscavage, N. E. Rogers, and R. E. Byrns. 1993. Oxidation of nitric oxide in aqueous solution to nitrite but not nitrate: comparison with enzymatically formed nitric oxide from L-arginine. *Proc. Natl. Acad. Sci. USA* **90**:8103–8107.

20. Blough, N. V., and O. C. Zafiriou. 1985. Reaction of superoxide with nitric oxide to form peroxynitrite in alkaline aqueous solution. *Inorg. Chem.* **24**:3504–3505.

21. Huie, R. E., and S. Padmaja. 1993. Reaction of NO with O_2^-. *Free Rad. Res. Commun.* **18**:195–199.

22. White, K. A., and M. A. Marletta. 1992. Nitric oxide synthase is a cytochrome P-450 type hemoprotein. *Biochemistry* **31**:6627–6631.

23. Stuehr, D. J., and M. Ikeda-Saito. 1992. Spectral characterization of brain and macrophage nitric oxide synthases. Cytochrome P-450-like hemoproteins that contain a flavin semiquinone radical. *J. Biol. Chem.* **267**:20547–20550.

24. Tayeh, M. A., and M. A. Marletta. 1989. Macrophage oxidation of L-arginine to nitric oxide, nitrite, and nitrate: tetrahydrobiopterin is required as a cofactor. *J. Biol. Chem.* **264**:19654–19658.

25. McMillan, K., D. S. Bredt, D. J. Hirsch, S. H. Snyder, J. E. Clark, and B. S. S. Masters. 1992. Cloned and expressed rat cerebellar nitric oxide synthase contains stoichiometric amounts of heme, which binds carbon monoxide. *Proc. Natl. Acad. Sci. USA* **89**:11141–11145.

26. Klatt, P., K. Schmidt, and B. Mayer. 1992. Brain nitric oxide synthase is a hemoprotein. *Biochem. J.* **288**:15–17.

27. Busconi, L., and T. Michel. 1993. Endothelial nitric oxide synthase. N-terminal myristoylation determines subcellular localization. *J. Biol. Chem.* **268**:8410–8413.

28. Abu-Soud, H. M., M. Loftus, and D. J. Stuehr. 1995. Subunit dissociation and unfolding of macrophage NO synthase: relationship between enzyme structure, prosthetic group binding and catalytic function. *Biochemistry* **34**:11167–11175.

29. Lamas, S., P. A. Marsden, G. K. Li, P. Tempst, and T. Michel. 1992. Endothelial nitric oxide synthase: molecular cloning and characterization of a distinct constitutive enzyme isoform. *Proc. Natl. Acad. Sci. USA* **89**:6348–6352.

30. Abu-Soud, H. M., and D. J. Stuehr. 1993. Nitric oxide synthases reveal a role for calmodulin in controlling electron transfer. *Proc. Natl. Acad. Sci. USA* **90**:10769–10772.

31. Cho, H. J., Q.-W. Xie, J. Calaycay, R. A. Mumford, K. M. Swederek, T. D. Lee, and C. Nathan. 1992. Calmodulin is a subunit of nitric oxide synthase from macrophages. *J. Exp. Med.* **176**:599–604.

32. Griscavage, J. M., J. M. Fukuto, Y. Komori, and L. J. Ignarro. 1994. Nitric oxide inhibits neuronal nitric oxide synthase by interacting with the heme prosthetic group. Role of tetrahydrobiopterin in modulating the inhibitory action of nitric oxide. *J. Biol. Chem.* **269**:21644–21649.

33. Giovanelli, J., K. L. Campos, and S. Kaufman. 1991. Tetrahydrobiopterin, a cofactor for rat cerebellar nitric oxide synthase, does not function as a reactant in the oxygenation of arginine. *Proc. Natl. Acad. Sci. USA* **88**:7091–7095.

34. Hobbs, A. J., J. M. Fukuto, and L. J. Ignarro. 1994. Formation of free nitric oxide from L-arginine by nitric oxide synthase: Direct enhancement of generation by superoxide dismutase. *Proc. Natl. Acad. Sci. USA* **91**:10992–10996.

35. Fukuto, J. M., A. J. Hobbs, and L. J. Ignarro. 1993. Conversion of nitroxyl (HNO) to nitric oxide (NO) in biological systems: The role of physiological oxidants and relevance to the biological activity of HNO. *Biochem. Biophys. Res. Commun.* **196**:707–713.

36. Abu-Soud, H. M., J. Wang, D. L. Rousseau, J. M. Fukuto, L. J. Ignarro, and D. J. Stuehr. 1995. Neuronal nitric oxide synthase self-inactivates by forming a ferrous-nitrosyl complex during aerobic catalysis. *J. Biol. Chem.* **270**:22997–23006.

37. Griscavage, J. M., N. E. Rogers, M. P. Sherman, and L. J. Ignarro. 1993. Inducible nitric oxide synthase from a rat alveolar macrophage cell line is inhibited by nitric oxide. *J. Immunol.* **151**:6329–6337.

38. Buga, G. M., J. M. Griscavage, and L. J. Ignarro. 1993. Negative feedback regulation of endothelial cell function by nitric oxide. *Circ. Res.* **73**:808–812.

39. Weiner, C. P., I. Lizasoain, S. A. Baylis, R. G. Knowles, I. G. Charles, and S. Moncada. 1994. Induction of calcium-dependent nitric oxide synthases by sex hormones. *Proc. Natl. Acad. Sci. USA* **91**:5212–5216.

40. Herdegen, T., S. Brecht, B. Mayer, J. Leah, W. Kummer, R. Bravo, and M. Zimmermann. 1993. Long lasting expression of JUN and KROX transcription factors and nitric oxide synthase in intrinsic neurons of the rat brain following axotomy. *J. Neurosci.* **13**:4130–4135.

41. Förstermann, U., and H. Kleinert. 1995. Nitric oxide synthase: expression and expressional control of the three isoforms. *Naunyn-Schmiedebergs Arch. Pharmacol.* **352**:351–364.

42. Nishida, K., J. P. Navas, A. A. Fisher, S. P. Dockery, M. Uematsu, R. M. Nerem, R. W. Alexander, and T. J. Murphy. 1992. Molecular cloning and characterization of the constitutive bovine aortic endothelial cell nitric oxide synthase. *J. Clin. Invest.* **90**:2092–2096.

43. Marsden, P. A., H. H. Heng, S. W. Scherer, R. J. Stewart, A. V. Hall, X. M. Shi, L. C. Tsui, and K. T. Schappert. 1993. Structure and chromosomal localization of the human constitutive endothelial nitric oxide synthase gene. *J. Biol. Chem.* **268**:17478–17488.

44. Förstermann, U., J. E. Kuk, M. Nakane, and J. S. Pollock. 1993. The expression of nitric oxide synthase is downregulated by tumor necrosis factor (TNF-alpha). *Naunyn-Schmiedebergs Arch. Pharmacol. [Suppl.]* **347**:R61.

45. Venema, R. C., K. Nishida, R. W. Alexander, and T. J. Murphy. 1994. Organization of the bovine gene encoding the endothelial nitric oxide synthase. *Biochim. Biophys. Acta* **1218**:413–420.

46. Park, S. K., H. L. Lin, and S. Murphy. 1994. Nitric oxide limits transcriptional induction of nitric oxide synthase in CNS glial cells. *Biochem. Biophys. Res. Commun.* **201**:762–768.

47. Sheffler, L. A., D. A. Wink, G. Melillo, and G. W. Cox. 1995. Exogenous nitric oxide regulates IFN-γ plus lipopolysaccharide-induced nitric oxide synthase expression in mouse macrophages. *J. Immunol.* **155**:886–894.

48. Colasanti, M., T. Persichini, M. Menegazzi, S. Mariotto, E. Giordano, C. M. Caldarera, V. Sogos, G. M. Lauro, and H. Suzuki. 1995. Induction of nitric oxide synthase mRNA expression. Suppression by exogenous nitric oxide. *J. Biol. Chem.* **270**:26731–26733.

49. Peng, H.-B., P. Libby, and J. K. Liao. 1995. Induction and stabilization of IκB-α by nitric oxide mediates inhibition of NF-κB. *J. Biol. Chem.* **270**:14214–14219.

50. Merrill, J. E., L. J. Ignarro, M. P. Sherman, J. Melinek, and T. E. Lane. 1993. Microglial cell cytotoxicity of oligodendrocytes is mediated through nitric oxide. *J. Immunol.* **151**:2132–2141.

51. Mitrovic, B., F. C. Martin, A. C. Charles, L. J. Ignarro, P. A. Anton, F. Shanahan, and J. E. Merrill. 1994. Neurotransmitters and cytokines in CNS pathology. *Prog. Brain Res.* **103**:319–330.

52. Mitrovic, B., L. J. Ignarro, H. V. Vinters, M. A. Akers, I. Schmid, C. Uittenbogaart, and J. E. Merrill. 1995. Nitric oxide induces necrotic but not apoptotic cell death in oligodendrocytes. *Neuroscience* **65**:531–539.

53. Nathan, C. 1992. Nitric oxide as a secretory product of mammalian cells. *FASEB J.* **6**:3051–3064.

54. Kilbourn, R. G., O. W. Griffith, and S. S. Gross. 1993. Pathogenetic mechanisms of septic shock. *N. Engl. J. Med.* **329**:1427–1428.

55. Kilbourn, R. G., and O. W. Griffith. 1992. Overproduction of nitric oxide in cytokine-mediated and septic shock. *J. Natl. Cancer Inst.* **84**:827–831.

56. Moncada, S., and A. Higgs. 1993. The L-arginine-nitric oxide pathway. *N. Engl. J. Med.* **329**:2002–2012.

57. Henry, Y., M. Lepoivre, J.-C. Drapier, C. Ducrocq, J. L. Boucher, and A. Guissani. 1993. EPR characterization of molecular targets for NO in mammalian cells and organelles. *FASEB J.* **7**:1124–1134.

58. Beckman, J. S., and J. P. Crow. 1993. Pathological implications of nitric oxide, superoxide and peroxynitrite formation. *Biochem. Soc. Trans.* **21**:330–334.

59. Stamler, J. S., D. I. Simon, J. A. Osborne, M. E. Mullins, O. Jaraki, T. Michel, D. J. Singel, and J. Loscalzo. 1992. S-Nitrosylation of proteins with nitric oxide: synthesis and characterization of biologically active compounds. *Proc. Natl. Acad. Sci. USA* **89**:444–448.

60. Gopalkrishna, R., Z. H. Chen, and U. Gundimeda. 1993. Nitric oxide and nitric oxide-generating agents induce a reversible activation of protein kinase C activity and phorbol ester binding. *J. Biol. Chem.* **268**:27180–27185.

61. Kröncke, K. D., K. Fehsel, T. Schmidt, F. T. Zenke, I. Dasting, J. R. Wesener, H. Bettermann, K. D. Breunig, and V. Kolb-Bachofen. 1994. Nitric oxide destroys zinc-sulfur inducing zinc release from metallothionein and inhibition of the zinc finger type yeast transcription activator LAC9. *Biochem. Biophys. Res. Commun.* **200**:1105–1110.

62. Wink, D. A., and J. Laval. 1994. The Fpg protein, a DNA repair enzyme, is inhibited by the biomediator nitric oxide *in vitro* and *in vivo. Carcinogenesis* **15**:2125–2129.

63. Wink, D. A., K. S. Kasprzak, C. M. Maragos, R. K. Elespuru, M. Misra, T. M. Dunams, T. A. Cebula, W. H. Koch, A. W. Andrews, J. S. Allen, and L. K. Keefer. 1991. DNA deaminating ability and genotoxicity of nitric oxide and its progenitors. *Science* **254**:1001–1003.

64. Routledge, M. N., D. A. Wink, L. K. Keefer, and A. Dipple. 1994. DNA sequence changes induced by two nitric oxide donor drugs in the supF assay. *Chem. Res. Toxicol.* **7**:628–632.

65. Tannenbaum, S. R., J. S. Wishnok, T. deRojas-Walker, S. Tamir, and H. Ji. 1994. DNA damage and cytotoxicity from nitric oxide, p. 24. *In Abstract Book of the First International Conference on Biochemistry and Molecular Biology of Nitric Oxide.* UCLA Sunset Village, Los Angeles, CA.

66. Villa, L. M., E. Sales, V. M. Darley-Usmar, M. W. Radomski, and S. Moncada. 1994. Peroxynitrite induces both vasodilation and impaired vascular relaxation in the isolated perfused rat heart. *Proc. Natl. Acad. Sci. USA* **91**:12383–12387.

67. Moro, M. A., V. M. Darley-Usmar, D. A. Goodwin, N. G. Read, R. Zamora-Pino, M. Feelisch, M. W. Radomski, and S. Moncada. 1994. Paradoxical fate and biological action of peroxynitrite on human platelets. *Proc. Natl. Acad. Sci. USA* **91**:6702–6706.

68. Wu, M., K. A. Pritchard, P. M. Kaminski, R. P. Fayngersh, T. H. Hintze, and M. S. Wolin. 1994. Involvement of nitric oxide and nitrosothiols in relaxation of pulmonary arteries to peroxynitrite. *Am. J. Physiol.* **266**:H2108–H2113.

69. Liu, S., J. S. Beckman, and D. D. Ku. 1994. Peroxynitrite, a product of superoxide and nitric oxide, produces coronary vasorelaxation in dogs. *J. Pharmacol. Exp. Ther.* **268**:1114–1121.

70. Beckman, J. S., T. W. Beckman, J. Chen, P. A. Marshall, and B. A. Freeman. 1990. Apparent hydroxyl radical production by peroxynitrite: implications for endothelial injury from nitric oxide and superoxide. *Proc. Natl. Acad. Sci. USA* **87**:1620–1624.

71. Radi, R., J. S. Beckman, K. M. Bush, and B. A. Freeman. 1991. Peroxynitrite oxidation of sulfhydryls. The cytotoxic potential of superoxide and nitric oxide. *J. Biol. Chem.* **266**:4244–4250.

72. Ischiropoulos, H., L. Zhu, J. Chen, M. Tsai, J. C. Martin, C. D. Smith, and J. S. Beckman. 1992. Peroxynitrite-mediated tyrosine nitration catalyzed by superoxide dismutase. *Arch. Biochem. Biophys.* **298**:431–437.

73. Miles, A. M., D. S. Bohle, P. A. Glassbrenner, B. Hansert, D. A. Wink, and M. B. Grisham. 1996. Modulation of superoxide-dependent oxidation and hydroxylation reactions by nitric oxide. *J. Biol. Chem.* **271**:40–47.

74. Daghigh, F., J. M. Fukuto, and D. E. Ash. 1994. Inhibition of rat liver arginase by an intermediate in NO biosynthesis, N^G-hydroxy-L-arginine: implications for the regulation of nitric oxide biosynthesis by arginase. *Biochem. Biophys. Res. Commun.* **202**:174–180.

75. Boucher, J. L., J. Custot, S. Vadon, M. Delaforge, M. Lepoivre, J. P. Tenu, A. Yapo, and D. Mansuy. 1994. N^W-Hydroxy-L-arginine, an intermediate in the L-arginine to nitric oxide pathway, is a strong inhibitor of liver and macrophage arginase. *Biochem. Biophys. Res. Commun.* **203**:1614–1621.

76. Morgan, D. M. L. 1994. Polyamines, arginine and nitric oxide. *Biochem. Soc. Trans.* **22**:879–883.

77. Chénais, B., A. Yapo, M. Lepoivre, and J.-P. Tenu. 1993. N^W-Hydroxy-L-arginine, a reactional intermediate in nitric oxide biosynthesis, induces cytostasis in human and murine tumor cells. *Biochem. Biophys. Res. Commun.* **196**:1558–1565.

78. Lenardo, M. J., and D. Baltimore. 1989. NF-κB: A pleiotropic mediator of inducible and tissue-specific gene control. *Cell* **58**:227–229.

79. Baeuerle, P. A. 1991. The inducible transcription factor NF-κB: regulation by distinct protein subunits. *Biochim. Biophys. Acta* **1072**:63–80.

80. Schreck, R., B. Meier, D. N. Männel, W. Dröge, and P. A. Baeuerle. 1992. Dithiocarbamates as potent inhibitors of nuclear factor κB activation in intact cells. *J. Exp. Med.* **175**:1181–1194.

81. Sherman, M. P., E. E. Aeberhard, V. Z. Wong, J. M. Griscavage, and L. J. Ignarro. 1993. Pyrrolidine dithiocarbamate inhibits induction of nitric oxide synthase activity in rat alveolar macrophages. *Biochem. Biophys. Res. Commun.* **191**:1301–1308.

82. Mülsch, B., R. Schray-Utz, P. I. Mordvintcev, S. Hauschild, and R. Busse. 1993. Diethyldithiocarbamate inhibits induction of macrophage NO synthase. *FEBS Lett.* **321**:215–218.

83. Lowenstein, C. J., E. W. Alley, P. Raval, A. M. Snowman, S. H. Snyder, S. W. Russell, and W. J. Murphy. 1993. Macrophage nitric oxide synthase gene: two upstream regions mediate induction by interferon γ and lipopolysaccharide. *Proc. Natl. Acad. Sci. USA* **90**:9730–9734.

84. Xie, Q.-W., R. Whisnant, and C. Nathan. 1993. Promoter of the mouse gene encoding calcium-independent nitric oxide synthase confers inducibility by interferon-γ and bacterial lipopolysaccharide. *J. Exp. Med.* **177**:1779–1784.

85. Xie, Q.-W., Y. Kashiwabara, and C. Nathan. 1994. Role of transcription factor NF-κB/rel in induction of nitric oxide synthase. *J. Biol. Chem.* **269**:4705–4708.

86. Schmidt, K. N., P. Amstad, P. Cerutti, and P. A. Baeuerle. 1995. The roles of hydrogen peroxide and superoxide as messengers in the activation of transcription factor NF-κB. *Chem. & Biol.* **2**:13–22.

87. Beg, A. A., T. S. Finco, P. V. Nantermet, and A. S. Baldwin. 1993. Tumor necrosis factor and interleukin-1 lead to phosphorylation and loss of IκB-α: a mechanism for NF-κB activation. *Mol. Cell. Biol.* **13**:3301–3310.

88. Liou, H. C., and D. Baltimore. 1993. Regulation of the NF-κB/rel transcription factor and IκB inhibitor system. *Curr. Opin. Cell. Biol.* **5**:477–487.

89. Baeuerle, P. A., and T. Henkel. 1994. Function and activation of NF-κB in the immune system. *Annu. Rev. Immunol.* **12**:141–179.

90. Siebenlist, U., G. Franzoso, and K. Brown. 1994. Structure, regulation and function of NF-κB. *Annu. Rev. Cell Biol.* **10**:405–455.

91. Schreck, R., and P. A. Baeuerle. 1991. A role of oxygen radicals as second messengers. *Trends Cell Biol.* **1**:39–42.

92. Schreck, R., K. Albermann, and P. A. Baeuerle. 1992. NF-κB: an oxidative stress-responsive transcription factor of eukaryotic cells. *Free Rad. Res. Commun.* **17**:221–237.

93. Meyer, M., R. Schreck, J. M. Müller, and P. A. Baeuerle. 1993. Redox control of gene expression by eukaryotic transcription factors NF-κB, AP-1 and SRF, pp. 217–235. *In* C. Pasquier (ed.), *Oxidative Stress on Cell Activation and Viral Infection*. Birkhäuser Verlag AG, Basel.

94. Schreck, R., P. Rieber, and P. A. Baeuerle. 1991. Reactive oxygen intermediates as apparently widely used messengers in the activation of the NF-κB transcription factor and HIV-1. *EMBO J.* **10**:2247–2258.

95. Meyer, M., R. Schreck, and P. A. Baeuerle. 1993. H_2O_2 and antioxidants have opposite effects on activation of NF-κB and AP-1 in intact cells: AP-1 as secondary antioxidant-responsive factor. *EMBO J.* **12**:2005–2015.

96. Suzuki, Y. J., M. Mizuno, and L. Packer. 1994. Signal transduction for nuclear factor-κB activation. Proposed location of antioxidant-inhibitable step. *J. Immunol.* **153**:5008–5015.

97. Schmidt, K. N., E. B. Traenckner, B. Meier, and P. A. Baeuerle. 1995. Induction of oxidative stress by okadaic acid is required for activation of transcription factor NF-κB. *J. Biol. Chem.* **270**:27136–27142.

98. Brown, K., S. Gerstberger, L. Carlson, G. Franzoso, and U. Siebenlist. 1995. Control of IκB-α proteolysis by site-specific, signal-induced phosphorylation. *Science* **267**:1485–1488.

99. Traenckner, E. B.-M., H. L. Pahl, T. Henkel, K. N. Schmidt, S. Wilk, and P. A. Baeuerle. 1995. Phosphorylation of human IκB-α on serines 32 and 36 controls IκB-α proteolysis and NF-κB activation in response to diverse stimuli. *EMBO J.* **14**:2876–2883.

100. Traenckner, E. B.-M., S. Wilk, and P. A. Baeuerle. 1994. A proteasome inhibitor prevents activation of NF-κB and stabilizes a newly phosphorylated form of IκB-α that is still bound to NF-κB. *EMBO J.* **13**:5433–5441.

101. Palombella, V. J., O. J. Rando, A. L. Goldberg, and T. Maniatis. 1994. The ubiquitin-proteasome pathway is required for processing the NF-κB1 precursor protein and the activation of NF-κB. 1994. *Cell* **78**:773–785.

102. Henkel, T., T. Machleidt, I. Alkalay, M. Kronke, Y. Ben-Neriah, and P. A. Baeuerle. 1993. Rapid proteolysis of IκB-α in response to phorbol ester, cytokines and lipopolysaccharide is a necessary step in the activation of NF-κB. *Nature* **365**:182–185.

103. Beg, A. A., and A. S. Baldwin, Jr. 1993. The IκB proteins: multifunctional regulators of Rel/NF-κB transcription factors. *Genes Dev.* **7**:2064–2070.

104. Radomski, M. W., R. M. Palmer, and S. Moncada. 1990. Glucocorticoids inhibit the expression of an inducible, but not the constitutive, nitric oxide synthase in vascular endothelial cells. *Proc. Natl. Acad. Sci. USA* **87**:10043–10047.

105. Di Rosa, M., M. Radomski, R. Carnuccio, and S. Moncada. 1990. Glucocorticoids inhibit the induction of nitric oxide synthase in macrophages. *Biochem. Biophys. Res. Commun.* **172**:1246–1252.

106. Geller, D. A., A. K. Nussler, S. M. Di, C. J. Lowenstein, R. A. Shapiro, S. C. Wang, R. L. Simmons, and T. R. Billiar. 1993. Cytokines, endotoxin and glucocorticoids regulate the expression of inducible nitric oxide synthase in hepatocytes. *Proc. Natl. Acad. Sci. USA* **90**:522–526.

107. Gilbert, R. S., and H. R. Herschman. 1993. Macrophage nitric oxide synthase is a glucocorticoid-inhibitable primary response gene in 3T3 cells. *J. Cell. Physiol.* **157**:128–132.

108. Mukaida, N., M. Morita, Y. Ishikawa, N. Rice, S. Okamoto, T. Kasahara, and K. Matsushima. 1994. Novel mechanism of glucocorticoid-mediated gene repression: nuclear factor-κB is target for glucocorticoid-mediated interleukin 8 gene repression. *J. Biol. Chem.* **269**:13289–13295.

109. Ray, A., and K. E. Prefontaine. 1994. Physical association and functional antagonism between the p65 subunit of transcription factor NF-κB and the glucocorticoid receptor. *Proc. Natl. Acad. Sci. USA* **91**:752–756.

110. Kleinert, H., C. Euchenhofer, I. Ihrig-Biedert, and U. Förstermann. 1996. Glucocorticoids inhibit the induction of nitric oxide synthase II by down-regulating cytokine-induced activity of transcription factor nuclear factor-κB. *Mol. Pharmacol.* **49**:15–21.

111. Auphan, N., J. A. DiDonato, C. Rosette, A. Helmberg, and M. Karin. 1995. Immunosuppression by glucocorticoids: inhibition of NF-κB activity through induction of IκB synthesis. *Science* **270**:286–290.

112. Aeberhard, E. E., S. A. Henderson, N. S. Arabolos, J. M. Griscavage, F. E. Castro, C. T. Barrett, and L. J. Ignarro. 1995. Nonsteroidal anti-inflammatory drugs inhibit expression of the inducible nitric oxide synthase gene. *Biochem. Biophys. Res. Commun.* **208**:1053–1059.

113. Kopp, E., and S. Ghosh. 1994. Inhibition of NF-κB by sodium salicylate and aspirin. *Science* **265**:956–959.

114. Vane, J. R., J. A. Mitchell, I. Appleton, A. Tomlinson, D. Bishop-Bailey, J. Croxtall, and D. A. Willoughby. 1994. Inducible isoforms of cyclooxygenase and nitric oxide synthase in inflammation. *Proc. Natl. Acad. Sci. USA* **91**:2046–2050.

115. M. Stefanovic-Racic, K. Meyers, C. Meschter, J. W. Coffey, R. A. Hoffman, and C. H. Evans. 1994. N-Monomethyl-arginine, an inhibitor of nitric oxide synthase, suppresses the development of adjuvant arthritis in rats. *Arthritis Rheum.* **37**:1062–1069.

116. Weinberg, J. B., D. L. Granger, D. S. Pisetsky, M. F. Seldin, M. A. Misukonis, S. N. Mason, A. M. Pippen, P. Ruiz, E. R. Wood, and G. S. Gilkeson. 1994. The

role of nitric oxide in the pathogenesis of spontaneous murine autoimmune disease: increased nitric oxide production and nitric oxide synthase expression in MRL-1pr/1pr mice, and reduction of spontaneous glomerulonephritis and arthritis by orally administered N^G-monomethyl-L-arginine. *J. Exp. Med.* **179**:651–660.

117. Ialenti, A., S. Moncada, and M. Di Rosa. 1993. Modulation of adjuvant arthritis by endogenous nitric oxide. *Br. J. Pharmacol.* **110**:701–705.

118. McCartney-Francis, M., J. B. Allen, D. E. Mizell, J. E. Albina, Q.-W. Xie, C. F. Nathan, and S. M. Wahl. 1993. Suppression of arthritis by an inhibitor of nitric oxide synthase. *J. Exp. Med.* **178**:749–754.

119. Lin, Y.-C., K. Brown, and U. Siebenlist. 1995. Activation of NF-κB requires proteolysis of the inhibitor IκB-α: signal-induced phosphorylation of IκB-α alone does not release active NF-κB. *Proc. Natl. Acad. Sci. USA* **92**:552–556.

120. Griscavage, J. M., M. P. Sherman, and L. J. Ignarro. 1992. Inhibition of nitric oxide and superoxide formation in activated rat alveolar macrophages by serine protease and NO synthase inhibitors. *FASEB J.* **6**:A1205.

121. Kilbourn, R., and G. Lopez-Berestein. 1990. Protease inhibitors block the macrophage-mediated inhibition of tumor cell mitochondrial respiration. *J. Immunol.* **144**:1042–1049.

122. Jorens, P. G., F. J. Van Overveld, H. Bult, P. A. Vermiere, and A. G. Herman. 1991. L-Arginine-dependent production of nitrogen oxides by rat pulmonary macrophages. *Eur. J. Pharmacol.* **200**:205–208.

123. Griscavage, J. M., S. Wilk, and L. J. Ignarro. 1995. Serine and cysteine proteinase inhibitors prevent nitric oxide production by activated macrophages by interfering with transcription of the inducible NO synthase gene. *Biochem. Biophys. Res. Commun.* **215**:721–729.

124. Kim, H., H. S. Lee, K. T. Chang, T. H. Ko, K. J. Baek, and N. S. Kwon. 1995. Chloromethyl ketones block induction of nitric oxide synthase in murine macrophages by preventing activation of nuclear factor-kappa B. *J. Immunol.* **154**:4741–4748.

125. Griscavage, J. M., S. Wilk, and L. J. Ignarro. In press. Inhibitors of the proteasome pathway interfere with induction of nitric oxide synthase in macrophages by blocking activation of nuclear factor-kappa B. *Proc. Natl. Acad. Sci. USA.* **93**:3308–3312.

126. Figueiredo-Pereira, M. E., K. A. Berg, and S. Wilk. 1994. A new inhibitor of the chymotrypsin-like activity of the multicatalytic proteinase complex (20S proteasome) induces accumulation of ubiquitin-protein conjugates in a neuronal cell. *J. Neurochem.* **63**:1578–1581.

127. Chen, F., D. C. Kuhn, S.-C. Sun, L. J. Gaydos, and L. M. Demers. 1995. Dependence and reversal of nitric oxide production on NF-κB in silica and lipopolysaccharide-induced macrophages. *Biochem. Biophys. Res. Commun.* **214**:839–846.

128. Lander, H. M., P. Sehajpal, D. M. Levine, and A. Novogrodsky. 1993. Activation of human peripheral blood mononuclear cells by nitric oxide-generating compounds. *J. Immunol.* **150**:1509–1516.

129. Lander, H. M., J. S. Ogiste, S. F. Pearce, R. Levi, and A. Novogrodsky. 1995. Nitric oxide-stimulated guanine nucleotide exchange on p21ras. *J. Biol. Chem.* **270**:7017–7020.

130. Rogers, N. E., and L. J. Ignarro. 1992. Constitutive nitric oxide synthase from cerebellum is reversibly inhibited by nitric oxide formed by L-arginine. *Biochem. Biophys. Res. Commun.* **189**:242–249.

131. Assreuy, I. Q., F. Q. Cunha, F. Y. Liew, and S. Moncada. 1993. Feedback inhibition of nitric oxide synthase activity by nitric oxide. *Br. J. Pharmacol.* **108**:833–837.

132. Rengasamy, A., and R. A. Johns. 1993. Regulation of nitric oxide synthase by nitric oxide. *Mol. Pharmacol.* **44**:124–128.

133. Wang, J., D. I. Rousseau, H. M. Abu-Soud, and D. J. Stuehr. 1994. Heme coordination of NO in NO synthase. *Proc. Natl. Acad. Sci. USA* **91**:10512–10516.

134. Burstyn, J. N., A. E. Yu, E. A. Dierks, B. K. Hawkins, and J. H. Dawson. 1995. Studies of the heme coordination and ligand binding properties of soluble guanylyl cyclase (sGC): characterization of Fe^{II}sGC and Fe^{II}sGC(CO) by electronic absorption and magnetic circular dichroism spectroscopies and failure of CO to activate the enzyme. *Biochemistry* **34**:5896–5903.

135. Stone, J. R., and M. A. Marletta. 1996. Spectral and kinetic studies on the activation of soluble guanylate cyclase by nitric oxide. *Biochemistry* **35**:1093–1099.

136. Stone, J. R., and M. A. Marletta. 1995. Heme stoichiometry of heterodimeric soluble guanylate cyclase. *Biochemistry* **34**:14668–14674.

137. Buga, G. M., R. Singh, S. Pervin, N. E. Rogers, D. A. Schmitz, C. P. Jenkinson, S. D. Cederbaum, and L. J. Ignarro. In press. Arginase activity in endothelial cells: inhibition by N^G-hydroxyarginine during high-output nitric oxide production. *Am. J. Physiol.* **271**:H1988–H1988.

138. Maragos, C. M., D. Morley, D. A. Wink, T. M. Dunams, J. E. Saavedra, A. Hoffman, A. A. Bove, L. Issac, J. A. Hrabie, and L. K. Keefer. 1991. Complexes of ·NO with nucleophiles as agents for the controlled biological release of nitric oxide. *J. Med. Chem.* **34**:3242–3247.

2

Peroxynitrite and Cell Signaling

Nathan Spear, Alvaro G. Estévez, Rafael Radi, and Joseph S. Beckman

2.1 Introduction

Highly reactive oxidants such as the hydroxyl radical are generally assumed to cause random cellular damage, which presumably results in cellular necrosis. However, signal transduction pathways in cells respond to many extracellular signals including oxidative stress. Oxidants have now been implicated in activating apoptosis, suggesting that cells respond to a certain threshold of oxidative stress by activating cell death pathways. Coordination of the complex and interlocking signaling pathways in eukaryotic cells is essential for proliferation, differentiation, and cell death. Biological oxidants can both activate and inactivate signaling pathways involving tyrosine kinases, transcription factors,[1,2] oxidation of key cellular thiols,[3] and calcium homeostasis. Depending on the interplay between intracellular signaling pathways, moderate exposure to certain oxidants may either promote cell proliferation, induce apoptosis, or cause frank necrosis. Oxidants produced during inflammation are critical for defense against foreign invasion, but are also likely to interact with growth factors and cytokines in a more elusive and poorly understood role in the initiation of wound healing and repair of tissue.

Peroxynitrite ($ONOO^-$) is a major oxidative product from activated macrophages and neutrophils resulting from the diffusion-limited reaction of superoxide and nitric oxide. It has been implicated in the oxidative damage detected in many pathological conditions.[4-7] Peroxynitrite is a powerful oxidant whose complex reactivity favors the selective attack and modification of key chemical moieties involved in cellular signal transduction. Peroxynitrite is particularly reactive with zinc-fingers, important in transcription factors and kinase regulatory regions and with iron/sulfur centers important for cell respiration. Peroxynitrite readily modifies solvent-accessible tyrosines in proteins to irreversibly form 3-nitrotyrosine, which cannot be phosphorylated. Nitration may weakly mimic phosphorylation and activate some enzymes. A further consequence of nitration is the pronounced effect on the assembly of cytoskeletal proteins, which disrupts the translocation of cellular signaling events between cellular compartments.

The chemistry of peroxynitrite is complex due to its ability to participate in one- and two-electron oxidations and reactions with catalysts *in vivo*, but provides insight into what types of cellular signaling processes might be affected by peroxynitrite. The ability of peroxynitrite to affect cellular signaling is complicated because peroxynitrite can produce nitrosothiols and nitric oxide in low yield (typically less than 1% of added peroxynitrite). However, minuscule concentrations of nitric oxide are highly effective at activating guanylate cyclase, which allows the secondary products derived from peroxynitrite to have profound physiological effects that are not directly related to oxidative attack by peroxynitrite. For example, we have found that the addition of submicromolar concentrations of peroxynitrite to vascular rings causes relaxation through a cyclic GMP (cGMP)-dependent process.[8]

2.1 Chemistry of Nitric Oxide

The production of peroxynitrite is dependent on the generation of nitric oxide, a small hydrophobic molecule that is an important local signaling molecule between cells. Nitric oxide synthase is a calmodulin dependent enzyme that oxidizes 1 mol of L-arginine by five electrons to 1 mol each of L-citruline and nitric oxide, using 1.5 mol of NADPH and 2 mol of oxygen as cosubstrates.[9] Nitric oxide has a total of 15 electrons, which requires nitric oxide to have an unpaired electron in its highest occupied molecular orbital. Although nitric oxide is a free radical, it is not inherently reactive. In fact, its reactivity is similar in chemical terms to that of molecular oxygen, which itself has two unpaired electrons. Both molecular oxygen and nitric oxide become far more damaging by secondary reactions, which produce more potent oxidants.

The principal biological target of nitric oxide in its signal transduction pathways is the ferrous heme-iron of guanylate cyclase.[10] Cyclic GMP produced by guanylate cyclase activates protein kinase G, which prosphorylates other signal mediators and initiates a cascade of events.[11,12] When these events occur in the vascular endothelium, the result is the relaxation of smooth muscle cells in the arterial wall.[13,14] In the brain, nitric oxide production is involved in neuronal plasticity and important for some forms of learning.[15,16]

The excessive production of nitric oxide has been linked to toxicities associated with injuries such as ischemia/reperfusion and in the tumoricidal activity of activated macrophages.[17–20] Neuronal toxicity mediated by excessive *N*-methyl-D-aspartate (NMDA) receptor activation has also been linked to nitric oxide production.[21,22] What is commonly overlooked is whether nitric oxide itself or an oxidant derived from nitric oxide is the proximal cause of toxicity. Comprehending the chemical nature of nitric oxide is key in understanding the mechanisms by which nitric oxide acts as an intercellular messenger or as a mediator of toxicity.

2.3 Nitrogen Dioxide Chemistry

Much of the apparent toxicity and reactivity resulting from the addition of milli-molar concentrations of nitric oxide to cultured cells or model systems *in vitro* results from the reaction of molecular oxygen with nitric oxide to form nitrogen dioxide (NO_2). The overall reaction is

$$2NO + O_2 \rightarrow 2NO_2 \qquad (2.1)$$
$$\text{rate} = 2K_3[NO]^2[O_2] \quad K = 2 \times 10^6 \text{ M}^{-2}\text{s}^{-1}$$

the rate of which depends on the square of nitric oxide concentration. Nitrogen dioxide is a strong one-electron oxidant with a reduction potential of +0.99 V. At fairly low concentrations, nitrogen dioxide can initiate lipid peroxidation by abstracting a hydrogen atom from the bis-allylic methylene carbon of polyunsaturated fatty acids.[23] In addition, NO_2 can oxidize thiols and amino acids and it can slowly nitrate aromatic compounds such as tyrosine.[24,25] As shown in Figure 2.1 A, nitration of tyrosine requires that one nitrogen dioxide oxidizes tyrosine

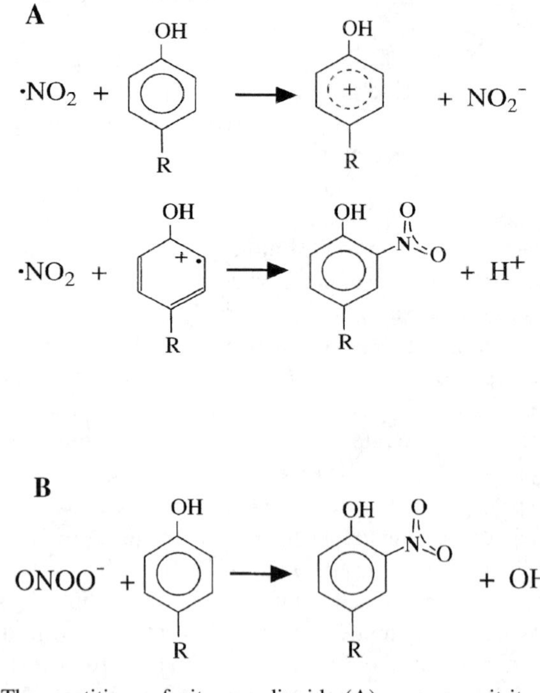

Figure 2.1. The reactitions of nitrogen dioxide (A) or peroxynitrite (B) with tyrosine results in the formation of 3-nitrotyrosine.

to the tyrosyl radical, which reacts with another nitrogen dioxide radical to form nitrotyrosine.[25] Nitrogen dioxide may also react with nitric oxide to form N_2O_3, another strong oxidant and nitrosylating species.[26] Although many *in vitro* studies have demonstrated that nitric oxide reacts with oxygen to form nitrogen dioxide, the concentration of nitric oxide needed for this reaction to proceed at a significant rate *in vivo* cannot generally compete with diffusion to red blood cells and removal by reaction with oxyhemoglobin. The production of nitric oxide requires two oxygens that must diffuse from red blood cells *in vivo* to the nitric oxide synthase. Therefore, nitric oxide must also be able to diffuse back to a red blood cell where it will be rapidly consumed by reaction with oxyhemoglobin.

2.4 Peroxynitrite Chemistry

Because nitric oxide is no more reactive than oxygen, much of its toxicity is mediated by forming more potent oxidants. Superoxide (O_2^-) rapidly reacts with nitric oxide to form the potent oxidant peroxynitrite. Superoxide is a radical anion formed when molecular oxygen is reduced by one electron. Because it is negatively charged, superoxide is not usually a strong oxidant at physiologic pH. However, superoxide is effective at oxidizing iron/sulfur centers where the positive charge of the iron facilitates reaction with the negatively charged superoxide. For example, aconitase is readily inactivated by superoxide. Production of superoxide has been linked to many pathological and toxicological conditions,[27,28] although the toxic effects have most often been attributed to the production of the hydroxyl radical through redox cycling with transition metals.

Interest in the possible consequences of generating O_2^- in biological systems began when McCord and Fridovich discovered the enzymatic activity of erythrocuprein.[29] The enzyme was renamed superoxide dismutase (SOD) because it catalyzes the dismutation of superoxide to oxygen and hydrogen peroxide. The importance of SOD is demonstrated by the fact that it is present in many tissues at concentrations as high as 10 μM. The enzyme contains a copper atom at its active site, which reacts with superoxide at an exceptionally fast rate (rate constant = 2×10^9 M^{-1} s^{-1}).[30] Given the high intracellular concentrations of SOD, the reaction between nitric oxide and superoxide is the only other biological substrate produced in high enough concentrations to compete for superoxide *in vivo*. However, this is only favorable when nitric oxide concentrations approach those of SOD in the cell, as happens in many pathological circumstances. The amount of peroxynitrite formed depends on the competition for superoxide between SOD and nitric oxide.

$$Cu^{++}\text{-SOD} + O_2^- \rightarrow Cu^+\text{-SOD} + O_2 \quad k_d = 2 \times 10^9 \ M^{-1} \ s^{-1} \qquad (2.2)$$
$$\text{reaction rate} = K_d[Cu^{++}\text{-SOD}][O_2^-]$$

$$NO + O_2^- \rightarrow ONOO^- \quad K_f = 6.7 \times 10^9 \ M^{-1} \ s^{-1} \qquad (2.3)$$
$$\text{reaction rate} = K_f[NO][O_2^-]$$

The normal range of NO needed for activation of guanylate cyclase is in the range of 10^{-9} to 10^{-7} M.[15,31] If the concentration of superoxide is constant, the rate of reaction between superoxide and SOD would be about 40-fold faster than with nitric oxide under normal conditions. However, the concentration of nitric oxide has been measured as high as 4×10^{-6} M during ischemia/reperfusion of rat brain.[13,32]. At these concentrations, more than half the superoxide produced will react with nitric oxide to form peroxynitrite rather than with SOD.

A major factor contributing to the toxicity of peroxynitrite is its stability as an anion. At alkaline pH, peroxynitrite anion exists in the *cis* conformation and is relatively stable. Peroxynitrite can be stored for several months in 0.3-M NaOH at −80°C. This stability results from the delocalization of the negative charge over all four atoms when peroxynitrite is in the *cis* conformation.

Peroxynitrous acid has a pK_a of 6.8. At physiological pH, about 20% of peroxynitrite is protonated (ONOOH), which allows isomerization to either the *cis* or *trans* conformations. Many of the compounds that react with peroxynitrite are oxidized by one electron and have zero-order reaction rates with respect to the target molecule. Reactions involving the hydroxyl radical often have similar characteristics to those seen with peroxynitrite, and ONOOH can decompose through a pathway that behaves like HO· and NO_2.[33] However, kinetic and thermodynamic considerations indicate that complete homolytic fission of the O—O bond is less likely to occur than the formation of an excited intermediate with the reactivity of hydroxyl radical.[34] Only about 30% of peroxynitrous acid reacts via the hydroxyl radical like pathway, the rest decomposes to nitric acid.

Although *cis*-peroxynitrite is stable it does react directly with sulfhydryl compounds.[35] The second-order rate constant for the reaction between peroxynitrite and thiols is in the range of 1.2 to 6×10^3 M^{-1} s^{-1}, which is about a thousand times faster than the rate of thiol oxidation by H_2O_2.

The fact that *cis*-peroxynitrite reacts directly with thiols could make it a relatively selective oxidant of cellular sulfhydryls *in vivo*. The major low molecular weight thiol is glutathione (GSH).[36] Several cellular processes are dependent on GSH, such as maintenance of protein sulfhydryls and the removal of hydroperoxides and electrophiles. Depletion of GSH has been shown to be toxic to cells and tissues and is thought to be due to impairment of these critical cellular processes.[37] In addition to the reaction with GSH, peroxynitrite may also react with proteins, including many involved in signal transduction, that contain cysteine residues essential for their proper function.

One type of reaction that is fairly unique to peroxynitrite is the nitration of phenolic compounds like tyrosine (Fig. 2.1B). The nitrated tyrosine product most often observed is 3-nitrotyrosine. Using antibodies raised against nitrated keyhole limpet hemocyanin, we and others have seen specific staining of nitrated proteins in a number of diseases and pathologies, including the endothelium and plaques in atherosclerotic vessels, synovial joints in rheumatoid arthritis, septic lung and heart, ischemic brain, and spinal chords of patients with amyotrophic lateral sclerosis.[38-40] The presence of nitrated proteins in these different pathological conditions indirectly implicates a role for peroxynitrite in the toxicities associated with those conditions. While other species such as NO_2 are capable of nitrating aromatic compounds, we have found that nitrotyrosine is a reasonable marker for peroxynitrite because the production of nitrotyrosine by NO_2 would be severely limited by the rate at which NO_2 is formed under physiological conditions. In addition, NO_2-mediated nitration requires two NO_2 molecules, which is also a limiting factor because there are many other cellular components that would compete with aromatic compounds reacting with NO_2 (Fig. 2.1A). Finally, NO_2 is far more reactive with thiols and ascorbate than tyrosine, which diminishes the relevance of nitration by this pathway greatly *in vivo*.

The addition of the nitro group to tyrosine lowers the pK_a of the hydroxyl group to approximately 7.5, which introduces a negative charge on the protein (Fig. 2.2). Thus, nitration could alter the conformation of the protein and alter its ability to recognize and bind to other proteins or it might inhibit its enzymatic activity. The negatively charged tyrosine residue might even be recognized as being phosphorylated and result in the unregulated activation of signaling pathways. Nitration of purified proteins by tetranitromethane has long been used to probe the function of tyrosines in enzyme function and shown to disrupt phosphorylation of receptors as well as to affect the polymerization of actin.

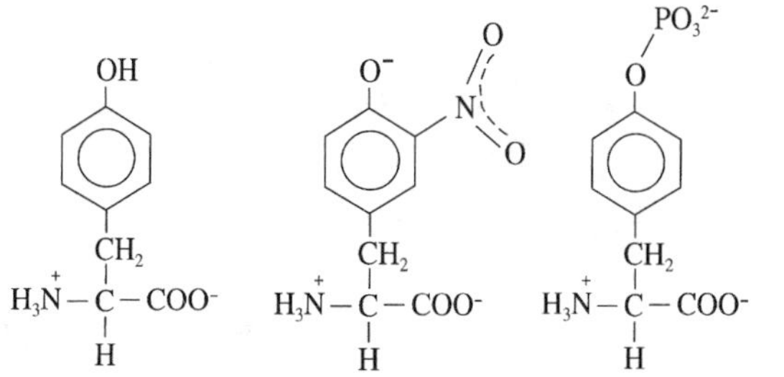

Figure 2.2. Chemical structures of tyrosine, 3-nitrotyrosine, and phosphotyrosine.

Peroxynitrite alone is able to nitrate tyrosine; however, in the presence of catalysts common to biological systems the rate of nitration can be enhanced many fold. Our group has demonstrated that bicarbonate enhances peroxynitrite-dependent luminol chemiluminescence and killing of bacteria. At that time, we proposed that the enhancement of luminol chemiluminescence by bicarbonate was dependent on the formation of a peroxynitrite-bicarbonate intermediate[41,42]. Recent work done by Lymar and Hurst[43] has established that peroxynitrite reacts with CO_2 through the mechanism shown below

$$ONOO^- + CO_2 \rightarrow ONOOCO_2^- \qquad (2.4)$$

One of the specific effects of carbon dioxide on peroxynitrite reactivity is the enhancement of tyrosine nitration.[44-46] The $ONOOCO_2^-$ adduct is very unstable with a half-life of approximately 3 ms, and reacts with tyrosine with a rate constant of greater than 2×10^5 M^{-1} s^{-1}.[45] Because the concentration of CO_2 is high in many physiological fluids, formation of $ONOOCO_2^-$ may be important when considering the consequences of generating $ONOO^-$ *in vivo*.

Transition metals are another important component of biological systems that can catalyze the reaction between tyrosine and peroxynitrite.[47] Low-molecular-weight complexes such as ferric-EDTA can react with peroxynitrite to form a nitronium ion like species as shown in Figure 2.3A. The first step in the reaction

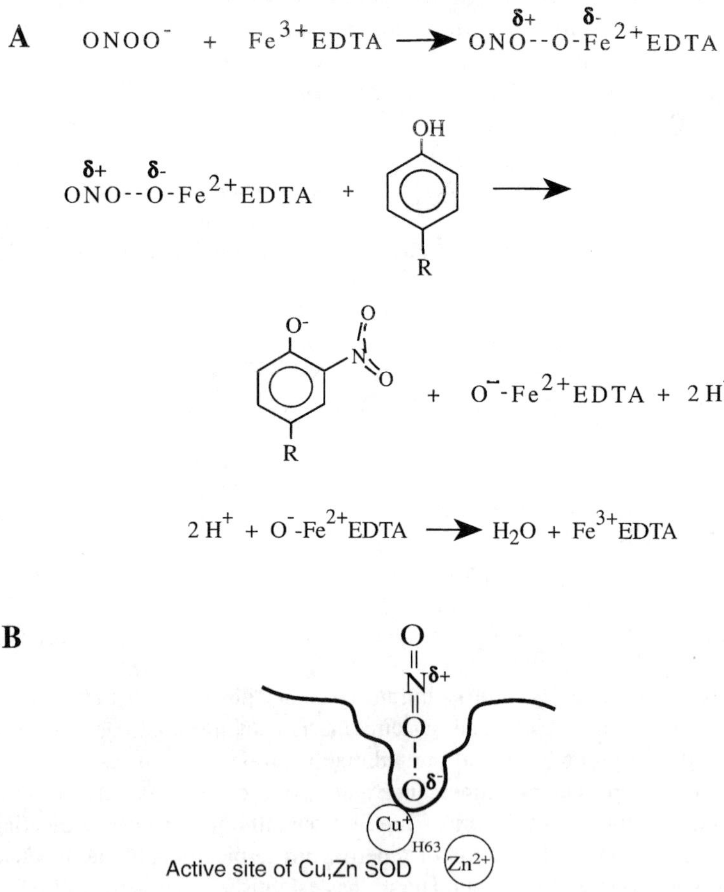

Figure 2.3. (2.3A) Proposed mechanism for the catalysis of peroxynitrite-dependent nitration by Fe-EDTA. (2.3B) Peroxynitrite reaction with the active site of Cu/Zn SOD results in the formation of a nitrating species.

is the addition of the negatively charged ONOO⁻ to the positively charged ferric ion. The electron density then shifts from the nitrogen atom to the iron, which could lead to the heterolytic cleavage of the O—O bond and the formation of NO_2^+. The production of NO_2^+ is not a necessity as the ONOO-Fe^{3+} complex itself may have the ability to act as a nitrating agent.

Peroxynitrite not only reacts with low-molecular-weight metal complexes but also reacts with metalloproteins. Hemeproteins like myeloperoxidase, iron-sulfur proteins, and copper proteins are a few of the different types of metalloproteins that react with peroxynitrite. The reaction between bovine-Cu,Zn SOD and peroxynitrite was one of the first studied.[48] Treatment of bovine SOD with ONOO⁻ resulted in the formation of a chromagen, which absorbed in the 430-nm region and was later identified as a nitrated tyrosine residue of the protein. Subsequent research has shown that the nitration of low-molecular-weight phenolic compounds catalyzed by SOD is one of the fastest known for ONOO⁻ (~10^5 M^{-1} s^{-1}). The mechanism of catalysis by SOD is similar to that shown for Fe^{3+}-EDTA (Fig. 2.3B). It is interesting to note that the amount of low-molecular-weight iron or copper (in the millimolar range) needed to detect a significant enhancement of the rate of nitration would probably not occur *in vivo*. However, the amount of SOD needed to enhance nitration is well within the physiological range of approximately 10 μM. This suggests that the differences in the reactions between ONOO⁻ and low-molecular-weight copper complexes or the copper of SOD must be due to differences in the environment of the two copper atoms. Enhancement of nitration by bovine SOD occurs not only with purified proteins but also homogenates of rat brain or heart. Because SOD can catalyze the nitration of proteins in a tissue homogenate containing many different targets for peroxynitrite, it is also likely to catalyze nitrotyrosine formation *in vivo* where there would also be a myriad of targets for peroxynitrite. We believe that this is the toxic gain-of-function occurring in mutations to Cu,Zn SOD linked with amyotrophic lateral sclerosis (ALS). The ALS mutations increase the probability that the mutated protein loses zinc, which increases the efficiency of nitrating tyrosines by peroxynitrite. Motor neurons seem to be particularly susceptible because they contain high concentrations of neurofilament L, which is a particularly favorable target for nitration by peroxynitrite. Abnormal assemblies of neurofilaments are a hallmark of ALS and neurofilaments are nitrated in ALS patients. This provides strong evidence that structural proteins are major targets of tyrosine nitration *in vivo*.

Although much of the chemistry pertaining to the oxidation and nitration reactions of peroxynitrite is known, there is still much to be learned about the effect that these reactions have on an intact living cell. Production of NO contributes to neurodegeneration that occurs after activation of the NMDA receptor. Two studies have shown that Cu/Zn SOD inhibits neuronal damage induced by NO donors or NMDA-receptor-stimulated NO production in cells in culture.[49–50] Lipton et al. also demonstrated that peroxynitrite was toxic to cultured cortical neurons.[50]

These findings suggest that the radical-radical reaction of NO with O_2^- contributes to neurodegeneration, but the mechanism of cell death was not determined.

2.5 Peroxynitrite and Apoptosis

Several recent publications have shown that treatment of human umbilical vein endothelial cells, HL60 cells or PC12 cells with peroxynitrite caused a large increase in programmed cell death or apoptosis.[51-53] Apoptosis is a tightly regulated cellular process that is essential for removing unwanted cells during development and for maintaining the homeostasis of cell populations.[54] Organisms can also use apoptosis as a defense mechanism whereby they can remove potentially dangerous cells, such as tumor cells or virus-infected cells.[55,56] If the regulation of apoptosis is compromised the consequences are serious, since many regulatory and inhibitor genes involved in apoptosis have been identified as oncogenes.

Several morphological and biochemical changes characterize cells undergoing apoptosis. Early events are condensation of the nuclear chromatin and cytoplasm as well as a ruffling of cellular membranes.[57] One common but not essential biochemical marker of apoptosis is the fragmentation of chromosomal DNA by specific nucleases.[58] The fragmented DNA and the contents of the cell are packed into membrane-bound apoptotic bodies, which are then phagocytosed by neighboring cells. An important aspect of apoptosis is the lack of an inflammatory response during the process.[57]

Agents that modulate the initiation of apoptosis include extracellular growth factors, hormones, viral infection, ionizing radiation, and chemotherapeutic agents.[54,59-61] Recent work has implicated an apoptotic mechanism in peroxynitrite-mediated toxicity.[51-53] The extracellular addition of peroxynitrite to PC12 cells resulted in a delayed cell death that was characteristic of apoptosis. Cells treated with peroxynitrite showed cytoplasmic shrinkage, nuclear condensation, and blebbing of cellular membranes. Cleavage of chromosomal DNA was also detected with terminal-deoxynucleotide transferase-mediated dUTP-digoxigenin nick end labeling (TUNEL).[52] Although peroxynitrite can directly damage DNA and might artifactually cause TUNEL staining, TUNEL staining was not found in peroxynitrite-treated PC12 cells until several hours after treatment and was prevented by the endonuclease inhibitor aurintricarboxylic acid. In summary, these morphological and biochemical changes are classical indicators of apoptosis.

Interestingly, the toxicity of peroxynitrite is affected differently when cells are exposed to different growth factors (Fig. 2.4). Pretreatment of PC12 cells with nerve growth factor (NGF) for 2 h protected against peroxynitrite. In contrast, PC12 cells pretreated with acidic and basic fibroblast growth factor (FGF-1 and FGF-2) stimulated cell death almost twofold.

The contrasting results with NGF and FGF-1 and -2 were unexpected. Previous work has shown that treating PC12 cells with NGF results in the extension of neurites and differentiation of the naive cells to a neuronal phenotype.[60] Treating

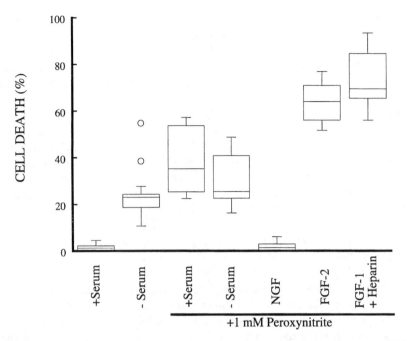

Figure 2.4. The effect of growth factors on the survival of PC12 cells after peroxynitrite treatment. The two leftmost groups were maintained in either the presence or absence of serum. Cultures of PC12 cells were also maintained in either the presence or absence of serum following exposure to 1-mM peroxynitrite. Separate cultures of PC12 cells were incubated (2 h) in the presence of serum with NGF (50 ng/ml), FGF-2 (50 ng/ml), or FGF-1 (50 ng/ml) plus heparin (50 U/ml) and then exposed to 1-mM peroxynitrite. Cell death was evaluated by a method employing staining with fluorescein diacetate and propidium iodide. The median and medial quartiles for the percentage cell death are given by the middle line and box. The whiskers or error bars give the maximum and minimum and the outliers are shown by the open circles. Adapted from Ref. 52.

the cells with FGF-2 also results in neurite outgrowth; however, there are small differences in kinetic and morphological features as compared to NGF.[61] Differences in the signaling mechanisms of these growth factors might explain why NGF can protect cells from peroxynitrite-induced apoptosis, while another can potentiate apoptosis (FGF). Our recent unpublished results suggest that NFG is protective in PC12 cells by activating a phosphoinositol-3-phosphate (PI-3) kinase pathway, while FGF increases toxicity by activating a p21Ras-dependent pathway.

2.6 Cell Signaling Pathways

Growth factors bind to distinct transmembrane receptors and activate intrinsic protein-tyrosine kinase activity in their intracellular domains. In particular, NGF binds to two different receptors on PC12 cells. Both the high-affinity NGF receptor (TrkA) and the low-affinity receptor (p75) are able to bind NGF and are able to regulate important signaling cascades.[64] On binding NGF, TrkA dimerizes and several tyrosine residues are autophosphorylated. Phosphorylated tyrosine residues of TrkA are recognized by proteins containing SH2 domains that recognize specific phosphotyrosines.[65] Through these interactions many different signaling pathways can be activated or inhibited. Some of the important signaling molecules activated by NGF binding to TrkA are the small GTP-binding protein p21Ras, phospholipase C-γ (PLC-γ), and PI-3 kinase (Fig. 2.5).[65-67] Peroxynitrite-dependent nitration of any of the tyrosine residues involved in controlling these pathways could potentially block phosphorylation and signal propagation.

Although it is well accepted that PI-3 kinase activity is essential for the proper function of many signaling pathways in PC12 cells, there is some controversy about the mechanism of PI-3 kinase activation by TrkA. Some investigators have proposed the direct binding of the p85 regulatory subunit of the enzyme to the phosphorylated receptor while others have suggested the presence of an adaptor protein.[67-68] Regardless of which mechanism is correct, it is certain that the proper function of PI-3 kinase is dependent on its interaction with phosphorylated tyrosine residues. Recent work has shown that stimulation of PI-3 kinase activity by NGF is essential for neurite extension in naive PC12 cells and for the rescue of PC12 cells from serum withdrawal.[69,70] Both of the studies mentioned above

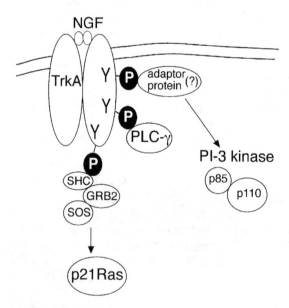

Figure 2.5. Diagram of NGF binding to the TrkA receptor and some of the different signaling pathways that are activated. SHC and GRB2 are adaptor proteins and SOS is a nucleotide exchange factor.

were able to determine the involvement of PI-3 kinase by using wortmannin, a specific inhibitor of the enzyme.

We have found that the combination of peroxynitrite and serum withdrawal induced similar levels of apoptosis as did either treatment alone, which suggested that the induction of apoptosis occurred in a common population of cells.[52] This finding led us to hypothesize that NGF was exerting its protective effect against peroxynitrite-induced apoptosis by stimulating PI-3 kinase activity. To test the hypothesis, we treated PC12 cells with wortmannin before and during the pretreatment with NGF. Cells treated with wortmannin and NGF had the same level of cell death as that detected in cells treated with peroxynitrite alone (unpublished observations). These results support the hypothesis that NGF protects PC12 cells from peroxynitrite-induced apoptosis by a mechanism dependent on the activation of PI-3 kinase. It is possible that peroxynitrite could be inhibiting PI-3 kinase activity by nitrating key tyrosine residues needed for the proper activation and function of the enzyme and that NGF maintains PI-3 kinase activity by inducing the phosphorylation of those same tyrosines and protects them from being nitrated.

Fibroblast growth factor activates many of the same signaling mechanisms as NGF, such as p21Ras and PI-3 kinase, and causes PC12 cells to undergo differentiation.[63,71,72] The fact that PC12 cells respond to FGF-2 indicates the presence of a functional FGF receptor that, once activated, is able to initiate signaling cascades. Although FGF-2 has been shown to activate PI-3 kinase, the level of activation was much less than that detected after NGF treatment.[71] Furthermore, Blumberg et al. demonstrated that there were many more phosphotyrosine proteins associated with PI-3 kinase after NGF treatment than after FGF-2 treatment.[68]

Some of these factors might explain why FGF does not protect PC12 cells from peroxynitrite as does NGF, but the reason for the potentiation of cell death seems to be related to the activation of p21Ras. When peroxynitrite-induced cell death was measured after FGF-1 (plus heparin to facilitate binding to its receptor) pretreatment, the level of cell death increased almost twofold.[52] However, a PC12 cell line expressing a dominant inhibitory mutant of p21Ras had the same level of cell death when pretreated with FGF-1 (plus heparin) as did cells treated with peroxynitrite alone (unpublished data). This suggests that enhancement of peroxynitrite-induced cell death is contingent on the activation of p21Ras. Thus, the toxicity of peroxynitrite can be either enhanced or decreased by the differential effects of growth factors on intracellular signaling pathways.

2.7 Transcription Factors

Binding of growth factors to their receptors ultimately results in the activation of specific transcription factors that regulate the transcription of genes with the correct regulatory sequences. Increased transcription of certain genes results in the activation of different cellular processes and the expression of distinct morphologies. Many transcription factors have cysteine residues that are required for the DNA binding function of the protein. For example, AP-1 has cysteine residues (Fig. 2.6A) that are critical for recognizing the 12-O-tetradecanoylphorbol-13-acetate (TPA) responsive element.[73] *In vitro* binding studies of AP-1 demonstrated that the DNA binding of the transcription factor was modulated by altering the oxidation state of these cysteine residues.[73] There is only one redox sensitive cysteine in the fos and jun proteins that form AP-1 and no

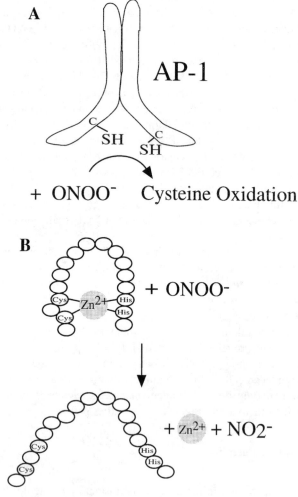

Figure 2.6. Oxidation of cysteine residues (2.6A) and zinc-finger moieties (2.6B) of transcription factors by peroxynitrite.

disulfides were detected in the oxidized protein[73]. This led to the hypothesis that the mechanism for the redox regulation of AP-1 activity depends on the oxidation of cysteine to the sulfenic acid oxidation state. Oxidation of sulfhydryls to sulfenic acids can also activate transcription factors[74]. For example, bacteria respond to hydrogen peroxide-induced oxidative stress through the OxyR transcription factor[75]. Similar to AP-1, OxyR has a redox sensitive cysteine that is thought to be oxidized to a sulfenic acid which changes the protein's conformation and DNA binding activity[75]. Therefore, peroxynitrite could affect cell signaling at the level of the transcription factor by oxidizing a cysteine residue to a sulfenic acid which could either inhibit or enhance DNA binding. Transcription factors are not the only cellular proteins with cysteines that are essential for the proper function of the cell's signaling mechanisms. The oxidation of protein thiols also effects the homeostasis of calcium and cytoskeletal organization.[3] The oxidation of these protein sulfhydryls by peroxynitrite could cause an increase in the calcium levels of the cytoplasm, which would have severe repercussions on the proper transduction of cellular signals.

Recent progress in determining how cells detect oxidative stress has shown that proteins with iron-sulfur clusters activate gene transcription after exposure to different oxidants[76]. Certain bacterial iron-sulfur proteins, such as the soxR protein, act as transcription factors that regulate the bacteria's response to oxidative stress. The soxR protein is a 2Fe-2S iron-sulfur protein which acts as a transcriptional activator of the soxS gene after exposure to superoxide or nitric oxide[77-79]. The crucial step in the activation of soxR is thought to involve oxidation of the iron-sulfur center which allows the protein to enhance transcription of soxS[78]. The increase in soxS protein causes several genes such as those encoding Mn-SOD and glucose-6-phosphate dehydrogenase to be activated[80]. Peroxynitrite is known to react rapidly with iron-sulfur proteins and could activate transcription factors with redox sensitive iron-sulfur clusters[81].

Transcription factors containing zinc-finger moieties are another major class of transcription factors that can be adversely effected by peroxynitrite. The zinc atom is coordinated by cysteine and histidine residues resulting in the formation of a structural motif known as a zinc-finger which is critical for the DNA binding activity of the protein (Fig. 2.6B). The zinc-fingers form the tertiary structures of the different transcription factors allowing them to recognize specific DNA sequences. Crow et al. demonstrated that the zinc-thiolate moiety of yeast alcohol dehydrogenase is particularly sensitive to oxidation by peroxynitrite.[82] The rate constant for the reaction between peroxynitrite and the zinc-thiolate of alcohol dehydrogenase is one of the fastest for peroxynitrite (10^5 M^{-1} s^{-1}). The relatively fast reaction with zinc-thiolates suggests that they may be a significant target for peroxynitrite. We have also recently found that zinc-histidine ligands are readily inactivated by peroxynitrite (unpublished results).

Small alterations in zinc-finger motifs can completely inhibit the DNA binding activity of the glucocorticoid receptor, which implies that any type of oxidation of a zinc-finger could have deleterious consequences.[83] The destruction of zinc-finger moieties could destroy the ability of a protein to recognize the correct DNA sequence or it might abolish DNA binding all together. Many other proteins that participate in cell signaling, including protein kinase C and Raf, also contain zinc-finger moieties that might be oxidized by peroxynitrite. Therefore, zinc-finger containing proteins of the cell's signaling pathways must also be considered when trying to determine what critical targets might be adversely effected by peroxynitrite.

Zinc-fingers and nitration of tyrosine residues involved in phosphorylation are two logical targets for peroxynitrite. We have begun to realize that structural proteins are affected by tyrosine nitration. Structural proteins such as actin can constitute up to 8 to 30% of total soluble protein and be millimolar in concentration in cells. When disassembled, many of the tyrosines involved in hydrophobic intersubunit contacts become exposed to nitration by peroxynitrite. The negative charge conferred by nitration can then disrupt the proper assembly of these subunits into polymers. Only a few subunits need to be modified to disrupt the assembly of a structure incorporating thousands of subunits. Thus, structural proteins are both abundant targets for peroxynitrite and the damaging effects can have dominant functional consequences. While we usually do not consider structural proteins as part of signal transduction, they are intimately involved in moving various activated signaling molecules between compartments. Many cytoplasmic receptors are translocated to the nucleus via the cytoskeletal network. Disruption of the cytoskeletal network is as central to the apoptotic process as nuclear condensation and fragmentation.

Acknowledgments

We thank Dr. J. Anthony Thompson and Dr. Gail Johnson for their support and for many helpful discussions. This work was supported by National Institutes of Health grants NS 24338 (JSB), NS33291 (JSB), and HL48676 (JSB).

References

1. Meyer, M., R. Schreck, and P. A. Baeuerle. 1993. H_2O_2 and antioxidants have opposite effects on activation of NF-κB and AP-1 in intact cells: AP-1 as secondary antioxidant-responsive factor. *EMBO J.* **12**:2005–2015.

2. Brumell, J. H., A. L. Burkhardt, J. B. Bolen, and S. Grinstein. 1996. Endogenous reactive oxygen intermediates activate tyrosine kinases in human neutrophils. *J. Biol. Chem.* **271**:1455–1461.

3. Orrenius, S., M. J. Burkitt, G. E. N. Kass, J. M. Dypbukt, and P. Nicotera. 1992. Calcium ions and oxidative cell injury. *Ann. Neurol.* **32**:S33–S42.

4. Beckman, J. S. 1991. The double edged role of nitric oxide in brain function and superoxide-mediated pathology. *J. Devel. Physiol.* **15**:53–59.

5. Dawson, V. L., T. M. Dawson, E. D. London, D. S. Bredt, and S. H. Snyder. 1991. Nitric oxide mediates glutamate neurotoxicity in primary cortical cultures. *Proc. Natl. Acad. Sci. USA* **88**:6368–6371.

6. Ischiropoulos, H., L. Zhu, and J. S. Beckman. 1992. Peroxynitrite formation from macrophage-derived nitric oxide. *Arch. Biochem. Biophys.* **298**:446–451.

7. Matheis, G., M. P. Sherman, G. D. Buckberg, D. M. Haybron, H. H. Young, and L. J. Ignarro. 1992. Role of L-arginine-nitric oxide pathway in myocardial reoxygenation injury. *Am. J. Physiol.* **262**:H616–H620.

8. Tarpey, M. M., J. S. Beckman, H. Ischiropoulos, J. S. Gore, and T. A. Brock. 1995. Peroxynitrite stimulates vascular smooth muscle cell cyclic GMP synthesis. *FEBS Lett.* **364**:314–318.

9. Palmer, R. M. J., D. S. Ashton, and S. Moncada. 1988. Arginine is the source of endothelial-derived nitric oxide. *Nature (London)* **333**:664–666.

10. Ignarro, L. J., J. B. Adams, P. M. Horwitz, and K. S. Wood. 1986. Activation of soluble guanylate cyclase by NO-hemoproteins involves NO-heme exchange. *J. Biol. Chem.* **261**:4997–5002.

11. Nairn, A. C., and P. Greengard. 1983. Cyclic GMP-dependent protein phosphorylatin in mammalian brain. *Fed. Proc.* **42**:3107–3113.

12. Robertson, B. E., R. Schubert, J. Hescheler, and M. T. Nelson. 1993. cGMP-dependent protein kinase activates Ca-activated K channels in cerebral artery smooth muscle cells. *Am. J. Physiol.* **265**:C299–C303.

13. Moncada, S., A. G. Herman, and P. M. Vanhouette. 1987. Endothelium-derived relaxing factor is identified as nitric oxide. *Trends Pharmacol. Sci.* **8**:365–368.

14. Ignarro, L. J. 1990. Biosynthesis and metabolism of endothelium-derived nitric oxide. *Annu. Rev. Pharmacol. Toxicol.* **30**:535–560.

15. Shibuki, K., and D. Okada. 1991. Endogenous nitric oxide release required for long-term synaptic depression in the cerebellum. *Nature (London)* **349**:326–329.

16. Dinerman, J. L., T. M. Dawson, M. J. Schell, A. Snowman, and S. H. Snyder. 1994. Endothelial nitric oxide synthase localized to hippocampal pyramidal cells: implications for synaptic plasticity. *Proc. Natl. Acad. Sci. USA* **91**:4214–4218.

17. Hibbs, J. B., Jr., R. R. Taintor, and Z. Vavrin. 1987. Macrophage cytotoxicity: Role of L-arginine deminiase and imino nitrogen oxidation to nitrite. *Science* **235**:173–235.

18. Hibbs, J. B., Jr., R. R. Taintor, Z. Vavrin, and E. M. Rachlin. 1988. Nitric oxide: a cytotoxic activated macrophage effector molecule. *Biochem. Biophys. Res. Commun.* **157**:87–94.

19. Nowicki, J. P., D. Duval, H. Poignet, and B. Scatton. 1991. Nitric oxide mediates neuronal death after focal cerebral ischemia in the mouse. *Eur. J. Pharmacol.* **204**:339–340.

20. Chen, J., K. A. Conger, M. J. Tan, and J. S. Beckman. 1994. Nitroarginine reduces infarction after middle cerebral artery occlusion in rats, pp. 264–272. *In* A. Hartman, F. Yatsu, and W. Kuschinsky (eds.), *Cerebral Ischemia and Basic Mechanisms,* Springer-Verlag, Berlin.

21. Lafon-Cazal, M., M. Culcasi, F. Gaven, S. Pietri, and J. Bockaert. 1993. Nitric oxide, superoxide and peroxynitrite: putative mediator of NMDA-induced cell death in cerebellar granule cells. *Neuropharmacology* **32**:1259–1266.

22. Lafon-Cazal, M., S. Pietri, M. Culcasi, and J Bockaert. 1993. NMDA-dependent superoxide production and neurotoxicity. *Nature* **364**:535–537.

23. Pryor, W. A., and J. W. Lightsey. 1981. Mechanisms of nitrogen dioxide reactions: initiation of lipid peroxidation and the production of nitrous acid. *Science* **214**:435–437.

24. Pryor, W. A., D. F. Church, C. K. Govindan, and G. Crank. 1982. Oxidation of thiols by nitric oxide and nitrogen dioxide: synthetic utility and toxicological implications. *J. Org. Chem.* **147**:156–158.

25. Prütz, W. A., H. Mönig, J. Butler, and E. J. Land. 1985. Reactions of nitrogen dioxide in aqueous model systems: oxidation of tyrosine units in peptides and proteins. *Arch. Biochem. Biophys.* **243**:125–134.

26. Wink, D. A., R. W. Nims, J. F. Darbyshire, D. Christodoulou, I. Hanbauer, G. W. Cox, F. Laval, J. Laval, J. A. Cook, M. C. Krishna, W. G. DeGraff, and J. B. Mitchell. 1994. Reaction kinetics for nitrosation of cysteine and glutathione in aerobic nitric oxide solutions at neutral pH. Insights into the fate and physiological effects of intermediates generated in the NO/O$_2$ reaction. *Chem. Res. Toxicol.* **7**:519–525.

27. Sawyer, D. T., and J. Valentine. 1981. How super is superoxide? *Acct. Chem. Res.* **14**:393–400.

28. Fridovich, I. 1986. Biological effects of the superoxide radical. *Arch. Biochem. Biophys.* **247**:1–11.

29. McCord, J. M., and I. Fridovich. 1969. Superoxide dismutase: an enzymic function for erythrocuprein (hemocuprein). *J. Biol. Chem.* **244**:6049–6055.

30. Klug, D., J. Rabani, and I. Fridovich. 1972. A direct demonstration of the catalytic action of superoxide dismutase through the use of pulse radiolysis. *J. Biol. Chem.* **247**:4839–4842.

31. Shibuki, K. 1990. An electrochemical microprobe for detecting nitric oxide release in brain tissue. *Neurosci. Res.* **9**:69–76.

32. Malinski, T., F. Bailey, Z. G. Zhang, and M. Chopp. 1993. Nitric oxide measured by a porphyrinic microsensor in rat brain after transient middle cerebral artery occlusion. *J. Cereb. Blood Flow Metab.* **13**:355–358.

33. Beckman, J. S., T. W. Beckman, J. Chen, P. M. Marshall, and B. A. Freeman. 1990. Apparent hydroxyl radical production from peroxynitrite: implications for endothelial injury by nitric oxide and superoxide. *Proc. Natl. Acad. Sci. USA* **87**:1620–1624.

34. Koppenol, W. H., J. J. Moreno, W. A. Pryor, H. Ischiropoulos, and J. S. Beckman. 1992. Peroxynitrite: a cloaked oxidant formed by nitric oxide and superoxide. *Chem. Res. Toxicol.* **5**:834–842.

35. Radi, R., J. S. Beckman, K. M. Bush, and B. A. Freeman. 1991. Peroxynitrite-mediated sulfhydryl oxidation: the cytotoxic potential of superoxide and nitric oxide. *J. Biol. Chem.* **266**:4244–4250.

36. Kosower, N. S. 1978. The glutathione status of cells. *Int. Rev. Cytol.* **54**:109–160.

37. Reed, D. J. 1990. Glutathione: toxicological implications. *Annu. Rev. Pharmacol. Toxicol.* **30**:603–631.

38. Beckman, J. S., Y. Z. Ye, P. Anderson, J. Chen, M. A. Accavetti, M. M. Tarpey, and C. R. White. 1994. Extensive nitration of protein tyrosines observed in human atherosclerosis detected by immunohistochemistry. *Biol. Chem. Hoppe-Seyler* **375**:81–88.

39. Ye, Y. Z., M. Strong, Z.-Q. Huang and J. S. Beckman. 1996. Antibodies that recognize nitrotyrosine. *Methods Enzymol.* **269**:201–209.

40. Haddad, I., G. Pataki, P. Hu, C. Galliani, J. S. Beckman, and S. Matalon. 1994. Quantitation of nitrotyrosine levels in lung sections of patients and animals with acute lung injury. *J. Clin. Invest.* **94**:2407–2413.

41. Zhu, L., C. Gunn, and J. S. Beckman. 1992. Bactericidal activity of peroxynitrite. *Arch. Biochem. Biophys.* **298**:452–457.

42. Radi, R., P. Cosgrove, J. S. Beckman and B. A. Freeman. 1993. Peroxynitrite-induced luminol chemiluminescence. *Biochem. J.* **290**:51–57.

43. Lymar, S. V., and J. K. Hurst. 1995. Rapid reaction between peroxonitrite ion and carbon dioxide: implications for biological activity. *J. Am. Chem. Soc.* **111**:8867–8868.

44. Uppu, R. M., G. L. Squadrito, and W. A. Pryor. 1996. Acceleration of peroxynitrite oxidations by carbon dioxide. *Arch. Biochem. Biophys.* **327**:335–343.

45. Lymar, S. V., Q. Jiang, and J. K. Hurst. Mechanism of carbon dioxide catalyzed oxidation of tyrosine by peroxynitrite. *Biochemistry.* **35**:7855–7861.

46. Denicola, A., B. A. Freeman, M. Trujillo and R. Radi. 1996. Peroxynitrite reaction with carbon dioxide/bicarbonate: kinetics and influences on peroxynitrite-mediated oxidations. *Arch. Biochem. Biophys.* **333**:49–58.

47. Beckman. J. S., H. Ischiropoulos, L. Zhu, M. van der Woerd, C. Smith, J. Chen, J. Harrison, J. C. Martin, and M. Tsai. 1992. Kinetics of superoxide dismutase and iron catalyzed nitration of phenolics by peroxynitrite. *Arch. Biochem. Biophys.* **298**:438–445.

48. Ischiropoulos, H., L. Zhu, J. Chen, H. M. Tsai, J. C. Martin, C. D. Smith, and J. S. Beckman. 1992. Peroxynitrite-mediated tyrosine nitration catalyzed by superoxide dismutase. *Arch. Biochem. Biophys.* **298**:431–437.

49. Dawson, V., T. Dawson, D. Bartley, G. Uhl, and S. Snyder. 1993. Mechanisms of nitric oxide-mediated neurotoxicity in primary brain cultures. *J. Neurosci.* **13**:2651–2661.

50. Lipton, S. A., Y. B. Choi, Z. H. Pan, S. Z. Lei, H. S. V. Chen, N. J. Sucher, J. Loscalzo, D. J. Singel, and J. S. Stamler. 1993. A redox-based mechanism for the neuroprotective and neurodestructive effects of nitric oxide and related nitroso-compounds. *Nature* **364**:626–631.

51. Bonfoco, E., D. Krainc, M. Ankarcrona, P. Nicotera, and S. A. Lipton. 1995. Apoptosis and necrosis: two distinct events induced, respectively, by mild and intense

insults with N-methyl-D-aspartate or nitric oxide/superoxide in cortical cell cultures. *Proc. Natl. Acad. Sci. USA* **92**:7162–7166.

52. Estévez, A. G., R. Radi, L. Barbeito, J. T. Shin, J. A. Thompson, and J. S. Beckman. 1995. Peroxynitrite-induced cytotoxicity in PC12 cells: evidence for an apoptotic mechanism differentially modulated by neurotrophic factors. *J. Neurochem.* **65**:1543–1550.

53. Lin, K.-T., J.-Y. Xue, M. Nomen, B. Spur, and P.Y.-K. Wong. 1995. Peroxynitrite-induced apoptosis in HL-60 cells. *J. Biol. Chem.* **270**:16487–16490.

54. Raff, M. C., B. A. Barres, J. F. Burne, H. S. Coles, Y. Ishizaki, and M. D. Jacobson. 1993. Programmed cell death and the control of cell survival: lessons from the nervous system. *Science* **262**:695–700.

55. Williams, G. T. 1991. Programmed cell death: apoptosis and oncogenesis. *Cell* **65**:1097–1098.

56. Vaux, D. L., G. Haecker, and A. Strasser. 1994. An evolutionary perspective on apoptosis. *Cell* **76**:777–779.

57. Kerr, J. F. R., C. M. Winterford, and B. V. Harmon. 1994. Apoptosis. Its significance in cancer and cancer therapy. *Cancer* **73**:2013–2026.

58. Wyllie, A. H., M. J. Arends, R. G. Morris, S. W. Walker, and G. Evan. 1992. The apoptosis endonuclease and its regulation. *Semin. Immunol.* **4**:389–397.

59. Searle, J., T. A. Lawson, P. J. Abbott, B. Harmon, and J. F. R. Kerr. 1975. An electron-microscope study of the mode of cell death induced by cancer-chemotherapeutic agents in populations of proliferating normal and neoplastic cells. *J. Pathol.* **116**:129–138.

60. Kyprianou, N., H. F. English, N. E. Davidson, and J. T. Isaacs. 1991. Programmed cell death during regression of the MCF-7 human breast cancer following estrogen ablation. *Cancer Res.* **51**:162–166.

61. Stephens, L. C., K. K. Ang, T. E. Schultheiss, L. Milas, and R. E. Meyn. 1991. Apoptosis in irradiated murine tumors. *Radiat. Res.* **127**:308–316.

62. Greene, L. A., and A. S. Tischler. 1976. Establishment of a noradrenergic clonal line of rat adrenal pheochromocytoma cells which respond to nerve growth factor. *Proc. Natl. Acad. Sci. USA* **73**:2424–2428.

63. Rydel, R. E., and L. A. Greene. 1987. Acidic and basic fibroblast growth factors promote stable neurite outgrowth and neuronal differentiation in cultures of PC12 cells. *J. Neurosci.* **7**:3639–3653.

64. Kaplan, D. R., D. Martin-Zance, and L. F. Parada. 1991. Tyrosine phosphorylatin and tyrosine kinase activity of the *trk* proto-oncogene product induced by NGF. *Nature* **350**:158–160.

65. Obermeier, A., R. A. Bradshaw, K. Seedorf, A. Choidas, J. Schlessinger, and A. Ullrich. 1994. Neuronal differentiation signals are controlled by nerve growth factor receptor/Trk binding sites for SHC and PLC-γ. *EMBO J.* **13**:1585–1590.

66. Muroya, K., S. Hattori, and S. Nakamura. 1992. Nerve growth factor induces rapid accumulation of the GTP-bound form of P21[ras] in rat pheochromocytoma PC12 cells. *Oncogene* **7**:277–281.

67. Soltoff, S. P., S. L. Rabin, L. C. Cantley, and D. R. Kaplan. 1992. Nerve growth factor promotes the activation of phosphatidylinositol 3-kinase and its association with the *trk* tyrosine kinase. *J. Biol. Chem.* **267**:17472–17477.

68. Blumberg, D., M. J. Radeke, and S. C. Feinstein. 1995. Specificity of nerve growth factor signaling: differential patterns of early tyrosine phosphorylation events induced by NGF, EGF, and bFGF. *J. Neurosci. Res.* **41**:628–639.

69. Kimura, K., S. Hattori, Y. Kabuyama, Y. Shizawa, J. Takayanagi, S. Nakamura, S. Toki, Y. Matsuda, K. Onodera, and Y. Fukui. 1994. Neurite outgrowth of PC12 cells is suppressed by wortmannin, a specific inhibitor of phosphatidylinositol 3-kinase. *J. Biol. Chem.* **269**:18961–18967.

70. Yao, R., and G. M. Cooper. 1995. Requirement for phosphatidylinositol-3 kinase in the prevention of apoptosis by nerve growth factor. *Science* **267**:2003–2006.

71. Raffioni, S., and R. A. Bradshaw. 1992. Activation of phosphatidylinositol 3-kinase by epidermal growth factor, basic fibroblast growth factor, and nerve growth factor in PC12 pheochromocytoma cells. *Proc. Natl. Acad. Sci. USA* **89**:9121–9125.

72. Wang, J.-K., G. Gao, and M. Goldfarb. 1994. Fibroblast growth factor receptors have different signaling and mitogenic potentials. *Mol. Cell. Biol.* **14**:181–188.

73. Abate, C., L. Patel, I. F. J. Rauscher, and T. Curran. 1990. Redox regulation of Fos and Jun DNA-binding. *Science* **249**:1157–1161.

74. Clairborne, A., H. Miller, D. Parsonage and R. P. Ross. 1993. Protein-sulfenic acid stabilization and function in enzyme catalysis and gene regulation. *FASEB.* **7**:1483–1490.

75. Storz, G., L. A. Tartaglia and B. N. Ames. 1990. Transcriptional regulator of oxidative stress-inducible genes: direct activation by oxidation. *Science.* **248**:189–194.

76. Rouault, T. A. and R. D. Klausner. 1996. Iron-sulfur clusters as biosensors of oxidants and iron. *TIBS.* **21**:174–177.

77. Hidalgo, E., J. M. Bollinger, Jr., T. M. Bradley, C. T. Walsh and B. Demple. 1995. Binuclear [2Fe-2S] clusters in the Escherichia coli soxR protein and role of the metal centers in transcription. *J. Biol. Chem.* **270**:20908–20914.

78. Hidalgo, E. and B. Demple. 1994. An iron-sulfur center essential for transcriptional activation by the redox sensing soxR protein. *EMBO J.* **13**:138–146.

79. Nonushiba, T., T. de Rojas-Walker, J. S. Wishnok, S. R. Tannenbaum, B. Demple. 1993. Activation by nitric oxide of an oxidative-stress response that defends Escherichia coli against activated macrophages. *Proc. Natl. Acad. Sci. USA.* **90**:9993–9997.

80. Demple, B. 1991. Regulation of bacterial oxidative stress genes. *Annu. Rev. Genet.* **25**:315–337.

81. Castro, L., M. Rodriguez and R. Radi. Aconitase is readily inactivated by peroxynitrite, but not by its precursor, nitric oxide. *J. Biol. Chem.* **269**:29409–29415.

82. Crow, J. P., J. S. Beckman, and J. M. McCord. 1995. Sensitivity of the essential zinc-thiolate moiety of yeast alcohol dehydrogenase to hypochlorite and peroxynitrite. *Biochemistry* **34**:3544–3552.

83. Ray, A., K. S. LaForge, and P. B. Sehgal. 1991. Repressor to activator switch by mutations in the first Zn finger of the glucocorticoid receptor: is direct DNA binding necessary? *Proc. Natl. Acad. Sci. USA* **88**:7086–7090.

3

Nitric Oxide Signaling Through Mitochondrial Calcium Release

Christoph Richter and Renato Laffranchi

3.1 Summary

Free calcium ions are important for the regulation of many enzyme systems, including those responsible for muscle contraction, nerve impulse transmission, neuronal activity, blood clotting, and modulation of hormone action. Intracellular Ca^{2+} can act as a second messenger in a variety of signaling systems. To serve these functions, fine-tuning of the intracellular Ca^{2+} concentration is essential. This is achieved by binding of Ca^{2+} to nonmembraneous proteins and by membrane-bound transport systems. Mitochondrial Ca^{2+} uptake and release participate directly in the regulation of the cytoplasmic Ca^{2+} concentration, and nitric oxide and its congeners modulate these mitochondrial activities in several ways. Thus, nitric oxide (nitrogen monoxide, NO·) binds to cytochrome oxidase and thereby prevents respiration and the maintenance of the mitochondrial membrane potential ($\Delta\psi$). The deenergization of mitochondria results in Ca^{2+} release from and prevention of Ca^{2+} uptake by mitochondria. Peroxynitrite ($ONOO^-$), the product formed by the reaction of NO· with superoxide (O_2^-), stimulates mitochondrial Ca^{2+} release with preservation of $\Delta\psi$. The NO·-dependent modulation of Ca^{2+} transport by mitochondria may be used in cell signaling, as exemplified by glucose-triggered insulin secretion from pancreatic beta cells. Insulin secretion, which occurs in two phases, requires an increase in the cytoplasmic free Ca^{2+} concentration. Beta cells produce nitric oxide, and release insulin in response to glucose during the first phase in an nitric oxide synthase-sensitive manner. When NO· is added to rat insulinoma cells, mitochondria become deenergized and lose their Ca^{2+}. The resulting increase in the cytoplasmic free Ca^{2+} then triggers insulin secretion. Since the mitochondrial energy state seems to be under the control of nitric oxide it is expected that future studies will identify other Ca^{2+}-dependent signaling systems, which are modulated by the action of nitric oxide and its congeners on mitochondria.

3.2 Cellular Calcium

Free calcium ions play a prominent role in the regulation of many enzyme systems, including those responsible for muscle contraction, nerve impulse transmission, neuronal activity, blood clotting, and modulation of hormone action. Specifically the intracellular Ca^{2+} serves the fundamental function of an intracellular messenger, and, therefore, its fine-tuning is required. The concentration of Ca^{2+}, which in resting cells is kept around 0.1 μM, can rapidly increase by up to one order of magnitude when transducing hormonal messages.

The intracellular Ca^{2+} concentration is regulated by binding to nonmembraneous proteins, and by membrane-bound transport systems. Besides buffering the intracellular Ca^{2+} concentration, the soluble proteins also participate in signal processing and are therefore called "Ca^{2+}-modulated proteins." Examples are troponin C, parvalbumin, and calmodulin. The main burden of regulation is, however, carried by Ca^{2+}-specific proteins, which transport the ion across membranes. Transport proteins were found in the plasma membrane, the endoplasmic (sarcoplasmic) reticulum, in mitochondria, and recently also in the nucleus. The transport systems differ in affinity and capacity for Ca^{2+} transport, and probably also in their importance for the maintenance of physiological and the development of pathological states of the cell.

3.3 Insulin Secretion

In mammals, the islets of Langerhans synthesize polypeptide hormones that regulate the storage and utilization of metabolic fuels. Fuel homeostasis is in large part controlled by the two hormones, insulin and glucagon.[1] The release of insulin is much better understood because in insulin-dependent diabetes mellitus this hormone is missing. Insulin secretion from islet beta cells is affected by a variety of factors. The most important is glucose, since the response to many other effectors either requires or is potentiated by the presence of this sugar. However, many other signals are involved, for example, the parasympathetic and enteric nervous system, other hormones, and other products of digestion.

When glucose is raised in the plasma after digestion of a meal, many signaling events are switched on in a beta cell before insulin secretion occurs. First of all, glucose is taken up by the beta cell through the GLUT-2 isoform of glucose transporters.[2] Expression of this transporter seems to be necessary but not sufficient for glucose-stimulated insulin secretion. Once taken up, glucose is phosphorylated by glucokinase.[3] This enzyme is the rate-limiting or pace-setting step in glucose metabolism. Glucose-6-phosphate is further metabolized in the glycolytic pathway. Glycogen synthesis or the pentose shunt are not physiologically relevant because they use only 1 to 7% or 2%, respectively, of the total glucose flux. Subsequent oxidative phosphorylation leads to ATP production and therefore an increase in the ATP:ADP ratio. This results in inhibition of the ATP-sensitive K$^+$ channels (K$_{ATP}$ channels) and therefore an increase in the cytosolic K$^+$ concentration, which leads to plasma membrane depolarization and causes an activation of voltage-gated, dihydropyridine-sensitive Ca^{2+} channels of the L class.[4,5] Thereafter, cytosolic Ca^{2+} increases and triggers insulin secretion. The influx of Ca^{2+} seems to be an absolute requirement for glucose-induced insulin secretion, since on removal of extracellular Ca^{2+}, insulin secretion no longer occurs.

Recently, evidence has been presented that glucose not only stimulates Ca^{2+} influx but also induces release of Ca^{2+} from intracellular stores by a xanthine-sensitive but ryanodine-insensitive, depolarization-dependent mechanism in mouse islets.[6] In human beta cells, glucose also induces Ca^{2+} release from intracellular stores.[7] Glucose raises cyclic ADPribose (cADPR), which induces Ca^{2+} release from microsomes of pancreatic islets and insulin secretion from digitonin-permeabilized islets.[8] Kato et al. reported that overexpression of CD38, a human lymphocyte antigen with both cADPR cyclase and cADPR hydrolase activity, results in enhanced insulin secretion in mice.[9]

With regard to time, glucose induces the following changes within 40 s: a rise in the ATP:ADP ratio, an increase in the reduction of pyridine nucleotides, and inhibition of the K$_{ATP}$ channels. Within 2 min [Ca^{2+}]$_i$ increases, coinciding roughly with the onset of insulin secretion.[10]

Glucose-stimulated insulin secretion is biphasic. The first phase starts within 2 min. It has a duration of 5 to 15 min and a peak after about 5 min. The second phase can last for a long time and is characterized by pulsatile oscillations. Each insulin pulse has a corresponding [Ca^{2+}]$_i$ oscillation.[11] There are oscillations of different frequencies. The slower ones are superimposed by faster oscillations.[12]

Exocytosis of insulin into the extracellular space requires ATP and is regulated by Ca^{2+}.[13] The Rab3A protein; a GTP-binding protein; and SNAP-25, a protein attached to the plasma membrane by palmitoylation, are also involved in the exocytotic process.[14,15]

Epinephrine, an agonist of adrenergic receptors, blocks insulin exocytosis. This occurs by activation of the heterotrimeric G proteins G$_i$ and G$_0$ and subsequent direct inhibition of insulin exocytosis without generation of a diffusable second messenger. Functionally, the G proteins are near the final, Ca^{2+}-dependent step of exocytosis.[16]

3.4 Mitochondria and Cellular Calcium Homeostasis

3.4.1 Traditional View

The importance of mitochondria as buffers of cytosolic Ca^{2+} under physiologic conditions has until recently been thought to be minor because the mitochondrial Ca^{2+} affinity and uptake rate are much smaller than those of the endo- or sarcoplasmic reticulum (for review see Ref. 17). This view was corroborated by electron probe X-ray microanalysis experiments, which show that mitochondria in situ contain much less Ca^{2+} than previously deduced from measurements with isolated mitochondria. Indeed, it is now generally accepted that mitochondria normally do not contain more than about 1 to 2 nmol of Ca^{2+} per milligram of protein. On the other hand, isolated mitochondria can accumulate enormous amounts of Ca^{2+} in the presence of inorganic phosphate due to the formation of insoluble calcium phosphate complexes in the matrix. Deposits of Ca^{2+} complexes in mitochondria in situ were also observed in electron microscopy and electron probe X-ray microanalyses when the plasma membrane had been permeabilized. It therefore appears that mitochondria take up and buffer the cytosolic Ca^{2+} when its concentration increases to levels that allow the operation of the mitochondrial low-affinity uptake system. The traditional view, therefore, sees mitochondria as safety devices against toxic increases of cytosolic Ca^{2+}, for example, when the plasma membrane Ca^{2+} pump is compromised. This is important, since net influx of Ca^{2+} across the plasma membrane appears to be a frequent and early final common mechanism by which some cell types are killed.

Ca^{2+} uptake occurs via the "uniporter" electrogenically, i.e., without charge compensation. Despite much effort, no protein engaged in Ca^{2+} uptake has been identified with certainty. Best studied are glycoproteins with M_r of 33,000 to 42,000, which bind Ca^{2+} in a ruthenium red- and La^{3+}-sensitive fashion. Ca^{2+} uptake is modulated by removal from or addition of glycoprotein to mitochondria, and antiglycoprotein antibodies inhibit Ca^{2+} uptake.

3.4.2 Modern View

Recently, convincing evidence has been accumulated that mitochondria also participate in a physiologic regulation of Ca^{2+} homeostasis. This advance in our understanding of mitochondria is based mainly on novel experimental approaches. They comprise, alone or in combination, the use of new, mainly fluorescent, probes; measurements at the single-cell level; and molecular biological techniques.

An electron probe microanalysis[18] first indicated significant Ca^{2+} sequestration by mitochondria after physiologic Ca^{2+} pulses. A turning point for our understanding of the importance of mitochondria in cellular Ca^{2+} handling was the introduction of recombinant aequorin into this field by the Pozzan group.[19-22] The coelenterate photoprotein aequorin emits light on Ca^{2+} binding and had already been widely used for measuring cytosolic Ca^{2+}. Pozzan and colleagues targeted functional aequorin into the mitochondrial matrix by putting a leader sequence specifying import into mitochondria onto the protein. This enabled them to measure Ca^{2+} in the mitochondrial matrix of intact cells. The initial report[19] provided the first direct measure of mitochondrial Ca^{2+} changes induced by physiologic stimuli that increase cytosolic Ca^{2+}. Both kinetics and amplitude of mitochondrial Ca^{2+} changes were much faster and larger than previously suggested.[23] Importantly, increases in mitochondrial Ca^{2+} were only transient. This, on the one hand, confirmed the previously held view that mitochondria play only a minor role as buffers of cytosolic Ca^{2+}. On the other hand, it suggested that mitochondria respond very dynamically to cytosolic Ca^{2+} transients. Subsequent work[20] showed that only those mitochondria close to inositol trisphosphate (IP_3)-gated channels transiently accumulate Ca^{2+} in living cells exposed to stimuli that generate IP_3. This result agrees with the *in vitro* observation that the mitochondrial Ca^{2+} uptake system has a low affinity for Ca^{2+} (see above). There seems to be a microheterogeneity among the mitochondria of the same cells, about 30% of them sensing the domains of high cytosolic Ca^{2+} concentrations.[21] While the above-mentioned studies used nonexcitable cells in which Ca^{2+} transients were evoked by release of Ca^{2+} from intracellular stores, another report[22] showed that stimulated Ca^{2+} influx into excitable cells raises the mitochondrial Ca^{2+} as efficiently as internal Ca^{2+} mobilization in nonexcitable cells. In the latter study, it was also shown that Ca^{2+} gradients exist between the cytosol and mitochondria at rest, that the mitochondrial Ca^{2+} increases to 6 to 12 μM on cell stimulation, and that the increase is very rapid. The high speed may be consistent with a rapid sensing by mitochondria of transient cytosolic Ca^{2+} increases. These data suggest that the mitochondrial Ca^{2+} uptake system enables an immediate response to cytosolic Ca^{2+} changes. In addition, calculations indicated that at peak stimulation the Ca^{2+} flux into mitochondria may account for as much as 20% of the total Ca^{2+} flux into the cell.

Also in acinar cells, both lacrimal and pancreatic[24,25] mitochondria accumulate Ca^{2+}, as shown with more traditional techniques employing cell fractionation and mitochondrial inhibitors. At 0.1-μM Ca^{2+} (a value close to the cytosolic Ca^{2+} concentration) the mitochondrial Ca^{2+} pool contained 26% of the total Ca^{2+}, started to increase when Ca^{2+} was raised above 0.5 μM, and reached 49% at 1-μM Ca^{2+}.[25] Similarly, mitochondria sequester Ca^{2+} in smooth muscle cells,[26] in sympathetic neurons,[27] and in dorsal root ganglion neurons.[28] Ca^{2+} flux into mitochondria was also indirectly followed in differentiated N1E-115 neuroblastoma cells[29] by measuring $\Delta\psi$ changes fluorimetrically.

Mitochondrial Ca^{2+} uptake and release has recently been found to be also important for intracellular Ca^{2+} waves. Frequency-modulated and spacially organized cytosolic Ca^{2+} oscillations are believed to be important in signal transduction.[30,31] As discussed in some detail later, the mitochondrial Ca^{2+} content influences the redox state of mitochondrial pyridine nucleotides. Hajnoczky et al.[32] measured NAD(P)H, reduced flavin adenine dinucleotide ($FADH_2$), intramitochondrial Ca^{2+} changes with the fluorescent indicator Rhod 2, and cytosolic Ca^{2+} oscillations in single hepatocytes. The latter, induced by IP_3, were efficiently transmitted to the mitochondria as mitochondrial Ca^{2+} oscillations, which, in turn, caused changes in the mitochondrial redox state. IP_3-induced Ca^{2+} wave activity in *Xenopus laevis* oocytes[33] was strengthened by respiratory substrate, increasing Ca^{2+} wave amplitude, velocity, and interwave period. These findings reveal that the $\Delta\psi$-driven Ca^{2+} uptake is a major factor in the regulation of IP_3-induced Ca^{2+} release and demonstrate a physiological role of mitochondria in Ca^{2+}-dependent cell signalling.

With isolated rat liver mitochondria and a conventional isotope technique the Gunter group[34] recently identified a rapid Ca^{2+} mode in mitochondria during physiologic-type Ca^{2+} pulses.

3.4.3 Integration of Old and New Findings

Following extensive study of isolated organellis in the 1960s and 1970s, interest in mitochondrial Ca^{2+} transport decreased when other cellular Ca^{2+} transport systems were discovered that seemed to be more important for cellular Ca^{2+} handling. The recent advent of new technologies brought a revival of the interest in mitochondria as regulators of Ca^{2+}-dependent processes. Note that so far, the early *in vitro* studies and the recent in situ studies of mitochondrial Ca^{2+} transport are in excellent agreement, which provides hope for a further understanding of mitochondria in their native environment.

3.4.4 Mitochondrial Calcium and Beta Cells

Important steps within several metabolic processes are activated by intramito-chondrial Ca^{2+}.[35] Pyruvate dehydrogenase and α-ketoglutarate dehydrogenase are controlled by micromolar Ca^{2+} changes, whereas isocitrate dehydrogenase requires more than 3 μM Ca^{2+} for activation.[36] The rate of electron transport in liver mitochondria may also be enhanced by Ca^{2+},[37] and the mitochondrial ATPase and adenine nucleotide translocator are modulated by Ca^{2+}.[38,39]

Some cell types have a particularly high demand for reducing equivalents in mitochondria because they require them for biosynthetic pathways (e.g., adrenal glomerulosa cells, which synthesize aldosterone), or because they rely almost exclusively on oxidative phosphorylation as a source of ATP (beta cells).[40] In single adrenal glomerulosa cells, Ca^{2+} influx into cells caused an Amytal (blocker of site I of the respiratory chain)-inhibitable reduction of pyridine nucleotides, indicating an activation of the mitochondrial dehydrogenases by Ca^{2+}.[41] Indeed, it was previously observed that inhibition of mitochondrial Ca^{2+} uptake by ruthenium red prevented aldosterone production in glomerulosa cells.[42] A similar coupling between hormone- or glucose-induced cytosolic Ca^{2+} oscillations and changes in the NADPH redox state, measured fluorimetrically, was also reported for liver and beta cells.[43,44] Other observations were made by Malaisse and coworkers.[45-47] When measuring the redox state of mitochondrial pyridine nucleotides of pancreatic islets indirectly by metabolite analysis they observed that glucose induced a rapid decrease in the mitochondrial $NADH/NAD^+$ ratio. This was interpreted to reflect an increased clearance of mitochondrial NADH through the respiratory chain and the malate-aspartate shuttle.

3.5 Oxidative Stress in Mitochondria

It is now well established that reactive oxygen species (ROS) such as superoxide (O_2^-), hydrogen peroxide (H_2O_2), the hydroxyl radical ($OH\cdot$), and singlet oxygen are products of normal metabolism (for reviews, see Refs. 48 and 49). In the early 1970s, Chance and coworkers reported that hydrogen peroxide (H_2O_2) is produced by the respiratory chain. The discovery of superoxide dismutase in mitochondria[50] prompted the search for O_2^- in mitochondria. Loschen et al.[51] were the first to show mitochondrial O_2^- production and to document that the constitutive mitochondrial H_2O_2 originates from O_2^-.

Mitochondria consume about 90% of the body's oxygen, and are the richest source of ROS since about 1 to 2% of oxygen metabolized by mitochondria is converted to O_2^- by several sites in the respiratory chain and matrix.[48] The steady-state concentrations of mitochondrial O_2^- and H_2O_2, the predominant precursors of the highly reactive OH·, are estimated to be in the picomolar and nanomolar ranges, respectively. In general, the main O_2^- generators in mitochondria are the ubiquinone radical and NADH dehydrogenase. However, according to Nohl and Jordan[52] and Nohl,[53] cytochrome b rather than ubisemiquinone is a site of O_2^- production. Superoxide is formed by autoxidation of the reduced components of the respiratory chain. This explains the increased O_2^- production with increased oxygen pressure, the stimulation by rotenone and antimycin, and the higher O_2^- production in state 4 compared to state 3 respiration. The inhibition of ROS production in mitochondria by cytochrome c depletion, cyanide, or myxothiazol is due to the inhibition of ubisemiquinone formation from ubiquinol, which requires the transfer of one electron to cytochrome c_1.

There is some variation among different tissues and species in the sites of ROS production. In the rat brain, O_2^- production is associated with the activity of dihydroorotate dehydrogenase rather than succinate dehydrogenase.[54] In heart mitochondria, a NADH-dependent component in the outer membrane contributes to H_2O_2 formation.[55] In insects, NADH dehydrogenase is not a source of H_2O_2.[56]

In addition to normal ROS production in mitochondria, reactive oxygen is formed in large amounts in the presence of certain compounds (e.g., so-called "redox cyclers," such as alloxan) and during some pathological states. Thus, alloxan, menadione, adriamycin (doxorubicin), rotenone, methylphenylpyridinium, tetrachloro-dibenzo-p-dioxin, elevated Ca^{2+}, or tumor necrosis factor α (TNF-α) stimulate ROS production by mitochondria, as does ischemia/reperfusion, ethanol consumption, and sepsis (for reviews see Refs. 57 through 59).

3.6 Nitric Oxide

3.6.1 Biology of Nitric Oxide

NO (nitrogen monoxide; note that here NO indicates nitric oxide independent of its redox state, whereas NO·, NO+, and NO− [see below] refer to the nitrogen monoxide radical, nitrosonium ion, and nitroxyl anion, respectively) presently receives enormous attention. It mediates beneficial responses such as maintenance of blood pressure, inhibition of platelet aggregation, tumoricidal activities, or destruction of foreign invaders in the immune response, and is probably of major importance in long-term memory.

NO· is synthesized by nitric oxide synthase (NOS). The family of NOS (reviewed in Refs. 60 and 61) consists of three isoenzymes, the neuronal constitutive (ncNOS), the endothelial constitutive (ecNOS), and the inducible (iNOS) isoform. The constitutive isoforms are strictly Ca^{2+}-dependent, whereas iNOS is not, although calmodulin is necessary for its activity. Activation of both cNOSs occurs via small Ca^{2+}-transients, while iNOS, once expressed, remains active as long as substrates are available. The constitutive isoforms show a typical interaction with calmodulin controlled by Ca^{2+}. In contrast, iNOS forms a tight complex with calmodulin at very low Ca^{2+} concentrations and thus appears to be Ca^{2+}-independent. The location of NOSs in cells is not clear; ecNOS is N-myristoylated, which targets it to the Golgi,[62] while ncNOS and iNOS seem to be cytosolic proteins. NOS activity in skeletal muscle was very recently shown to be membrane-, possibly, mitochondria-associated,[63] and very recent immunochemical studies showed NOS associated with the inner membrane of rat liver and brain mitochondria[64] and the outer membrane of avian ciliary ganglion mitochondria.[65]

Next to its beneficial action, NO can also inhibit T-cell proliferation, inhibit viral replication in macrophages, suppress protein synthesis, and mediate cell lysis by multiple mechanisms (reviewed in Refs. 60 and 61). The dichotomy of NO is in part due to a broad array of redox species with distinctive properties and reactivities: NO^+ (nitrosonium), NO·, and NO^- (nitroxyl anion) (reviewed in Refs. 66 and 67; see also Ref. 68), and the ability of NO· to combine with O_2^- to yield peroxynitrite $(ONOO^-)$.[69]

Peroxynitrite is an efficient oxidant of thiols.[70] Its *in vivo* formation was recently shown in macrophages and other cells. Peroxynitrite production is associated with the activation and expression of iNOS and implicated in the pathophysiology of diseases such as acute endotoxemia, inflammatory bowel disease, neurological disorders, and atherosclerosis. Inhibition by SOD, the O_2^- scavenger, of NO·-mediated cytotoxicity suggests that $ONOO^-$ may contribute to the NO-mediated biological effects (see Ref. 71).

Nitroxyl anion is formed from NO· by reduced SOD.[72] It is another NO congener that oxidizes thiols. Nitroxyl anion is, like $ONOO^-$, neuroprotective at *N*-methyl-D-aspartate receptors because these compounds lead to disulfide formation at critical thiols of the redox modulator site of the receptor,[73,74] which inhibits Ca^{2+} entry into the cell.

3.6.2 Nitric Oxide Synthase, Nitric Oxide, and Insulin Secretion

Fifteen years ago it was reported that sodium nitroprusside, now known to release NO, increased cyclic GMP (cGMP) in rat islets and caused insulin secretion.[75] Since it was subsequently established that NO stimulates guanylate cyclase, the same group[76] studied glucose- and arginine-stimulated insulin release and glucose-induced changes in cGMP in rat islets and RINm5F insulinoma cells. Both arginine and glucose increased islet cGMP levels. NOS inhibition reduced glucose- and arginine-induced insulin release and glucose-induced increases in cGMP. This suggested a regulation of the cGMP level in beta cells by NO (but see later cited Ref. 80).

Histochemical studies and protein immunoblots[77] indicated that beta-cell-derived HIT-T15 cells contain cNOS and produce NO from arginine in response to glucose. Secretion of insulin from these cells was preventable by NOS inhibition. Also in islets of rat pancreas cNOS was detected[77] but it is not clear whether in this organ NOS staining could be ascribed to beta cells. In fact, we were unable to detect any NOS immunochemically in beta cell of rats (R. Laffranchi and C. Richter, 1996). This does not rule out a physiological role of NO in beta cells because NOS was found in alpha cells, from where it conceivably exerts a paracrine effect.

Aqueous NO induced intracellular Ca^{2+} mobilization and secretory activity in rat pancreatic beta cells.[78] The fact that cells responded to NO in the absence of extracellular Ca^{2+}, and the inhibition by 10-μM thapsigargin, known to inhibit the microsomal Ca^{2+} uptake system quite specifically at nanomolar concentrations, were taken as evidence that NO mobilizes Ca^{2+} from microsomal pools and acts as a signaling molecule in beta-cell insulin secretion.

The precise effect of NO and cGMP on insulin secretion, however, remains controversial.[79,80] As seen below, NO seems to be involved in insulin secretion at least during its initial phase, possibly with mitochondria as targets.

3.6.3 Nitric Oxide in Mitochondria

3.6.3.1 Nitric Oxide as a Mitochondrial Poison

Excessive and prolonged production of oxidized nitrogen species inhibits the mitochondrial enzymes complex I and complex II of the electron transport chain, and cis-aconitase, an enzyme of the citric acid cycle.[81] The target of nitric oxide in these enzymes is the catalytically active nonheme iron-sulfur center. Nitric oxide binding results in loss of iron and permanent inhibition of mitochondrial respiration, which can lead to cellular dysfunction. NO· has been compared with $ONOO^-$ as to its inhibitory capacity in mitochondria. Aconitase is the principle site of inhibition by $ONOO^-$ (and O_2^-), while this enzyme is resistant to NO·.[82,83]

3.6.3.2 Nitrogen Monoxide (NO·) as a Regulator of Cytochrome Oxidase

The most-cited and best understood physiological target of NO· is guanyl cyclase. NO· binds to and stimulates it, and thus controls cell functions via cGMP, cGMP-gated channels, cGMP-dependent protein kinases, and phosphodiesterases. However, NO· also binds to cytochrome oxidase and reversibly inhibits respiration, as seen with the isolated enzyme, submitochondrial particles, mitochondria, hepatocytes, brain nerve terminals, and astrocytes.[84-91] Cytochrome oxidase inhibition is competitive with oxygen due to binding of NO· to the oxygen binding site heme a_3 and to Cu_B^+ of the reduced enzyme.[89,92] Why the inhibition is transient is not clear at the moment but several findings point to consumption of NO· as the underlying reason.[93] Thus, cytochrome oxidase can reduce NO·,[94] and NO· can combine with O_2 to form NO_x and with O_2^- to form $ONOO^-$.

Concentrations of NO· measured in a range of biologic systems are similar to those shown to inhibit cytochrome oxidase and mitochondrial respiration *in vitro,* and inhibition of NO· synthesis results in a stimulation of respiration in many systems. It was, therefore, recently proposed that NO· exerts a good part of its physiological and pathological effects on cells by inhibiting cytochrome oxidase.[89,95]

3.6.3.3 Nitric Oxide Synthase in Mitochondria

As mentioned above, there is evidence that NOS is associated with mitochondria. This offers exciting new insights into the biology of NO. For example, if an enzyme were stimulated by Ca^{2+} and were located in the matrix or at the inner side of the inner mitochondrial membrane, this would provide a self-regulating system for mitochondrial Ca^{2+} homeostasis in which Ca^{2+} uptake by mitochondria would lead to NO· formation. NO· could promote Ca^{2+} release (see below) by collapsing $\Delta\psi$ via inhibition of cytochrome oxidase.

3.6.3.4 Other Targets of Nitric Oxide in Mitochondria

Even if NO· were not formed inside mitochondria it may have a profound impact on the organelles, since it is uncharged and can easily traverse membranes. For example, extramitochondrially formed NO· could combine in mitochondria with O_2^- and form $ONOO^-$, which could then stimulate Ca^{2+} release from mitochondria with maintenance of $\Delta\psi$ (see below). Another possibility would be a NO·-catalyzed autoADPribosylation of mitochondrial NAD^+-binding proteins.[96,97] Also, mitochondria are rich in glutathione and contain key sulfhydryl enzymes such as the adenine nucleotide translocator or creatine kinase, which are putative targets of nitric oxide congeners.

3.7 Reactive Oxygen and Nitrogen Species as Regulators of Mitochondrial Ca^{2+} Homeostasis

3.7.1 Mitochondrial Ca^{2+} Release

In principle, Ca^{2+} can leave mitochondria in three ways: by nonspecific leakage through the inner membrane, by reversal of the influx carrier, and by a Na^+-dependent or -independent release pathway.[17,98] Only the latter two are physiologically relevant because they operate when $\Delta\psi$ is high. The Na^+-dependent pathway predominates in mitochondria of heart, brain, skeletal muscle, adrenal cortex, brown fat, and most tumor tissue. The Na^+-independent pathway is important in liver, kidney, lung, and smooth muscle mitochondria, probably exchanges Ca^{2+} with H^+, and is linked to the redox state of mitochondrial pyridine nucleotides. Compounds causing their oxidation (and hydrolysis) promote Ca^{2+} release from intact mitochondria. This release has been reviewed.[99–101]

3.7.2 Prooxidant-Induced, NAD$^+$-Linked Ca^{2+} Release

3.7.2.1 Specific Release from Energized Mitochondria

Ca^{2+} release from mitochondria was first associated with the oxidation of their pyridine nucleotides by Lehninger's group,[102] who showed that enzymatic oxidation of NAD(P)H by acetoacetate or oxaloacetate promotes release whereas reduction of NAD(P)$^+$ by β-hydroxybutyrate in the presence of rotenone prevents it. Shortly thereafter it was reported[103,104] that hydroperoxides such as *t*-butylhydroperoxide or hydrogen peroxide promote mitochondrial pyridine nucleotide oxidation followed by hydrolysis of NAD$^+$ to ADPribose and nicotinamide, and by Ca^{2+} release. Since then, many prooxidants were identified that stimulate Ca^{2+} release from intact liver, heart, brain, and kidney mitochondria. Ca^{2+} release is secondary to pyridine nucleotide oxidation and hydrolysis, and requires protein monoADPribosylation. The release is specific, occurs from energized mitochondria with preserved $\Delta\psi$, and does not engage formation of a "pore" or "permeability transition" (reviewed in Refs. 59 and 105).

Cyclosporine A (CSA) and its derivatives are useful for the study of mitochondrial Ca^{2+} handling. A number of them were tested as inhibitors of pyridine nucleotide hydrolysis, Ca^{2+} release, and the matrix-located peptidyl-prolyl *cis-trans* isomerase.[106] There is an impressive positive correlation between the extent of inhibition of these three parameters by the different cyclosporine derivatives. This strongly suggests that the ADPribose-dependent Ca^{2+} release engages peptidyl-prolyl *cis-trans* isomerase, and further documents the specificity of this pathway.

3.7.2.2 NAD$^+$ Hydrolysis Is Under the Control of Vicinal Thiols

NAD$^+$ hydrolysis, and therefore ADPribosylation and Ca^{2+} release, are under the control of vicinal thiols: phenylarsine oxide, which reversibly forms a five-membered ring with vicinal thiols, promotes the Ca^{2+}-dependent intramitochondrial NAD$^+$ hydrolysis, and thereby the specific Ca^{2+} release.[107] Gliotoxin, a fungal metabolite carrying a disulfide moiety also promotes the Ca^{2+}-dependent intramitochondrial NAD$^+$ hydrolysis and thereby the specific Ca^{2+} release, but is inactive when its sulfurs are reduced or methylated.[108] Thus, intramitochondrial, Ca^{2+}-dependent NAD$^+$ hydrolysis is prevented when some vicinal thiols are in the reduced SH-form, and occurs when they are connected, either by a cross-linking reagent or by oxidation to the disulfide form.

3.7.3 Peroxynitrite (ONOO⁻), a Thiol Oxidant, Stimulates the Specific Mitochondrial Ca^{2+} Release Pathway

Since ONOO⁻ oxidizes thiols, and since vicinal thiols control the specific mitochondrial Ca^{2+} release pathway it was tested whether ONOO⁻ is able to activate it. Peroxynitrite indeed induces Ca^{2+} release from rat liver mitochondria.[109] This release occurs (1) with preservation of $\Delta\psi$, (2) when mitochondrial pyridine nucleotides are oxidized but not when they are reduced, (3) parallel to NAD^+ hydrolysis, (4) in a CSA-inhibitable manner, (5) without inhibition of respiration, and (6) without entry of extramitochondrial solutes such as sucrose into mitochondria. These findings allow the conclusion that ONOO⁻ can mobilize mitochondrial Ca^{2+} by stimulating the specific, ADPribose-dependent release pathway.

3.7.4 Nitrogen Monoxide (NO·) Inhibits Cytochrome Oxidase and Causes Ca^{2+} Release from Deenergized Mitochondria

3.7.4.1 Studies with Hepatic Mitochondria

NO· at submicromolar, physiologically relevant concentrations potently deenergizes isolated mitochondria.[86] Deenergization is observed when mitochondria utilize respiratory substrates such as pyruvate *plus* malate, succinate, or ascrobate *plus* tetramethyl-phenylenediamine, but not when mitochondria are energized with ATP. Deenergization is due to a transient inhibition of cytochrome oxidase. Its extent and duration is determined by the concentration of NO· and oxygen, and the kind of respiratory substrate (Fig. 3.1). Importantly, cytochrome oxidase is particularly sensitive to NO· at oxygen concentrations below 30 μM[110] (Fig. 3.2), i.e., at intracellular oxygen tensions. The NO·-induced changes of the mitochondrial energy state are transient, and are paralleled by release and reuptake of mitochondrial Ca^{2+} (Fig. 3.3). These findings reveal a direct action of NO· on the mitochondrial respiratory chain and suggest that NO· exerts some of its physiological and pathological effects by deenergizing mitochondria.

3.7.4.2 Studies with Hepatocytes

Also in freshly prepared hepatocytes NO· deenergizes mitochondria.[87] Deenergization is reversible at low, but longer-lasting at higher NO· concentrations. The drop and the recovery of $\Delta\psi$ are accompanied by a rise and fall of cytosolic Ca^{2+} levels. NO at higher concentrations, provided by nitrosoglutathione in combination with dithiothreitol, kills hepatocytes. Killing is reduced when the cytosolic Ca^{2+} is chelated, or when Ca^{2+} "cycling" (Ca^{2+} release followed by reuptake) by

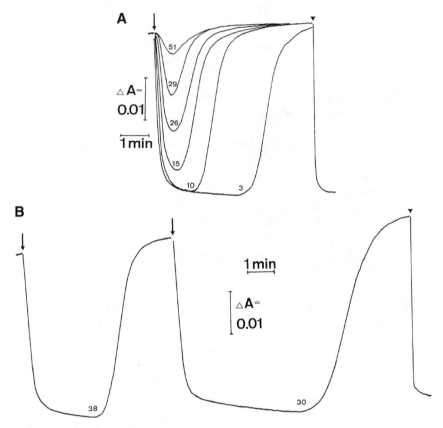

Figure 3.1. Transient deenergization by NO of rat liver and brain mitochondria. Rat liver (panels A to C) or brain (panel D) mitochondria were incubated at 1 mg of protein/ ml in buffer in the presence of 5-mM ethylene glycol bis(β-aminoethyl ether)-N,N,N',N'-tetraacetic acid and 10 μM safranine at various oxygen tensions. The membrane potential was measured by following safranine absorption at 511 to 533 nm. (A) After energization with potassium succinate (2.5 mM) in the presence of rotenone (5 μM), 2 μl of NO-saturated buffer was added (arrow) (final NO concentration, 1.3 μM). The numbers indicate the oxygen tension (in mm Hg) in the mitochondrial suspension. (B) Mitochondria were energized with 0.6-M ascorbate plus 10-mM tetramethyl-phenylenediamine in the presence of 5-μM rotenone. Twice μl of NO-saturated buffer (arrows) were added (final NO concentration 1.3 μM each), and the oxygen tension was measured as in panel A. (C) Mitochondria were energized with 1.5-mM ATP in the presence of 5-μM rotenone at 30 mm Hg oxygen tension. At the first two arrows, 10-μl NO-saturated buffer (final NO concentration, 6.5 μM each), at the third arrow, 20 μl of NO-saturated buffer (final NO concentration, 13 μM) were added. (D) Rat brain mitochondria were incubated in the presence of 5-mM pyruvate plus 2.5-mM malate at an oxygen tension of 28 mm Hg. At each arrow, 1-μl NO-saturated buffer (final NO concentration 0.65 μM, each) was added. At the triangles, 1-μM carbonyl cyanide *m*-chlorophenylhydrazone (panels A, B, and D) or 1.7-μg oligomycin/mg of mitochondrial protein (panel C) were added. (Taken from Schweizer and Richter)[86] with permission.)

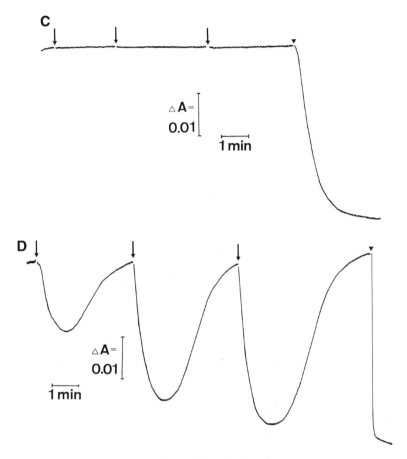

Figure 3.1. Continued

mitochondria is prevented by CSA. Apparently NO can kill cells by releasing Ca^{2+} from mitochondria and thereby flooding the cytosol with Ca^{2+}.

3.8 Nitrogen Monoxide (NO·) Stimulates Insulin Secretion by Inducing Ca^{2+} Release from Mitochondria

3.8.1 Studies with Added NO·

At lower micromolar concentration NO· stimulates an epinephrine-sensitive insulin secretion from cells of the beta cell line, INS-1 (Fig. 3.4).[111] Insulin secretion is paralleled by a reversible decrease of $\Delta\psi$, and by an increase of the cytosolic Ca^{2+}. When mitochondria had been depleted of their Ca^{2+} by uncoupling and blockade of the respiratory chain prior to NO· addition, no NO·-induced insulin

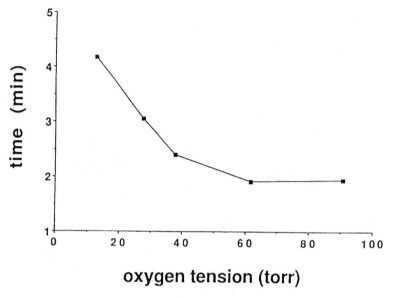

oxygen tension (torr)

Figure 3.2. Transient deenergization of isolated rat liver mitochondria by NO. Succinate-energized rat liver mitochondria were exposed to 0.5-µM NO under various oxygen tensions, and the mitochondrial membrane potential was measured as shown in Fig. 3.1. The time required to regain 50% of the absorbance change after deenergization is plotted against the oxygen tension. (Taken from Richter et al.[110] with permission.)

secretion was seen. Conversely, when the cells were treated with the mitochondrial poisons insulin secretion was stimulated. Chelation of intracellular, but not of extracellular Ca^{2+} prevents the NO·-induced insulin secretion.[111] Similar results were obtained with islets of Langerhans isolated from rat pancreas. These data show that NO· can stimulate insulin secretion by deenergizing mitochondria and thereby triggering mitochondrial Ca^{2+} release. Therefore, NO· may play a role in modulating physiological insulin secretion from beta cells (see also the next section).

3.8.2 First Phase of Physiological Insulin Secretion from Beta Cells Depends on Nitric Oxide

As outlined above, insulin secretion occurs in at least two phases and requires an increase in the cytosolic Ca^{2+} content. We have found (R. Laffranchi, M. Reinecke, I. David, C. Richter, and G.A. Spinas, manuscript submitted) that NO participates in glucose-stimulated insulin secretion. The action of NO predominates in the initial phase of secretion.

Islets of Langerhans were isolated from the pancreas of newborn rats (Zur-SD) and precultured in RPMI-1640 for 5 to 6 days. Thereafter, 50 islets were

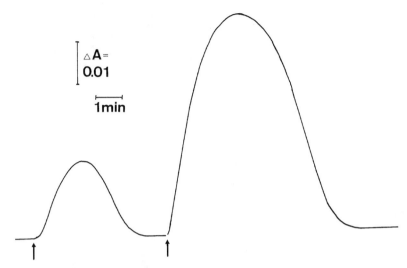

Figure 3.3. Ca^{2+} uptake and release by rat liver mitochondria exposed to NO. Mitochondria at 1 mg of protein/ml were incubated in buffer in the presence of 50 μM of the Ca^{2+} indicator arsenazo III at an oxygen tension of 30 mmHg. Ca^{2+} uptake and release were measured by following the absorption at 675 to 685 nm. After energization with potassium succinate (2.5 mM) in the presence of rotenone (5 μM), twice 2 μl of NO-saturated buffer (final NO concentration 1.3 μM) were added (arrows). The mitochondrial Ca^{2+} content was 10 nmol/mg of protein as determined with arsenazo III after addition of carbonyl cyanide *m*-chlorophenylhydrazone. (Taken from Schweizer and Richter,[86] with permission.)

transferred to a perfusion chamber (300 μl) and perifused with Krebs-Ringer buffer at a flow rate of 500 μl/min. After an equilibration period of 45 min in Krebs-Ringer buffer plus 1.6-mM glucose 0.5-mM L-methylarginine (L-NMMA, a NOS inhibitor), 0.5-mM D-methylarginine (D-NMMA, not capable of NOS inhibition), or 50-μM trifluoperazine (TFP, a calmodulin antagonist), respectively, were added to the buffer, or the buffer was left unchanged (control). Glucose concentration was kept at 1.6 mM for another 10 min, was then raised to 16 mM for 40 min and was finally reduced to 1.6 mM for 15 min.

The increase in glucose concentration resulted in a typical biphasic insulin secretion from control islets. The NO synthase inhibitor L-NMMA caused a marked decrease of the first phase of insulin secretion: area under the curve (AUC, a measure of the secreted insulin, in picograms) 142.1 ± 31.0 pg/islet × 10 min versus 285.1 ± 39.4 pg/islet × 10 min in control islets ($p < 0.01$). The second phase was not influenced by L-NMMA: AUC 764.8 ± 97.7 pg/islet × 30 min and 805.8 ± 256.4 pg/islet × 30 min in control and L-NMMA-exposed islets, respectively. Perifusion with the inactive enantiomer D-NMMA showed no significant difference as compared with control islets. Addition of TFP to the perifusate resulted in an inhibition of both phases of insulin release: first phase

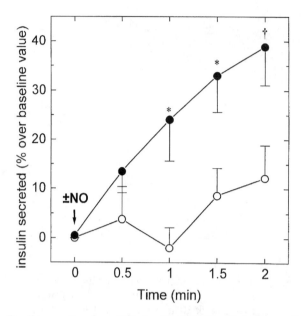

Figure 3.4. Insulin secretion from INS-1 cells. NO (2 μM) in form of NO-saturated buffer was added (time 0 on the abscissa) 1 min after suspending the 1.5×10^7 cells. NO treatment is shown as closed symbols; controls (no NO, or nitrogen-purged buffer added) are shown as open symbols. Insulin accumulation was measured at the time points indicated. Secreted insulin is expressed as percentage of base line values at time 0 (48.9 ± 8.0 ng/ml and 44.2 ± 2.8 ng/ml in controls and NO-treated cells, respectively). Data are mean ± standard error of the mean (SEM) of five to eight experiments. *$P < 0.03$; †$P < 0.05$ as compared to controls, respectively. (Taken from Laffranchi et al.,[111] with permission).

AUC 291.1 ± 27.1 pg/islet × 10 min in controls versus 48.3 ± 15.5 pg/islet × 10 min with TFP ($p < 0.05$); second phase AUC 686.7 ± 155.1 pg/islet × 30 min in controls versus 254.6 ± 77.6 pg/islet × 30 min with TFP ($p < 0.05$).

The finding that TFP inhibits the first phase of insulin secretion is in line with the inhibition of the first phase by the specific NO synthase inhibitor ʟ-NMMA, because the constitutive isoform of NO synthase is a calmodulin-dependent enzyme. Thus, these results show that endogenously formed NO is to a considerable extent involved in the signal transduction of the first phase of glucose-stimulated insulin secretion.

How could a modulation of physiological insulin secretion from beta cells come about? It is conceivable that NO raises the cytosolic Ca^{2+} content by deenergizing mitochondria and thereby causing their Ca^{2+} to be released before the Ca^{2+} channels of the plasma membrane are opened. Alternatively, by deenergizing mitochondria NO may prevent Ca^{2+} uptake into these organelles when the plasma membrane channels are opened. Detailed time-course studies at the single-cell level will provide answers to these crucial questions.

3.9 Conclusion and Outlook

Both nitric oxide and Ca^{2+} are used in biology as messenger molecules. Recent findings show that mitochondria are of central importance for an interplay between these two messenger molecules. By transiently deenergizing mitochondria, NO· can allow relatively large increases in cytoplasmic Ca^{2+} levels to occur. Since in many short-term cell signaling systems Ca^{2+} enters the cell via the plasma membrane or is mobilized from intracellular stores such as the endoplasmic reticulum, the action of NO· on cytochrome oxidase can be expected to be of general relevance in short-term cell signaling. For example, it will be interesting to investigate the influence of NO· on mitochondria of glutamate-excitable neurons, in which on stimulation, Ca^{2+} enters the cell and NO· is produced.

It should also be kept in mind that $ONOO^-$ mobilizes Ca^{2+} from mitochondria without deenergizing them. This may be important for the maintenance of intramitochondrial Ca^{2+} homeostasis, which is crucial for cellular fuel homeostasis. In this way, $ONOO^-$ may participate at the mitochondrial level in long-term cell signaling systems.

References

1. Unger, R. H. 1981. The milieu interieur and the islets of Langerhans. *Diabetologia* **20**:1–11.

2. Thorens, B., H. K. Sarkar, H. R. Kaback, and H. F. Lodish. 1988. Cloning and functional expression in bacteria of a novel glucose transporter present in liver, intestine, kidney, and beta-pancreatic islet cells. *Cell* **55**:281–290.

3. Matschinsky, F. M., and J. E. Ellerman. 1968. Metabolism of glucose in the islets of Langerhans. *J. Biol. Chem.* **243**:2730–2736.

2. Misler, S., L. C. Falke, K. Gillis, and M. L. McDaniel. 1986. A metabolite-regulated potassium channel in rat pancreatic B cells. *Proc. Natl. Acad. Sci. USA* **83**:7119–7123.

5. Prentki, M., and F. M. Matschinsky. 1987. Ca^{2+}, cAMP, and phospholipid-derived messengers in coupling mechanisms of insulin secretion. *Physiol. Rev.* **67**:1185–1248.

6. Roe, M. W., M. E. Lancaster, R. J. Mertz, J. F. Worley III, and I. D. Dukes. 1993. Voltage-dependent intracellular calcium release from mouse islets stimulated by glucose. *J. Biol. Chem.* **268**:9953–9956.

7. Rojas, E., P. B. Carroll, C. Ricordi, A. C. Boschero, S. S. Stojilkovic, and I. Atwater. 1994. Control of cytosolic free calcium in cultured human pancreatic beta-cells occurs by external calcium-dependent and -independent mechanisms. *Endocrinology* **134**:1771–1781.

8. Takasawa, S., K. Nata, H. Yonekura, and H. Okamoto. 1993. Cyclic ADP-ribose in insulin secretion from pancreatic beta cells. *Science* **259**:370–373.

9. Kato, I., S. Takasawa, A. Akabane, O. Tanaka, H. Abe, T. Takamura, Y. Suzuki, K. Nata, H. Yonekura, T. Yoshimoto, and H. Okamoto. 1995. Regulatory role of CD38 (ADP-ribosyl cyclase/cyclic ADP-ribose hydrolase) in insulin secretion by glucose in pancreatic β cells. *J. Biol. Chem.* **270:**30045–30050.

10. Smith, P. A., P. Rorsman, and F. M. Ashcroft. 1989. Modulation of dihydropyridine-sensitive Ca^{2+} channels by glucose metabolism in mouse pancreatic beta-cells. *Nature* **342:**550–553.

11. Bergsten, P., E. Grapengiesser, E. Gylfe, A. Tengholm, and B. Hellman. 1994. Synchronous oscillations of cytoplasmic Ca^{2+} and insulin release in glucose-stimulated pancreatic islets. *J. Biol. Chem.* **269:**8749–8753.

12. Bergsten, P. 1995. Slow and fast oscillations of cytoplasmic Ca^{2+} in pancreatic islets correspond to pulsatile insulin release. *Am. J. Physiol.* **268:**E282–E287.

13. Burgess, T. L., and R. B. Kelly. 1987. Constitutive and regulated secretion of proteins. *Annu. Rev. Cell. Biol.* 3:243–293.

14. Olszewski, S., J. T. Deeney, G. T. Schuppin, K. P. Williams, B. E. Corkey, and C. J. Rhodes. 1994. Rab3A effector domain peptides induce insulin exocytosis via a specific interaction with a cytosolic protein doublet. *J. Biol. Chem.* **269:**27987–27991.

15. Sadoul, K., J. Lang, C. Montecucco, U. Weller, R. Regazzi, S. Catsicas, C. B. Wollheim, and P. A. Halban. 1995. SNAP-25 is expressed in islets of Langerhans and is involved in insulin release. *J. Cell Biol.* **128:**1019–1028.

16. Lang, J., I. Nishimoto, T. Okamoto, R. Regazzi, C. Kiraly, U. Weller, and C. B. Wollheim. 1995. Direct control of exocytosis by receptor-mediated activation of the heterotrimeric GTPases G_i and G_0 or by the expression of the active G_α subunits. *EMBO J.* **14:**3635–3644.

17. Carafoli, E. 1987. Intracellular calcium homeostasis. *Annu. Rev. Biochem.* **56:**395–433.

18. Wendt-Gallitelli, M. F., and G. Isenberg. 1991. Total and free myoplasmic calcium during a contraction cycle: x-ray microanalysis in guinea-pig ventricular myocytes. *J. Physiol. (London)* **435:**349–372.

19. Rizzuto, R., A. W. M. Simpson, M. Brini, and T. Pozzan. 1992. Rapid changes of mitochondrial Ca^{2+} revealed by specifically targeted recombinant aequorin. *Nature* **358:**325–327.

20. Rizzuto, R., M. Brini, M. Murgia, and T. Pozzan. 1993. Microdomains with high Ca^{2+} close to the IP_3-sensitive channels that are sensed by neighboring mitochondria. *Science* **262:**744–747.

21. Rizzuto, R., C. Bastianutto, M. Brini, M. Murgia, and T. Pozzan. 1994. Mitochondrial Ca^{2+} homeostasis in intact cells. *J. Cell Biol.* **126:**1183–1194.

22. Rutter, G. A., J.-M. Theler, M. Murgia, C. B. Wollheim, T. Pozzan, and R. Rizzuto. 1993. Stimulated Ca^{2+} influx raises mitochondrial free Ca^{2+} to supramicromolar levels in a pancreatic β-cell line. Possible role in glucose and agonist-induced insulin secretion. *J. Biol. Chem.* **268:**22385–22390.

23. Miyata, H., H. S. Silverman, S. J. Sollott, E. G. Lakatta, M. D. Stern, and R. G. Hansford. 1991. Measurement of mitochondrial free Ca^{2+} concentration in living single rat cardiac myocytes. *Am. J. Physiol.* **261**:H1123–H1134.

24. Bird, G. S. J., J. F. Obie, and J. W. Putney, Jr. 1992. Functional homogeneity of the non-mitochondrial Ca^{2+} pool in intact mouse lacrimal acinar cells. *J. Biol. Chem.* **267**:18382–18386.

25. Toescu, E. C., J. M. Gardner, and O. H. Petersen. 1993. Mitochondrial Ca^{2+} uptake at submicromolar $[Ca^{2+}]_i$ in permeabilized pancreatic acinar cells. *Biochem. Biophys. Res. Commun.* **192**:854–859.

26. Mix, T. C. H., R. M. Drummond, R. A. Tuft, and F. S. Fay. 1994. Mitochondria in smooth muscle sequester Ca^{2+} following stimulation of cell contraction. *Biophys. J.* **66**:A97.

27. Friel, D. D., and R. W. Tsien. 1994. An FCCP-sensitive Ca^{2+} store in bullfrog sympathetic neurons and its participation in stimulus-evoked changes in $[Ca^{2+}]_i$. *J. Neurosci.* **14**:4007–4024.

28. Werth, J. L., and S. A. Thayer. 1994. Mitochondria buffer physiological calcium loads in cultured rat dorsal root ganglion neurons. *J. Neurosci.* **14**:348–356.

29. Loew, L. M., W. Carrington, R. A. Tuft, and F. S. Fay. 1994. Physiological cytosolic Ca^{2+} transients evoke concurrent mitochondrial depolarization. *Proc. Natl. Acad. Sci. USA* **91**:4340–4344.

30. Rapp, P. E., and M. J. Berridge. 1981. The control of transepithelial potential oscillations in the salivary gland of *Calliphora erythocephala*. *J. Exp. Biol.* **93**:119–132.

31. Rooney, T. A., E. Sass, and A. P. Thomas. 1990. Agonist-induced cytosolic calcium oscillations originate from a specific locus in single hepatocytes. *J. Biol. Chem.* **265**:10792–10796.

32. Hajnoczky, G., L. D. Robb-Gaspers, M. B. Seitz, and A. P. Thomas. 1995. Decoding of cytosolic calcium oscillations in the mitochondria. *Cell* **82**:415–424.

33. Jouaville, L. S., F. Ichas, E. L. Holmuhamedov, P. Camacho, and J. D. Lechleiter. 1995. Synchronization of calcium waves by mitochondrial substrates in *Xenopus laevis* oocytes. *Nature* **377**:438–441.

34. Sparagna, G. C., K. K. Gunter, S.-S. Sheu, and T. E. Gunter. 1995. Mitochondrial calcium uptake from physiological-type pulses of calcium. A description of the rapid uptake mode. *J. Biol. Chem.* **270**:27510–27515.

35. McCormack, J. G., A. P. Halestrap, and R. M. Denton. 1990. Role of calcium ions in regulation of mammalian intramitochondrial metabolism. *Physiol. Rev.* **70**:391–425.

36. McCormack, J. G., and R. M. Denton. 1985. Hormonal control of intramitochondrial Ca^{2+}-sensitive enzymes in heart, liver and adipocyte tissue. *Biochem. Soc. Trans.* **13**:664–667.

37. Halestrap, A. P., and E. J. Griffiths. 1993. Mitochondrial pyrophosphate metabolism in health and disease, pp. 365–377. *In* L. H. Lash and D. P. Jones (eds.), *Mitochondrial Dysfunction*. Academic Press, San Diego.

38. Yamada, E. W., and N. J. Huzel. 1988. The calcium-binding ATPase inhibitor protein from bovine heart mitochondria. Purification and properties. *J. Biol. Chem.* **263:**11498–11503.

39. Brown, G. C. 1992. Control of respiration and ATP synthesis in mammalian mitochondria and cells. *Biochem. J.* **284:**1–13.

40. Sekine, N., V. Cirulli, R. Regazzi, L. J. Brown, E. Gine, J. Tamarit-Rodriguez, M. Girotti, S. Marie, M. J. MacDonald, C. B. Wollheim, and G. A. Rutter. 1994. Low lactate dehydrogenase and high mitochondrial glycerol phosphate dehydrogenase in pancreatic β-cells. Potential role in nutrient sensing. *J. Biol. Chem.* **269:**4895–4902.

41. Pralong, W.-F., L. Hunyady, P. Varnai, C. B. Wollheim, and A. Spät. 1992. Pyridine nucleotide redox state parallels production of aldosterone in potassium-stimulated adrenal glomerulosa cells. *Proc. Natl. Acad. Sci. USA* **89:**132–136.

42. Capponi, A. M., M. F. Rossier, E. Davies, and M. B. Valloton. 1988. Calcium stimulates steroidogenesis in permeabilized bovine adrenal cortical cells. *J. Biol. Chem.* **263:**16113–16117.

43. Pralong, W.-F., C. Bartley, and C. B. Wollheim. 1990. Single islet β-cell stimulation by nutrients: relationship between pyridine nucleotides, cytosolic Ca^{2+} and secretion. *EMBO J.* **9:**53–60.

44. Pralong, W.-F., A. Spät, and C. B. Wollheim. 1994. Dynamic pacing of cell metabolism by intracellular Ca^{2+} transients. *J. Biol. Chem.* **269:**27310–27314.

45. Ramirez, R., F. Malaisse-Lagae, A. Sener, and W. J. Malaisse. 1995. The coupling of metabolic to secretory events in pancreatic islets: the mitochondrial redox state. *Med. Sci. Res.* **23:**249–250.

46. Ramirez, R., A. Sener, and W. J. Malaisse. 1995. Hexose metabolism in pancreatic islets: effect of D-glucose on the mitochondrial redox state. *Mol. Cell. Biochem.* **142:**43–48.

47. Ramirez, R., A. Sener, and W. J. Malaisse. 1995. Hexose metabolism in pancreatic islets: regulation of the mitochondrial NADH/NAD$^+$ ratio. *Biochem. Mol. Med.* **55:**1–7.

48. Chance, B., H. Sies, and A. Boveris. 1979. Hydroperoxide metabolism in mammalian organs. *Physiol. Rev.* **59:**527–605.

49. Cadenas, E. 1989. Biochemistry of oxygen toxicity. *Annu. Rev. Biochem.* **58:**79–110.

50. Weisiger, R. A., and I. Fridovich. 1973. Mitochondrial superoxide dismutase. Site of synthesis and intramitochondrial localization. *J. Biol. Chem.* **248:**4793–4796.

51. Loschen, G., A. Azzi, C. Richter, and L. Flohe. 1974. Superoxide radicals as precursors of mitochondrial hydrogen peroxide. *FEBS Lett.* **42:**68–72.

52. Nohl, H., and W. Jordan. 1986. The mitochondrial site of superoxide formation. *Biochem. Biophys. Res. Commun.* **138:**533–539.

53. Nohl, H. 1991. Formation of reactive oxygen species associated with mitochondrial respiration, pp. 108–116. *In* K. J. P. Davies (ed.), *Oxidative Damage and Repair.* Pergamon Press, Oxford.

54. Forman, H. J., and J. Kennedy. 1976. Dihydroorotate-dependent superoxide production in rat brain and liver. A function of the primary dehydrogenase. *Arch. Biochem. Biophys.* **217**:411–421.

55. Nohl, H. 1987. A novel superoxide radical generator in heart mitochondria. *FEBS Lett.* **214**:269–273.

56. Sohal, R. S. 1993. Aging, cytochrome oxidase activity, and hydrogen peroxide release by mitochondria. *Free Rad. Biol. Med.* **14**:583–588.

57. Richter, C., and B. Frei. 1988. Ca^{2+} release from mitochondria induced by prooxidants. *Free Rad. Biol. Med.* **4**:365–375.

58. Richter, C., and G. E. N. Kass. 1991. Oxidative stress in mitochondria: Its relationship to cellular Ca^{2+} homeostasis, cell death, proliferation, and differentiation. *Chem. Biol. Interact.* **77**:1–23.

59. Richter, C., and M. Schweizer. In press. Oxidative stress in mitochondria. *In* J. Scandalios (ed.), Oxidative Stress and the Molecular Biology of Antioxidant Defenses, Cold Spring Harbor Laboratory Press, Cold Spring Harbor, NY.

60. Kröncke, K.-D., K. Fehsel, and V. Kolb-Bachofen. 1995. Inducible nitric oxide synthase and its product nitric oxide, a small molecule with complex biological activities. *Biol. Chem. Hoppe-Seyler* **376**:327–343.

61. Snyder, S. H. 1995. No endothelial NO. *Nature* **377**:196–197.

62. Sessa, W. C., G. Garcia-Cardenas, J. Liu, A. Keh, J. S. Pollock, J. Bradley, S. Thiru, I. M. Braverman, and K. M. Desai. 1995. The Golgi association of endothelial nitric oxide synthase is necessary for the efficient synthesis of nitric oxide. *J. Biol. Chem.* **270**:17641–17644.

63. Kobzik, L., B. Stringer, J. L. Balligand, M. B. Reid, and J. S. Stamler. 1995. Endothelial type nitric oxide synthase in skeletal muscle fibers: mitochondrial relationships. *Biochem. Biophys. Res. Commun.* **211**:375–381.

64. Bates, T. E., A. Loesch, G. Burnstock, and J. B. Clark. 1995. Immunochemical evidence for a mitochondrially located nitric oxide synthase in brain and liver. *Biochem. Biophys. Res. Commun.* **213**:896–900.

65. Nichol, K. A., N. Chan, D. F. Davey, and M. R. Bennett. Location of nitric oxide synthase in the developing avian ciliary ganglion. *J. Auton. Nerv. Syst.* **51**:91–102.

66. Stamler, J. S., D. J. Singel, and J. Loscalzo. 1992. Biochemistry of nitric oxide and its redox-activated forms. *Science* **258**:1898–1902.

67. Stamler, J. S. 1994. Redox signaling: nitrosylation and related target interactions of nitric oxide. *Cell* **78**:931–936.

68. Lipton, S. A., Y. B. Choi, Z. H. Pan, S. Z. Lei, H. S. V. Chen, N. J. Sucher, J. Loscalzo, D. J. Singel, and J. S. Stamler. 1993. A redox-based mechanism for the neuroprotective and neurodestructive effects of nitric oxide and related nitroso-compounds. *Nature* **364**:626–632.

69. Beckman, J. S., T. W. Beckman, J. Chen, P. A. Marshall, and B. A. Freeman. 1990. Apparent hydroxyl radical production by peroxynitrite: Implications for endothelial injury from nitric oxide and superoxide. *Proc. Natl. Acad. Sci. USA* **87**:1620–1624.

70. Radi, R., J. S. Beckman, K. M. Bush, and B. A. Freeman. 1991. Peroxynitrite oxidation of sulfhydryls. The cytotoxic potential of superoxide and nitric oxide. *J. Biol. Chem.* **266**:4244–4250.

71. Lin, K.-T., J.-Y. Xue, M. Nomen, B. Spur, and P. Y.-K. Wong. 1995. Peroxynitrite-induced apoptosis. *J. Biol. Chem.* **270**:16487–16490.

72. Murphy, M. E., and H. Sies. 1991. Reversible conversion of nitroxyl anion to nitric oxide by superoxide dismutase. *Proc. Natl. Acad. Sci. USA* **88**:10680–10684.

73. Lipton, S. A., D. J. Singel, and J. S. Stamler. 1994. Redox-activated states of nitric oxide determine neuronal protection *versus* neuronal injury, pp. 183–189. *In* H. Takagi, N. Toda, and R. D. Hawkins (eds.), *Nitric Oxide: Roles in Neuronal Communication and Neurotoxicity.* Japan Scientific Societies Press and CRC Press, Tokyo.

74. Lipton, S. A., W.-K. Kim, P. V. Rayudu, W. Asaad, D. R. Arnelle and J. S. Stamler. 1995. Singlet and triplet nitroxyl anion (NO⁻) lead to NMDA receptor downregulation and neuroprotection. *Endothelium* **3**:S44 (abstract 174).

75. Laychock, S. G. 1981. Evidence for guanosine 3′,5′-monophosphate as a putative mediator of insulin secretion from isolated rat islets. *Endocrinology* **108**:1197–1205.

76. Laychock, S. G., M. E. Modica, and C. T. Cavanaugh. 1991. L-arginine stimulates cyclic guanosine 3′,5′-monophosphate formation in rat islets of Langerhans and RINm5F insulinoma cells: evidence for L-arginine:nitric oxide synthase. *Endocrinology* **129**:3043–3052.

77. Schmidt, H. H. H. W., T. D. Warner, K. Ishii, H. Sheng, and F. Murad. 1992. Insulin secretion from pancreatic B cells caused by L-arginine-derived nitrogen oxides. *Science* **255**:721–723.

78. Willmott, N. J., A. Galione, and P. A. Smith. 1995. Nitric oxide induces intracellular Ca²⁺ mobilization and increases secretion of incorporated 5-hydroxytryptamine in rat pancreatic β-cells. *FEBS Lett.* **371**:99–104.

79. Corbett, J. A., M. A. Sweetland, J. L. Wang, J. R. Lancaster, Jr., and M. L. McDaniel. 1993. Nitric oxide mediates cytokine-induced inhibition of insulin secretion by human islets of Langerhans. *Proc. Natl. Acad. Sci. USA* **90**:1731–1735.

80. Jones, P. M., S. J. Persaud, T. Bjaaland, J. D. Pearson, and S. L. Howell. 1992. Nitric oxide is not involved in the initiation of insulin secretion from rat islets of Langerhans. *Diabetologia* **35**:1020–1027.

81. Dawson, T. M., V. L. Dawson, D. A. Bartley, G. R. Uhl, and S. H. Snyder. 1992. A novel neuronal messenger molecule in brain: the free radical, nitric oxide. *J. Neurosci.* **13**:297–311.

82. Castro, L., M. Rodriguez, and R. Radi. 1994. Aconitase is readily inactivated by peroxynitrite, but not by its precursor, nitric oxide. *J. Biol. Chem.* **269**:29409–29415.

83. Hausladen, A., and I. Fridovich. 1994. Superoxide and peroxynitrite inactivate aconitases, but nitric oxide does not. *J. Biol. Chem.* **269**:29405–29408.

84. Brown, G. C., and C. E. Cooper. 1994. Nanomolar concentrations of nitric oxide reversibly inhibit synaptosomal respiration by competing with oxygen at cytochrome oxidase. *FEBS Lett.* **356**:295–298.

85. Carr, G. J., and S. J. Ferguson. 1990. Nitric oxide formed by nitrite reductase of *Paracoccus denitrificans* is sufficiently stable to inhibit cytochrome oxidase activity and is reduced by its reductase under aerobic conditions. *Biochim. Biophys. Acta* **1017**:57–62.

86. Schweizer, M., and C. Richter. 1994. Nitric oxide potently and reversibly deenergizes mitochondria at low oxygen tension. *Biochem. Biophys. Res. Commun.* **204**:169–175.

87. Richter, C., V. Gogvadze, R. Schlapbach, M. Schweizer, and J. Schlegel. 1994. Nitric oxide kills hepatocytes by mobilizing mitochondrial calcium. *Biochem. Biophys. Res. Commun.* **205**:1143–1150.

88. Cleeter, M. W. J., J. M. Cooper, V. M. Darley-Usmar, S. Moncada, and A. H. V. Schapira. 1994. Reversible inhibition of cytochrome c oxidase, the terminal enzyme of the mitochondrial respiratory chain, by nitric oxide. Implications for neurodegenerative diseases. *FEBS Lett.* **345**:50–54.

89. Torres, J., V. Darley-Usmar, and M. T. Wilson. 1995. Inhibition of cytochrome c oxidase in turnover by nitric oxide: mechanism and implications for control of respiration. *Biochem. J.* **312**:169–173.

90. Brown, G. C., J. P. Bolaños, S. J. R. Heales, and J. B. Clark. 1995. Nitric oxide produced by activated astrocytes rapidly and reversibly inhibits cellular respiration. *Neurosci. Lett.* **193**:201–204.

91. Takehara, Y., T. Kanno, T. Yoshioka, M. Inoue, and K. Utsumi. 1995. Oxygen-dependent regulation of mitochondrial energy metabolism by nitric oxide. *Arch. Biochem. Biophys.* **323**:27–32.

92. Brudvig, O. W., O. H. Stevens, and O. I. Chan. 1980. Reactions of nitric oxide with cytochrome oxidase. *Biochemistry* **19**:5275–5285.

93. Clarkson, R. B., S. W. Norby, A. Smirnov, S. Boyer, N. Vahidi, R. W. Nims, and D. A. Wink. 1995. Direct measurement of the accumulation and mitochondrial conversion of nitric oxide within chinese hamster ovary cells using an intracellular electron paramagnetic resonance technique. *Biochem. Biophys. Acta* **1243**:496–502.

94. Zhao, X. J., V. Sampath, and W. W. Caughey. 1995. Cytochrome c oxidase catalysis of the reduction of nitric oxide to nitrous oxide. *Biochem. Biophys. Res. Commun.* **212**:1054–1060.

95. Brown, G. C. 1995. Nitric oxide regulates mitochondrial respiration and cell functions by inhibiting cytochrome oxidase. *FEBS Lett.* **369**:136–139.

96. McDonald, L. S., and J. Moss. 1993. Stimulation by nitric oxide of an NAD linkage to glyceraldehyde-3-phosphate dehydrogenase. *Proc. Natl. Acad. Sci. USA* **90**:6238–6241.

97. Brüne, B., S. Dimmeler, L. M. Y. Vedia and E. G. Lapetina. 1994. Nitric oxide—a signal for ADP-ribosylation of proteins. *Life Sci.* **54**:61–70.

98. Crompton, M. 1985. The regulation of mitochondrial calcium transport in heart. *Curr. Top. Membr. Transp.* **25**:231–276.

99. Richter, C., and B. Frei. 1988. Ca^{2+} release from mitochondria induced by prooxidants. *Free Rad. Biol. Med.* **4**:365–375.

100. Richter, C., and G. E. N. Kass. 1991. Oxidative stress in mitochondria: its relationship to cellular Ca^{2+} homeostasis, cell death, proliferation, and differentiation. *Chem. Biol. Interact.* **77**:1–23.

101. Richter, C. 1992. Mitochondrial calcium transport, pp. 349–358. *In* N. Neuberger and L. L. M. Van Deenen (eds.), *New Comprehensive Biochemistry Volume Molecular Mechanisms in Bioenergetics,* L. Ernster (ed.). Elsevier, Amsterdam.

102. Lehninger, A. L., A. Vercesi, and E. A. Bababunmi. 1978. Regulation of Ca^{2+} release from mitochondria by the oxidation-reduction state of pyridine nucleotides. *Proc. Natl. Acad. Sci. USA* **75**:1690–1694.

103. Lötscher, H. R., K. H. Winterhalter, E. Carafoli, and C. Richter. 1979. Hydroperoxides can modulate the redox state of mitochondrial pyridine nucleotides and the calcium balance in rat liver mitochondria. *Proc. Natl. Acad. Sci. USA* **76**:4340–4344.

104. Lötscher, H. R., K. H. Winterhalter, E. Carafoli, and C. Richter. 1980. Hydroperoxide-induced loss of pyridine nucleotides and release of calcium from rat liver mitochondria. *J. Biol. Chem.* **255**:12579–12583.

105. Richter, C., and J. Schlegel. 1993. Mitochondrial Ca^{2+} release induced by prooxidants. *Toxicol. Lett.* **67**:119–127.

106. Schweizer, M., J. Schlegel, D. Baumgartner, and C. Richter. 1993. Sensitivity of mitochondrial peptidyl-prolyl *cis-trans* isomerase, pyridine nucleotide hydrolysis, and Ca^{2+} release to cyclosporine A and related compounds. *Biochem. Pharmacol.* **45**:641–646.

107. Schweizer, M., P. Durrer, and C. Richter. 1994. Phenylarsine oxide stimulates the pyridine nucleotide-linked Ca^{2+} release from rat liver mitochondria. *Biochem. Pharmacol.* **48**:967–973.

108. Schweizer, M., and C. Richter. 1994. Gliotoxin stimulates Ca^{2+} release from intact rat liver mitochondria. *Biochemistry* **33**:13401–13405.

109. Schweizer, M., and C. Richter. In press. Peroxynitrite stimulates the pyridine nucleotide-linked Ca^{2+} release from intact rat liver mitochondria. *Biochemistry* **35**:4524–4528.

110. Richter, C., V. Gogvadze, R. Laffranchi, R. Schlapbach, M. Schweizer, M. Suter, P. Walter, and M. Yaffee. 1995. Oxidants in mitochondria: from physiology to diseases. *Biochim. Biophys. Acta* **1271**:67–74.

111. Laffranchi, R., V. Gogvadze, C. Richter, and G. A. Spinas. 1995. Nitric oxide (nitrogen monoxide, NO) stimulates insulin secretion by inducing calcium release from mitochondria. *Biochem. Biophys. Res. Commun.* **217**:584–591.

4

Phospholipase A₂ Activation: An Early Manifestation of Oxidative Stress

Alex Sevanian and Julia Rashba-Step

4.1 General Considerations

Phospholipase A₂ (PLA₂) is recognized as an integral component of the oxidant stress response system. This enzyme is well poised among the cellular oxidant stress response components due to its inherent function in maintaining cell membrane phospholipid composition and modifying the fatty acyl profile according to the functional demands of the cell. The "fluidity" and organization of membrane phospholipids is manipulated through PLA₂ and associated acyl-transferase enzyme activities that *tailor* phospholipids according to the cells needs. This modulation of membrane phospholipid structure can be viewed as fundamentally linked to the other important role of many cell-associated PLA₂s in regulating production of eicosanoids and related autocoids. The rate of eicosanoid production has been shown to be dependent on the ambient lipid peroxide concentrations, and this so-called "peroxide tone" affects not only enzymes that convert unsaturated fatty acids to autocoid products (represented largely by lipoxygenases and cyclooxygenases) but also PLA₂ activity, which provides the fatty acids to these enzymes. Accordingly, oxidant-mediated PLA₂ activity should be considered as a mediator of processes such as inflammation,[1] the immune response,[2] and NADPH oxidase activation.[3] Oxidative stress and resulting cell injury, as found, for example, after tissue ischemia, causes perturbations in eicosanoid synthesis when tensions are high or when oxygen is reintroduced to ischemic cells.[4,5] Marked increases in the levels of free fatty acids, attributable to PLA₂, have been reported in ischemic tissues or after ischemia and reperfusion[5,6] as well as in a variety of other forms of oxidant stress where increased free radical formation is demonstrated.[7,8]

Increases in PLA₂ catalytic activity are usually evident after membrane disruption, as occurs after lipid peroxidation. This has been shown in model membranes as well as biologic membranes[9] and intact cells. Many PLA₂, including those that have no known role in autocoid production, respond similarly to disruptions in phospholipid structural organization. The use of chaotropic agents[10] or oxidation of phospholipids in bilayer, micellar, and other nonbilayer structures, usually results in higher PLA_2 hydrolytic activity toward the component phospholipids. Oxidation of specific membrane-associated proteins, such as receptors or transducer components, produces transient increases in catalytic activity. Broadening research in the area of cell signal transduction biochemistry has begun to clarify the role of oxidants in mediating or modulating a host of signaling events. This chapter highlights aspects of the fields of PLA_2 enzymology and membrane biochemistry that describe points of interaction between oxidants, oxidative modification of phospholipids, or derived fatty acids, and signaling pathways involving PLA_2 activity. Emphasis is placed on the major PLA_2 classes that have been studied in terms of their catalytic activities toward oxidized membranes, or the modulation of their activities by oxidant action on cells or cell membranes. The PLA_2 of interest include those listed in the Table 4.1 below, and discussion is limited primarily to the "calcium-dependent" enzymes.

Table 4.1.   Types of PLA$_2$ of Interest.

Type	Structural features	Regulated by	Representative enzymes
I	14 kDa, his-48, 14 cys, lys-56-acyl residues 25-52 Ca^{2+}-binding loop	Ca^{2+}, acylation	Pancreatic, *N. naja* venom
II	14 kDa, his-48, 14 cys, lys-11-acyl COOH-terminus extension ending in cys residues 25-52 Ca^{2+}-binding loop	Ca^{2+}, acylation	Platelet, *C. adamanteus C atrox*, synovial fluid
III	14 kDa, 10 cys, similar structure but evolutionarily divergent from I and II	Ca^{2+}, mellitin	Bee venom
IV	85 kDa, unrelated to I–III, Ca^{2+}-binding domain homologous to PKC	Ca^{2+}, phosphorylation	Platelet, macrophage endothelial

Adapted from Mayer and Marshall.[105]

4.2   PLA$_2$ Activity in Relation to Membrane Structure

The bilayer arrangement of phospholipids represents one of several possible membrane organizational states. Phospholipid mobility is also quite varied among these organizational states and can assume a range of motion from highly ordered to nearly isotropic as defined within the limits attainable in the physiological ranges of pH and temperature.[11] The mobility and structural arrangements of phospholipids are most divergent with minimal cooperativity of motion at the phase transition temperature. Under such circumstances, pronounced irregularities can exist within the membrane matrix,[12] and barrier properties of the membrane break down.[13] Since the structural organization of membrane phospholipids appears to be quite dynamic and their physical arrangements evanescent, the conventional view of the membrane "bilayer" is likely one of several fleeting structural arrangements. The bilayer configuration is generally held to be the most resistant to PLA_2 hydrolysis, although factors such as ionic surface charge can have a large effect on hydrolytic susceptibility of the bilayer.[14] Moreover, addition of other agents such as diacylglycerol[15] or detergents[16] leads to increased isotropic motion of phospholipids, increased ionic permeability, and susceptibility to PLA_2 hydrolysis. Phospholipids entering nonbilayer domains tend to experience rapid structural transitions and transbilayer movements. The microenvironments exhibiting these rapid and divergent movements are postulated to be the origins of membrane fusion processes[17] and sites of enhanced PLA_2 hydrolytic activity.[18] The interface between various phase domains represents "defects" that are characterized by such properties as leakiness to ions and structural instability, manifested by vesicle fusion.[19] Hydrolytic activity of PLA_2 is maximal at the phase transition temperature and the defective regions are regarded as sites of initial hydrolysis.[20,21] It is not uncommon to find that phospholipids in disordered arrangements are manyfold more susceptible to PLA_2 than the same phospholipids in ordered bilayers. For further details on the relationship between bilayer structure and PLA_2 activity see an article by Burack and Biltonen.[22]

4.2.1 Effects of Lipid Peroxidation on Membrane Structure

Disruptions in membrane structure by oxidation of phospholipids also makes them more susceptible to the action of PLA_2.[23,24] Evidence supporting this comes from studies with artificial membrane as well as from numerous biological membrane preparations. Peroxidation of membrane lipids via free radical reactions commonly leads to formation of hydroperoxides as primary products.[25] Formation of hydroperoxides not only produces species that inactivate proteins[26] but alters membrane physical properties, due at least in part to changes in phospholipid orientational distribution and dynamics—assessed by various biophysical techniques. This alteration in membrane structure appears to be sufficient to activate extracellular phospholipases.[27] Mechanisms responsible for these events are very rapid since hydrolysis of phospholipids can occur within seconds after treatments of cells or artificial membranes with oxidants. *In vitro* studies suggest that PLA_2 is inherently "primed" to undergo large increases in catalytic activity based on the nature of its substrates and that this activation does not require posttranscriptional enzyme modification, aside from modifications during the catalytic process. Although the identity of the PLA_2 involved were not specified with regards to studies using cells or cell membranes, it appears that the stimulation in hydrolytic activity takes place almost immediately after an oxidant challenge.[28,29]

Phospholipid monolayers have been used to study the molecular conformations assumed in the presence of hydroperoxy- and hydroxy- groups in phosphatidylcholine Monolayer films of phospholipids were used to study the interaction of PLA_2 at the lipid-water interface, particularly in terms of the lipid characteristics, the influence of surface pressure on the molecular surface area and lipid packing, and accessibility of the enzyme to the sn-2 ester bond of phospholipids.[30] In most cases, surface pressures below 35 nM/m permit sufficient intermolecular spacing (area/molecule) for PLA_2 to penetrate and degrade the phospholipid monolayer. Both hydroperoxy-(PLPCOOH) and hydroxy (PLPCOH)-phosphatidylcholines derived from 1- palmitoyl, 2,linoleoyl-phosphatidylcholine (PLPC) showed a pronounced susceptibility to hydrolysis at surface pressures below 35 nM/m as compared to unoxidized phosphatidylcholine, resulting in faster rates of surface pressure decrease during the course of hydrolysis.[31] The order of pressure decrease rate was PLPCOOH > PLPCOH > PLPC. As the penetration behavior of PLA_2 was identical for all substrates studied, the results indicated that the more polar oxidized phospholipids assume an enlarged molecular surface area produced by the pulling of the oxidized fatty acid toward the lipid-water interface.

Oxidation of aqueous dispersions of phospholipids increases the gel to liquid crystalline phase transition temperatures.[32] The modified thermotropic properties of oxidized membrane phospholipids are indicative of altered structural organization and of the formation of multiphasic arrangements. Accumulation of polar residues within the monolayer, or membrane, is expected to alter the thermotropic behavior of the matrix as insertion of oxygen substantially increases the dipole character of atom pairs comprising the paraffin chains. The conformational changes imposed by oxidized phospholipids are consistent with the reported increase in PLA_2 activity at membrane sites having packing defects.[20] Moreover, the polar nature of the oxidized free fatty acids released by PLA_2 would lead to a faster expulsion to the aqueous phase, facilitating or prolonging the catalytic activity of the enzyme that is subject to product inhibition.[16] Oxidation of membrane phospholipid bilayers causes many forms of disruption, likely including factors such as calcium binding, phospholipid packing irregularities and phase separations. McLean et al.[27] measured the rates of PLA_2 activity in oxidized vesicles independent of interfacial factors using lipid substrate solubilized with Triton X-100. The rate of hydrolysis was dramatically increased in peroxidized dispersions as compared to unoxidized controls. Moreover, the rates of hydrolysis of unoxidized phospholipids in the peroxidized dispersions was unchanged indicating a hydrolytic preference for oxidized substrates in mixed phospholipids. Sevanian and Kim[33] described different effects in terms of substrate specificity in that they found that lipid peroxidation induced hydrolysis of intact unsaturated phospholipids along with the peroxidized phospholipids. The order of hydrolytic preference was 20:4 = 22:6 > 18:2 >> 18:1 > 18:1 in oxidized liposomes, however, hydrolysis of all unoxidized lipids was about 10-fold less than oxidized phospholipids. These two findings are consistent in the sense that PLA_2 shows a marked preference for oxidized substrates but an absolute preference may not be possible since unoxidized phospholipids coexist with peroxidized phospholipids in PLA_2-susceptible domains. The formation of these domains can be attributed to the presence of oxidized lipids.[34] As pointed out later, similar findings have been reported with biological membranes and intact cells where phospholipid peroxidation appears to induce release of intact as well as peroxidized fatty acids. A protective role of PLA_2 was postulated based on these observations.

The membrane-disruptive effect of oxidized phospholipids appears analogous to the action of detergents such as bile salts, which were used to elicit enhanced hydrolytic activity by converting bilayer phospholipid arrangements to mixed micelle conformations.[10] Formation of oxidized phospholipids among a bilayer of unoxidized phospholipids is thought to create *heterogeneities* in physical organization leading to compositional phase separation.[22] These lateral phase separations have been proposed to cause aggregation of fatty acids producing domains with high anionic charge density to which PLA_2 adheres, undergoes a conformational change (for which several bound states have been proposed[22]) and enters the catalytically active state. This process is referred to as binding activation.[35] In addition, lipid peroxidation products in membranes induce pore formation, behaving like "ionophores"[36] or impairing ion pumps.[37] This may cause influx of extracellular calcium or release calcium from internal stores,[38] which triggers PLA_2 activity.

Utilizing model membrane bilayers containing small amounts of oxidized phospholipids (~5 mol%), the kinetic analysis of PLA_2 showed that the enzyme was operating in a *closed system* where either depletion of substrate at the interface takes place during catalysis and/or accumulation of products results in end product inhibition. Thus, the rate of substrate replenishment to an otherwise closed system influences the extent to which enzymatic activity is sustained. Peroxidized phospholipids were found to facilitate greater enzyme substrate interactions, either on a specific vesicle or among vesicles.[39] Although the effects of peroxidation on PLA_2 activity may be explained in other ways, a prevalent opinion is that preferential hydrolysis of oxidized phospholipids occurs at domains of structural heterogeneity that are created by lipid peroxidation products.[27] PLA_2 binding to domains enriched in oxidized phospholipids followed by increased rates of hydrolysis were proposed to occur by enhanced lateral migration of substrates or by membrane fusion with greater rates of substrate influx into the peroxidized domains. The kinetics of influx resembled the kinetics of lipid flux during membrane lipid peroxidation.[40] Recognition of structural imperfections by virtually all forms of PLA_2 suggests that the cell-associated enzymes may provide a "housekeeping" role, removing damaged lipids and thereby minimizing formation of heterogeneous domains. Employing fluorescence lifetime techniques as well as angle-resolved fluorescence dynamics (AFD) and electron spin resonances (ESR) measurements on model membranes, it was found that small amounts of PLPC-hydroxides or PLPC-hydroperoxides strongly decrease the molecular orientational order but do not affect the reorientational dynamics (=fluidity) of the membrane lipids.[34] The two major lifetime components described for diphenylhexatriene (DPH) are found to be significantly different for oxidized phospholipid-containing vesicles as compared to unoxidized ones. The major fluorescence signal component is shifted to longer lifetimes, while the minor component assumes greater fractional intensity with broadened lifetime distributions.[41] This indicates that the dielectric environment of DPH changes within domains enriched with lipid peroxidation products and such domains are preferred sites for PLA$_2$ attack.

Studies of PLA$_2$ activation in relation to changes in vesicle membrane fluidity (using trimethylammonio (TMA)-DPH) revealed altered physical properties during the course of hydrolysis.[42] TMA fluorescence rapidly and progressively decreased on PLA$_2$ activation after a short lag period reflecting transitions from a bilayer structure due to the formation of hydrolysis products. These products may include the small amounts of fatty acids utilized for enzyme autoacylation, dimerization and activation as described for the group I and group II PLA$_2$.[43] Comparison of the kinetics of hydrolysis between unoxidized and oxidized vesicles reveals that a lag period is not evident for oxidized vesicles.[39] Possibilities that may account for this include (1) oxidized phospholipids or small amounts of oxidized free fatty acids already exist in the peroxidized vesicles such that a lag period usually required for their release is circumvented. This occurs during the early stages of PLA$_2$-induced hydrolysis of zwitterionic phospholipids in contrast to anionic vesicles where binding and hydrolysis are rapid. As anionic fatty acid products accumulate at the membrane interface, there is a rapid progression of hydrolysis as more enzyme partitions to the interface. (2) The oxidized fatty acids (or derived aldehydes) may acylate, for example, the lysine residues in the NH$_2$ terminal helix, or at lysine 56 near the catalytic region (histidine 48) of the group I PLA$_2$,[44] resulting in enzyme activation. The oxidized vesicles allow more effective calcium binding at the interface[39] and the calcium binding loop encompassing the catalytic site of the calcium-dependent enzymes.[45] (3) The structure of the vesicle is disrupted such that ability of the enzyme to access substrates is enhanced as described above. The amount of lipid peroxidation required to achieve enhanced catalytic rates has been investigated by incorporating increasing amounts of oxidized phospholipid (PLPCOOH) into PLPC vesicles. Figure 4.1 relates the extent of snake venom (type 1) PLA$_2$ activity to the content of PLPCOOH in vesicles. Utilizing a fixed 10-min incubation interval, the extent of hydrolysis increases substantially only when the content of PLPCOOH exceeds approximately 2 mol% of total phospholipids (i.e., ~40 nmol/mg phospholipid vesicle). Beyond this level of oxidized phospholipids, the extent of hydrolysis increases sharply to a maximum of 20 mol%, after which catalytic activity either does not change or decreases. It is important to note that PLA$_2$ activity was measured based on the hydrolysis of total PLPC substrate, consisting of the release of oxidized and unoxidized fatty acids. Note also that catalytic activation by lipid peroxidation requires peroxide concentrations very close to the amounts of peroxide needed to stimulate propagation of lipid peroxidation using a variety of catalysts (vide infra). Recent analyses show that a direct antioxidant role is unlikely for PLA$_2$ as the kinetics for eliminating lipid peroxides from the membrane matrix are inadequate compared to their direct reduction by peroxidases

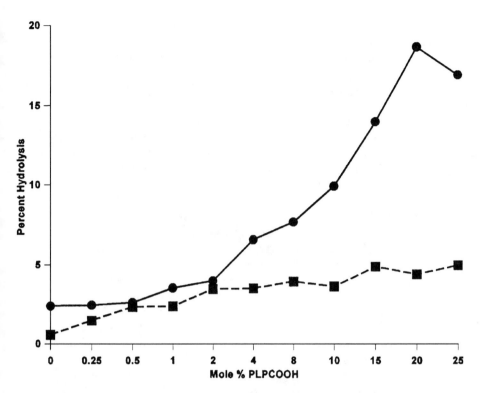

Figure 4.1. The extent of snake venom *(Crotalus adamanteus)* PLA₂ activity toward large unilamellar vesicles comprised of 1-palmitoyl,2-linoleoyl-phosphatidylcholine as a function of the content of phospholipid hydroperoxides. Vesicles were prepared with various amounts of 1-palmitoyl, 2-[13]hydroperoxylinoleoyl-phosphatidylcholine added to 1-palmitoyl, 2-linoleoyl-phosphatidylcholine in the amounts indicated along the x axis as mole percent of total vesicle phospholipid. The extent of PLA₂ activity is expressed as the percentage of total phospholipid hydrolysis during a fixed 10 min incubation (at 37°C) interval in Tris-buffer (pH 7.6) containing 1-mM calcium and 0.25 nmol/mg phospholipid.

such as phospholipid hydroperoxide glutathione peroxidase (PHGPX).[46] However, an indirect role is plausible based on the capacity of PLA₂ to maintain oxidatively modified lipids at levels below that required for autocatalytic propagation reactions. Although the molar concentrations of peroxides can not be directly compared between cell membranes, lipoproteins and artificial lipid preparations, several studies show that a range of 5 to 25 nmol lipid peroxide per mg lipid (or approximately 2 mol% lipid) is required to stimulate the autocatalytic phase of lipid peroxidation in unsaturated lipid systems.[47–49] If the concentration of lipid peroxides is based on the total content of oxidizable lipid, then similar amounts of peroxide (~2 mol%) induce propagation of lipid peroxidation and activation of PLA$_2$. This relationship may either be coincidental or there could be an important underlying role for PLA$_2$ in curtailing membrane lipid peroxidation by limiting the amounts of lipid peroxides that can accumulate in a membrane.

4.2.2 Modulation of PLA$_2$ Activity by Calcium

Peroxidation of phosphatidylcholine vesicles has been shown to increase the calcium-binding potential to a degree comparable to addition of the anionic phospholipid, dioleoylphosphatidic acid.[39] A similarity was found between the calcium binding profiles and PLA$_2$ activity, and under conditions where approximately 5% of the phospholipids were peroxidized, the effective calcium concentration required for half-maximal activity was less than one-half that required for unoxidized vesicles. Indeed, significant activity was found at micromolar calcium concentrations for oxidized vesicles, whereas comparable activity in unoxidized vesicles required nearly millimolar calcium. Peroxidation of vesicle phospholipids markedly increased the rate and extent of hydrolysis, even in the presence of anionic phospholipid or detergent. Since deoxycholate is known to induce vesicle fusion such that a larger proportion of enzyme is associated with a fewer number of enlarged vesicles, and since peroxidation also induces vesicle fusion,[50] it was postulated that a combination of structural and calcium binding affinity changes are produced in membranes following lipid peroxidation that evoke an additive effect on PLA$_2$ activity.

Membrane structural alterations during lipid peroxidation represent one possible mechanism by which the activity of PLA$_2$ is increased, although signaling processes initiated by changes in intracellular calcium have a pronounced effect on the postranslational modification, translocation, and activation of PLA$_2$. For many of the PLA$_2$, enhanced activity appears to be secondary to increases in intracellular calcium and/or calcium association with membranes. This stimulation may be attributable to calcium-induced phase separation of phospholipids leading to formation of structural interfaces that are sites of enhanced hydrolytic activity.[51] Phospholipids containing anionic moieties would also readily coordinate calcium, thus bridging the active site of PLA$_2$ in proximity to the sn-2 fatty ester group. These phase separations have been described for oxidized membrane phospholipids (as noted above), but formation of secondary polar products of lipid peroxidation, such as carbonyls, would also provide anionic groups capable of binding calcium. Note that the so-called absolute requirements stated for calcium are not as strict as many reports imply since other divalent ions,[52] and even high salt concentrations,[53] can support catalytic activity, however, calcium is most effective at the usual intracellular ion concentration range. For the low-molecular-weight enzymes there is strong dependence for calcium binding to the amino acid loop region encompassing the catalytic site, where Tyr-28, Gly-30, Gly-32, and Asp-49 coordinate calcium in proximity to the His-48-active site.[54] In addition, the coordination of calcium is required for binding to the substrate as interfacial binding and catalysis have an absolute requirement for calcium. The calcium-mediated catalysis is described in two steps: (1) enzyme binding to the interface, via an interface recognition site distinct topologically from the catalytic site[45] and (2) binding of the phospholipid substrate to the active site of the interface-bound enzyme to produce the enzyme substrate complex. The intracellular calcium-dependent PLA_2 reportedly translocates from the cytosol to membranes at calcium concentrations ranging from 300 nM to 10 μM[55] and activation of other signaling pathways, such as mediated by G-proteins,[56] appears to enhance the calcium mediated activation of PLA_2. This suggests that calcium is necessary but not sufficient for full activation of PLA_2 by various stimuli. As pointed out in the next section, the extent of PLA_2 activation is not dependent on the peak increase in intracellular calcium levels since other factors impact the extent of enzyme activity. Nevertheless, the rise in calcium is an obligatory feature of the signaling process, and in the case of oxidant stimulation the extent of the calcium mobilization or rise required for activation may be small, although this is influenced by the degree of oxidative damage.

In the case of calcium-independent PLA_2, no apparent requirement exists for binding or catalysis, indicating a divergence in enzyme structure that enables catalytic activity toward micellar substrates but not anionic substrates. This is reminiscent of serine esterases such as lecithin cholesterol acyltransferase (LCAT) and lipoprotein lipase,[57,58] which have an active site amino acid triad consisting of serine, histidine, and aspartate.[59] Peroxidized lipids have also been found to be good substrates for the acidic calcium-independent PLA_2, which actively releases the peroxidized fatty acids.[60] A major form of the calcium-independent enzymes is the plasmalogen selective PLA_2, prevalent in sarcolemma.[61] Studies with this enzyme showed that catalytic activity responded to the structure organization of plasmenyl- and phosphatidylcholine components in membranes and to oxidant action on the myocardial phospholipids.[62] It was also shown that oxidants produced following ischemia/reperfusion or by inflammation readily attacked the susceptible vinyl-ether linkages in plasmalogen phosphatides forming oxidation products that were excellent substrates for calcium-independent PLA_2. A related observation was made using lung and liver preparations from rats fed vitamin E and selenium deficient diets. Together with marked increases in lipid peroxidation products, there was a rise in calcium-independent PLA_2 activity, whereas calcium-dependent PLA_2 was minimally affected.[63] This enzymatic activity is associated with cytosolic fractions in several tissues and displays high inducible activity toward acyl-unsaturated as well as saturated phospholipids.[64]

4.3 Signaling of PLA_2 Activity

To this point the properties of the group I and group II PLA_2 have been largely addressed, all of which have a similar primary sequence and three-dimensional structure. These enzymes are either extracellular or cell associated but are distinct from the type IV PLA$_2$, or calcium-independent PLA$_2$, which are intracellular and show no sequence homology to the other enzymes. The type II and type IV PLA$_2$ also share roles in cell signaling pathways as their synthesis or activation appears to be mediated by receptor-associated and oxidant-modulated pathways. Type II PLA$_2$ synthesis and secretion is stimulated following exposures to cytokines such as interleukin-1β (IL-1β) and tumor necrosis factor alpha (TNF-α).[65] Some of the low-molecular-weight PLA$_2$ (types I and II) also appear to contribute to signaling events associated with release of biologically active fatty acids or autocoid precursors,[66] or these PLA$_2$ are formed and released from cells in response to cytokine stimuli.[67,68] For example, IL-1, TNFα, and bradykinin induce synthesis and secretion of synovial fluid PLA$_2$, which is responsible for prostaglandin E$_2$ (PGE$_2$) formation during rheumatoid arthritis.[69,70] Participation in cell signaling and inflammatory events can occur by release of fatty acid autocoid precursors such as linoleic acid. Formation of 13-hydroxyoctadecenoic acid (13-HODE) through the concerted action of PLA$_2$ and 15-lipoxygenase[71] is reported to have a modulatory effect on mitogenesis by suppression of protein kinase C (PCK-β) activity.[72] This effect is associated with incorporation of 13-HODE into diacylglycerol indicating that reacylation mechanisms are required to produce specific PKC regulatory diacylglycerols. Utilization of other unsaturated fatty acids by lipoxygenases expands the potential pathways by which oxidant activation of many PLA$_2$ contribute to cell responses downstream from the generation of oxyradical species.

4.3.1 Convergence of Housekeeping and Signaling Pathways

In addition to the stimulatory effects of oxyradical generating systems, "oxidant" activation of PLA$_2$ is readily afforded by lipoxygenase-mediated peroxidation of membrane phospholipids. In this regard, activation of specific lipoxygenases serves as the "oxidant stimulus" for PLA$_2$. This is well known for the type IV- and 15-lipoxygenases, which have measurable activity toward phospholipid vesicles,[73] lipoproteins,[74] and the activities of which are markedly enhanced toward deoxycholate/phospholipid micelles.[75] The oxidation process appears to follow an interfacial kinetic reaction analogous to the dual phospholipid model proposed for PLA$_2$.[76] Similar activity has been reported for 15-lipoxygenase whose action is inhibited by tocopherol analogs[77] and for the lipoxygenase, which readily oxidizes the phospholipids in reticulocyte organelles.[78] Lipoxygenase is converted from its resting (ferrous) state to the active (ferric) state through an oxidative reaction utilizing lipid hydroperoxide formed either by autoxidation or enzymatically.[79] This activation may occur either directly through gradual accumulation of hydroperoxides (via the enzymes own action), by oxidant chemical or other catalyst-induced autoxidation, or by disruption of membranes from which lipoxygenase substrates are derived. Membrane phospholipid disruption represents a potentially important means by which lipid autoxidation is facilitated[80] and lipoxygenase activity is enhanced.[81,82] The activation of lipoxygenase has been observed when membrane phospholipid structure becomes disordered, permitting the enzyme to oxidize unsaturated fatty acyl moieties in phospholipids directly[82] or attack substrates that are released through activation of phospholipases.[83] The "programmed" activation of lipoxygenases represents an enzymatically regulated mechanism for generating messengers derived by the PLA_2-mediated release of the oxidized fatty acids from membranes. This programming may originate and ultimately be regulated by coordinate activation of these enzymes as described for CD28 receptor mediated kinase signaling of inflammatory cell proliferation.[84] Moreover, conversion of arachidonic acid by lipoxygenase mediates $cPLA_2$ (also known as 85-kDa PLA_2) activity apparently through a leukotriene receptor dependent pathway.[85] This latter finding points to a possible positive feedback mechanism for modulating $cPLA_2$-mediated arachidonic acid release wherein release of arachidonic acid and its conversion to leukotrienes (or its peroxidation) promotes further activation of $cPLA_2$.

In most of the literature to date, PLA_2 activation by lipid peroxidation or oxidant challenge appears to involve calcium-dependent enzyme(s). Alternately, the signaling events producing enhanced PLA_2 activity are calcium-dependent. In a landmark study, PLA_2 activity in mitochondria was found to increase approximately 60% after lipid peroxidation in the absence of measurable changes in calcium concentrations.[86] This suggested that small changes in calcium levels associated with membranes or calcium-independent PLA_2 were involved in the hydrolytic response. Calcium mobilization and phospholipid degradation are held to be interrelated with the exception of the calcium-independent PLA_2 whose activation may be modulated by other mechanisms. Membrane damage via lipid peroxidation perturbs calcium homeostasis, which serves as an important initial signal for PLA_2 activity.[87] This may be viewed as part of the *housekeeping* function of PLA_2, since the removal of oxidized fatty acids from phospholipids represents part of the maintenance of membrane structure and dynamics. In addition, the release of specific lipid peroxidation products appears to be contribute to the mimicry of oxidants in cell signaling pathways. Examples include oxidant mediated release of arachidonic acid from phosphatidylinositol (PI), or release of inositol triphosphate (IP_3) or diacylglycerol (DAG) as factors activating kinases.[88] Perhaps best studied is the diacylglycerol-mediated increase in the affinity of PKC for calcium, lowering the calcium requirement for PKC and PLA_2 activation.[89] Oxidant-mediated calcium signaling induces hydrolysis of membrane phospholipids in a manner characteristic of PLA_2 activation following inotropic stimuli.[87] Analogous to the action of natural agonists, peroxide-mediated release of calcium proceeds via a discharging of cell pools, making cells unresponsive to a second stimulus. The rise in calcium levels following oxidant challenge is frequently lower than that produced by receptor-mediated events[87,90,91] and involves extracellular as well as intracellular sources, however, the extent and duration of calcium elevation depends on the degree of oxidant challenge.[92]

Enhanced calcium transport through calcium channels as well as the inhibition of ion pumps occurs in the presence of oxidants, including lipid peroxides. Tert-Butyl hydroperoxide (tBOOH) at 200 μM induced an increase in intracellular calcium in hepatocytes without evident formation of IP$_3$.[93] The increase in calcium was independent of membrane lipid peroxidation but appeared to involve formation of mixed disulfides with oxidized glutathione (GSSG), or other oxidized thiols associated with IP$_3$ receptors.[91,94] These observation indicate that cellular thiol states may be important determinants for oxidant-mediated induction of calcium signaling. A similar effect of linoleic acid hydroperoxide (LOOH) was noted after treatment of rabbit aortic endothelial cells.[95] At a dose range of 0.1–0.4 μmol/10^6 cells, there was a rapid but transient increase in intracellular free calcium concentrations derived from both intra- and extracellular sources. Moreover, the sustained elevations in calcium appeared to be due to diminished reuptake of calcium into intracellular pools. Taken together, these findings indicate that lipid peroxidation can trigger measurable and functionally effective increases in intracellular calcium levels that stimulate events leading to PLA$_2$ activation. The tBOOH-induced PLA$_2$ activation is also proposed to involve activation of a serine esterase,[96] probably through the action of other regulators of cPLA$_2$ such as calcium, PKC (see below), or PLA$_2$ activating protein.[97]

Studies with α-tocopherol[98] and tocopherol analogs[99] indicate that peroxide-induced activation of PLA$_2$ requires membrane lipid peroxidation since antioxidant inhibition of peroxidation prevents enzyme activation. However, these studies also suggest that the antioxidants inhibit signaling events downstream from membrane associated calcium-mediated events. These include direct redox interactions with PKC, which in turn, regulates PLA$_2$ activity.[100] Activation of PLA$_2$ may also take place by proteolytic signaling events that are well known to be stimulated by oxidants[101] or by elevations in intracellular calcium concentrations. Treatment of endothelial cells with H$_2$O$_2$ increased serine esterase activity which correlated with activation of PLA$_2$.[102] Although exposure to H$_2$O$_2$ is widely known to cause elevations in intracellular calcium levels and this is accompanied by enhanced PLA$_2$ activity, only part of the PLA$_2$ activity appears to be calcium-dependent.[103] Our lab recently showed that peroxide-induced hydrolysis of cell phospholipids was only partly calcium-dependent and that either calcium-independent PLA$_2$ or compromised reacylation reactions could account for the calcium-independent hydrolytic activity.[95] The latter possibility will be discussed below in greater detail. Antiproteases have been reported to block activation of hydrolysis, whereas PLA$_2$ activity was enhanced following treatments of cell homogenates with trypsin.[102] These findings indicate that at least some cell-associated PLA$_2$ are regulated by alternate mechanisms involving endogenous inhibitors[104] or transitions from proenzyme to active enzyme states that are influenced by oxidants.

Regulation of PLA$_2$ activity by phosphorylation appears limited to the type IV PLA$_2$ (i.e., cPLA$_2$),[105] and this mechanism of catalytic regulation provides an important link to the cell signaling pathways utilizing protein kinases. However, little is known about the effects of oxidants or lipid peroxidation on state of $cPLA_2$ phosphorylation and associated catalytic activity. The type IV PLA_2 contains a 45 amino acid domain homologous to PKC that is involved in calcium binding and interfacial association with membrane lipids, but not required for catalysis.[106,107] It also contains at least two distinct phosphorylation sites that are substrates for tyrosine and serine/threonine kinases.[108,109] This latter feature appears to be important for cell signaling processes affecting the release of arachidonic acid. There are scattered reports showing that presumptive $cPLA_2$ activity is increased following exposure of intact cells to oxidants,[110,111] however, the sites of oxidant action causing stimulated PLA_2 activity remain obscure.

Activation of $cPLA_2$ has been shown to be mediated by PKC, and this can occur under conditions that do not require marked elevations in $[Ca^{2+}]_i$, as described previously.[112] PKC activation by oxidants and lipid peroxides has been described[113] and oxidant-mediated activation was shown to alter the Ca^{2+} requirement of the enzyme, converting it to a Ca^{2+}-independent form.[114] Indeed, the lipid (membrane) bound form of PKC becomes active at low (physiological) Ca^{2+} concentrations.[115] Gopalakrishna and Anderson[114] described the transient activation of PKC following addition of H_2O_2, altering the kinetic behavior of the enzyme to an isoform that is Ca^{2+}/phospholipid-independent. This activation is thought to take place by direct oxidization of specific vicinal thiol in the regulatory domain of PKC, which is reported to affect the reversible activation/inactivation of the enzyme.[116] Studies using lipid peroxides showed a similar activation of PKC although very high peroxide concentrations were required to stimulate enhanced phosphorylating activity.[117] As recently reported, however, induction of membrane lipid peroxidation after adding low concentrations of peroxides could amplify the stimulatory action on PKC.[95] It was suggested that fatty acid hydroperoxides activate PKC by satisfying the requirement for phosphatidylserine (PS) and that this effect occurred in the presence or absence of Ca^{2+},[117] enabling PKC to bind to membranes to a greater extent. Activation of PKC appears to be an initial *kinase* step in the cascade of phosphorylation reactions leading to $cPLA_2$ phosphorylation and activation.[109,118] The enhanced PKC activity produced after oxidant exposures, and presumably the phosphorylation of $cPLA_2$ by PKC, appears to be insufficient for stimulated $cPLA_2$ activity[119,120] unless accompanied by small increases in intracellular calcium concentrations.[120] In recent studies (unpublished), we found that $cPLA_2$ activity increased approximately 50% in the presence α,β-PKC isozymes and 1-mM calcium when incubated with an *in vitro* system consisting of unsaturated PC/PS vesicles. This was accompanied by nearly 100% phosphorylation of the $cPLA_2$.

Figure 4.2 depicts a mechanism for PKC activation and binding to membranes containing lipid peroxidation products. This first step is hypothetical and proposes a mechanism by which interaction of the enzyme requires association with membranes containing peroxidized lipids. Although depicted as translocation, peroxi-

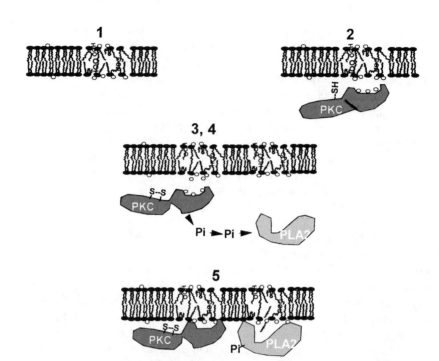

Figure 4.2. Activation of cPLA$_2$ via PKC shown following peroxidation of membrane phospholipids. The process is depicted in five steps. Step 1 involves peroxidation of phospholipid to form phospholipid hydroperoxides, as indicated by the -OOH groups. This produces local disruption in phospholipid structure and packing. In addition, an increased calcium binding takes place at these domains (as indicated by open circles representing calcium ions). In addition, the intracellular calcium concentrations also increase. Step 2 depicts one of two possible effects on PKC, in the example shown, there is interaction of cytosolic PKC with peroxides formed in membranes and with calcium liberated from membrane stores (or from external sources), and bound to PKC-calcium-binding domains. Alternatively, the oxidized phospholipids and increased calcium levels may activate cytosolic PKC, causing translocation to the membrane, binding, and activation. Steps 3 and 4 show the oxidation of vicinal thiols (shown as -SH in step 2 and disulfide bridges in steps 3 and 4) by the peroxidation products interacting with PKC. This results in activation of PKC and a phosphorylation cascade to PLA$_2$ in the cytosol (cPLA$_2$). The intermediacy of specific kinases mediating phosphorylation of cPLA$_2$, specifically at the serine-505 residue) is depicted in Figure 4.3. Step 5 shows the phosphorylation of cPLA$_2$ and its translocation and binding to the membrane, particularly at oxidatively damaged loci and their associated calcium ions. Also shown is the translocation and association of active PKC with the membrane. The activated cPLA$_2$ then hydrolyzes phospholipids in the oxidized domains, releasing hydroperoxides as well as degrading its natural substrate, arachidonic acid-containing phosphatidylcholine.

dation membrane lipids may enhance PKC-membrane interactions and transiently increase its phosphorylating activity or increase PKC activity in general. The key events consist of phosphorylation, translocation, and hydrolytic activation of cPLA$_2$, however, there may be several phosphorylation steps involving kinases such as *raf,* MEK, and ERK (MAPkinase), the latter of which causes the greatest phosphorylation/activation of cPLA$_2$[118] (see below for details). The key steps in this scheme are shown as (1) binding of calcium to the peroxidized membrane (The calcium binding to peroxidized domains may be based on an increased membrane anionic charge.[121]); (2) translocation or enhanced association of PKC with peroxidized-calcium containing domains; (3) interaction of membrane peroxides with the regulatory domain of PKC (depicted by -SH); (4) catalytic activation of PKC leading to a phosphorylation cascade to cPLA$_2$ (or possibly direct cPLA$_2$-phosphorylation—see Fig. 4.3); (5) translocation of phosphorylated cPLA$_2$ to the membrane, its binding to peroxidized-calcium containing domains followed by hydrolysis of phospholipids in these domains. It was shown that cPLA$_2$ activity moves from the cytosolic to membrane fraction when cells were lysed in calcium-containing buffers.[122,123] Studies with purified cPLA$_2$ from COS-1 cell fractions showed that as calcium concentrations increased from 100 nM (typical for the resting cells) to the 300 nM (found in activated cells), cPLA$_2$ disappeared from the cytosol and was found associated with the membrane fraction.[124]

According to this scheme, lipid peroxides can function as another class of lipid activators for PKC,[125] with the potential for activating PLA$_2$ through several possible steps in the kinase signaling cascade. Formulation of key events described above relies on the reported effects of antioxidants, calcium chelators, or buffering agents, as well as effects of sulfhydryl reducing agents in terms of reversing oxidant inactivation of PKC.[95,96,110,114,126] The inhibitory effect of vitamin E on lipid hydroperoxide (LOOH)-induced phospholipid hydrolysis and increased [Ca^{2+}]$_i$, for example, indicates that propagation of cellular lipid peroxidation is needed to elicit the hydrolytic response to LOOH.[127] It appears that low levels of LOOH may be relatively ineffective at stimulating phospholipid hydrolysis directly, however, conditions that allow for propagation of lipid peroxidation induce or amplify the cellular response. Although inhibition of lipid peroxidation is a plausible mechanism by which vitamin E prevents lipid peroxide-induced hydrolysis, other possible actions of vitamin E, including direct inhibition of PKC activity or its translocation,[100] as well as the direct inhibition of PLA$_2$,[99] should be considered. In this respect, unsaturated fatty acids such as arachidonic acid may also activate PKC directly[128] or participate in PKC mediated events such as c-*fos* expression during oxidant stimulation of cell growth, which requires PLA$_2$-dependent arachidonic acid release and its metabolism by lipoxygenase-P-450 monooxygenase pathways.[129]

Recently, c-PLA$_2$ was shown to be phosphorylated by PKC and by microtubule-associated protein 2 kinase (MAP kinase), two major serine/threonine kinases,[118] resulting in enhanced catalytic activity. Studies with recombinantly expressed

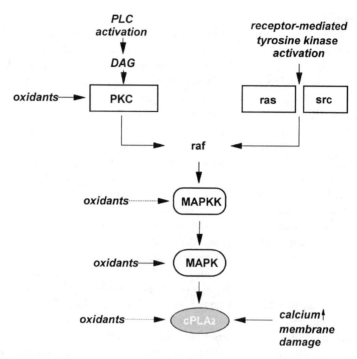

Figure 4.3. The proposed phosphorylation pathway involved in CPLA$_2$ activation and possible sites of oxidant action. Two initial sites of activation are represented. The pathway initiated from the right side of the scheme involves receptor-mediated activation of tyrosine kinases, such as found with platelet-derived growth factor (PDGF), leading to activation of tyrosine kinases *ras* and *src*. Activation of *raf* by these tyrosine kinases leads to phosphorylation and activation of MAPkinase kinase *(MAPKK)*. Alternately, *raf* activation may take place via a separate route beginning with receptor mediated stimulation of phospholipase C (PLC) and formation of DAG, as shown on the left side of the scheme. DAG formation serves to activate PKC, and this is mimicked by direct activation using phorbol esters such as TPA. This alternate pathway may also be stimulated by oxidants acting on the PKC regulatory domain, where the activated PKC phosphorylates *raf*. Activation of *raf* serves as a point of convergence between oxidant vs receptor-mediated triggering of the phosphorylation cascade. Steps subsequent to *raf* involve MAPKK- (or MEK)-mediated phosphorylation of MAPkinase (MAPK) which phosphorylates cPLA$_2$ at serine-505, resulting in full catalytic activation and phospholipid hydrolysis. Alternative sites for kinase activation are proposed at MAPKK and MAPK. In addition, increased cPLA$_2$ activity may be produced by oxidant-mediated peroxidation of membranes, representing a structurally based, calcium-dependent stimulation of phospholipid hydrolysis.

fragments of cPLA₂ showed that PKC and P42-MAP-kinase act at different phosphorylation sites. Phosphorylation of multiple sites is required for activation of the enzyme[130] and one of these sites (i.e., serine-505) must be phosphorylated for enzyme activation,[118,130] while phosphorylation of c-PLA₂ in the c-terminal region is required for translocation to the membrane.[123] These phosphorylating events, together with calcium-mediated binding to membranes, appear to coordinately stimulate phospholipid hydrolysis with release of arachidonic acid. Regulation of cPLA₂ activity by PKC, and the intermediacy of downstream kinases that directly phosphorylate cPLA₂ (i.e., MAPkinase), is indicated by complete inhibition of agonist-induced (or peroxide-induced) hydrolytic activity by PKC inhibitors[95] or by PKC down-regulation via prolonged treatments with phorbol ester.[96] Recent evidence also indicates that oxidation of PKC increases the activity of the unbound, calcium/phospholipid-independent form of the enzyme[114] that could facilitate PLA₂ phosphorylation, translocation, or hydrolytic activity without marked or prolonged changes in intracellular free calcium concentrations. H₂O₂ induced a rapid increase in cPLA₂ phosphorylation in smooth muscle cells that corresponded to the activation of PKC and MAPkinase.[131] There is mounting evidence that the MAP family kinases are stimulated in response to oxidants. Examples include cell exposure to linoleic acid hydroperoxides,[132] X irradiation, and H₂O₂ treatment[133] and glutathione depletion.[134] Activation of p42 MAPkinase requires phosphorylation at two residues, thr-183 and tyr-185, as catalyzed by MAPkinase kinase (MAPKK), which appears also to be modulated by oxidants.[135] Although stimulation of phosphorylating activity may be one means by which oxidants evoke these kinases, inhibition of dephosphorylating enzymes can also lead to the same increase in phosphorylation and activation. Inhibition of tyrosine phosphatases by the oxidizing agents has been reported.[136,137,138] Since all protein-tyrosine phosphatases have cysteine residues in their active site,[139] oxidation of these cysteines would cause inhibition of tyrosine phosphatases and account for the observed stimulation of tyrosine phosphorylation. Likewise, CL100 phosphatase has been shown to inactivate p42 MAPkinase by dephosphorylating thr-183 and tyr-185 at similar rates. In addition, the rapid inactivation of p42 MAPkinase is catalyzed by protein phosphatase 2A and a tyrosine phosphatase distinct from CL100.[140] These enzymes may also be inactivated by oxidants. Guy et al.[141] proposed that oxidant-induced phosphorylation is due to the inactivation of a redox-sensitive protein phosphatase. Such inactivation is compatible with a stimulated phosphorylation activity as catalyzed by PKC.

Figure 4.3 incorporates the proposed sites at which oxidants (particularly lipid peroxides or derived products) can stimulate the recognized components of the kinase cascade leading to cPLA₂ phosphorylation. The main components are thought to be the three kinases proximate to cPLA₂ consisting of *raf, MAPkinase kinase* (or MEK), and *MAP kinase* (or ERK). Signaling to MAP kinase can coverage via cytokine (agonist)-receptor-mediated activity involving *ras* and *src* kinsase, or via PKC. The latter may be a primary site for oxidant activation, which in turn phosphorylates and activates *ras*. This is likely to be an important site for oxidant activation since PKC inhibition blocks $cPLA_2$ phosphorylation, as discussed above. However, alternative sites for activation may include oxidant interaction with MAPkinase kinase or MAPkinase, and full phosphorylation and stimulation of $cPLA_2$ hydrolytic activity may require activation of these latter kinases.

4.3.2 Reacylation Following Lipid Peroxidation—Role of PLA₂

Hydrolysis of oxidized phospholipids by PLA_2 leads has two significant outcomes: (1) metabolism and elimination of the released fatty acids and (2) a decrease in membrane disruption which is afforded by repair of the oxidized PL. The former includes reduction of hydroperoxides to nonreactive alcohols by glutathione peroxidases (GSH-Pdx), involving the cytosolic- and membrane-associated sele-noenzymes.[142,143] The concerted action of PLA_2 and GSH-Pdx eliminates the prooxidant lipid peroxides and primes the resulting lysophospholipids for repair by a series of reacylation enzymes at the expense of fatty acyl coenzyme A (CoA). Although phospholipid hydrolysis is widely reported to be stimulated by lipid peroxidation, there is conflicting evidence that reacylation reactions may be either inhibited or enhanced depending on the extent of oxidant stress and energy deprivation.[27,28] Indeed, under severe oxidant stress or prolonged ischemia there evidence for enhanced lipid peroxidation[144] and loss in reacylation activities such that fatty acid reincorporation into phospholipids does not keep up with hydrolysis, even if hydrolytic activity is not appreciably enhanced.[145] Activation of phospholipid deacylation/reacylation reactions is an important feature of membrane repair after lipid peroxidation. Ischemia-reperfusion-induced release of fatty acids displays a pattern similar to that driven by free radical peroxidation of membranes. This is based on studies showing that antioxidants such as nordihy-droguaiaretic acid[146] and ascorbate[147] protected cells and prevented PLA_2 activation. Also, a similar pattern of lipid peroxidation is found for cells or tissues treated with oxidants or submitted to ischemia and reperfusion.[148,149] Since oxidants appear to produce hydrolysis of phospholipids that is calcium-dependent and -independent,[103] other functional deficits aside from PLA_2 must be involved in the hydrolytic process. The most commonly reported effect is inhibition of reacylation reactions or the energy stores associated with reincorporation of fatty acids into phospholipids.[103,150,151]

Insufficient information is presently available to evaluate the contribution of PLA_2 activation versus inhibition of phospholipid reacylation following oxidant exposure and peroxidation of membrane phospholipids. It is clear that high doses of oxidants, such as H_2O_2 at concentrations > 1 mM, produce marked depletions in cellular ATP levels, which are proposed to mediate increases in intracellular calcium levels along with increased free fatty acid concentrations.[103,150] Although PLA_2 activity appears to be increased under these circumstances, accumulation of free fatty acids is more plausibly explained by the strong inhibition of fatty acid CoA synthetase.[152] Exposures to *t*BOOH or hyperoxia stimulate reacylation of erythrocyte membrane phospholipids.[153,154] Fatty acid hydroperoxides were found to inhibit reacylation of neuronal phospholipids,[155,156] indicating a dual effect of lipid peroxidation on the deacylation-reacylation cycle. Pacifici et al.[127] proposed that deacylation (PLA$_2$) and reacylation reactions are acutely increased after lipid peroxidation, but with progressive accumulation of peroxidation products reacylation reactions become inhibited purportedly due to formation of inhibitory complexes with thiol moieties of CoA, which prevents fatty acid activation.[156] This may be the primary lesion causing accumulation of fatty acids and lysophospholipids following ischemica-reperfusion where differential degradation of unsaturated mitochondrial phospholipids is found.[157] It appears that during the cycle of hydrolysis and reacylation there are certain phospholipids, such as phosphatidylethanolamine (PE), PS, and PI that are resistant to degradation, while others, such as PC and cardiolipin are rapidly hydrolyzed. It is known that PE and PS tend to be localized on the cytosolic side of the leaflet bilayer, while PC is more evenly distributed on both sides of the membrane bilayer.[158] The susceptibility of PC to oxidation[127] and alterations in its distribution via increased transbilayer movements in oxidized membranes[18,32] may account for enhanced PLA$_2$ attack of PC and its prominent involvement in reacylation reactions following oxidant challenge.[127]

Oxidative mechanisms contributing to free fatty acid accumulation and altered reacylation of phospholipids can also include effects of lysophospholipids and free fatty acids on the signaling cascade leading to PLA$_2$ activation. There is mounting evidence that some forms of PLA$_2$ strongly determine the incorporation of specific fatty acids into phospholipids. This has been shown for the calcium-independent PLA$_2$, which regulates incorporation of arachidonic acid into macrophage membrane phospholipids by providing lysophospholipid as an acyl acceptor.[159] The calcium-dependent c-PLA$_2$ also appears to have a similar action, but in this case acyl transferase reactions are catalyzed where lysophospholipid is produced and the liberated arachidonic acid is transferred to glycerol to form monoacylglycerol.[160] Moreover, cPLA$_2$ can further degrade the formed lysophospholipid. It is possible that the relative amounts of lysophospholipid present can modulate the activity of cPLA$_2$ by regulating PKC, either via the signaling pathway described above or by stimulating formation of reactive oxygen species that interact with PKC.[161] Lysophosphatidylcholine was shown to act synergistically with DAG to stimulate PKC activity *in vitro*[162] and to impair endothelial cell responsiveness to receptor-mediated stimuli by activating PKC.[163] These reactions implicate a regulatory cycle involving cPLA$_2$ formation or degradation of lysophospholipid, the levels of which, in turn, modulate the activity of cPLA$_2$.

4.4 Conclusions

In summary, membrane lipid peroxidation can induce PLA$_2$ activity through two interrelated processes. The first is calcium-dependent and mediated by signaling events originating at the plasma membrane (or intracellular membranes). This involves, at least in part, signal transduction proteins and membrane derived messengers, hence mimicking receptor mediated events. These processes can occur at low calcium concentrations, particularly at sites where oxidation of regulatory sulfhydryls takes place or where calcium binding is enhanced. The second process is calcium-independent and mediated by aberrant physiologic processes evoked by oxidative stress. Under conditions where membranes are not severely damaged and excessive calcium leakage does not occur, peroxide-induced structural alterations can enhance phospholipid hydrolytic susceptibility under conditions requiring lower calcium concentrations. The cytosolic high-molecular-weight PLA_2 ($cPLA_2$) found in several cell types may account for this enhanced activity as it requires micromolar calcium concentrations for full activity. Peroxide-mediated stimulation of $cPLA_2$ involves PKC activation and enzyme phosphorylation. These conditions are similar to the redox regulation described for $cPLA_2$ activation, its phosphorylation, and its translocation to the membrane in Ca^{2+}-dependent process. Although $cPLA_2$ activation depends on PKC, recent findings also indicate that tyrosine kinases may be involved and that protein-tyrosine kinase-mediated phosphorylation amplifies PLA_2 activating stimuli, perhaps by means of lipoxygenase catalyzed eicosanoid synthesis.[164] Based on these studies a direct action of oxidants on PKC and mitogen activated protein kinase (MAP-kinase or tyrosine kinase) is plausible. Identification of central phosphorylating enzyme systems that are modulated by oxidants and also stimulate PLA_2 activity demonstrates an intrinsic role for reactive oxygen species in autocoid production and phospholipid turnover.

Acknowledgments

The authors acknowledge the support of grant HL45206 from the National Institutes of Health during the period that this review was written. Support was also provided by a gift from Pfizer Pharmaceutical Group.

References

1. Axelrod, J. 1990. Receptor-mediated activation of phospholipase A_2 and arachidonic acid release in signal transduction. *Biochem. Soc. Trans.* **18**:503–507.

2. Shimidzu, T., and L. S. Wolfe. 1990. Arachidonic acid and signal transduction. *J. Neurol.* **55**:1–15.

3. Dana, R., H. L. Malech, and R. Levy. 1994. The requirement for phospholipase A_2 for activation of the assembled NADPH oxidase in human neutrophils. *Biochem. J.* **297**:217–223.

4. Werns, S. W., M. J. Shea, and B. R. Lucachesi. 1985. Free radicals in ischemic myocardial injury. *Free Rad. Biol. Med.* **1**:103–110.

5. Jones, R. L., J. C. Miller, P. K. Williams, K. R. Chien, J. T. Willerson, and L. M. Buja. 1987. The relationship between arachidonate release and calcium overloading during ATP depletion in cultured neonatal rat cardiac myocytes. *Fed. Proc.* **46:**1152. [See also 1987. *Basic Res. Cardiol.* **82**(suppl. 1):121.]

6. Kajiyama, K., D. F. Pauly, H. Hughes, S. B. Yoon, M. L. Entman, and J. B. McMillin-Woods. 1987. Protection by verapamil of mitochondrial glutathione equilibrium and phospholipid changes during reperfusion of ischemic canine myocardium. *Circ. Res.* **61:**301–310.

7. Sevanian, A. 1988. Phospholipase and lipid peroxidation, pp. 77–95. *In* C. K. Chow (ed.), *Cellular Antioxidant Defense Mechanisms,* Vol. II. CRC Press, Boca Raton, FL.

8. Arroyo, C. M., J. H. Kramer, R. H. Leiboff, G. W. Mergner, B. F. Dickens, and W. B. Weglicki. 1987. Spin trapping of oxygen and carbon-centered free radicals in ischemic canine myocardium. *Free Rad. Biol. Med.* **3:**313–316.

9. Butler, M., and L. G. Abood. 1982. Use of phospholipase A to compare phospholipid organization in synaptic membranes, myelin and liposomes. *J. Membrane Biol.* **66:**1–7.

10. Borgstrom, B. 1993. Phosphatidylcholine as substrate for human pancreatic phospholipase A_2. Importance of the physical state of the substrate. *Lipids* **28:**371–375.

11. Sklar, L. A. 1984. Fluorescence polarization studies of membrane fluidity: where do we go from here? pp. 99–117. *In* M. Kates and L. A. Manson (eds.), *Biomembranes,* Vol. 12. Plenum Press, New York.

12. Haest, C. W. M., J. de Gier, G. A. van Es, A. J. Verkleij, and L. L. M. van Deenen. 1972. Fragility of the membrane barrier of *Eschericia coli. Biochim. Biophys. Acta* **288:**43–48.

13. Noordham, P. C., A. Killian, R. F. M. Oude Elferink, and J. de Gier. 1982. Comparative study of the properties of saturated phosphatidylethanolamine and phosphatidylcholine bilayers: barrier characteristics and susceptibility to phospholipase A_2 degradation. *Chem. Phys. Lipids* **31:**191–196.

14. Dawson, R. M. C. 1982. Phospholipid structure as a modulator of intracellular turnover. *J. Am. Oil Chem. Soc.* **59:**401–407.

15. Dawson, R. M. C., R. F. Irvine, J. Bray, and P. J. Quinn. 1984. Long chain unsaturated diacylglycerols cause a perturbation in the structure of phospholipid bilayers, rendering them susceptible to phospholipase attack. *Biochem. Biophys. Res. Commun.* **125:**836–840.

16. Berg, O. G., Y. Bao-Zhu, J. Rogers, and M. K. Jain. 1991. Interfacial catalysis by phospholipase A_2: Determination of the interfacial kinetic rate constants. *Biochemistry* **30:**7283–7297.

17. Chernomordik, L. V., M. M. Kozlov, G. B. Melikyan, I. G. Abidor, V. S. Markin, and Y. A. Chizmadzhev. 1985. The shape of lipid molecules and monolayer membrane fusion. *Biochim. Biophys. Acta* **812:**643–650.

18. Noordham, P. C., C. J. A. van Echfeld, B. De Kruijff, and J. de Gier. 1981. Rapid transbilayer movement of phosphatidylcholine in unsaturated phosphatidylethanolamine containing model membrane. *Biochim. Biophys. Acta* **646:**483–487.

19. Gast, K., D. Zirmer, A. M. Ladhoff, J. Schreiber, R. Koelsch, K. Kretschemer, and J. Lasch. 1982. Autooxidation-induced fusion of lipid vesicles. *Biochim. Biophys. Acta* **686**:99–103.

20. Upreti, G. C., and M. K. Jain. 1980. Action of phospholipase A_2 on unmodified phosphatidylcholine bilayers: organization defects are preferred sites of action. *J. Membrane Biol.* **55**:113–120.

21. Kensil, C. R., and E. A. Dennis. 1985. Action of cobra venom phospholipase A_2 on large unilamellar vesicles. *Lipids* **20**:80–84.

22. Burack, R. W., and R. L. Biltonen. 1994. Lipid bilayer heterogeneities and modulation of phospholipase A_2 activity. *Chem. Phys. Lipids* **73**:209–222.

23. Sevanian, A. 1988. Phospholipase A_2: a secondary membrane antioxidant, pp. 84–99. *In* A. Sevanian (ed.), *Lipid Peroxidation in Biological Systems.* American Oil Chemists' Society, Champaign.

24. Cikryt, P., S. Feurstein, and A. Wendel. 1982. Selenium and non-selenium-dependent glutathione peroxidases in mouse liver. *Biochem. Pharmacol.* **31**:2873–2877.

25. Terao J. 1990. Reactions of lipid hydroperoxides, pp. 219–238. *In* C. Vigo-Pelfrey (ed.), *Membrane Lipid Oxidation,* Vol. I. CRC Press, Boca Raton, FL.

26. Fruebis, J., S. Parthasarathy, and D. Steinberg. 1992. Evidence for a concerted reaction between lipid hydroperoxides and polypeptides. *Proc. Natl. Acad. Sci. USA* **89**:10588–10592.

27. McLean, L. R., K. A. Hagaman, and W. S. Davidson. 1993. Role of lipid structure in the activation of phospholipase A_2 by peroxidized phospholipids. *Lipids* **28**:505–509.

28. Weglicki, W. B., B. F. Dickens, and I. Tong-Mak. 1984. Enhanced liposomal phospholipid degradation and lysophospholipid produced due to free radicals. *Biochem. Biophys. Res. Commun.* **124**:229–233.

29. Boyer, C. S., G. L. Bennenberg, A. Ryrfeldt, and P. Moldeus. 1993. Investigations into the mechanisms of oxidant-mediated arachidonic acid mobilization. *Toxicologist* **13**:337–341.

30. Demel, R. A., W. S. M. van Kessel, R. F. A. Zwaal, B. Roelofsen, and L. M. van Deenen. 1975. Relation between various phospholipase actions on human red cell membranes and the interfacial phospholipid pressure in monolayers. *Biochim. Biophys. Acta* **406**:97–107.

31. van den Berg, J. M., J. A. Op den Kamp, B. H. Lubin, and F. A. Kuypers. 1994. Conformational changes in oxidized phospholipids and their preferential hydrolysis by phospholipase A_2: a monolayer study. *Biochemistry* **32**:4962–4967.

32. Coolbear, K. P., and K. M. W. Keough. 1983. Lipid oxidation and gel to liquid crystalline transition temperatures of synthetic polyunsaturated mixed-acid phosphatidylcholine. *Biochim. Biophys. Acta* **732**:531–535.

33. Sevanian, A., and E. Kim. 1985. Phospholipase A_2 dependent release of fatty acids from peroxidized membranes. *Free Rad. Biol. Med.* **1**:263–27.

34. Wratten, M. L., G. van Ginkel, A. A. van't Veld, A. Bekker, E. van Faassen, and A. Sevanian. 1992. Structural and dynamic effects of oxidatively modified phospholipids in unsaturated lipid membranes. *Biochemistry* **31**:10901–10907.

35. Jain, M. K., and O. Berg. 1989. The kinetics of interfacial catalysis by phospholipase A₂ and regulation of interfacial activation: hopping versus scooting. *Biochim. Biophys. Acta* **1002:**127–132.

36. Kim, R. S., and F. S. LaBella. 1988. The effect of linoleic and arachidonic derivatives on calcium transport in vesicles from cardiac sarcoplasmic reticulum. *J. Cell. Mol. Cardiol.* **20:**119–130.

37. Menshikova, E. V., V. B. Ritov, A. A. Shvedova, N. Elsayed, M. H. Karol, and V. E. Kagan. 1995. Pulmonary microsomes contain a Ca²⁺-transport system sensitive to oxidative stress. *Biochim. Biophys. Acta* **1228:**165–174.

38. Lotscher, H. R., K. H. Winterhalter, E. Carafoli, and C. Richter. 1979. Hydroperoxides can modulate the redox state of pyridine nucleotides and the calcium balance in rat liver mitochondria. *Proc. Natl. Acad. Sci. USA* **76:**4340–4344.

39. Salgo, M. G., F. P. Corongiu, and A. Sevanian. 1993. Enhanced interfacial catalysis and hydrolytic specificity of phospholipase A₂ towards peroxidized phosphatidylcholine. *Arch. Biochem. Biophys.* **304:**123–132.

40. Volkov, E. I., and A. T. Mustafin. 1985. Mathematical model of lipid peroxidation in membranes. *Proc. Natl. Acad. Sci. USSR* **6:**805–821.

41. Wratten, M. L., E. Gratton, M. Van de Ven, and A. Sevanian. 1989. DPH lifetime distributions in vesicles containing phospholipid hydroperoxides. *Biochem. Biophys. Res. Commun.* **164:**169–175.

42. Bell, J. D., and R. L. Biltonen. 1989. The temporal sequence of events in the activation of phospholipase A₂ by lipid vesicles. Studies with monomeric enzyme from *Agkinstodon piscivorus*. *J. Biol. Chem.* **264:**12194–12200.

43. Cho, W. Q., A. G. Tomasselli, R. L. Heinrikson, and F. J. Kezdy. 1988. The chemical basis for interfacial activation of monomeric phospholipase A₂. *J. Biol. Chem.* **263:**11237–11241.

44. Tomasselli, A. G., J. Hui, J. Fisher, A. Zucker, and H. Neely. 1989. Dimerization and activation of porcine pancreatic phospholipase A₂ via substrate level activation of lysine 56. *J. Biol. Chem.* **264:**10041–10047.

45. Heinrikson, R. L. 1991. Dissection and sequence analysis of phospholipase A₂. *Methods Enzymol.* **197:**201–214.

46. Antunes, F., A. Salvador, and R. E. Pinto. 1995. PHGPx and phospholipase A₂/GPx: Comparative importance on the reduction of hydroperoxides in rat liver mitochondria. *Free Rad. Biol. Med.* **19:**669–677.

47. Thomas, J. P., B. Kalyanaraman, and A. W. Girotti. 1994. Involvement of preexisting lipid hydroperoxides in Cu²⁺-stimulated oxidation of low-density lipoprotein. *Arch. Biochem. Biophys.* **315:**244–254.

48. Kim, E., and A. Sevanian. 1991. Hematin and peroxide catalyzed peroxidation of phospholipid liposomes. *Arch. Biochem. Biophys.* **288:**324–330.

49. Thomas, C. E., and R. L. Jackson. 1991. Lipid hydroperoxide involvement in copper-dependent and independent oxidation of LDL. *J. Pharmacol. Exp. Ther.* **256:**1182–1188.

50. Sevanian, A., M. L. Wratten, L. L. McLeod, and E. Kim. 1988. Lipid peroxidation and phospholipase A_2 activity in liposomes composed of unsaturated phospholipids: a structural basis for enzyme activation. *Biochim. Biophys. Acta* **961**:316–327.

51. Borovjagin, V. L., J. A. Vergara, and T. J. McIntosh. 1982. Morphology of the intermediate stages in the lamellar to hexagonal lipid-phase transition. *J. Membrane Biol.* **69**:199–208.

52. Reynolds, L. J., L. L. Hughes, A. I. Louis, R. M. Kramer, and E. A. Dennis. 1993. Metal ion and salt effects on the phospholipase A_2, lysophospholipase, and transacylase activities of human cytosolic phospholipase A_2. *Biochim. Biophys. Acta* **1167**:272–280.

53. Jain, M. K., and M. H. Gelb. 1991. Phospholipase A_2-catalyzed hydrolysis of vesicles: uses of interfacial catalysis in the scooting mode. *Methods Enzymol.* **197**:112–125.

54. Dijkstra, B. W., J. Drenth, K. H. Kalk, and P. J. Vandermaalen. 1978. Three-dimensional structure and disulfide bond connections in bovine pancreatic phospholipase A_2. *J. Mol. Biol.* **124**:53–60.

55. Cockroft, S. 1992. G-protein-regulated phospholipase C, D and A_2-mediated signaling in neutrophils. *Biochim. Biophys. Acta* **111**:135–160.

56. Burch, R. M., A. Luini, and J. Axelrod. 1986. Phospholipase A_2 and phospholipase C are activated by distinct GTP-binding proteins in response to alpha 1-adrenergic stimulation in FRTLP thyroid cells. *Proc. Nat. Acad. Sci. USA* **83**:7201–7205.

57. Wion, K. L., T. G. Kirchgessner, A. J. Lusis, M. C. Schotz, and R. M. Lawn. 1987. Human lipoprotein lipase complementary DNA sequence. *Science* **235**:1638–1641.

58. Martin, G. A., S. J. Busch, G. D. Meredith, A. D. Gardin, D. J. Blankenship, S. J. Mao, A. E. Rechtin, C. W. Woods, M. M. Racke, and M. P. Schafer. 1988. Isolation and cDNA sequence of human postheparin plasma hepatic triglyceridelipase. *J. Biol. Chem.* **263**:10907-10914.

59. Fielding, C. J., and P. E. Fielding. 1995. Molecular physiology of reverse cholesterol transport. *J. Lipid Red.* **36**:211–228.

60. Gamache, D. A., A. A. Fawzy, and R. C. Franson. 1987. Preferential hydrolysis of peroxidized phospholipid by lysosomal phospholipase C. *Biochim. Biophys. Acta* **958**:116–124.

61. Ford, D. A., and R. W. Gross. 1989. Plasmenylethanolamine is the major storage depot for arachidonic acid in rabbit vascular smooth muscle and is rapidly hydrolyzed after angiotensin II stimulation. *Proc. Natl. Acad. Sci. USA* **86**:3479–3483.

62. Scherrer, L. A., and R. W. Gross. 1989. Subcellular distribution, molecular dynamics and catabolism of plasmalogens in myocardium. *Mol. Cell. Biochem.* **88**:97–105.

63. Kuo, C.-F., S. Cheng, and J. R. Burgess. 1995. Deficiency of vitamin E and selenium enhances calcium-independent phospholipase A_2 activity in rat lung and liver. *J. Nutr.* **125**:1419–1429.

64. Burgess, J. R., and C.-F. Kuo. In press. Increased calcium-independent phospholipase A_2 activity in vitamin E and selenium deficient rat lung, liver and spleen cytosol is time-dependent and reversible. *J. Nutr. Biochem.*

65. van den Bosch, H., C. Schalkwijk, J. Pfeilschifter, and F. Marki. 1992. The induction of cellular group II phospholipase A_2 by cytokines and its prevention by dexamethasone, pp. 1–10. *In* N. G. Bazan (ed.), *Neurobiology of Essential Fatty Acids.* Plenum Press, New York.

66. Fonteh, A. N., D. A. Bass, L. A. Marshall, M. Seeds, J. M. Samet, and F. H. Chilton. 1994. Evidence that secretory phospholipase A_2 plays a role in arachidonic acid release and eicosanoid biosynthesis by mast cells. *J. Immunol.* **152:**5438–5446.

67. Seilhamer, J. J., W. Pruzanski, P. Vadas, S. Plant, J. A. Miller, J. Kloss, and L. K. Johnson. 1989. Cloning and recombinant expression of phospholipase A_2 present in rheumatoid arthritic synovial fluid. *J. Biol. Chem.* **264:**5335–5338.

68. Crowl, R. M., T. J. Stoller, R. R. Conroy, and C. R. Stoner. 1991. Induction of phospholipase A_2 gene expression in human hepatoma cell by mediators of the acute phase response. *J. Biol. Chem.* **266:**2647–2651.

69. O'Neill, L. A., and G. P. Lewis. 1989. Interleukin-1 potentiates bradykinin and TNF alpha-induced PGE2 release. *Eur. J. Pharmacol.* **166:**131–137.

70. Gilman, S. C., J. Chang, P. R. Zeigler, J. Uhl, and E. Mochan. 1988. Interleukin-1 activates phospholipase A_2 in human synovial cells. *Arthritis Rheum.* **31:**126–130.

71. Glasgow, W. C., and T. E. Eling. 1994. Structure-activity relationship for potentiation of EGF-dependent mitogenesis by oxygenated metabolites of linoleic acid. *Arch. Biochem. Biophys.* **311:**286–292.

72. Cho, Y., and V. A. Ziboh. 1994. 13-hydroxyoctadecanoic acid reverses epidermal hyperproliferation via selective inhibition of protein kinase C-β activity. *Biochem. Biophys. Res. Commun.* **201:**286–292.

73. Kuhn, H., J. Belkner, and R. Wiesner. 1990. Oxygenation of biological membranes by the pure reticulocyte lipoxygenase. *J. Biol. Chem.* **265:**18351–18361.

74. Belkner, J., R. Wiesner, J. Rathman, J. Barnett, E. Sigal, and H. Kuhn. 1993. Oxygenation of lipoproteins by mammalian lipoxygenases. *Eur. J. Biochem.* **213:**251–261.

75. Roveri, A., M. Maiorino, and F. Ursini. 1994. Enzymatic and immunological measurements of soluble and membrane-bound phospholipid-hydroperoxide glutathione peroxidase. *Methods Enzymol.* **233:**202–212.

76. Hendrickson, H. S., and E. A. Dennis. 1984. Kinetic analysis of the dual phospholipid model for phospholipase A_2 action. *J. Biol. Chem.* **259:**5734–5739.

77. Arai, H., A. Nagao, J. Terao, T. Suzuki, and K. Takama. 1995. Effect of d-α-tocopherol analogues on lipoxygenase-dependent peroxidation of phospholipid-bile salt micelles. *Lipids* **30:**135–140.

78. Kroschwald, P., A. Kroschwald, H. Kuhn, P. Ludwig, B. J. Thiele, M. Hohne, T. Schewe, and S. M. Rapoport. 1989. Occurrence of the erythroid cell specific arachidonate 15-lipoxygenase in human reticulocytes. *Biochem. Biophys. Res. Commun.* **160:**954–960.

79. Naidu, K. A., and A. P. Kulkarni. 1994. Lipoxygenase: a non-specific oxidative pathway for xenobiotic metabolism. *Prostaglandins Leukot. Essent. Fatty Acids* **50:**155–159.

80. Mowri, H., S. Nojima, and K. Inoue. 1984. Effect of lipid composition of liposomes on their sensitivity to peroxidation. *J. Biochem.* **95**:551–558.

81. Green, F. A. 1989. Generation and metabolism of lipoxygenase products in normal and membrane-damaged cultured human keratinocytes. *J. Invest. Dermatol.* **93**:486–491.

82. Schnurr, K., H. Kuhn, S. M. Rapoport, and T. Schewe. 1995. 3,5-Di-t-butyl-4-hydroxytoluene (BHT) and probucol stimulate selectively the reaction of mammalian 15-lipoxygenase with biomembranes. *Biochim. Biophys. Acta* **1254**:66–72.

83. Sparrow, C. P., S. Parthasarathy, and D. Steinberg. 1988. Enzymatic modification of low density lipoprotein by purified lipoxygenase plus phospholipase A_2 mimics cell-mediated oxidative modification. *J. Lipid Res.* **29**:745–753.

84. Los, M., H. Schenk, K. Hexel, P. A. Baeuerle, and W. Droge. 1995. IL-2 gene expression and NFkB activation through CD28 requires reaction oxygen production by lipoxygenase. *EMBO J.* **14**:3731–3740.

85. Wijkander, J., J. T. O'Flaherty, A. B. Nixon, and R. L. Wykle. 1995. 5-lipoxygenase products modulate the activity of the 85-kDa phospholipase A_2 in human neutrophils. *J. Biol. Chem.* **270**:26543–26549.

86. Yasuda, M., and T. Fujita. 1977. Effect of lipid peroxidation on phospholipase A_2 activity of rat liver mitochondria. *Jpn. J. Pharmacol.* **27**:429–435.

87. Elliott, S. J., G. Meszaros, and W. P. Schilling. 1992. Effect of oxidant stress on calcium signaling in vascular endothelial cells. *Free Rad. Biol. Med.* **13**:635–650.

88. Shasby, D. M., M. Yorek, and S. S. Shasby. 1988. Exogenous oxidants initiate hydrolysis of endothelial cell inositol phospholipids. *Blood* **72**:491–499.

89. Nishizuka, Y. 1992. Intracellular signaling by hydrolysis of phospholipids and activation of protein kinase C. *Science* **258**:607–612.

90. Roveri, A., M. Coassin, M. Maiorino, A. Zamburlini, F. T. M. van Amsterdam, E. Ratti, and F. Ursini. 1992. Effect of hydrogen peroxide on calcium homeostasis in smooth muscle cells. *Arch. Biochem. Biophys.* **297**:265–270.

91. Koshita, M., K. Miwa, and T. Oba. 1993. Sulfhydryl oxidation induces calcium release from fragmented sarcoplasmic reticulum even in the presence of glutathione. *Experientia* **49**:282–284.

92. Hoyal, C. R., E. Gozal, H. Zhou, K. Foldenauer, and H. J. Forman. 1996. Modulation of the rat alveolar macrophage respiratory burst by hydroperoxides is calcium dependent. *Arch. Biochem. Biophys.* **326**:166–171.

93. Rooney, T. A., D. C. Renard, E. J. Sass, and A. P. Thomas. 1991. Oscillatory cytosolic calcium waves independent of stimulated inositol 1,4,5-trisphosphate formation in hepatocytes. *J. Biol. Chem.* **266**:12272–12282.

94. Renard, D. C., M. B. Seitz, and A. P. Thomas. 1992. Oxidized glutathione causes sensitization of calcium release to inositol 1,4,5-trisphosphate in permeabilized hepatocytes. *Biochem. J.* **284**:507–512.

95. Sweetman, L. L., N. Zhang, H. Peterson, L. L. McLeod and A. Sevanian. 1995. Effect of linoleic acid hydroperoxide on endothelial cell calcium homeostasis and arachidonic acid release. *Arch. Biochem. Biophys.* **323**:97–107.

96. Chakraborti, S., J. R. Michael, G. H. Gurtner, S. S. Ghosh, G. Dutta, and A. Merker. 1993. Role of a membrane-associated serine esterase in the oxidant activation of phospholipase A_2 by t-butyl hydroperoxide. *Biochem J.* **292:**585–589.

97. Bomalski, J. S., M. Fallon, S. T. Crooke, P. C. Meunier, and M. A. Clark. 1990. Identification and isolation of phospholipase A_2 activating protein in human rheumatoid arthritis synovial fluid: induction of eicosanoid synthesis and an inflammatory response in joints injected in vivo. *J. Lab. Clin. Med.* **116:**814–818.

98. Douglas, C. E., A. C. Chan, and P. C. Choy. 1986. Vitamin E inhibits platelet phospholipase A_2. *Biochim. Biophys. Acta* **876:**639–642.

99. Pentlund, A. P., A. R. Morrison, S. C. Jacobs, L. L. Aruza, J. S. Hebert, and L. Packer. 1992. Tocopherol analogs suppress arachidonic acid metabolism with phospholipase inhibition. *J. Biol. Chem.* **267:**15528–15584.

100. Chatelain, E., D. Boscoboinik, G. Bartoli, V. E. Kagan, F. K. Gey, L. Packer, and A. Azzi. 1993. Inhibition of smooth muscle cell proliferation and protein kinase C activity by tocopherols and tocotrienols. *Biochim. Biophys. Acta* **1176:**83–89.

101. Davies, K. J. A., A. G. Weise, A. Sevanian, and E. Kim. 1990. Repair systems in oxidative stress, pp. 123–141. *In Molecular Biology of Aging,* UCLA Symposium on Molecular and Cellular Biology, Alan R. Liss, New York.

102. Chakraborti, S., and J. R. Michael. 1991. Involvement of a serine esterase in oxidant mediated activation of phospholipase A_2 in pulmonary endothelium. *FEBS Lett.* **281:**185–187.

103. Rice, K. L., P. G. Duane, S. L. Archer, D. P. Gilboe, and D. E. Niewoehner. 1992. H_2O_2 injury causes Ca^{2+}-dependent and independent hydrolysis of phosphatidylcholine in alveolar epithelial cells. *Am. J. Physiol.* **263:**L430–L438.

104. Ballou, L. R., and W. Y. Cheung. 1983. Marked increase of human platelet phospholipase A_2 activity in vitro and demonstration of an endogenous inhibitor. *Proc. Natl. Acad. Sci USA* **80:**5203–5207.

105. Mayer, R. J., and L. A. Marshall. 1993. New insights on mammalian phospholipase A_2(s); comparison of arachidonyl-selective and non-selective enzymes. *FASEB J.* **7:**339–348.

106. Wijkander, J., and R. Sundler. 1992. Macrophage arachidonate-mobilizing phospholipase A_2: role of Ca^{2+} for membrane binding but not for catalytic activity. *Biochem. Biophys. Res. Commun.* **184:**118–124.

107. Clark, J. D., L.-L. Lin, R. W. Kriz, C. S. Ramesha, L. A. Sultzman, A. Y. Lin, N. Milona, and J. L. Knopf. 1991. A novel arachidonic acid-selective cytosolic PLA_2 contains a Ca^{2+}-dependent translocation domain with homology to PKC and GAP. *Cell* **65:**1043–1051.

108. Tsujishita, Y., Y. Asaoka, and Y. Nishizuka. 1994. Regulation of phospholipase A_2 in human leukemia cell lines: its implication for intracellular signaling. *Proc. Natl. Acad. Sci. USA* **91:**674–6278.

109. Qiu, Z.-H., and C. C. Leslie. 1994. Protein kinase C-dependent and -independent pathways of mitogen-activated protein kinase activation in macrophages by stimuli that activate phospholipase A_2. *J. Biol. Chem.* **269:**19480–19487.

110. Goldman, R., R. Ferber, R. Meller, and U. Zor. 1994. A role for reactive oxygen species in zymosan and beta-glucan induced protein tyrosine phosphorylation and phospholipase activation in murine macrophages. *Biochim. Biophys. Acta* **1222:**265–276.

111. Chakraborti, S., G. H. Gurtner, and J. R. Michael. 1989. Oxidant-mediated activation of phospholipase A_2 in pulmonary endothelium. *Am. J. Physiol.* **257:**L430–L437.

112. Xing, M., and R. Mattera. 1992. Phosphorylation-dependent regulation of phospholipase A_2 by G-proteins and Ca in HL granulocytes. *J. Biol. Chem.* **267:**25966–25975.

113. Taher, M. M., J. G. N. Garcia, and V. Natarajan. 1993. Hydroperoxide-induced diacylglycerol formation and protein kinase C activation in vascular endithelial cells. *Arch. Biochem. Biophys.* **303:**260–266.

114. Gopalakrishna, R., and W. B. Anderson. 1989. Ca^{2+}- and phospholipid-independent activation of protein kinase C by selective oxidative modification of the regulatory domain. *Proc. Natl. Acad. Sci. USA* **86:**6758–6762.

115. Bazzi, M. D., and G. L. Nelsestuen. 1988. Constitutive activity of membrane-inserted protein kinase C. *Biochem. Biophys. Res. Commun.* **152:**336–343.

116. Gopalakrishna, R., and W. B. Anderson. 1991. Reversible oxidative activation and inactivation of protein kinase C by the mitogen/tumor promoter periodate. *Arch. Biochem. Biophys.* **285:**382–387.

117. O'Brian, C. A., N. E. Ward, B. Weistein, A. W. Bull, and L. J. Marnett. 1988. Activation of rat brain protein kinase C by lipid oxidation product. *Biochem. Biophys. Res. Commun.* **155:**1374–1380.

118. Nemenoff, R. A., S. Winitz, N.-X. Qian, V. van Putten, G. L. Johnson, and L. E. Heasley. 1993. Phosphorylation and activation of a high molecular weight form of phospholipase A_2 by p42 microtubule-associated protein 2 kinase and protein kinase C. *J. Biol. Chem.* **268:**1960–1964.

119. Wijkander, J., and R. Sundler. 1991. An 100-kDa arachidonate-mobilizing phospholipase A_2 in mouse spleen and the macrophage cell line J774. Purification, substrate interaction and phosphorylation by protein kinase C. *Eur. J. Biochem.* **202:**873–880.

120. Chakraborty, S., J. R. Michael, and T. Sanyal. 1992. Defining the role of protein kinase C in calcium-ionophore-(A23187)-mediated activation of phospholipase A_2 in pulmonary endothelium. *Eur. J. Biochem.* **206:**965–972.

121. Vladimirov, Y. A., V. I. Olenev, T. B. Suslova, and Z. P. Cheremisina. 1980. Lipid peroxidation in mitochondrial membranes. *Adv. Lipid Res.* **17:**173–249.

122. Kramer, R. M., C. Hession, B. Johansen, G. Hayes, P. McGray, E. P. Chow, R. Tizard, and R. B. Pepinsky. 1989. Structure and properties of a human non-pancreatic phospholipase A_2. *J. Biol. Chem.* **264:**5768–5775.

123. Channon, J. Y., and C. C. Leslie. 1990. A calcium-dependent mechanism for associating a soluble arachidonoyl-hydrolyzing phospholipase A_2 with membrane in the macrophage cell line RAW 264. *J. Biol. Chem.* **265:**5409–5413.

124. Clark, J. D., L.-L. Lin, R. W. Kriz, C. S. Ramesha, L. A. Sultzman, A. Y. Lin, N. Milona, and J. L. Knopf. 1995. A novel arachidonic acid-selective cytosolic PLA₂ contains a Ca-dependent translocation domain with homology to PKC and GAP. *Cell* **65**:1043–1051.

125. Bell, R. M., and D. J. Burns. 1991. Lipid activation of protein Kinase C. *J. Biol. Chem.* **266**:4661–4664.

126. Zor, U., E. Ferber, P. Gergely, K. Scucs, V. Dombradi, and R. Goldman. 1993. Reactive oxygen species mediate phorbol ester-regulated tyrosine phosphorylation and phospholipase A₂ activation: potentiation by vanadate. *Biochem. J.* **295**:879–888.

127. Pacifici, E. H. K., L. L. McLeod, and A. Sevanian. 1994. Lipid hydroperoxide-induced peroxidation and turnover of endothelial cell phospholipids. *Free Radic. Biol. Med.* **17**:297–309.

128. Murakami, K., S. Y. Chan, and A. Routtenberg. 1986. Protein kinase C activation by cis-fatty acid in the absence of Ca and phospholipids. *J. Biol. Chem.* **261**:15424–15429.

129. Rao, G. N., B. Lassegue, K. K. Griendling, R. W. Alexander, and B. C. Berk. 1993. Hydrogen peroxide-induced c-*fos* expression is mediated by arachidonic acid release: role of protein kinase C. *Nucleic Acids Res.* **21**:1259–1263.

130. Lin, L.-L., M. Wartmen, A. Y. Lin, J. L. Knopf, A. Seth, and R. J. Davis. 1993. cPLA₂ is phosphorylated and activated by MAP kinase. *Cell* **72**:269–278.

131. Rao, G. N., et al. 1995. Hydrogen peroxide activation of cytosolic phospholipase A₂ in vascular smooth muscle cells. *Biochim. Biophys. Acta.* **1265**:67–72.

132. Rao, G. N., R. W. Alexander, and M. S. Runge. 1995. Linoleic acid and its metabolites, hydroxyoctadecadienoic acids, stimulate c-*fos,* c-*jun* and c-*myc* mRNA expression, mitogen-activated protein kinase activation, and growth in rat aortic smooth muscle cells. *J. Clin. Invest.* **96**:842–847.

133. Stevenson, M. A., et al. 1994. X-irradiation, phorbol esters, and H₂O₂ stimulate mitogen-activated protein kinase activity in NIH-3T3 cells through the formation of reactive intermediates. *Cancer Res.* **54**:12–15.

134. Russo, T., N. Zabrano, F. Esposito, R. Ammendola, F. Cimino, M. Fiscella, J. Jackman. 1995. A p53-independent pathway for activation of WAF1/CIP1 expression following oxidative stress. 1995. *J. Biol. Chem.* **270**:29386–29391.

135. Lander, H. M., J. S. Ogiste, K. K. Teng, and A. Novogradsky. 1995. p21ras as a common signaling target of reactive free radicals and cellular redox status. *J. Biol. Chem.* **270**:21195–21198.

136. Garcia-Morales, P., Y. Minami, E. Luong, R. D. Klausner, and L. E. Samelson. 1990. Tyrosine phosphorylation in T cells is regulated by phosphatase activity: studies with phenylarsine oxide. *Proc. Natl. Acad. Sci. USA* **87**:9255–9259.

137. Hadari, Y. R., B. Geiger, O. Nadiv, I. Sabanay, C. T. Robers, Jr., D. LeRoith, and Y. Zick. 1993. Hepatic tyrosine-phosphorylated proteins identified and localized following in vivo inhibition of protein tyrosine phosphatases: effects of H₂O₂ an vanadate administration into rat livers. *Mol. Cell. Endocrinol.* **97**:9–17.

138. Sullivan, S. G., D. T.-Y. Chiu, M. Errasfa, J. M. Wang, J.-S. Qi, and A. Stern. 1994. Effects of H_2O_2 on protein tyrosine phosphatase activity in HER14 cells. *Free Rad. Biol. Med.* 16:399–403.

139. Fischer, E. H., H. Charbonneau, and N. K. Tonks. 1991. Protein tyrosine phosphatases: a diverse family of intracellular and transmembrane enzymes. *Science* **253**:401–406.

140. Alessi, D. R., N. Gomez, G. Moorhead, T. Lewis, and P. Cohen. 1995. Inactivation of p42 MAPKinase by protein phosphatase 2A and a protein tyrosine phosphatase, but not CL100, in various cell lines. *Curr. Biol.* **5**:283–295.

141. Guy, G. R., J. Cairns, S. Ng, and Y. H. Tan. 1993. Inactivation of a redox-sensitive protein phosphatase during the early events of tumor necrosis factor/interleukin-1 signal transduction. *J. Biol. Chem.* **268**:2141–2148.

142. Van Kuijk, F. J. G. M., A. Sevanian, G. Handelman, and E. A. Dratz. 1987. A new role for phospholipase A_2. *Trends Biochem. Sci.* **12**:31–34.

143. Ursini, F., and A. Bindoli. 1987. The role of selenium peroxidases in the protection against damage of membranes. *Chem. Phys. Lipids* **44**:255–276.

144. Romaschin, A. D., I. Rebeyka, G. J. Wilson, and D. A. G. Mickle. 1987. Conjugated dienes in ischemic and reperfused myocardium: an in vivo chemical signature of oxygen free radical mediated injury. *J. Mol. Cell. Cardiol.* **19**:289–302.

145. McLeod, L. L. M., and A. Sevanian. Submitted for publication. Lipid alterations in an endothelial cell model of ischemia and reperfusion. *Free Rodic. Biol. Med.*

146. Robison, T., A. Sevanian, and H. Forman. 1990. Inhibition of arachidonic acid release by nordihydroguaiaretic acid and its antioxidant action in rat alveolar macrophages and Chinese hamster lung fibroblasts. *Toxicol. Appl. Pharmacol.* **105**:113–122.

147. Sciamanna, M. A., J. Zinkel, A. Y. Fabi, and C. P. Lee. 1992. Ischemic injury to rat forebrain mitochondria and protection by ascorbate. *Arch. Biochem. Biophys.* **305**:215–224.

148. Takayama, F., T. Egashira, and Y. Yamanaka. 1994. The multiple hydroperoxides of choline phospholipids occurring in plasma after ischemia-reperfusion in rate liver. *J. Toxicol. Sci.* **19**:97–106.

149. Block, F., M. Kunkel, and K. H. Sontag. 1995. Posttreatment with EPC-K1, an inhibitor of lipid peroxidation and of phospholipase A_2 activity, reduces functional deficits after global ischemia in rats. *Brain Res. Bull.* **36**:257–260.

150. Duane, P. G., K. L. Rice, D. E. Charbonneau, M. B. King, D. P. Gilboe, and D. E. Niewoehner. 1991. Relationship of oxidant-mediated cytotoxicity to phospholipid metabolism in endothelial cells. *Am. J. Respir. Cell Mol. Biol.* **4**:408–416.

151. Malis, C. D., and J. V. Bonventre. 1986. Mechanism of calcium potentiation of oxygen free radical injury to renal mitochondria. *J. Biol. Chem.* **261**:14201–14206.

152. Hornberger, W., and H. Patscheke. 1990. Primary stimuli of icosanoid release inhibit arachidonoyl-CoA synthetase and lysophospholipid acyltransferase. *Eur. J. Biochem.* **187**:175–181.

153. Lubin, B. G., S. B. Shohet, and D. G. Nathan. 1972. Changes in fatty acid metabolism after erythrocyte peroxidation: stimulation of a membrane repair process. *J. Clin. Invest.* **51:**338–345.

154. Dise, C. A., and D. B. P. Goodman. 1986. t-Butyl hydroperoxide alters fatty acid incorporation into erythrocyte membrane phospholipid. *Biochim. Biophys. Acta* **859:**69–78.

155. Dise, C., J. Clark, C. Lambersten, and D. Goodman. 1987. Hyperbaric hyperoxia reversibly inhibits erythrocyte phospholipid fatty acid turnover. *J. Appl. Physiol.* **62:**533–538.

156. Zaleska, M., and D. Wilson. 1989. Lipid hydroperoxides inhibit reacylation of phospholipids in neuronal membranes. *J. Neurochem.* **52:**255–260.

157. Sun, D., and D. D. Gilboe. 1994. Ischemia-induced changes in cerebral mitochondrial free fatty acids, phospholipids and respiration in the rat. *J. Neurochem.* **62:**1921–1928.

158. Lubin, B., D. Chiu, J. Bastacki, B. Roelofsen, and L. L. M. van Deenen. 1981. Abnormalities in membrane phospholipid organization in sickled erythrocytes. *J. Clin. Invest.* **67:**1643–1649.

159. Balsinde, J., I. D. Bianco, E. J. Ackermann, K. Conde-Frieboes, and E. A. Dennis. 1995. Inhibition of calcium-independent phospholipase A₂ prevents arachidonic acid incorporation and phospholipid remodeling in P388D1 macrophages. *Proc. Natl. Acad. Sci. USA* **92:**8527–8531.

160. Hanel, A. M., and M. H. Gelb. 1995. Multiple enzymatic activities of the human cytosolic 85-kDa phospholipase A₂: hydrolytic reactions and acyl transfer to glycerol. *Biochemistry* **34:**7807–7818.

161. Ohara, Y., T. E. Peterson, B. Zheng, J. F. Kuo, and D. G. Harrison. 1994. Lysophosphatidylcholine increases vascular superoxide anion production via protein kinase C activation. *Arterioscler. Thromb.* **14:**1007–1013.

162. Oishi, K., R. L. Raynor, P. A. Charp, and J. F. Kuo. 1988. Regulation of protein kinase C by lysophospholipids. *J. Biol. Chem.* **263:**6865–6871.

163. Kugiyama, K., M. Ohgushi, S. Sugiyama, T. Murohara, K. Fukunaga, E. Miyamoto, and H. Yasue. 1992. Lysophosphatidylcholine inhibits surface receptor-mediated intracellular signals in endothelial cells by a pathway involving protein kinase C activation. *Circ. Res.* **71:**1422–1428.

164. Glaser, K. B., A. Sung, J. Bauer, and B. M. Weichman. 1993. Regulation of eicosanoid biosynthesis in the macrophage. Involvement of protein tyrosine phosphorylation and modulation by selective protein tyrosine kinase inhibitors. *Biochem. Pharmacol.* **45:**711–717.

5

Oxidant-Mediated Activation of Phospholipases C and D

Viswanathan Natarajan, William M. Scribner, and Suryanarayana Vepa

5.1 Introduction

The response of mammalian cells to external stimuli involves transduction of signals across the plasma membrane into the cell cytoplasm and nucleus. Binding of hormones, neurotransmitters, growth factors, cytokines, and a variety of agonists to specific receptors results in the activation of effector enzymes, which catalyze the generation of second messengers. For example, activation of adenylate cyclase results in hydrolysis and conversion of ATP to cyclic AMP (cAMP), the first second-messenger system reported in mammalian cells.[1] Similarly, stimulation of phospholipases A_2, C, and D generates arachidonic acid, diacylglycerol (DAG), and phosphatidic acid (PA), respectively.[2] These lipid-derived second messengers are further metabolized or converted to oxygenated derivatives of arachidonate or lysophosphatidic acid (LPA). In addition to agonists, reactive oxygen species (ROS) or oxidants such as hydrogen peroxide (H_2O_2), superoxide anion, hydroxyl radical, nitric oxide (NO), or peroxynitrite also modulate cell-signaling pathways in mammalian cells.[3,4] Oxidants have been implicated in the pathophysiology of several human diseases (Table 5.1) and are known to cause

Table 5.1. A Partial List of Disorders that Likely Involve Oxidative Stress.

Alzheimer's disease	Myocardial infarction
Asthma	Pulmonary hypertension
Atherosclerosis	Parkinson's disease
Autoimmune diseases	Radiation injury
ARDS	HIV infection
Emphysema	Vasculitis

ARDS, adult respiratory distress syndrome; HIV, human immunodeficiency virus.

cell injury and damage.[4–6] Although the mechanism(s) of oxidant-induced human disorders are not well understood, they may involve modulation of signal transduction pathways. Several reviews on the activation and regulation of phospholipases A_2, C, and D (PLA_2, PLC, and PLD) by agonists have been published.[7–10] Only in the last decade has the role of oxidants as bioregulators gained increasing attention. Hence, this chapter focuses on the oxidant-mediated activation and regulation of PLC and PLD and the physiological significance of this stimulation in mammalian cells.

5.2 Phospholipase C in Cell Signaling

5.2.1 Phospholipase C

PLC is a key player in cell signaling involving lipid-derived second messengers.[11] PLC catalyzes the hydrolysis of phosphatidylinositol-4,5-bisphosphate (PIP_2) to generate inositol(1,4,5)trisphosphate (IP_3) and DAG. IP_3 regulates release of Ca^{2+} from the endoplasmic reticulum, whereas DAG is an endogenous, positive activator of protein kinase C (PKC).[12] DAG can also be phosphorylated to PA by the enzymatic action of DAG kinase.[13] PA is the immediate precursor for de novo biosynthesis of acidic phospholipids and possesses mitogenic properties.[14,15] Based on the phospholipid substrates utilized, PLC can be broadly categorized as PIP_2-PLC, PI-PLC (phosphatidylinositide specific), or PC-PLC (phosphatidylcholine specific). The reactions catalyzed by each of these PLCs are given in Figure 5.1. Of the three PLC enzymes, PIP_2-PLC and PI-PLC have been well characterized and exists in multiple isoforms (β, γ, δ).[16] All the three isoforms are single polypeptides of molecular weights 150, 145, and 85 kDa, respectively. So far more than 10 mammalian PI-PLC enzymes belonging to the three isotypes have been identified (four β, two γ, four δ). All the isoenzymes of PLC share two regions of high homology designated as X (60%) and Y (40%). The length of spacer region separating the X and Y regions is varied among the three types. Common structural features in PLC include SH2 and SH3 domains and a plekstrin homology (PH) domain.[17,18] The PH domain is the most recently identified domain present in all PLC types and has been implicated in protein-protein interactions.[19] SH2 and SH3 domains are small modules of protein structures (100 and 50 amino acids, respectively) and control protein-protein interactions involving tyrosine phosphorylated residues and proline-rich sequences.[20] Very little is known about the structural features of PC-PLC.

5.2.2 Activation and Regulation of PLC

A variety of ligands and extracellular stimuli activate PI-PLC, and isotype-specific activation is a rule rather than an exception. For example, a variety of polypeptide growth factors (platelet-derived growth factor [PDGF], epidermal growth factor

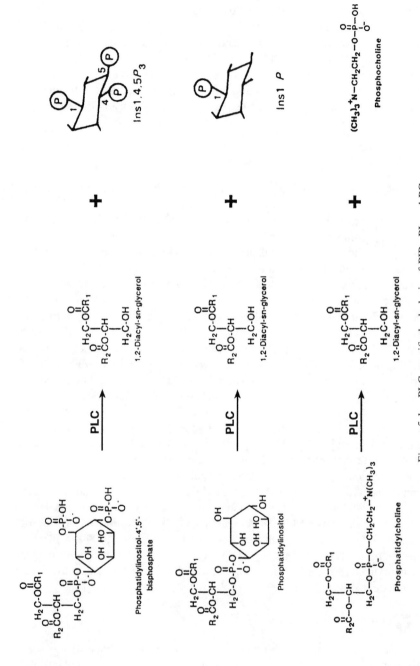

Figure 5.1. PLC-specific hydrolysis of PIP$_2$, PI, and PC.

[EGF], and fibroblast growth factor [bFGF]) activate PLC-γ,[16] while bradykinin, vasopressin, thromboxane A_2, bombesin, angiotensin II, histamine, acetylcholine, adrenergic activators, and thyroid-stimulating hormone stimulate PLC-β2.[21] The ligands involved in the activation of PLC-δ are currently not well studied.

The regulation of PI-PLC isoenzymes is very intricate and tightly controlled. The two isoenzymes of PLC, PLC-β and PLC-γ, differ greatly in their modes of regulation. PLC-β activation and regulation involves heterotrimeric G proteins,[21,22] whereas PLC-γ activation involves tyrosine phosphorylation and interaction with kinases.[16,17] For example, β1 activation requires Gqα,[22,23] whereas β2 employs G$\beta\gamma$.[22] Thus cell-specific expression of PLC isoenzymes and G proteins contribute to the complexity of PLC activation and regulation. Furthermore, cross talk between β and γ isoenzymes has been proposed based on the activation of PLC-γ by activators of PLC-β.[24]

The activators of PLC-γ employ a different mechanism to achieve stimulation and regulation. Growth factors such as PDGF bind to receptors, leading to receptor dimerization and activation of intrinsic tyrosine kinase activity.[25] The receptor activation results in phosphorylation of many proteins including PLC-γ and association of PLC-γ to the membrane.[25] Apart from the receptor tyrosine kinase mediated activation of PLC-γ, the stimulation can also be achieved by nonreceptor tyrosine kinase pathways. The agents that activate PLC-γ through nonreceptor pathways include ligands for T-cell antigen receptors[16,26] and cytokine receptors. The nonreceptor tyrosine kinases activated in this pathway include the members of the *src* family, *syk* and *jak/tyk*. The association of PLC-γ to these kinases probably involves protein scaffolding and may represent a significant part of activation mechanism.

5.2.3 Oxidant-Induced Activation of PLC

ROS such as H_2O_2, superoxide anion, hydroxyl radical, hypochlorous acid, nitric oxide, or peroxynitrite are normal metabolites of all mammalian cells utilizing O_2 for respiration. As these ROS are toxic to the cell, do not normally accumulate, and are detoxified by a variety of defense mechanisms operative within the cell. Most mammalian cells convert or metabolize the ROS to nontoxic derivatives using antioxidants, enzymes, metal-ion catalyzed reactions, and intracellular thiol agents[27] (Fig. 5.2). Conditions that offset this balance between ROS generated and metabolized can lead to an accumulation of ROS within cells.[6] As ROS are chemically and biochemically active, their interaction with cellular components causes lipid peroxidation, modification of proteins and carbohydrates, and DNA fragmentation. These oxidant-mediated alterations in cellular components lead to cell injury, changes in permeability, influx of calcium, and ultimately cell death. However, the amount of oxidants that accumulate or produced by polymorphonuclear leukocytes vary, depending on the mode of activation. Recently,

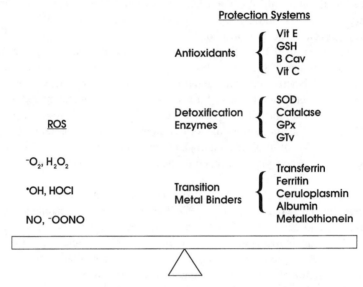

DEFENSE MECHANISMS AGAINST OXIDATIVE STRESS

Figure 5.2. Cellular defense mechanisms against accumulation of ROS.

awareness has increased of the role of oxidants as bioregulators in cellular processes such as mitogenesis, cell proliferation, apoptosis, etc., as evidenced by the increasing frequency of reports on the influence of oxidants on signaling pathways.

Brief exposure of endothelial cells (ECs) to H_2O_2, results in an alteration of permeability that is preceded by an increase in inositol phosphates and DAG, derived from PLC-catalyzed hydrolysis of inositol phospholipids.[28] A similar activation of PLC by H_2O_2 was also observed in MDCK cells.[29] The increased PLC-mediated generation of DAG results in PKC activation and decreased electrical resistance across the monolayers. Whisler et al.[30] observed that H_2O_2-simulated CD3/TCR-mediated signaling in T cells by increased IP_3 production and mobilization of intracellular Ca^{2+}. Furthermore, it was observed that H_2O_2-mediated PKC activation was dependent on PLC stimulation as inhibition of PLC by 4-bromophenacylbromide resulted in ~ 50% reduction in PKC activation. It is known that H_2O_2 increases tyrosine phosphorylation in a variety of cells including ECs and T cells.[31] As PLC-γ is regulated by tyrosine phosphorylation, mediated by either receptor or nonreceptor tyrosine kinases, it is tempting to speculate that the PLC subtype stimulated by H_2O_2, or other oxidants, is PLC-γ. This speculation gains further support from the observation that H_2O_2 increases tyrosine phosphorylation of the EGF receptor, the insulin receptor, and also *src* family of nonreceptor kinases.[32] Qin et al.[33] observed that H_2O_2 (10 mM) increased tyrosine phosphorylation of PLC-γ1 in permeabilized blood lymphocytes and suggested the involvement of p72[syk] tyrosine kinase in the phosphorylation of PLC-γ1.

In a study aimed at understanding the mechanism or mechanisms responsible for H_2O_2-induced contraction of pulmonary arteries, pretreatment of the arteries with NCDC (2-nitro-4-carboxyphenyl-*N,N*-diphenyl-piperazine), a PLC and serine protease inhibitor, attenuated contractions, suggesting a plausible link between H_2O_2-dependent activation of PLC and pulmonary artery contraction.[34] Schachter et al.[35] detected a four- to fivefold increase in phosphoinositide turnover with 1-mM H_2O_2 in a vascular smooth muscle cell line from embryonic rat aorta. In contrast to these results, Vercellotti et al.[36] reported that in human umbilical vein endothelial cells, H_2O_2-attenuated thrombin-mediated increase in PLC activity. However, the mode of this attenuation is not known. Furthermore, cardiac sarcolemmal PLC activity *in vitro* was inhibited by H_2O_2 and hypochlorous acid, whereas it was apparently not affected by oxygen free radicals.[37] Oxidation of SH groups in PLC or one of its regulatory component was suggested to be a possible cause for this inhibition.

Ethanol-induced hepatotoxicity is mediated by a cytochrome P-450, Cyp 2E1, and involves the formation of radical intermediates and ROS.[38] A role for lipid peroxidation was demonstrated in ethanol-induced hepatotoxicity and in a recent study Nanji et al.[39] demonstrated increased PLC activity in ethanol-fed rats. It is quite possible that free radicals and ROS formed during ethanol exposure were responsible for the observed increase in PLC activity.

An oxidative burst in eosinophils and neutrophils results in the generation of superoxide anion and H_2O_2.[40] Inhibition of PLC attenuated the formation of ROS, implicating a role for PLC in the oxidative burst. Involvement of PLC was also shown in the oxidative burst of cultured plant cells.[41] The agents, which elicit H_2O_2 formation in suspension culture of soybean cells, also increased phosphoinositide turnover. In contrast to these results, Gordon et al.[42] demonstrated that exogenously added bacterial PLC-inhibited superoxide and H_2O_2 generation induced by phorbol myristate acetate (PMA) and fMet-Leu-Phe (fMLP).[42]

Other conditions that enhance the formation of ROS include hypoxia, hyperoxia, and anoxia. Hypoxia caused increases in PLC activity in cat carotid body[43] and also enhanced IP_3 turnover of cardiac myocytes in response to alpha-adrenergic stimulation.[44] Activation of PLC along with PLA_2 was shown to have a role in hypoxic cell injury.[45]

In a remarkable study, Finkel and colleagues have demonstrated that activation of smooth muscle cells by PDGF increased intracellular H_2O_2 production.[46] In this study, most of the functional responses of PDGF were blocked by attenuating the production of H_2O_2. It is known that PDGF activates PLC-γ and it will be interesting to investigate whether PLC activation causes formation of H_2O_2 or H_2O_2 production leads to activation of PLC.

The effect of H_2O_2 and hydroperoxides on bovine pulmonary artery endothelial cell (BPAEC), DAG, and PKC was investigated. In unstimulated BPAECs, PKC was primarily localized in the cytosol and H_2O_2 treatment resulted in a dose- (10^{-5} to 10^{-3} M) and time- (15 to 60 min) dependent translocation of PKC from the cytosol to the membrane.[47] Similar activation of PLC was also observed with linoleic acid hydroperoxide and not with cumene or t-butyl hydroperoxide. In addition to PKC activation, H_2O_2 also caused increased production (two- to three-fold) of inositol phosphates and DAG. A similar increase in cytosolic PKC specific activity without altering membrane bound PKC activity was observed in hepatocytes exposed to quinones.[48] In contrast to these data, Von Ruecker et al.[49] showed that H_2O_2 plus Cu^{2+} increased membrane associated PKC in rat hepatocytes. Similarly, *in vitro* activation of PKC by lipoxin A^{50} and 5-hydroxy-eicosatetraenoic acid was observed.[51] Treatment of porcine pulmonary artery ECs with NO_2 resulted in PKC activation and increased [^{3}H]DAG production.[52] These results suggest oxidant-induced production of DAG and PKC activation.

5.2.4 Activation of PLC by Vanadate and Pervanadate

Vanadate, a potent phosphatase inhibitor, also acts as an oxidant and an insulin mimetic agent. Orthovanadate increased PLC activity, tyrosine phosphorylation, and translocation of PLC-γ in permeabilized mast cells.[53] Vanadate also activated PLC in permeabilized rat basophilic leukemia (RBL) cell ghosts, however, ATP was required for this activation and addition of high-energy phosphate compounds such as phosphoenolpyruvate and phosphocreatine potentiated this activation.[54] Tertrin-Clary et al.[55] demonstrated that in rat placental cells vanadate was a more potent activator of PLC than fluoride. A combination of vanadate and TPA activated PLC in rat peritoneal macrophages, and this activation of PLC was preceded by ROS formation, PKC activation, and protein tyrosine phosphorylation.[56] Vanadate combines with H_2O_2 in solution to form peroxovanadate ($V^{4+}OOH$), which is a much more potent inhibitor of tyrosine phosphatase than vanadate and H_2O_2. Peroxovanadate-induced platelet activation was associated with tyrosine phosphorylation of PLC-γ and increased PLC activity, as seen by increased inositol phosphate production.[57] Peroxovanadate-mediated signaling cascade in Jurkat human leukemic T cells included activation of tyrosine kinases *lck* and *fyn*, tyrosine phosphorylation of PLC-γ, and a rise in intracellular Ca^{2+} levels.[58] Grinstein and colleagues studied the effect of H_2O_2 and vanadate on Ca^{2+} homeostasis in granulocytic HL60 cells. Peroxides of vanadate induced transient increases in Ca^{2+} and increased production of inositol phosphates.[59] Immunoprecipitation studies demonstrated increased tyrosine phosphorylation of PLC-γ2 on treatment with vanadyl peroxides. Zick and coworkers studied the effect of H_2O_2 and vanadate by injecting the same into the portal vein of rat livers and identified PLC-γ as one of the proteins whose tyrosine phosphorylation was increased by these agents.[60] In BPAECs, addition of H_2O_2 (1 mM) plus vanadate (100 μM) greatly enhanced the generation of DAG, IP_3, and Ca^{2+} release. Immuno-precipitation of the cell lysates with anti-PLC-γ followed by Western blotting with antiphosphotyrosine antibody revealed increased tyrosine phosphorylation of PLC-γ as compared to H_2O_2 or vanadate alone (data not shown).

5.3 Activation and Regulation of Phospholipase D

5.3.1 Phospholipase D

PLD (E.C.3.1.4.4) catalyzes the hydrolysis of cellular phospholipids generating PA and a polar head group. The PA-generated by PLD can be dephosphorylated to DAG by PA phosphatase (PA Pase).[13] Although phosphatidylcholine (PC) is the preferred substrate for the crude mammalian PLD, the enzyme also hydrolyses other phospholipids such as phosphatidylethanolamine and phosphatidylinositol.[8,61–64] In addition to hydrolytic activity, PLD also exhibits transphosphatidylation property. In the presence of short-chain primary alcohols (C2 to C6), PA generated by PLD is enzymatically transferred to the primary alcohol, forming the corresponding phosphatidylalcohol (Fig. 5.3).[65,66] As PA is an important intermediate in the biosynthesis of acidic phospholipids, direct measurement of PA will not be a true index of PLD activation. However, formation of phosphatidyl alcohol is a reliable and reproducible marker for measuring PLD activity in intact cells and subcellular fractions.[8,61,67]

5.3.2 PLD Activation and Cell Signaling

Although PLD was first reported in plants by Hanahan and Chaikoff[68] in 1948, it took almost 40 years to recognize activation of PLD to external stimuli in mammalian cells. In 1987, Bocckino et al.[69] reported increased PA production in hepatocytes after stimulation with vasopressin. Since this initial report, activation of PLD by a variety of stimuli in mammalian cells has been reported.[3,8,9,61,70–73] Most of the agonists that activate PIP_2/PI-specific PLC also stimulate PLD. Stimulation of PLD generates PA, which can be subsequently converted to lysoPA by a phospholipase A1/A2 or to DAG by PA Pase. Diacylglycerol is an endogenous activator of classical (α,β,γ) and novel (δ,ε,η,θ) PKC isoenzymes and may also serve as a source of arachidonic acid for prostanoid production.[74] Thus, PLD activation, in response to an external stimulus, can produce DAG over a longer period of time and sustain cellular PKC activation. Phosphatidic acid, in the absence of DAG, can also activate PKC, suggesting a physiological role for PLD activation in mammalian cells.[75]

5.3.3 Regulation of Agonist-Induced PLD Activation

At least four possible mechanism(s) of regulation of agonist-induced PLD activation have been identified in mammalian cells. The regulation of PLD may involve (1) PKC activation, (2) changes in intracellular calcium, (3) GTP-binding proteins

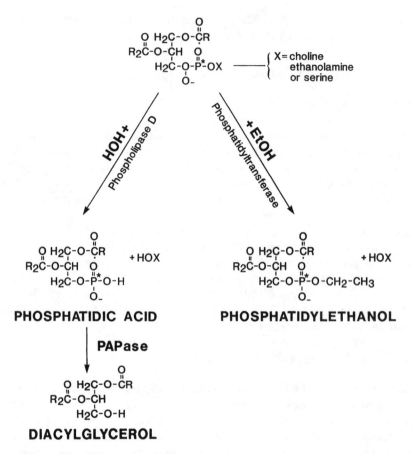

Figure 5.3. PLD-catalyzed hydrolysis and transphosphatidylation reactions.

including small molecular weight G-proteins, and (4) tyrosine kinases/protein tyrosine phosphatases.

5.3.3.1 Role of PKC in PLD Activation

In most of the mammalian systems agonist-induced PLD activation was dependent on PKC stimulation via PIP_2-PLC dependent pathway.[9] The agonist-mediated PLD activation was similar to TPA-induced PLD activation, suggesting a PKC dependent mechanism. Inhibition of PKC by pharmacological agents such as staurosporin, H-7, bisindolymaleimide, or calphostin C attenuated agonist- or TPA-induced PLD activation.[67] Additionally, down-regulation of PKC by prolonged exposure to TPA or Phorbol-12,13-dibutyrate (PDBu) blocked phorbol ester- or agonist-induced PLD activation.[67] These studies are consistent with the role of PKC in PLD activation. However, in lymphocytes, mast cells, and neutrophils, PKC inhibitors failed to attenuate fMLP-mediated PLD activation, suggesting PKC independent mechanism for PLD stimulation.[3] Furthermore, the mechanism of PKC-mediated regulation of PLD is unclear. Studies by Conricode et al.[76] suggested that PLD activation by PKC may not involve protein phosphorylation but may depend on protein-protein interaction. Recent studies in neutrophils by Lambeth and coworkers suggest that TPA-mediated PLD activation may involve phosphorylation of a plasma membrane protein.[77] However, there is conflicting evidence regarding the isoform(s) of PKC involved in PLD activation. Biochemical studies including over expression studies point out that PKC α and β isoforms may be the isoforms involved in the regulation of PLD.[77]

5.3.3.2 Role of G-Proteins in PLD Activation

Based on inhibition by cholera or pertussis toxin, fMLP, NaF, and other agonists are thought to utilize G_{i2} and/or G_{i3} G-proteins in PLD activation. Recent studies with *Clostridium botulinum* C3 exoenzyme indicate that the low-molecular-weight G-protein, rho, may be directly involved in PLD activation in HL-60 cells.[78,79] In human umbilical vein endothelial cells, pretreatment with cholera toxin (1 μg/ml, 1 h) potentiated the α-thrombin-induced PLD activation suggesting that Gs linked adenylate cyclase may be involved in the modulation of thrombin effect.[80]

5.3.3.3 Role of Calcium in PLD Activation

Agents that increase intracellular calcium such as thrombin, ATP, bradykinin, fMLP, and calcium ionophores stimulated PLD in a variety of mammalian cells including ECs. PLD activation, as measured by (^{32}P) phosphatidylethanol (PEt) formation, was attenuated by either depletion of extracellular calcium by EGTA or intracellular calcium by BAPTA-AM.[73] However, TPA-mediated (^{32}P)PEt accumulation was not affected by EGTA treatment, whereas BAPTA attenuated TPA-mediated PEt formation by 50%.[67,80] These data suggest that agonist-induced PLD activation in ECs involve PKC activation and is dependent on changes in intracellular calcium.

5.3.3.4 Tyrosine Kinases in Agonist-Induced PLD Activation

Signal transduction of growth factors such as PDGF, EGF, bFGF, and insulin involves increased protein tyrosine phosphorylation and PLD activation.[81-84] In osteoblast-like MC3TC-E1 cells[85] and human embryonic kidney cells expressing human m3 muscarinic acetylcholine receptor,[86] increased PLD activation via protein tyrosine phosphorylation in response to PGF2 alpha and carbachol, respectively, was observed.[87] The role of protein tyrosine phosphorylation in growth factor- and agonist-induced PLD activation was based on inhibitor studies with genistein, tyrophostin, ST 271, erbstatin, and other tyrosine kinase inhibitors.[87-90]

As protein tyrosine phosphorylation is a balance between tyrosine kinases and protein tyrosine phosphatases, inhibition of phosphatases with agents such as vanadate, phenylarsine oxide or diamide should potentiate protein tyrosine phosphorylation. In human neutrophils, PLD and not PLC stimulation by PAF, fMLP, and LTB4 was blocked by ST271, ST638, or erbstatin.[87] Furthermore, the PLD activation in permeabilized HL-60 cells was potentiated by pervanadate, an inhibitor of phosphatase, implicating tyrosine-kinase-dependent activation.[59] Recent studies in rat basophilic leukemia (RBL-2H3) cells suggest the involvement of tyrosine phosphorylation in immunoglobulin E- (IgE)-mediated PLD activation.[91] Similar to the growth-factor-dependent activation of PLD, cholecystokinin and caerulein-induced pancreatic messages were transduced via P13 kinase, tyrosine kinase, and PLD in pancreatic acini.[92] In U937 myeloid leukocytes, GTP[S]/Mg-ATP-activated PLD was synergistically regulated by G proteins and tyrosine kinases.[93] The mechanism(s) involved in tyrosine-kinase-dependent PLD activation by growth factors and other agonists is yet to be determined but may involve (1) phosphorylation of PLD by tyrosine kinases, (2) interaction between an intermediate tyrosine phosphorylated protein and PLD, and (3) phosphorylation of an inhibitory protein to PLD, causing PLD activation.

5.3.4 Oxidant-Induced PLD Activation

Oxidant-induced DAG accumulation in ECs may be due to activation of PIP_2/PLC, PC-PLC, PC/PLD, or all. A biphasic accumulation of DAG was observed in BPAEC exposed to H_2O_2.[47] The first phase of DAG formation was observed within 5 min of H_2O_2 treatment, whereas the second and sustained phase of DAG accumulation was seen after 10 min of H_2O_2 treatment.[3,47] This observation suggested that two different mechanisms are involved in the two phases of DAG production. It was hypothesized that the second phase of DAG accumulation was due to PLD activation, resulting in PA formation, followed by conversion of PA to DAG by PAPase.[3,94] Treatment of BPAECs with H_2O_2 or linoleic acid hydroperoxide resulted in enhanced (^{32}P)PEt formation, indicating PLD activation.[94] The oxidant-induced PLD activation was not associated with cytotoxicity and addition of ferrous chloride augmented H_2O_2-induced (^{32}P)PEt formation about twofold, suggesting a role for hydroxyl radical in the activation. Desferoxamine, a chelator of iron, also inhibited serum or Fe-induced activation of PLD. The oxidant-induced PLD activation was not attenuated by chelation of either extracellular calcium with EGTA or intracellular calcium with BAPTA. Furthermore, the PKC inhibitors staurosporin, bisindolylmaleimide, and calphostin C showed no effect on H_2O_2-induced PLD stimulation. These data support the notion that the oxidant-induced PLD activation was PKC- and Ca^{2+}-independent and different from that of agonist- or TPA-induced PLD activation in ECs. Similarly, noncytotoxic levels of H_2O_2 was shown to stimulate hydrolysis of phosphatidylcholine (PC) in NIH3T3 cells.[95] Interestingly, chronic treatment of

NIH3T3 cells with TPA blocked H_2O_2-induced PE and not PC hydrolysis, suggesting different isoforms of PLD involved in the activation.

In addition to H_2O_2 and linoleic acid hydroxperoxide, endothelial cell PLD was also stimulated by 4-hydroxynonenal (4-HNE), a major product of lipid peroxidation.[96] The 4-HNE-induced PLD activation was also PKC- and Ca^{2+}-independent and hydrolyzed both PC and PE as lipid substrates.[96] Oxidants like H_2O_2 and 4-HNE also activated PLD in other cell types including smooth muscle cells, fibroblasts, and neutrophils.[95,97] In HL-60 cells, pervanadate (H_2O_2 plus vanadate) increased PLD activity and this was linked to superoxide production and increased respiratory burst.[59] Oxidized low-density lipoprotein (LDL) (Ox-LDL) also activated PLD in rabbit femoral artery smooth muscle cells.[98] This activation of PLD was specific for Ox-LDL as native LDL or acetylated LDL had no effect. Also, PKC inhibitors and down regulation of PKC by TPA did not alter Ox-LDL-mediated PLD activation. These results suggest that oxidant and Ox-LDL-induced PLD activation were PKC-independent.

5.3.5 Role of Tyrosine Kinases in Oxidant-Induced PLD Activation

While considerable progress has been made in understanding the regulation of PLA_2 and PLC, very little is known about the regulation of PLD at a biochemical and molecular level. Although, PLD has been purified and cloned from plants and bacteria, only recently the human PLD1 gene from HeLa cells was cloned and a 120-kDa protein exhibiting activity was expressed in Cos-7 and Sf9 cells.[99] Hence, all the studies pertaining to PLD activation and regulation have been carried out in intact cells or reconstituted cell-free systems.

Two lines of studies provide evidence for the role of tyrosine kinases in oxidant-mediated PLD activation in mammalian cells. Inhibitors of tyrosine kinases and protein tyrosine phosphatases modulated oxidant-induced PLD activation. In ECs, H_2O_2-induced PLD activation was attenuated by genistein, tyrphostin, and herbimycin in a dose- and time-dependent manner (Fig. 5.4). The effect of tyrosine kinase inhibitors was highly specific for H_2O_2-, 4-HNE- or Ox-LDL-induced PLD activation as TPA- or bradykinin-mediated PLD activation was not blocked.[94,96] Exogenous addition of H_2O_2 to ECs also induced a time- and dose-dependent increase in protein tyrosine phosphorylation.[31] Several proteins (17 to 200 kDa) were identified as tyrosine phosphorylated proteins by Western blot analysis with antiphosphotyrosine antibodies.[94] A correlation between protein tyrosine phosphorylation and PLD activation was observed using varying concentrations of genistein (Fig. 5.5). A similar correlation between pervanadate-induced PLD activation and protein tyrosine phosphorylation was also observed in neutrophils.[59] These inhibitor studies point to a role for protein tyrosine phosphorylation in oxidant-induced PLD activation.

The potentiating effect of phosphatase inhibitors on oxidant-induced PLD activation offers the second line of evidence for tyrosine-kinase-dependent PLD

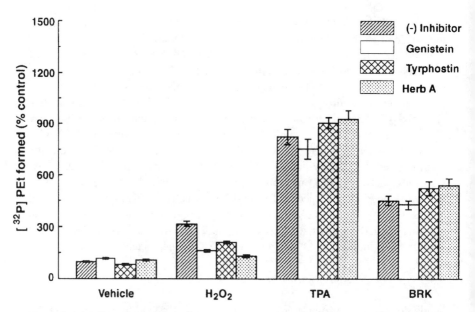

Figure 5.4. Effect of tyrosine inhibition on H$_2$O$_2$-induced PLD activation. [^{32}P]orthophosphate labeled BPAECs were pretreated with vehicle or vehicle containing genistein (100 µM), tyrphostin (10 µM), or herbimycin A (10 µM) for 60 min. Cells were washed and challenged with H$_2$O$_2$ (1 mM), TPA (100 µM), or bradykinin (BRK; 1 µM) for 30 min in the presence of 0.5% ethanol. Lipids were extracted under acidic conditions and [^{32}P]PEt was separated by TLC.[31] Values are represented as percentage control.

activation. Addition of vanadate to H$_2$O$_2$-enhanced PLD activation in ECs.[31] Other phosphatase inhibitors such as phenylarsine oxide and diamide also potentiate the H$_2$O$_2$-, 4-HNE-, and Ox-LDL-induced PLD activation.[100] Similar effects of vanadate on fMLP-, LTB4-, and H$_2$O$_2$-mediated PLD activation was observed in HL-60 cells.[87] In addition to PLD activation, addition of higher concentrations of vanadate (100 µM to 1 mM) also greatly enhanced protein tyrosine phosphorylation of EC proteins (Natarajan, V., Scribner, W. M., Vepa, S. 1996). However,

Figure 5.5. Correlation between H$_2$O$_2$-induced PLD activation and protein tyrosine phosphorylation. BPAECs were labeled with [^{32}P]orthophosphate[31] and pretreated with varying concentrations of genistein for 60 min. Unlabeled cells were used for the determination of protein tyrosine phosphorylation. H$_2$O$_2$-induced PLD activation was determined as indicated in Figure 5.4. H$_2$O$_2$-induced protein tyrosine phosphorylation was determined after sodium dodecylsulfate–polyacrylamide gel electrophoresis (SDS-PAGE) and Western blotting using antiphosphotyrosine antibody and detection with enhanced chemiluminescence (ECL).

A.

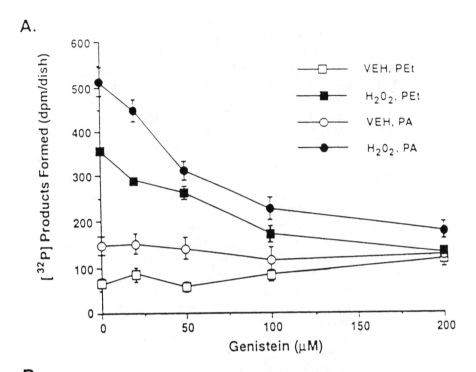

B.

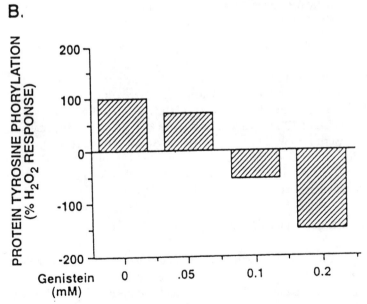

Figure 5.5

vanadate, at lower concentrations, showed only marginal increase in protein phosphorylation and PLD activation.

Vanadate not only functions as a phosphatase inhibitor but also, in the presence of H_2O_2, forms pervanadate (V^{4+}-OOH). Pervanadate is more cell permeable, as compared to vanadate, and is also a potent inhibitor of phosphatases.[101] Pervanadate is a potent insulin mimetic agent[102] and promotes insulin receptor protein tyrosine phosphorylation.[103] In neutrophils,[87] HL-60 cells,[59] and ECs,[31] pervanadate was shown to be more potent activator of PLD than H_2O_2. Furthermore, pervanadate increased protein tyrosine phosphorylation of ECs[104,105] and neutrophil proteins[87,106] and potentiated phosphorylation of polyGlu-Tyr in cell-free preparations. The pervanadate-induced PLD activation in neutrophils and ECs was inhibited by tyrosine kinase inhibitors, suggesting protein tyrosine phosphorylation as an important step in the activation of PLD.[3,59]

While there is increasing evidence for the involvement of tyrosine kinases in growth factor and oxidant-induced PLD activation, neither the mechanism of this activation nor the tyrosine-phosphorylated proteins responsible for PLD activation has been identified. Studies carried out by Vepa et al. in ECs suggest caveolin (22 kDa) and focal adhesion kinase (FAK) (125 kDa) as targets for H_2O_2-induced protein tyrosine phosphorylation.[104,105] Also, MAP-Kinase family (Erk1/Erk2) as target proteins for tyrosine phosphorylation in human neutrophils by granulocyte-macrophage colony-stimulating factor (GM-CSF) and in endothelial cells by H_2O_2 and pervanadate have been reported.[106-108] As microtubule-associated protein (MAP)-kinase has potential serine, threonine, and tyrosine phosphorylation sites, its activation may be critical in PLD activation.

5.3.6 Effect of Oxidants on PLD Activity in Vitro

While oxidants activated PLD in intact cells, an opposite effect was observed in cell-free preparations. Incubation of EC membrane preparations with H_2O_2 (0.1 to 10 mM) slightly inhibited hydrolysis of [^{3}H]PC to [^{3}H]PA in vitro. Under similar conditions, 4-HNE showed no effect on PLD in the membrane fractions (Natarajan, V., Scribner, W. M., Vepa, S. 1996). Addition of vanadate to H_2O_2 treated cell-free preparations inhibited PLD catalyzed hydrolysis of [^{3}H]PC to [^{3}H]PA (Fig. 5.6). It is possible that vanadate and DPV oxidize SH groups in membrane preparations, which are essential for PLD activity.

5.4 Physiologic Significance of PLC and PLD Activation

The physiological responses in which PLC has been implicated include a wide range of cellular events such as secretion, contraction, neural activity, fertilization, cell hypertrophy, proliferation, differentiation, and apoptosis.[11,18,21,109,110] PIP_2 hydrolysis by PLC is implicated in cell motility and changes in cell structure involving actin polymerization and rearrangement of the cytoskeletal network.[18,109] The significance of PLC-mediated signaling is further proven by the fact that

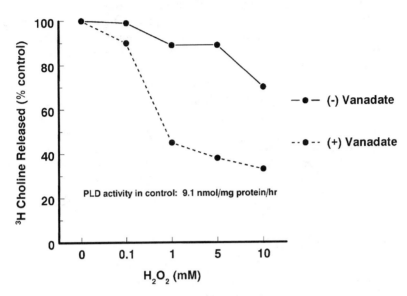

Figure 5.6. Effect of vanadate on H_2O_2-induced PLD activation in cell-free preparations. PLD activity in membrane fractions from BPAEC was measured using [^{3}H]PC as substrate and determining the release of [^{3}H]choline. PLD activity in untreated membranes was 9.1 nmol/mg protein/h.

both the substrate and the products have second-messenger roles. Activation of classical and novel PKC isotypes requires DAG and Ca^{2+} and PIP_2-PLC provides the initial burst of these metabolites for activation. IP_3-induced changes in intracellular Ca^{2+} can initiate a variety of cellular responses and modulate enzyme activities such as calmodulin, PLA_2, and PLD, Ca^{2+}-dependent kinases, and phosphatases. The physiological role for PLD activation in mammalian cells is yet to be clearly understood. Phosphatidic acid and lysoPA may have second-messenger functions and have been reported to be biologically active in eliciting a variety of cellular responses.[111] Phosphatidic acid has been suggested to promote Ca^{2+} entry,[112-114] inhibit adenylate cyclase,[115] exhibit mitogenic properties,[14,116-118] activate receptor and nonreceptor tyrosine kinases,[114,119] and be involved in stimulation-secretion coupling.[114] Generation of PA via the PLD pathway has been functionally linked to NADPH oxidase and superoxide generation in neutrophils and activation of PKC in mammalian cells.[15] Phosphatidic acid has been described to activate PIP_2-specific PLC both *in vitro* and in intact cells[120] and lysophosphatidic acid enhances tyrosine phosphorylation of focal adhesion kinase.[121] Phosphatidic acid is rapidly hydrolyzed to DAG by PA Pase, causing sustained accumulation of DAG and PKC activation.[13] Furthermore, DAG can be an important source of arachidonic acid for prostanoid synthesis.[12] Exposure of mammalian cells to agonists and subsequent activation of DAG lipase can result in mobilization of arachidonic acid and generation of prostanoids as a result of PLD-mediated generation of PA. The mechanism PA- or LPA-mediated signal transduction may involve receptors that are coupled to GTP-binding proteins and modulation of tyrosine kinases and phosphatases.[122] Exogenously added PA or LPA increases [³H]thymidine incorporation into DNA in fibroblasts and smooth muscle cells,[98,117] suggesting mitogenic effect. Ox-LDL is a risk factor in atherosclerosis and is involved in smooth muscle cell (SMC) proliferation. The mechanism of SMC proliferation is not well understood, but Ox-LDL-induced PLD activation and generation of PA may represent one of the initial signaling events. Recent studies in SMCs suggest that exogenously added Ox-LDL-enhanced incorporation of [³H]thymidine into DNA, which was also mimicked by PA or LPA.[98] Furthermore, Ox-LDL also increased intracellular free Ca^{2+} in SMCs. These results suggest that Ox-LDL in atherosclerotic plaques may enhance SMC proliferation through modulation of PLD and protein kinases.

5.5 Concluding Remarks

Oxidants are potent and rapid activators of PLC and PLD in mammalian cells, resulting in the generation of DAG and PA, respectively. The potential ramifica-

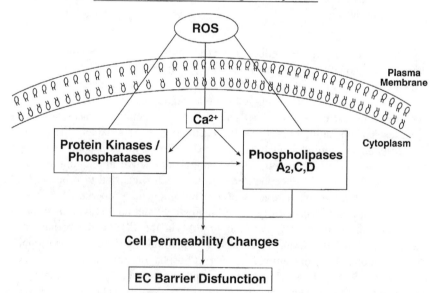

Figure 5.7. Scheme illustrating the role of ROS in modulating signal transduction pathways in endothelial cells.

tions of DAG and PA/LPA formation includes sustained activation of PKC, increase in intracellular calcium, and modulation of tyrosine kinases and phosphatases (Fig. 5.7). Alterations in the activation of protein kinases and generation of second messenger, in turn, can effect cell permeability, differentiation, proliferation, and apoptosis. Further understanding of oxidant-induced signal transduction pathways should provide new insights into cell functions under normal and diseased conditions.

Acknowledgments

We wish to thank Beverly Clark for her secretarial assistance and Rita Shamlal for her technical support. The work from V. Natarajan's laboratory is supported by grants from the National Institutes of Health (HL47671), K04HL03095; American Lung Association Career Investigator Award (VN), and an American Heart Association-Indiana Affiliate Postdoctoral Fellowship to Suryanarayana Vepa.

References

1. Hardman, J. G., G. A. Robison, and E. W. Sutherland. 1971. Cyclic nucleotides. [Review]. *Annu. Rev. Physiol.* **33**:311–336.

2. Exton, J. H. 1994. Messenger molecules derived from membrane lipids. [Review]. *Curr. Opin. Cell Biol.* **6**:226–229.

3. Natarajan, V. 1995. Oxidants and signal transduction in vascular endothelium [see comments]. [Review]. *J. Lab. Clin. Med.* **125**:26–37.

4. Saran, M., and W. Bors. 1989. Oxygen radicals acting as chemical messengers: a hypothesis. [Review]. *Free Rad. Res. Commun.* **7**:213–220.

5. Ward, P. A. 1991. Mechanisms of endothelial cell injury. [Review]. *J. Lab. Clin. Med.* **118**:421–426.

6. Brigham, K. L. 1986. Role of free radicals in lung injury. [Review]. *Chest* **89**:859–863.

7. Waite, M. 1987. *The Phospholipases,* Handbook of Lipid Research ed., Plenum Press, New York.

8. Billah, M. M., and J. C. Anthes. 1990. The regulation and cellular functions of phosphatidylcholine hydrolysis. [Review]. *Biochem. J.* **269**:281–291.

9. Liscovitch, M., and V. Chalifa. 1994. Signal-Activated Phospholipase D, pp. 31–63. *In Signal-Activated Phospholipases.* M. Liscovitch (ed.), Landes Co., Austin, TX.

10. Natarajan, V., W. M. Scribner, and V. Suryanarayana. (1996). Regulation of phospholipase D by tyrosine kinases. *Chem. Phys. Lipids.* **80**:103–116.

11. Berridge, M. J., and R. F. Irvine. 1984. Inositol trisphosphate, a novel second messenger in cellular signal transduction. [Review]. *Nature* **312**:315–321.

12. Nishizuka, Y. 1984. The role of protein kinase C in cell surface signal transduction and tumour promotion. [Review]. *Nature* **308**:693–698.

13. Brindley, D. N. 1984. Intracellular translocation of phosphatidate phosphohydrolase and its possible role in the control of glycerolipid synthesis. [Review]. *Prog. Lipid Res.* **23**:115–133.

14. Knauss, T. C., F. E. Jaffer, and H. E. Abboud. 1990. Phosphatidic acid modulates DNA synthesis, phospholipase C, and platelet-derived growth factor mRNAs in cultured mesangial cells. Role of protein kinase C. *J. Biol. Chem.* **265**:14457–14463.

15. Agwu, D. E., L. C. McPhail, S. Sozzani, D. A. Bass, and C. E. McCall. 1991. Phosphatidic acid as a second messenger in human polymorphonuclear leukocytes. Effects on activation of NADPH oxidase. *J. Clin. Invest.* **88**:531–539.

16. Rhee, S. G., and K. D. Choi. 1992. Multiple forms of phospholipase C isozymes and their activation mechanisms. [Review]. *Adv. Second Messenger Phosphoprotein Res.* **26**:35–61.

17. Rhee, S. G., and K. D. Choi. 1992. Regulation of inositol phospholipid-specific phospholipase C isozymes. [Review]. *J. Biol. Chem.* **267**:12393–12396.

18. Lee, S. B., and S. G. Rhee. 1995. Significance of PIP2 hydrolysis and regulation of phospholipase C isozymes. [Review]. *Curr. Opin. Cell Biol.* **7**:183–189.

19. Parker, P. J., B. A. Hemmings, and P. Gierschik. 1994. PH domains and phospholipases—a meaningful relationship? [Review]. *Trends Biochem. Sci.* **19**:54–55.

20. Koch, C. A., D. Anderson, M. F. Moran, C. Ellis, and T. Pawson. 1991. SH2 and SH3 domains: elements that control interactions of cytoplasmic signaling proteins. *Science* **252**:668–674.

21. Exton, J. H. 1994. Phosphoinositide phospholipases and G proteins in hormone action. [Review]. *Annu. Rev. Physiol.* **56**:349–369.

22. Rhee, S. G. 1994. Regulation of phosphoinositide-specific phospholipase C by a-protein. In M. Liscovitch (ed.), *Signal-Activated Phospholipases.* R. G. Landes Company, Austin, TX. pp. 1–12.

23. Gutowski, S., A. Smrcka, L. Nowak, D. G. Wu, M. Simon, and P. C. Sternweis. 1991. Antibodies to the alpha q subfamily of guanine nucleotide-binding regulatory protein alpha subunits attenuate activation of phosphatidylinositol, 4,5-bisphosphate hydrolysis by hormones. *J. Biol. Chem.* **266**:20519–20524.

24. Marrero, M. B., W. G. Paxton, J. L. Duff, B. C. Berk, and K. E. Bernstein. 1994. Angiotensin II stimulates tyrosine phosphorylation of phospholipase C-gamma 1 in vascular smooth muscle cells. *J. Biol. Chem.* **269**:10935–10939.

25. Claesson-Welsh, L. 1994. Platelet-derived growth factor receptor signals. [Review]. *J. Biol. Chem.* **269**:32023–32026.

26. Boulton, T. G., N. Stahl, and G. D. Yancopoulos. 1994. Ciliary neurotrophic factor/leukemia inhibitory factor/interleukin 6/oncostatin M family of cytokines induces tyrosine phosphorylation of a common set of proteins overlapping those induced by other cytokines and growth factors. *J. Biol. Chem.* **269**:11648–11655.

27. Heffner, J. E., and J. E. Repine. 1989. Pulmonary strategies of antioxidant defense. [Review]. *Am. Rev. Respir. Dis.* **140**:531–554.

28. Shasby, D. M., M. Yorek, and S. S. Shasby. 1988. Exogenous oxidants initiate hydrolysis of endothelial cell inositol phospholipids. *Blood* **72**:491–499.

29. Shasby, D. M., M. Winter, and S. S. Shasby. 1988. Oxidants and conductance of cultured epithelial cell monolayers: inositol phospholipid hydrolysis. *Am. J. Physiol.* **255**:C781–C788.

30. Whisler, R. L., Y. G. Newhouse, L. Beiqing, B. K. Karanfilov, M. A. Goyette, and K. V. Hackshaw. 1994. Regulation of protein kinase enzymatic activity in Jurkat T cells during oxidative stress uncoupled from protein tyrosine kinases: role of oxidative changes in protein kinase activation requirements and generation of second messengers. *Lymphokine Cytokine Res.* **13**:399–410.

31. Natarajan, V., S. Vepa, R. S. Verma, and W. M. Scribner. 1996. Inhibitors of tyrosine kinases and protein tyrosine phosphatases modulate hydrogen peroxide-induced activation of endothelial cell phospholipase D. *Am. J. Physiol.* **271**(15):L400–L408.

32. Yurchak, L. K., J. S. Hardwick, K. Amrein, K. Pierno, and B. M. Sefton. 1996. Stimulation of phosphorylation of Tyr-394 by hydrogen peroxide reactivates biologically inactive, non-membrane-bound forms of Lck. *J. Biol. Chem.* **271**:12549–12554.

33. Qin, S., T. Inazu, and H. Yamamura. 1995. Activation and tyrosine phosphorylation of p72^{syk} as well as calcium mobilization after hydrogen peroxide stimulation in peripheral blood lymphocytes. *Biochem. J.* **308**:347–352.

34. Sheehan, D. W., E. C. Giese, S. F. Gugino, and J. A. Russell. 1993. Characterization and mechanisms of H_2O_2-induced contractions of pulmonary arteries. *Am. J. Physiol.* **264**:H1542–H1547.

35. Schachter, M., K. L. Gallagher, M. K. Patel, and P. S. Sever. 1989. Oxidants and vascular smooth muscle. *Biochem. Soc. Trans.* **17**:1096–1097.

36. Vercellotti, G. M., S. P. Severson, P. Duane, and C. F. Moldow. 1991. Hydrogen peroxide alters signal transduction in human endothelial cells [see comments]. *J. Lab. Clin. Med.* **117**:15–24.

37. Meij, J. T., S. Suzuki, V. Panagia, and N. S. Dhalla. 1994. Oxidative stress modifies the activity of cardiac sarcolemmal phospholipase C. *Biochim. Biophys. Acta* **1199**:6–12.

38. Ingelman-Sundberg, M., and I. Johansson. 1984. Mechanisms of hydroxyl radical formation and ethanol oxidation by ethanol-inducible and other forms of rabbit liver microsomal cytochromes P-450. *J. Biol. Chem.* **259**:6447–6458.

39. Nanji, A. A., S. Zhao, R. G. Lamb, S. M. Sadrzadeh, A. J. Dannenberg, and D. J. Waxman. 1993. Changes in microsomal phospholipases and arachidonic acid in experimental alcoholic liver injury: relationship to cytochrome P-450 2E1 induction and conjugated diene formation. *Alcohol. Clin. Exp. Res.* **17**:598–603.

40. Perkins, R. S., M. A. Lindsay, P. J. Barnes, and M. A. Giembycz. 1995. Early signalling events implicated in leukotriene B4-induced activation of the NADPH oxidase in eosinophils: role of Ca^{2+}, protein kinase C and phospholipases C and D. *Biochem. J.* **310**:795–806.

41. Legendre, L., Y. G. Yueh, R. Crain, N. Haddock, P. F. Heinstein, and P. S. Low. 1993. Phospholipase C activation during elicitation of the oxidative burst in cultured plant cells. *J. Biol. Chem.* **268**:24559–24563.

42. Gordon, L. I., C. Schmeichel, S. Prachand, and S. A. Weitzman. 1990. Inhibition of polymorphonuclear leukocyte oxidative metabolism by exogenous phospholipase C. *Cell. Immunol.* **128**:503–515.

43. Pokorski, M., and R. Strosznajder. 1993. PO2-dependence of phospholipase C in the cat carotid body. *Adv. Exp. Med. Biol.* **337**:191–195.

44. Heathers, G. P., A. S. Evers, and P. B. Corr. 1989. Enhanced inositol trisphosphate response to alpha 1-adrenergic stimulation in cardiac myocytes exposed to hypoxia. *J. Clin. Invest.* **83**:1409–1413.

45. Kawaguchi, H., M. Shoki, K. Iizuka, H. Sano, Y. Sakata, and H. Yasuda. 1991. Phospholipid metabolism and prostacyclin synthesis in hypoxic myocytes. *Biochim. Biophys. Acta* **1094**:161–167.

46. Sundaresan, M., Z. X. Yu, V. J. Ferrans, K. Irani, and T. Finkel. 1995. Requirement for generation of H$_2$O$_2$ for platelet-derived growth factor signal transduction. *Science* **270**:296–299.

47. Taher, M. M., J. G. Garcia, and V. Natarajan. 1993. Hydroperoxide-induced diacylglycerol formation and protein kinase C activation in vascular endothelial cells. *Arch. Biochem. Biophys.* **303**:260–266.

48. Kass, G. E., S. K. Duddy, and S. Orrenius. 1989. Activation of hepatocyte protein kinase C by redox-cycling quinones. *Biochem. J.* **260**:499–507.

49. von Ruecker, A. A., B. G. Han-Jeon, M. Wild, and F. Bidlingmaier. 1989. Protein kinase C involvement in lipid peroxidation and cell membrane damage induced by oxygen-based radicals in hepatocytes. *Biochem. Biophys. Res. Commun.* **163**:836–842.

50. Hansson, A., C. N. Serhan, J. Haeggstrom, M. Ingelman-Sundberg, and B. Samuelsson. 1986. Activation of protein kinase C by lipoxin A and other eicosanoids. Intracellular action of oxygenation products of arachidonic acid. *Biochem. Biophys. Res. Commun.* **134**:1215–1222.

51. O'Flaherty, J. T., and J. Nishihira. 1987. 5-Hydroxyeicosatetraenoate promotes Ca2+ and protein kinase C mobilization in neutrophils. *Biochem. Biophys. Res. Commun.* **148**:575–581.

52. Patel, J. M., K. M. Sekharam, and E. R. Block. 1992. Oxidant and angiotensin II-induced subcellular translocation of protein kinase C in pulmonary artery endothelial cells. *J. Biochem. Toxicol.* **7**:117–123.

53. Atkinson, T. P., M. A. Kaliner, and R. J. Hohman. 1992. Phospholipase C-gamma 1 is translocated to the membrane of rat basophilic leukemia cells in response to aggregation of IgE receptors. *J. Immunol.* **148**:2194–2200.

54. Dreskin, S. C. 1995. ATP-dependent activation of phospholipase C by antigen, NECA, Na3VO4, and GTP-gamma-S in permeabilized RBL cell ghosts: differential augmentation by ATP, phosphoenolpyruvate and phosphocreatine. *Mol. Cell. Biochem.* **146**:165–170.

55. Tertrin-Clary, C., M. P. De La Llosa-Hermier, M. Roy, M. C. Chenut, C. Hermier, and P. De La Llosa. 1992. Activation of phospholipase C by different effectors in rat placental cells. *Cell. Signal.* **4**:727–736.

56. Goldman, R., and U. Zor. 1994. Activation of macrophage PtdIns-PLC by phorbol ester and vanadate: involvement of reactive oxygen species and tyrosine phosphorylation. *Biochem. Biophys. Res. Commun.* **199**:334–338.

57. Blake, R. A., T. R. Walker, and S. P. Watson. 1993. Activation of human platelets by peroxovanadate is associated with tyrosine phosphorylation of phospholipase C gamma and formation of inositol phosphates. *Biochem. J.* **290**:471–475.

58. Imbert, V., J. F. Peyron, D. Farahi Far, B. Mari, P. Auberger, and B. Rossi. 1994. Induction of tyrosine phosphorylation and T-cell activation by vanadate peroxide, an inhibitor of protein tyrosine phosphatases. *Biochem. J.* **297**:163–173.

59. Bourgoin, S., and S. Grinstein. 1992. Peroxides of vanadate induce activation of phospholipase D in HL-60 cells. Role of tyrosine phosphorylation. *J Biol. Chem.* **267**:11908–11916.

60. Hadari, Y. R., B. Geiger, O. Nadiv, I. Sabanay, C. T. Roberts, Jr., D. LeRoith, and Y. Zick. 1993. Hepatic tyrosine-phosphorylated proteins identified and localized following in vivo inhibition of protein tyrosine phosphatases: effects of H_2O_2 and vanadate administration into rat livers. *Mol. Cell. Endocrinol.* **97**:9–17.

61. Exton, J. H. 1990. Signaling through phosphatidylcholine breakdown. [Review]. *J. Biol. Chem.* **265**:1–4.

62. Liscovitch, M., J. K. Blusztajn, A. Freese, and R. J. Wurtman. 1987. Stimulation of choline release from NG108-15 cells by 12-O-tetradecanoylphorbol 13-acetate. *Biochem. J.* **241**:81–86.

63. Kiss, Z., and W. B. Anderson. 1989. Alcohols selectively stimulate phospholipase D-mediated hydrolysis of phosphatidylethanolamine in NIH 3T3 cells. *FEBS Lett.* **257**:45–48.

64. McNulty, S., G. S. Lloyd, R. M. Rumsby, R. M. Sayner, and M. G. Rumsby. 1992. Ethanolamine is released from glial cells in primary culture on stimulation with foetal calf serum and phorbol ester. *Neurosci. Lett.* **139**:183–187.

65. Kobayashi, M., and J. N. Kanfer. 1987. Phosphatidylethanol formation via transphosphatidylation by rat brain synaptosomal phospholipase D. *J. Neurochem.* **48**:1597–1603.

66. Gustavsson, L., and C. Alling. 1987. Formation of phosphatidylethanol in rat brain by phospholipase D. *Biochem. Biophys. Res. Comm.* **142**:958–963.

67. Natarajan, V., and J. G. Garcia. 1993. Agonist-induced activation of phospholipase D in bovine pulmonary artery endothelial cells: regulation by protein kinase C and calcium. *J. Lab. Clin. Med.* **121**:337–347.

68. Hanahan, D. J., and I. L. Chaikoff. 1948. On the nature of the phosphorus-containing lipides of cabbage leaves and their relation to a phospholipide-splitting enzyme contained in these leaves. *J. Biol. Chem.* **172**:191–198.

69. Bocckino, S. B., P. F. Blackmore, P. B. Wilson, and J. H. Exton. 1987. Phosphatidate accumulation in hormone-treated hepatocytes via a phospholipase D mechanism. *J. Biol. Chem.* **262**:15309–15315.

70. Shukla, S. D., and S. P. Halenda. 1991. Phospholipase D in cell signalling and its relationship to phospholipase C. [Review]. *Life Sci.* **48**:851–866.

71. Cockcroft, S. 1962. G-protein-regulated phospholipases C, D and A_2-mediated signalling in neutrophils. [Review]. *Biochim. Biophys. Acta* **1113**:135–160.

72. Billah, M. M. 1993. Phospholipase D and cell signaling. [Review]. *Curr. Opin. Immunol.* **5**:114–123.

73. Exton, J. H. 1994. Phosphatidylcholine breakdown and signal transduction. [Review]. *Biochim. Biophys. Acta* **1212**:26–42.

74. Nishizuka, Y. 1992. Intracellular signaling by hydrolysis of phospholipids and activation of protein kinase C. [Review]. *Science* **258**:607–614.

75. Stasek, J. E., Jr., V. Natarajan, and J. G. Garcia. 1993. Phosphatidic acid directly activates endothelial cell protein kinase C. *Biochem. Biophys. Res. Commun.* **191**:134–141.

76. Conricode, K. M., K. A. Brewer, and J. H. Exton. 1992. Activation of phospholipase D by protein kinase C. Evidence for a phosphorylation-independent mechanism. *J. Biol. Chem.* **267**:7199–7202.

77. Lopez, I., D. J. Burns, and J. D. Lambeth. 1995. Regulation of phospholipase D by protein kinase C in human neutrophils. Conventional isoforms of protein kinase C phosphorylate a phospholipase D-related component in the plasma membrane. *J. Biol. Chem.* **270**:19465–19472.

78. Malcolm, K. C., A. H. Ross, R. G. Qiu, M. Symons, and J. H. Exton. 1994. Activation of rat liver phospholipase D by the small GTP-binding protein RhoA. *J. Biol. Chem.* **269**:25951–25954.

79. Ohguchi, K., Y. Banno, S. Nakashima, and Y. Nozawa. 1996. Regulation of Membrane-bound Phospholipase D by Protein Kinase C in HL60 Cells. *J. Biol. Chem.* **271**:4366–4372.

80. Garcia, J. G., J. W. Fenton II, and V. Natarajan. 1992. Thrombin stimulation of human endothelial cell phospholipase D activity. Regulation by phospholipase C, protein kinase C, and cyclic adenosine 3'5'-monophosphate. *Blood* **79**:2056–2067.

81. Reynolds, N. J., H. S. Talwar, J. J. Baldassare, P. A. Henderson, J. T. Elder, J. J. Voorhees, and G. J. Fisher. 1993. Differential induction of phosphatidylcholine hydrolysis, diacylglycerol formation and protein kinase C activation by epidermal growth factor and transforming growth factor-alpha in normal human skin fibroblasts and keratinocytes. *Biochem. J.* **294**:535–544. [Published erratum appears in 1993 *Biochem. J.* **295**(Pt 3):903.]

82. Price, B. D., J. D. Morris, and A. Hall. 1989. Stimulation of phosphatidylcholine breakdown and diacylglycerol production by growth factors in Swiss-3T3 cells. *Biochem. J.* **264**:509–515.

83. Plevin, R., S. J. Cook, S. Palmer, and M. J. Wakelam. 1991. Multiple sources of sn-1,2-diacylglycerol in platelet-derived-growth-factor-stimulated Swiss 3T3 fibroblasts. Evidence for activation of phosphoinositidase C and phosphatidylcholine-specific phospholipase D. *Biochem. J.* **279**:559–565.

84. Randall, R. W., G. D. Spacey, and R. W. Bonser. 1993. Phospholipase D activation in PDGF-stimulated vascular smooth muscle cells. *Biochem. Soc. Trans.* **21**:352S.

85. Kozawa, O., A. Suzuki, and Y. Oiso. 1995. Tyrosine kinase regulates phospholipase D activation at a point downstream from protein kinase C in osteoblast-like cells. *J. Cell. Biochem.* **57**:251–255.

86. Schmidt, M., S. M. Huwe, B. Fasselt, D. Homann, U. Rumenapp, J. Sandmann, and K. H. Jakobs. 1994. Mechanisms of phospholipase D stimulation by m3 muscarinic acetylcholine receptors. Evidence for involvement of tyrosine phosphorylation. *Eur. J. Biochem.* **225**:667–675.

87. Uings, I. J., N. T. Thompson, R. W. Randall, G. D. Spacey, R. W. Bonser, A. T. Hudson, and L. G. Garland. 1992. Tyrosine phosphorylation is involved in receptor coupling to phospholipase D but not phospholipase C in the human neutrophil. *Biochem. J.* **281**:597–600. [Published erratum appears in 1992 *Biochem. J.* **283**(Pt 3):919.]

88. Wilkes, L. C., V. Patel, J. R. Purkiss, and M. R. Boarder. 1993. Endothelin-1 stimulated phospholipase D in A10 vascular smooth muscle derived cells is dependent on tyrosine kinase. Evidence for involvement in stimulation of mitogenesis. *FEBS Lett.* **322**:147–150.

89. Rivard, N., G. Rydzewska, J. S. Lods, and J. Morisset. 1995. Novel model of integration of signaling pathways in rat pancreatic acinar cells. *Am. J. Physiol.* **269**:G352–G362.

90. Kim, B. Y., S. C. Ahn, H. K. Oh, H. S. Lee, T. I. Mheen, H. M. Rho, and J. S. Ahn. 1995. Inhibition of PDGF-induced phospholipase D but not phospholipase C activation by herbimycin A. *Biochem. Biophys. Res. Commun.* **212**:1061–1067.

91. Lin, P., W. J. C. Fung, S. Li, T. Chen, B. Repetto, K. S. Huang, and A. M. Gilfillan. 1994. Temporal regulation of the IgE-dependent 1,2-diacylglycerol production by trosine kinase activation in a rat (RBL 2H3) mast-cell line. *Biochem. J.* **299**:109–114.

92. Rivard, N., G. Rydzewska, J. S. Lods, J. Martinez, and J. Morisset. 1994. Pancreas growth, tyrosine kinase, PtdIns 3-kinase, and PLD involve high-affinity CCK-receptor occupation. *Am. J. Physiol.* **266**:G62–G70.

93. Dubyak, G. R., S. J. Schomisch, D. J. Kushner, and M. Xie. 1993. Phospholipase D activity in phagocytic leucocytes is synergistically regulated by G-protein-and tyrosine kinase-based mechanisms. *Biochem. J.* **292**:121–128.

94. Natarajan, V., M. M. Taher, B. Roehm, N. L. Parinandi, H. H. Schmid, Z. Kiss, and J. G. Garcia. 1993. Activation of endothelial cell phospholipase D by hydrogen peroxide and fatty acid hydroperoxide. *J. Biol. Chem.* **268**:930–937.

95. Kiss, Z., and W. H. Anderson. 1994. Hydrogen peroxide regulates phospholipase D-mediated hydrolysis of phosphatidylethanolamine and phosphatidylcholine by different mechanisms in NIH 3T3 fibroblasts. *Arch. Biochem. Biophys.* **311**:430–436.

96. Natarajan, V., W. M. Scribner, and M. M. Taher. 1993. 4-Hydroxynonenal, a metabolite of lipid peroxidation, activates phospholipase D in vascular endothelial cells. *Free Rad. Biol. Med.* **15**:365–375. Published erratum appears in 1994, *Free Rad. Biol. Med.* **16**(2):295.]

97. Fialkow, L., C. K. Chan, S. Grinstein, and G. P. Downey. 1993. Regulation of tyrosine phosphorylation in neutrophils by the NADPH oxidase. Role of reactive oxygen intermediates. *J. Biol. Chem.* **268**:17131–17137.

98. Natarajan, V., W. M. Scribner, C. M. Hart, and S. Parthasarathy. 1995. Oxidized low density lipoprotein-mediated activation of phospholipase D in smooth muscle cells: a possible role in cell proliferation and atherogenesis. *J. Lipid Res.* **36**:2005–2016.

99. Hammond, S. M., Y. M. Altshuller, T.-C. Sung, S. A. Rudge, K. Rose, J. A. Engebrecht, A. J. Morris, and M. A. Frohman. 1995. Human ADP-ribosylation factor-activated phosphatidylcholine-specific phospholipase D defines a new and highly conserved gene family. *J. Biol. Chem.* **270**:29640–29648.

100. Natarajan, V., W. M. Scribner, and S. Vepa. Submitted for publication. Phosphatase inhibitors potentiate 4-hydroxynonenal-induced phospholipase D activation in vascular endothelial cells. *Am. J. Resp. Cell Mol. Biol.*

101. Posner, B. I., R. Faure, J. W. Burgess, A. P. Bevan, D. Lachance, G. Zhang-Sun, I. G. Fantus, J. B. Ng, D. A. Hall, B. S. Lum, and A. Shaver. 1994. Peroxovanadium compounds. A new class of potent phosphotyrosine phosphatase inhibitors which are insulin mimetics. *J. Biol. Chem.* **269**:4596–4604.

102. Heffetz, D., I. Bushkin, R. Dror, and Y. Zick. 1990. The insulinomimetic agents H_2O_2 and vanadate stimulate protein tyrosine phosphorylation in intact cells. *J. Biol. Chem.* **265**:2896–2902.

103. Heffetz, D., W. J. Rutter, and Y. Zick. 1992. The insulinomimetic agents H_2O_2 and vanadate stimulate tyrosine phosphorylation of potential target proteins for the insulin receptor kinase in intact cells. *Biochem. J.* **288**:631–635.

104. Vepa, S., W. M. Scribner, and V. Natarajan. In press. Activation of protein phosphorylation by oxidants in vascular endothelial cells: identification of tyrosine phosphorylation of caveolin. *Free Rad. Biol. Med.*

105. Vepa, S., W. M. Scribner, and V. Natarajan. 1995. Hydrogen peroxide induces tyrosine phosphorylation of FAK in endothelial cells, p. 181. Abst. 9th International Conference on Second Messengers and Phosphoproteins. Nashville, TN USA October 27–November 1.

106. Bourgoin, S., P. E. Poubelle, N. W. Liao, K. Umezawa, P. Borgeat, and P. H. Naccache. 1992. Granulocyte-macrophage colony-stimulating factor primes phospholipase D activity in human neutrophils in vitro: role of calcium, G-proteins and tyrosine kinases. *Cell. Signal.* **4**:487–500.

107. Gomez-Cambronero, J., J. M. Colasanto, C. K. Huang, and R. I. Sha'afi. 1993. Direct stimulation by tyrosine phosphorylation of microtubule-associated protein (MAP) kinase activity by granulocyte-macrophage colony-stimulating factor in human neutrophils. *Biochem. J.* **291**:211–217.

108. Scribner, W. M., S. Vepa, and V. Natarajan. 1993. Diperoxovanadate-induced activation of MAP kinase, p. 182. Abst. 9th International Conference on Second Messengers and Phosphoproteins. Nashville, TN, USA October 27–November 1.

109. Janmey, P. A. 1994. Phosphoinositides and calcium as regulators of cellular actin assembly and disassembly. [Review]. *Annu. Rev. Physiol.* **56**:169–191.

110. De Jonge, H. W., H. A. Van Heugten, and J. M. Lamers. 1995. Signal transduction by the phosphatidylinositol cycle in myocardium. [Review]. *J. Mol. Cell. Cardiol.* **27**:93–106.

111. Kroll, M. H., G. B. Zavico, and A. L. Schafer. 1989. Second messenger function of phosphatidic acid in platelet activation. *J. Cell. Physiol.* **139**:558–564.

112. van Corven, E. J., A. Groenink, K. Jalink, T. Eichholtz, and W. H. Moolenaar. 1989. Lysophosphatidate-induced cell proliferation: identification and dissection of signaling pathways mediated by G proteins. *Cell* **59**:45–54.

113. Moolenaar, W. H., R. L. van der Bend, E. J. van Corven, K. Jalink, T. Eichholtz, and W. J. van Blitterswijk. 1992. Lysophosphatidic acid: a novel phospholipid with hormone- and growth factor-like activities. [Review]. *Cold Spring Harbor Symp. Quant. Biol.* **57**:163–167.

114. Moolenaar, W. H. 1995. Lysophosphatidic acid, a multifunctional phospholipid messenger. [Review]. *J. Biol. Chem.* **270**:12949–12952.

115. Dunlop, M. E., and R. G. Larkins. 1989. Effects of phosphatidic acid on islet cell phosphoinositide hydrolysis, Ca2+, and adenylate cyclase. *Diabetes* **38**:1187–1192.

116. Moolenaar, W. H., W. Kruijer, B. C. Tilly, I. Verlaan, A. J. Bierman, and S. W. de Laat. 1986. Growth factor-like action of phosphatidic acid. *Nature* **323**:171–173.

117. Wood, C. A., L. Padmore, and G. K. Radda. 1993. The effect of phosphatidic acid on the proliferation of Swiss 3T3 cells. *Biochem. Soc. Trans.* **21**:369S.

118. Kaszkin, M., J. Richards, and V. Kinzel. 1992. Proposed role of phosphatidic acid in the extracellular control of the transition from G2 phase to mitosis exerted by epidermal growth factor in A431 cells. *Cancer Res.* **52**:5627–5634.

119. Bocckino, S. B., P. B. Wilson, and J. H. Exton. 1991. Phosphatidate-dependent protein phosphorylation. *Proc. Natl. Acad. Sci. USA* **88**:6210–6213.

120. Jackowski, S., and C. O. Rock. 1989. Stimulation of phosphatidylinositol 4,5-bisphosphate phospholipase C activity by phosphatidic acid. *Arch. Biochem. Biophys.* **268**:516–524.

121. Saville, M. K., A. Graham, K. Malarkey, A. Paterson, G. W. Gould, and R. Plevin. 1994. Regulation of endothelin-1- and lysophosphatidic acid-stimulated tyrosine phosphorylation of focal adhesion kinase (pp125fak) in Rat-1 fibroblasts. *Biochem. J.* **301**:407–414.

122. English, D., Y. Cui, and R. A. Siddiqui. (1996). Messenger functions of phosphatidic acid. *Chem. Phys. Lipids.* **80**:117–132.

6

Peroxide Tone in Eicosanoid Signaling

Richard J. Kulmacz and William E. M. Lands

6.1 Overview: Mediators or Markers?

The close association of lipid hydroperoxides (ROOH) and hydrogen peroxide (HOOH) with inflammatory/proliferative processes sets the context for examining whether those peroxides are intermediates without which the processes could not occur or are simply extraneous waste products. The peroxides and the processes are so intertwined that "oxidative stress" and signal transductions paradoxically seem at times to change their roles, as a cause produces an effect that then becomes a cause for further effects. The need to intervene successfully in a wide range of inflammatory/proliferative disorders makes it important to develop a functional understanding of the degree to which ambient levels of hydroperoxides influence those disorders. Those ambient levels constitute the peroxide tone discussed in this chapter.

Important among the intricate intra- and intercellular signals linked to peroxide tone are the eicosanoid cascades that provide a rich series of physiological and pathological modulators. The biosynthesis of eicosanoids from essential fatty acid precursors is strongly responsive to changes in peroxide tone, and a careful review of the fatty acid oxygenase reaction mechanisms provides useful examples for understanding and interpreting the paradoxical cause and effect roles of peroxides. The following discussion of peroxide tone in eicosanoid signaling notes some of the parallel recruitments of diverse major mediators that occur during the overall inflammatory/proliferative events. This diverse combination of autocrine and paracrine mediators creates an intricate network of signals for which it is difficult to identify with certainty any single initiating or triggering event. Rather, one's effort is perhaps better spent in discerning the role of peroxides in those processes that amplify or diminish the overall response and in developing intervention strategies to maintain a healthy balance. Peroxide tone is one important aspect of such a balance.

6.2 Peroxides in Cell Responses

During cell responses to stimuli, peroxides stimulate eicosanoid signaling as the cell converts from its quiescent, basal state to a transient activated state, altering the supply of the mediators indicated in Figure 6.1. The figure arranges the major types of signaling in four quadrants: cytokines and their protein kinase signaling cascades and activated gene transcription in the upper left; formation and removal of HOOH/ROOH in the upper right; phospholipase action, platelet-activating factor (PAF) synthesis and calcium ion in the lower left; and formation and release of eicosanoids in the lower right. The figure is illustrative only because the amounts of the indicated gene products differ qualitatively and quantitatively among different types of cells. Also, the signaling may be either acute or chronic, involving negative as well as positive feedback processes. The figure is designed

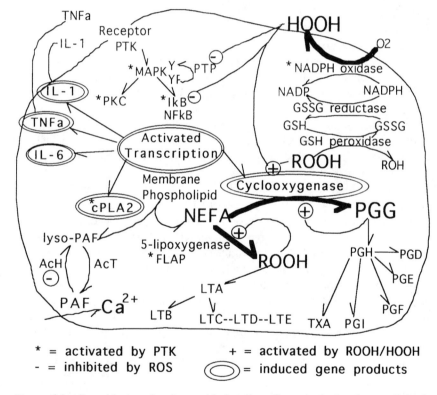

```
*  = activated by PTK        +  = activated by ROOH/HOOH
-  = inhibited by ROS      ◯  = induced gene products
```

Figure 6.1. Peroxide tone in eicosanoid signaling. Gene products whose activity is stimulated by phosphorylation by protein kinase are indicated by an asterisk (*); those inhibited by increased peroxide tone, by a minus sign (−); those activated by peroxide tone, by a plus (+); and those up-regulated by activated transcription are within double ovals.

to help readers identify direct and indirect steps by which peroxide tone may mediate events.

Cytokines are major extracellular inducers of an activated state, and their receptors trigger protein kinase signaling cascades that activate cellular enzymes by phosphorylation (noted by asterisks) and activate NF-κB for transcription of several important genes (noted by double ovals). The resultant gene products further enhance positive feedback processes, amplifying inflammatory and proliferative events.

Peroxides enhance eicosanoid formation by inhibiting (noted by minus signs) several regulatory proteins and by activating (noted by plus signs) fatty acid oxygenases. They inactivate protein tyrosine phosphatases, permitting accumulation of active tyrosine phosphate derivatives to give more kinase-mediated signals; they inactivate the inhibitor, IκB, to permit more NF-κB to enter the nucleus and activate transcription of inflammatory genes; and they inactivate PAF acetylhydrolase to give more PAF-dependent entry of Ca^{2+}, with enhanced phospholipase A_2 (PLA_2) and 5-lipoxygenase activating protein (FLAP) actions. All of these actions of peroxides enhance eicosanoid signaling.

One induced gene product, PLA_2, is further activated by protein kinase(s), and it catalyzes the accumulation of two types of product: the nonesterified fatty acid (NEFA) precursors of the eicosanoids and the various "lyso-phospholipids," of which a 1-O-alkyl form can be acetylated to form PAF. PAF acetyl hydrolase usually prevents the accumulation of appreciable PAF, but its rapid inactivation by peroxides permits PAF to accumulate to pathophysiologic levels. PAF is a very potent autacoid, which amplifies events, increasing intracellular calcium levels, phospholipase action, eicosanoid synthesis, and the production and release of more cytokines.

Eicosanoids, including prostaglandins, thromboxanes, and leukotrienes, are formed from accumulated NEFA by the action of cyclooxygenases and lipoxygenases, and that action requires the presence of sufficient peroxide tone. The various eicosanoids produced by different cells differ in their ability to enhance or suppress the release and action of various cytokines and to enhance or suppress further recruitment of more inflammatory cells. Thus, many tactics for decreasing harmful inflammatory and proliferative events decrease the accumulation of eicosanoids. Nonsteroidal anti-inflammatory drugs block cyclooxygenase action (thus diminishing the formation and action of prostanoids such as PGE_2 or thromboxane A_2 (TXA), whereas anti-inflammatory steroids induce formation of IκB (thus diminishing NFκB activation of genes for the formation and action of additional proinflammatory proteins such as PLA_2 or prostaglandin H synthase-2 (PGHS-2).

6.3 Hydroperoxides Required by Fatty Acid Oxygenases

Twenty-five years ago, while attempting to "trap" a putative hydroperoxide intermediate in the enzyme-catalyzed oxygenation of arachidonate to prostaglandin, Lands et al.[1] added glutathione peroxidase to convert any intermediate hydroperoxide to alcohol. Instead of accumulating the expected hydroxy derivative, the oxygenase reaction itself stopped, and the fatty acid substrate was recovered unchanged. This result led to subsequent demonstrations that all fatty acid oxygenases examined in detail require activation by hydroperoxide to catalyze the addition of molecular oxygen to polyunsaturated fatty acids (reviewed in Ref. 2). Thus, production of important signaling autacoids from essential fatty acids was possible only in the presence of some finite level of peroxide tone. Because peroxides were commonly regarded as harmful, toxic materials at that time, the biomedical research community was not enthusiastic in accepting hydroperoxides as essential for normal physiology. Nevertheless, the obvious essentiality of certain polyunsaturated fatty acids enhanced agreement that an oxidative conversion to prostaglandins (by some mechanism) provided health benefits.

Several unusual kinetic features of the oxygenase-catalyzed biosynthesis of eicosanoids made early understanding of the control of eicosanoid biosynthesis slow to develop. For example, the accelerative initial oxygenation kinetics indicated a need to accumulate some intermediate or to remove some interfering material.[3] This autoactivation is consistent with the initial hydroperoxide product of fatty acid oxygenation being the activating agent.[4,5] Subsequent examination of the lag phase and its elimination by added hydroperoxides confirmed the concept of an essential role for peroxide tone in catalysis by cyclooxygenase[6] and lipoxygenase.[7,8] Suppression of fatty acid oxygenation by adding glutathione peroxidase after the reaction was proceeding indicated that the activating material was accessible to removal by the added peroxidase. The inhibition by glutathione peroxidase could be reversed by adding *N*-ethylmaleimide to stop glutathione peroxidase removal of the ROOH activator. A lag phase during return of oxygenase activity confirmed that most of the activating mediator(s) had been removed, leaving unactivated oxygenase molecules that could subsequently "restart" in an autoactivating manner. Such results were significant in indicating that the enzyme was reversibly, rather than permanently, activated by hydroperoxide. Subsequently, the recognition of a glutathione peroxidase activity for phospholipid hydroperoxides has opened the question of whether membrane bound forms of activating ROOH might serve as local regulators of oxygenases, especially mammalian 15-lipoxygenase.[9]

The explosive autoactivation of the eicosanoid-forming oxygenases raised the prospect of forming ever-increasing amounts of eicosanoids, and it stimulated concerns for identifying some negative feedback regulation to prevent overproduction of the potent autacoids. This concern was enhanced by recognition that the water-soluble product rapidly diffused away from its site of synthesis by the enzyme localized on membranes. After accelerating to an optimal rate, oxygenation reaction rates customarily slow progressively, not by product inhibition, but by a reaction-catalyzed inactivation[10] similar to that seen with "suicide-substrates."[11,12] This phenomenon indicated that a "fail-safe" form of self-limiting eicosanoid production occurred, setting a maximum number of ROOH molecules produced per molecule of enzyme. However, that interpretation also was reluctantly received in the early 1970s, because it implied that living systems would expend appreciable energy to orchestrate the availability of a specific protein to form a needed autacoid from an essential fatty and then "wastefully" discard the essential catalyst. Now, biochemists, molecular biologists, and cell physiologists recognize the inevitable cost of tight regulatory control, with significant energy being expended on initiating needed actions at the same time that the cell begins to expend further energy to stop those actions before imbalances develop. It is within such counterpoised envelopes of opposing actions that living systems successfully adapt to their varying environments,[13] and energy considerations become secondary to those of control and survival. Retention of such an energy-expensive process during evolution is most probably an indication of the overall importance to survival of successfully controlling that process.

6.4 Fatty Acid Oxygenase Mechanisms

6.4.1 Cyclooxygenases

Early mechanisms proposed for conversion of arachidonate to prostaglandin G_2 (PGG_2) in the cyclooxygenase reaction[14,15] invoked a free radical pathway similar to those established for autooxidation of fatty acids. A radical-based mechanism was also suspected from the observation of free radicals during oxygenase catalysis by crude enzyme preparations.[15] The presence of isotope effects with specifically deuterated substrates[16] indicated that the rate-limiting step was abstraction of hydrogen from the fatty acid, presumably involving an oxidizing group on the enzyme. A requirement of the cyclooxygenase for hydroperoxide activators,[12] prompted the suggestion that the oxidant used for hydrogen abstraction is generated by interaction of the synthase with hydroperoxides.[17] It was later found that purified ovine PGHS-1 generated protein-linked free radicals during reaction with hydroperoxides.[18] The similarity between the spectrum of the initial hydroperoxide-induced PGHS radical, characterized by a wide doublet electron pragmatic resonance (EPR) signal, and the stable tyrosyl radical in ribonucleotide reductase led Ruf and his colleagues[19] to propose a detailed catalytic mechanism with tyrosyl radical as a key enzyme intermediate (Fig. 6.2). This mechanism was based on a heme-dependent peroxidase cycle[20] in which a tyrosyl radical was proposed to be formed by an internal electron transfer reaction in peroxidase compound I. The tyrosyl radical abstracts a hydrogen from position 13 of arachidonate to generate a fatty acyl radical, which then attacks molecular oxygen and rearranges to form a PGG_2 radical. In the final step, a hydrogen is transferred back to release PGG_2 and regenerate the tyrosyl radical. An essential aspect of this mechanism is the activation of the fatty acid rather than molecular oxygen. Indeed, the lack of inhibition of the cyclooxygenase by carbon monoxide[6,21] indicated that the heme does not bind molecular oxygen during the cyclooxygenase reaction.

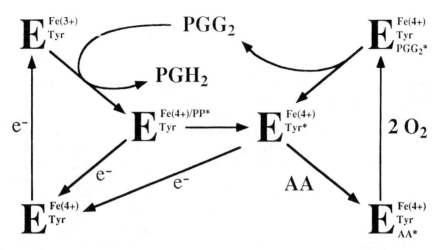

Figure 6.2. Hypothetical unitary mechanism for the peroxidase and cyclooxygenase catalysis by PGHS.[59] The peroxidase cycle is at the left and the cyclooxygenase cycle at the right. The oxidation state of heme-iron is indicated for each of the enzyme intermediates. Tyr and Tyr*, ground and free radical states of the active site tyrosine; PP*, protoporphyrin free radical; e⁻, reducing equivalent from cosubstrate; AA and AA*, arachidonic acid and arachidonyl free radical; PPG₂ and PGG₂ hydroperoxy free radicals.

Evaluation of the tyrosyl radical mechanism has been complicated by the fact that several different EPR radical signals have been observed during reaction of PGHS-1.[22] The wide doublet, assigned to a tyrosyl radical with a strained ring conformation,[23] is seen first in reactions of the native enzyme; the wide doublet changes to a wide singlet as the reaction progresses. If the enzyme is treated with indomethacin or any of several other cyclooxygenase inhibitors before reaction with hydroperoxide, neither the wide-doublet nor the wide-singlet EPR signals are observed, but instead a narrow singlet, which has been assigned to a tyrosyl radical with a relaxed ring conformation.[23,24] The wide-doublet tyrosyl radical is formed early in reaction of PGHS-1 with arachidonate[19,25-26] and it accumulates further when the fatty acid is exhausted,[26] consistent with its proposed role in cyclooxygenase catalysis (Fig. 6.2). At this point, the simplest interpretation is that there are at least three distinct PGHS-1 radicals: a wide doublet from native enzyme,[18,19] one narrow singlet from self-inactivated enzyme;[26,27] and another narrow singlet from enzyme treated with anti-cyclooxygenase agents.[24] These tyrosyl radical signals probably originate from different tyrosine residues. Understanding the role of tyrosyl radicals in the catalytic reaction helps interpret the inhibitory actions of phenolic agents described in Section 6.8.

Site-directed mutagenesis and chemical modification studies provided direct evidence that the several hydroperoxide-induced radicals in PGHS-1 are indeed tyrosyl radicals, and they identified tyrosine 385 (Y385) as the likely site of the wide-doublet radical in the native enzyme.[28-29] Crystallographic results place Y385 between the heme and the likely fatty acid site,[30] quite consistent with the sequential oxidation of heme, Y385, and fatty acid in the proposed mechanism (Fig. 6.2). The tyrosine residues giving rise to the narrow singlet radicals in inactivated and inhibited PGHS-1 have not yet been identified. The clear demarcation of separate active sites for the cyclooxygenase and peroxidase in the crystal structure[30] confirmed earlier conclusions from kinetic studies.[31]

Sequential incubation of PGHS-1 with hydroperoxide to generate the wide-doublet radical, and then with arachidonate, led to depletion of the tyrosyl radical and formation of a new radical species.[32] Parallel reactions with deuterated fatty acid showed this new radical to be an arachidonyl radical. The narrow singlet generated in indomethacin-treated PGHS-1 was not able to react with arachidonate, indicating that intact cyclooxygenase capacity was required for formation of the arachidonyl radical. These results demonstrated that the wide-doublet tyrosyl radical is chemically competent to oxidize arachidonate to a fatty acyl radical, the initial step in the proposed cyclooxygenase catalytic cycle (Fig. 6.2). This step links the peroxidase and cyclooxygenase activities, providing a specific molecular basis for the concept of symbiosis between the two PGHS catalytic activities.[33] The basic features of the combined peroxidase/cyclooxygenase mechanism in Figure 6.2 thus seem to be established for PGHS-1,[34] although some of the details remain controversial, such as whether the tyrosyl radical derives from compounds I or II.[35]

The recent discovery of the second PGHS isoform (reviewed in Ref. 36) provided an opportunity for independent testing of the mechanism. The cyclooxygenase activity of PGHS-2 also requires activation by hydroperoxide.[37,38] The residues making up the cyclooxygenase active site in PGHS-1[30] are almost completely conserved in PGHS-2; there is more variation between the two isoforms in the peroxidase site residues. Of particular interest, PGHS-2 retains a tyrosine residue in the position corresponding to Y385 in PGHS-1, and reaction of microsomal PGHS-2 with hydroperoxide generates a free radical signal.[39] This radical, which is formed almost quantitatively with homogeneous PGHS-2,[40] is slightly narrower than the wide singlet seen in PGHS-1 reactions, suggesting some differences in the radical environment in the two isoforms. Unlike the situation with PGHS-1, the line shape of the PGHS-2 wide singlet can be accounted for by a single tyrosyl radical with a partially strained ring conformation.[40] Another contrast with PGHS-1 is that the PGHS-2 wide singlet is the only radical observed throughout the reaction with hydroperoxide; it remains to be determined if a wide-doublet species can be observed at shorter reaction times.

6.4.2 Lipoxygenases

Lipoxygenases from plant and animal sources seem to have similar reaction mechanisms; most detailed mechanistic studies have been done with soybean lipoxygenase. As with the cyclooxygenase reaction, the free-radical-mediated autooxidation of polyunsaturated fatty acids has served as a basic paradigm. Free radicals are detected during lipoxygenase catalysis.[41,42] These radicals are associated with the lipid rather than the enzyme, and they have been assigned to metabolites resulting from abstraction of a hydrogen atom from the fatty acid by an enzyme oxidant.[43] All lipoxygenases examined so far have one atom of non-heme-iron as a cofactor, and the metal is believed to furnish the oxidizing potential for hydrogen abstraction. Crystallographic studies indicate an unusual iron coordination in soybean lipoxygenase, with ligands furnished by three histidine residues and the carboxyl terminus itself, confirming earlier results from mutagenic studies with the soybean enzyme and mammalian 5- and 12-lipoxygenases.[44,45]

Early investigators observed a pronounced induction phase in the reaction of soybean lipoxygenase, indicating autocatalytic behavior, and noted that the induction period could be shortened by addition of hydroperoxide.[7] Later studies showed that soybean lipoxygenase activity could be completely suppressed by a hydroperoxide scavenger, demonstrating an absolute requirement for a reversible activation by hydroperoxide.[8,46] Mammalian 5-lipoxygenase and 12-lipoxygenase also require hydroperoxide activator.[47,48,49] Thus, both plant and animal lipoxygenases resemble the cyclooxygenase in being converted from a latent to an active state by their own product. Such product activation provides an accelerative positive feedback that ensures vigorous synthesis once initiated.

The reaction of soybean lipoxygenase with hydroperoxide converts the iron from the ferrous to the ferric oxidation state,[50,51] and the proportion of enzyme in the active state correlates well with the proportion of ferric enzyme.[52] These observations fit well with the mechanism proposed originally by deGroot et al.,[42] shown in Figure 6.3. The initial step in this mechanism is oxidation of the latent enzyme by hydroperoxide to generate the active, ferric form. The ferric enzyme then abstracts a hydrogen from the fatty acid to generate a fatty acyl radical, which reacts with molecular oxygen to form the hydroperoxide and regenerate the ferric enzyme. The mechanism in Figure 6.3 provides a satisfactory account for much of the complex kinetic behavior of the soybean enzyme,[52] but the exact role of the iron in catalysis and the structure of the radical intermediates remain controversial.[42,53,54] The nature of the hydrogen abstraction step is particularly interesting because of the extremely large kinetic isotope effect observed.[55,56]

Another unresolved issue is the number of lipid bonding sites. Smith and Lands[8] and Cook and Lands[46] proposed separate sites for substrate fatty acid and activator hydroperoxide, whereas deGroot et al.[42] suggested that the substrate and activator compete for the same hydrophobic site. Kinetic measurements have so far been unable to distinguish between models with one and two sites.[52,57] Two cavities lined with hydrophobic residues are apparent from crystallographic data.[45] Arachidonate can be docked into one of these cavities with the 11,14 diene system positioned near the iron, consistent with the mechanism in Figure 6.3.

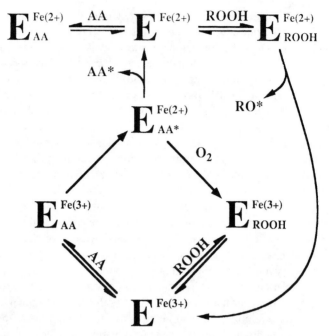

Figure 6.3. Hypothetical mechanism for aerobic lipoxygenase catalysis.[42] The oxidation state of the non-heme-iron is indicated for each enzyme intermediate. AA and AA*, arachidonic acid and arachidonyl free radical; ROOH, fatty acid hydroperoxide; RO*, fatty acid alkoxy free radical.

The second cavity is proposed to funnel oxygen to the active site, but an allosteric action of fatty acid or hydroperoxide bound in the second cavity is not ruled out.[45]

6.4.3 Common Mechanistic Features of Fatty Acid Oxygenases

The above discussion makes it clear that, despite the complete lack of structural resemblance between the PGHS isoforms and any of the animal or plant lipoxygenases, there are many parallels between the cyclooxygenase and lipoxygenase reactions and their mechanisms. Both enzymes are episodically stimulated to form products that participate in defensive events. In both reactions, iron is in a more oxidized state in the active enzyme than in the latent enzyme. Both reactions follow a chemical strategy of activating the fatty acid, rather than activating molecular oxygen. Both cyclooxygenase and lipoxygenase have a fundamentally autocatalytic nature, as the product hydroperoxide converts latent enzyme to a catalytically active form. Thus, both are regulated by peroxide tone. Finally, both types of enzyme exhibit inactivation during catalysis, limiting the amount of product that can be formed by a given amount of enzyme.

6.5 Amplification of Intracellular Hydroperoxides

Amplification of hydroperoxide levels is a consequence of the activated oxygenase site producing many molecules of product hydroperoxide, each of which can diffuse out and activate a latent oxygenase molecule nearby. This autocatalytic increase in peroxide tone is driven by the relative efficiency of the product hydroperoxide for activating latent oxygenase to produce more peroxide compared to its efficiency for being removed by peroxidase action. Such efficiencies are commonly defined by a K_m value, the steady-state concentration (peroxide tone) needed for a half-maximal rate. The isolated oxygenases *in vitro* can accumulate appreciable hydroperoxide in the reaction mixture and thereby develop maximal rates, but *in vivo* the peroxide tone is often maintained at a very low level by various peroxide-removing enzymes, even though they may be functioning inefficiently at a peroxide tone far below their optimal range.

Overall production and removal of hydroperoxides depends, in turn, on the molecular structures of the various substrate acids and their hydroperoxide products. Selectivities of the enzymes maintaining fatty acids in tissue lipids combine with typical U.S. dietary supply to make arachidonate, 20:4n-6, the most abundant polyunsaturated fatty acid to be released by PLA$_2$ from tissue glycerolipids. Arachidonate is more abundant than its precursors, 18:2n-6 and 20:3n-6, and also at higher levels than the competitive antagonist, 22:6n-3, or its precursor, 20:5n-3. Of the many polyunsaturated acids that might appear in tissues, only 20:3n-6, 20:4n-6, and 20:5n-3 are effective substrates for the cyclooxygenase, and 20:4n-6 and 20:5n-3 are principal substrates for the 5-lipoxygenase.

6.5.1 Amplification of Hydroperoxide Level by PGHS

The cyclooxygenase activity of PGHS is activated through the heme-dependent peroxidase activity that has kinetic characteristics comparable to those of "classic" hemoprotein peroxidases.[20,58,59] To examine in detail the balance between the cyclooxygenase and peroxidase of PGHS in handling the hydroperoxide, PGG$_2$, glutathione peroxidase was added to compete for hydroperoxide in the two hydroperoxide actions with PGHS-1.[60] Quantitative estimates of the peroxide tone requirements of the two PGHS catalytic activities were derived from their susceptibility to inhibition by the hydroperoxide scavenger. Interactions of PGG$_2$ with PGHS peroxidase were comparable to those with glutathione peroxidase, with half-maximal ROOH removal occurring at about 1 μM peroxide tone. However, a much lower peroxide tone (20 nM) was sufficient for half-maximal stimulation of ROOH formation by cyclooxygenase. As a result of the large difference in hydroperoxide requirements for the two activities, PGHS cyclooxygenase catalysis can amplify the hydroperoxide level by about a factor of 50 (from 20 to 1000 nM) without accumulating sufficient hydroperoxide to permit its rapid removal by the PGHS peroxidase.

The ability of PGHS to generate a significant pulse of hydroperoxide, even in the presence of ample reducing cosubstrates, has been confirmed by direct measurements.[34] In this situation, once some of the PGHS in a given cell has been activated, the increasing intracellular peroxide tone probably activates most of the PGHS within that cell, producing PGG_2 at near-maximal rates and further increasing the hydroperoxide tone. Only when the hydroperoxide level approaches micromolar levels does the appreciable peroxidase activity of the cell (glutathione peroxidase, PGHS peroxidase, etc.) become effective in preventing further increases in peroxide tone. In this manner, the lipid hydroperoxide tone was recognized as nearly nanomolar for basal states of negligible eicosanoid formation, whereas it was in the range of 10 to 100 nM for physiologic eicosanoid signaling, and in the range of 100 to 1000 nM for pathophysiologic signaling.[61-62] Hydrogen peroxide is 100-fold less efficient than lipid hydroperoxide as a cyclooxygenase activator,[63] but it appears in pericellular fluids at 100-fold greater levels than lipid hydroperoxides. Its possible role in cellular function is discussed in a later section.

6.5.2 Differential Activation of the PGHS Isoforms

The cyclooxygenase activity of PGHS-2, the isoform induced by proliferative and proinflammatory cytokines (reviewed in Ref. 36), is more resistant to inhibition by hydroperoxide scavengers than is the PGHS-1 cyclooxygenase.[37,38] Half-maximal activation of PGHS-2 cyclooxygenase required only 2-nM lipid hydroperoxide, about an order of magnitude lower than for PGHS-1. The greater efficiency of hydroperoxides in activating the PGHS-2 isoform permits the inducible isoform to exhibit rapid acceleration to vigorous catalytic rates from basal hydroperoxide levels below 10 nM, values too low to activate PGHS-1 effectively. Thus, tissues in which inflammatory cytokines have induced PGHS-2 can be predicted to be even more responsive to peroxide tone than normal basal tissues. Furthermore, there might be differential modulation of the two cyclooxygenase activities, even when both PGHS isoforms are in the same subcellular location. Such selective activation of PGHS-2 may contribute to the differential regulation of cyclooxygenase catalysis by the two isoforms observed in fibroblasts and macrophages.[64]

6.5.3 Differential Activation of PGHS and Lipoxygenases

The affinity for lipid hydroperoxide activator of the prototypical lipoxygenase, soybean lipoxygenase, was estimated to be in the range of 1 to 30 μM.[8,65] Activation of this lipoxygenase catalysis is thus considerably less efficient than activation of PGHS cyclooxygenase, and such levels of lipid hydroperoxide are not likely to occur in normal tissue conditions. As a consequence, PGHS may have a predominant role in initially amplifying hydroperoxide levels in cells that contain both types of fatty acid oxygenases. With limited action by cellular peroxidases, a pulse of hydroperoxide generated by activation of PGHS may be sufficient to activate lipoxygenases and give a form of "cross-stimulation." In this way, PGHS (especially PGHS-2) might serve as a crucial agent in transducing intercellular HOOH signals into intracellular hydroperoxide increases and subsequent signalling with prostaglandins and leukotrienes (see also Section 6.7). More detailed quantitative characterization of the half-maximal peroxide tone needed for 5-, 12-, and 15-lipoxygenase are needed to clarify the way in which these lipoxygenases are activated. Also, further quantitative studies with the phosphorylated 5-lipoxygenase activating protein (FLAP; see Fig. 6.1) are needed to interpret adequately the factors limiting 5-lipoxygenase amplification of hydroperoxides. Lipoxygenases also decompose hydroperoxides to various products, some representing important metabolic intermediates, and some probably side products.[66,67]

6.6 Relative k_{cat} Values and a Consequence of Eating Fish

Early studies with crude cyclooxygenase preparations found much lower rates of oxygenation with eicosapentaenoic acid (20:5n-3) compared to rates with the n-6 fatty acids, 20:3n-6 and 20:4n-6.[68] Those results led to a suggestion that the low cyclooxygenase k_{cat} for n-3 fatty acids from tissue lipids might permit useful competitive antagonism of n-6 eicosanoid biosynthesis and thereby diminish pathophysiological events that are mediated by excessive production of the common n-6 prostaglandins. After the 1975 recognition that cyclooxygenase mediates formation of the principal thrombogenic agent, thromboxane, attention turned to possible moderation of thrombotic situations by prior nutritional supplementation with fish oils rich in n-3 fatty acids. Later, discovery of lipoxygenase-mediated leukotriene biosynthesis extended this nutritional tactic to other inflammatory/proliferative diseases, such as atherosclerosis, arthritis, or asthma. Various studies showed physiological benefits of dietary supplies of the long-chain n-3 fatty acids (especially 20:5n-3 and 22:6n-3) that are abundant in fish (reviewed in Ref. 69 and 70), but detailed studies with more purified cyclooxygenase preparations showed surprising similarities in reaction kinetics with 20:5n-3 and 20:4n-6. An important question that remained unanswered was in what way the cyclooxygenase kinetic constants with 20:5n-3 differ from those with 20:4n-6.

Careful fitting of a mechanistic model to the observed cyclooxygenase kinetic patterns permitted calculation of the major kinetic constants involved in the enzyme-catalyzed events, and helped interpret similarities and differences for 20:5n-3 and 20:4n-6.[71] The two hydroperoxide products, PGG$_2$ and PGG$_3$, were equally effective as cyclooxygenase activators. However, a several-fold lower k_{cat} for 20:5n-3 compared to 20:4n-6 explained most of the differences in behavior of these two substrates. The slower generation of ROOH activator weakens the feedback-activation process, making reaction with the n-3 substrate much more vulnerable than reaction with the n-6 substrate to conditions that suppress ambient peroxide tone. Overall, the n-3 eicosanoids tend to be formed more slowly and also to be less potent physiologically than the n-6 eicosanoids. The situation with 20:5n-3 emphasizes that the auto-accelerative cyclooxygenase kinetics depend on both efficient activation by hydroperoxide and efficient oxygenation of the fatty acid substrate. In the case of 20:5n-3, the oxygenation reaction can start, but then it slows and stops, even while latent enzyme remains. The slow k_{cat} for 20:5n-3 hinders amplification of intracellular peroxide tone by PGHS-1, but the greater efficiency of ROOH with PGHS-2 probably permits the n-3 substrate to form appreciable prostanoids with the inducible PGHS-2 isoform more readily than with the constitutive PGHS-1 isoform. Further study is needed to determine whether activation rate or catalytic rate factors are major contributors to the selectivity differences between the two isoforms.[72] The k_{cat} is the principal difference between the major eicosanoid precursor in meat (20:4n-6) and that in fish and seafoods (20:5n-3), and this difference has a significant consequence in suppressing actions mediated by PGHS-1 in normal tissues at basal states. Careful evaluation of the kinetics of PGHS-2 will be needed to see if the 10-fold greater efficiency of peroxide activation of PGHS-2 precludes suppression by 20:5n-3 in inflamed tissues in which PGHS-2 has been induced.

A very low cyclooxygenase k_{cat} also helps interpret an unusual feature of acetylenic fatty acids. A limited number of positional isomers inactivate the cyclooxygenase irreversibly under conditions that resemble a self-catalyzed inactivation, except that no oxygenation reaction was detected.[73] The inactivation apparently depended on low levels of HOOH in the incubations,[74] generating an intermediate form of enzyme that interacts with specific acetylenic isomers to form a destructive transitional state. The fact that only certain positional isomers inactivated the enzyme indicates that a structurally specific conformation of liganded enzyme is vulnerable to a "suicide" transition. The active acetylenic isomers had spatial conformations that resembled a cis-ethylenic (n-6) structure, a common feature in many cyclooxygenase substrates. Retrospective consideration of earlier results[75] with only low peroxide tone indicate that 20:5n-3 also caused irreversible inactivation of the enzyme in a manner similar to the action of certain acetylenic isomers.

6.7 Inflammatory/Proliferative Events

An increased eicosanoid biosynthesis is commonly associated with the onset of inflammatory conditions, and such conditions are clearly associated with extracellular HOOH tone and intracellular ROOH tone. To monitor and interpret the role of peroxide tone in these pathophysiological situations, better analytical methods were developed to detect the very small amounts of ROOH predicted to occur by using one of the most sensitive detectors known for ROOH, PGHS-1.[76,77] (Now that PGHS-2 is known to be 10-fold more sensitive to ROOH activation, even more sensitive assays may be possible). With cultured macrophages exposed to phorbol esters, the activated protein kinases in these cells caused translocation and stimulation of NADPH oxidase to form comparatively high levels (5 to 50 μM) of reactive oxygen species that constitute a generally elevated peroxide tone, especially with the freely diffusing HOOH (see Fig. 6.1). In this model system, over 70% of the extracellular cyclooxygenase activator was shown to be HOOH, since its accumulation could be prevented by added catalase.[78] Formation and release of micromolar amounts of HOOH by phagocytic cells at inflammatory/proliferative foci serves as an intercellular signaling process for triggering PGHS action within nearby cells, with a consequent intracellular amplification of lipid hydroperoxide. With the cultured macrophages, some extracellular activator remaining after catalase treatment was destroyed by glutathione peroxidase, and it appeared to be nanomolar amounts of ROOH.[78] Nevertheless, the question whether the extracellular ROOH was a mediator or marker of the macrophage response was not fully addressed by this protocol. Insufficient evidence for ROOH-mediated intercellular signals has come from studies with added extracellular ROOH, although experiments with HOOH added to cells clearly showed its ability for intercellular stimulation of eicosanoid synthesis.

The 100-fold greater efficacy of ROOH over HOOH in promoting eicosanoid biosynthesis with PGHS-1 stimulated further attempts to discern a role for extracellular ROOH. Greater efflux of ROOH (about 0.5 nmol/ml) was measured in the plasma coming from the tissue region involved with major thoracic or abdominal surgical operations relative to that from uninvolved regions.[79] Similarly, plasma draining from regions of pulmonary or abdominal sepsis contained about 0.5 nmol/ml more hydroperoxide than that from uninvolved regions.[80] The combined results showed plasma ROOH to be a significant marker for regional pathology,[81] however, plasma ROOH appeared to be cleared rapidly on passage through capillary beds between arterial and venous systems. Thus, uncertainty remains about the participation of extracellular ROOH in mediating pathological states, even though intercellular signaling by HOOH and intracellular signaling by ROOH seems certain. Investigators need to consider developing a controlled study design that can give more conclusive evidence regarding whether or not extracellular ROOH serves as an intercellular signal and a causal factor in cell pathology rather than merely as a marker of the condition.

6.8 Antiinflammatory Drug Efficacy

Because so many pathological states are linked to excessive production of eicosanoids, biomedical research gives considerable attention to therapeutic mechanisms for suppressing eicosanoid production. The complex network of signals noted in Section 6.2 illustrates many areas in which drug development for diminution of pathophysiology might be explored. Historically, two major general strategies are (1) the use of steroids to suppress the inflammatory signal cascades that increase fatty acid substrate access to active oxygenase and (2) the use of nonsteroidal anti-inflammatory drugs (NSAIDs) to block oxygenase conversion of fatty acids to active eicosanoids. Three major classes of inhibitory mechanism were identified for NSAID action[82]: reversible antagonism of fatty acid substrate binding,[83] time-dependent inactivation of the oxygenase active site,[83] and antagonism of hydroperoxide activation.[84] Although various structural analogs of fatty acids (such as n-3 fatty analogs) are obvious candidates for competitive inhibition, their rapid metabolism by various lipid-metabolizing enzymes limits their therapeutic kinetics. Rather, poorly metabolized carboxylic acid derivatives that competitively bind to the fatty acid substrate site, such as ibuprofen (with a k_i of about 5 μM), have proved to be successful therapeutic agents. The presence of the arginine residue at position 120 of PGHS-1 is essential for these competitive carboxylic inhibitors to effectively interact with the enzyme.[85,86]

Surprisingly small structural modifications of the reversible inhibitors can provide ligands that bind very tightly to the substrate site and inactivate PGH synthase in a time-dependent and almost irreversible manner. This slowly reversible inactivation[87,88] has high structural specificity, and it occurs without forming a covalent derivative of the enzyme.[89] Thus it differs greatly from the inactivating irreversible acetylation[90] that occurs after acetylsalicylate (aspirin) binds to the substrate site. A residual activity reported for enzyme treated with time-dependent agents[88,91,92] may be due to the very slow reversal of inhibition during the experimental protocol. An opportunity remains to develop more structurally specific inactivators that bind tightly to the oxygenase substrate site. A large group of time-dependent agents without a carboxyl function (typified by NS398, SC58125, CGP28238, and DuP697) is now being developed for selective inhibition of the PGHS-2 isoform (reviewed by Hershman[36]). An important consequence of the poorly reversible inhibitors is that the PGHS remains inactive after the bulk of the drug has been cleared from the area, and restoration of eicosanoid biosynthesis depends principally on the synthesis of new oxygenase protein. In this case, somewhat more careful attention to drug dosage and pharmacokinetics is needed than for the reversible agents.

The third mechanism of NSAID action in inhibiting prostaglandin formation involves antagonising the hydroperoxide activation of the oxygenase activity. This mode of inhibition is reversible and can be overwhelmed by greater peroxide tone. The "deactivating" inhibitors in this category (such as acetamidophenol[84]) thus tend to be more effective in conditions of low than in high peroxide tone.[93] This phenomenon led Lands and Hanel[94] to propose this category of agents as especially suited to general analgesic actions where only slight increase in peroxide tone of hyperalgesic areas may be expected. The ability of a variety of phenolic agents to antagonize the activation of cyclooxygenase[68,84,95] seems related to the mediating role of the tyrosyl radical in the enzyme-catalyzed chain reaction.

Facile chain transfer from PGHS-1 to form a nonprotein phenoxy radical represents an opportunity to devise site-specific radical-trapping agents that return the activated PGHS oxygnease to its unactivated "ground state." This situation emphasizes the earlier interpretation[94] of a lower peroxide tone (and less hydroperoxide activation) occurring in hyperalgesic tissues permitting successful intervention with phenolic agents like acetamidophenol. In contrast, in an inflamed tissue where cytokines have induced PGHS-2, the 10-fold greater effectiveness of ROOH in activating PGHS-2 (2 nM for PGHS-2 versus 20 nM for PGHS-1[38]) may tend to make such chain-transfer, or radical-quenching, agents less effective.

Another remaining puzzle is the biphasic action of many phenolic compounds, stimulating the cyclooxygenase velocity at low concentrations, and increasing the number of catalytic events before self-inactivation.[96] Perhaps these radical-trapping agents can prevent some of the destructive side reactions of radical intermediates in the PGHS mechanisms. The phenolic antioxidant compounds also stimulate conversion of PGG_2 to prostaglandin H_2 (PGH_2) by the PGHS peroxidase.[21] The influence of molecular structure on cyclooxygenase stimulation by a series of phenolic compounds has been studied.[97] The stimulatory effect on the cyclooxygenase is unlikely to be due solely to the agents acting as stoichiometric reductants, because tryptophan was found to be recovered unoxidized.[21] A catalytic action on electron transfer within the cyclooxygenase cycle has been proposed[71] but as yet lacks direct experimental support.

6.9 Summary

The eicosanoids, prostaglandins, thromboxanes, and leukotrienes are not normally present at appreciable levels in unperturbed tissues, but they accumulate rapidly in a burst of synthesis following activation of a cell, from which they diffuse quickly and then are rapidly inactivated by metabolism in nearby cells. The oxygenase catalytic mechanism facilitates the transient nature of this response by an accelerative positive feedback due to the immediate product being an activating hydroperoxide and by a self-regulatory negative feedback of reaction-inactivation. Eicosanoid synthesis is enhanced by two enhancements of peroxide tone: intra- and intercellular diffusion of HOOH released from nearby activated cells, and intracellular ROOH produced by increased rates of eicosanoid formation. Further enhancement of eicosanoid synthesis comes from induction of PGHS-2, which is 10 times more sensitive to activation by peroxide tone than is the constitutive PGHS-1. The increased eicosanoids then facilitate recruitment of inflammatory cells, increasing the general peroxide tone in the region. The elevated HOOH and ROOH act on other regulatory proteins to further increase peroxide tone and eicosanoid synthesis. Thus, resolution of the inflamed, hyperresponsive state is likely to be best achieved with decreases in the amount of PGHS-2 as well as decreases in the ambient levels of HOOH and ROOH.

Acknowledgment

Recent research by RJK described in this chapter was supported by a grant from the National Institutes of Health, GM-52170.

References

1. Lands, W. E. M., R. E. Lee, and W. L. Smith. 1971. Factors regulating the biosynthesis of various prostaglandins. *Ann. N. Y. Acad. Sci.* **180**:107–122.

2. Lands, W. E. M. 1979. The biosynthesis and metabolism of prostaglandins. *Ann. Rev. Physiol.* **41**:633–652.

3. Cook, H. W., and W. E. M. Lands. 1976. A mechanism for the suppression of cellular biosynthesis of prostaglandins. *Nature* **260**:630–632.

4. Cook, H. W., and W. E. M. Lands. 1975. Evidence for an activating factor formed during prostaglandin biosynthesis. *Biochem. Biophys. Res. Commun.* **65**:464–471.

5. Hemler, M. E., G. Graff, and W. E. M. Lands. 1978. Accelerative autoactivation of prostaglandin biosynthesis by PGG2. *Biochem. Biophys. Res. Commun.* **85**:1325–1337.

6. Hemler, M. E., and W. E. M. Lands. 1980. Evidence for a peroxide-initiated free radical mechanism of prostaglandin biosynthesis. *J. Biol. Chem.* **255**:6253–6261.

7. Haining, J. L., and B. Axelrod. 1958. Induction period in the lipoxidase-catalyzed oxidation of linoleic acid and its abolition by substrate peroxide. *J. Biol. Chem.* **232**:193–202.

8. Smith, W. L., and W. E. M. Lands. 1972. Oxygenation of unsaturated fatty acids by soybean lipoxygenase. *J. Biol. Chem.* **247**:1038–1042.

9. Schnurr, K., J. Belkner, S. Ursini, T. Schewe, and K. Kuhn. 1996. The selenoenzyme phospholipid hydroperoxide glutathione peroxidase controls the activity of the 15-lipoxygenase with complex substrates and preserves the specificity of the oxygenase products. *J. Biol. Chem.* **271**:4653–4658.

10. Smith, W. L., and W. E. M. Lands. 1971. Stimulation and blockade of prostaglandin synthesis. *J. Biol. Chem.* **246**:6700–6704.

11. Smith, W. L., and W. E. M. Lands. 1970. The self-catalyzed destruction of lipoxygenase. *Biochem. Biophys. Res. Commun.* **41**:846–851.

12. Smith, W. L., and W. E. M. Lands. 1972. Oxygenation of polyunsaturated fatty acids during prostaglandin biosynthesis by sheep vesicular gland. *Biochemistry* **11**:3276–3285.

13. Marshall, P. J., R. J. Kulmacz, and W. E. M. Lands. 1987. Constraints on prostaglandin synthesis in tissues. *J. Biol. Chem.* **262**:3510–3517.

14. Samuelsson, B. 1965. On the incorporation of oxygen in the conversion of 8,11,14-eicosatrienoic acid to prostaglandin E_1. *J. Am. Chem. Soc.* **87**:3011–3013.

15. Nugteren, D. H., R. K. Beerthuis, and D. A. van Dorp. 1966. The enzymic conversion of all-*cis* 8,11,14-eicosatrienoic acid into prostaglandin E₁. *Rec. Trav. Chim. Pays-Bas* **85**:405–419.

16. Hamberg, M., and B. Samuelsson. 1967. On the mechanism of the biosynthesis of prostaglandins E₁ and F₁-alpha. *J. Biol. Chem.* **242**:5336–5343.

17. O'Brien, P. J., and A. Rahimtula. 1976. The possible involvement of a peroxidase in prostaglandin biosynthesis. *Biochem. Biophys. Res. Commun.* **70**:832–838.

18. Kulmacz, R. J., Tsai, A. -L., and G. Palmer. 1987. Heme spin states and peroxide-induced radical species in prostaglandin H synthase. *J. Biol. Chem.* **262**:10524–10531.

19. Karthein, R., R. Dietz, W. Nastainczyk, and H. H. Ruf. 1988. Higher oxidation states of prostaglandin H synthase: EPR study of a transient tyrosyl radical in the enzyme during the peroxidase reaction. *Eur. J. Biochem.* **171**:313–320.

20. Lambier, A.-M., C. M. Markey, H. B. Dunford, and L. J. Marnett. 1985. Spectral properties of the higher oxidation states of prostaglandin H synthase. *J. Biol. Chem.* **260**:14894–14896.

21. Ohki, S., N. Ogino, S. Yamamoto, and O. Hayaishi. 1979. Prostaglandin hydroperoxidase, an integral part of prostaglandin endoperoxide synthetase from bovine vesicular gland microsomes. *J. Biol. Chem.* **254**:829–836.

22. Smith, W. L., T. E. Eling, R. J. Kulmacz, L. J. Marnett, and A. Tsai. 1992. Tyrosyl radicals and their role in hydroperoxide-dependent activation and inactivation of prostaglandin endoperoxide synthase. *Biochemistry* **31**:3–7.

23. Barry, B. A., M. K. El-Deeb, P. O. Sandusky, and G. T. Babcock. 1990. Tyrosine radicals in photosystem II and related model compounds. *J. Biol. Chem.* **265**:20139–20143.

24. Kulmacz, R. J., G. Palmer, and A.-L. Tsai. 1991. Prostaglandin H synthase: perturbation of the tyrosyl radical as a probe of anticyclooxygenase agents. *Mol. Pharmacol.* **40**:833–837.

25. DeGray, J. A., G. Lassmann, J. F. Curtis, T. A. Kennedy, L. J. Marnett, T. E. Eling, and R. P. Mason. 1992. Spectral analysis of the protein-derived tyrosyl radicals from prostaglandin H synthase. *J. Biol. Chem.* **267**:23583–23588.

26. Tsai, A.-L., G. Palmer, and R. J. Kulmacz. 1992. Prostaglandin H synthase. Kinetics of tyrosyl radical formation and of cyclooxygenase catalysis. *J. Biol. Chem.* **267**:17753–17759.

27. Lassmann, G., R. Odenwaller, J. F. Curtis, J. A. DeGray, R. P. Mason, L. J. Marnett, and T. E. Eling. 1991. Electron spin resonance investigation of tyrosyl radicals of prostaglandin H synthase. *J. Biol. Chem.* **266**:20045–20055.

28. Tsai, A. L., L. C. Hsi, R. J. Kulmacz, G. Palmer, and W. L. Smith. 1994. Characterization of the tyrosyl radicals in ovine prostaglandin H synthase-1 by isotope replacement and site-directed mutagenesis. *J. Biol. Chem.* **269**:5085–5091.

29. Shimokawa, T., R. J. Kulmacz, D. L. DeWitt, and W. L. Smith. 1990.Tyrosine 385 of prostaglandin endoperoxide synthase is required for cyclooxygenase catalysis. *J. Biol. Chem.* **265**:20073–20076.

30. Picot, D., P. J. Loll, and R. M. Garavito. 1994. The X-ray crystal structure of the membrane protein prostaglandin H₂ synthase-1. *Nature* **367**:243–249.

31. Marshall, P. J., and R. J. Kulmacz. 1988. Prostaglandin H synthase: distinct binding sites for cyclooxygenase and peroxidase substrates. *Arch. Biochem. Biophys.* **266**:162–170.

32. Tsai, A.-L., R. J. Kulmacz, and G. Palmer. 1995. Spectroscopic evidence for reaction of prostaglandin H synthase-1 tyrosyl radical with arachidonic acid. *J. Biol. Chem.* **270**:10503–10508.

33. Kulmacz, R. J., J. F. Miller, Jr., and W. E. M. Lands. 1985. Prostaglandin H synthase: an example of enzymic symbiosis. *Biochem. Biophys. Res. Commun.* **130**:918–923.

34. Wei, C., R. J. Kulmacz, and A. L. Tsai. 1995. Comparison of branched-chain and tightly coupled reaction mechanisms for prostaglandin H synthase. *Biochemistry* **34**:8499–8512.

35. Bakovic, M., and H. B. Dunford. 1996. Reactions of prostaglandin endoperoxide synthase and its compound I with hydroperoxides. *J. Biol. Chem.* **271**:2048–2056.

36. Herschman, H. R. 1996. Prostaglandin synthase 2. *Biochim. Biophys. Acta* **1299**:125–140.

37. Capdevila, J. H., J. D. Morrow, Y. Y. Belosludtsev, D. R. Beauchamp, R. N. DuBois, and J. R. Falck. 1995. The catalytic outcomes of the constitutive and the mitogen inducible isoforms of prostaglandin H₂ synthase are markedly affected by glutathione and glutathione peroxidase(s). *Biochemistry* **34**:3325–3337.

38. Kulmacz, R. J., and L.-H. Wang. 1995. Comparison of hydroperoxide initiator requirements for the cyclooxygenase activities of prostaglandin H synthase-1 and -2. *J. Biol. Chem.* **270**:24019–24023.

39. Hsi, L. C., C. W. Hoganson, G. T. Babcock, and W. L. Smith. 1994. Characterization of a tyrosyl radical in prostaglandin endoperoxide synthase-2. *Biochem. Biophys. Res. Commun.* **202**:1592–1598.

40. Xiao, G., A.-L. Tsai, G. Palmer, W. C. Boyar, P. J. Marshall, and R. J. Kulmacz. 1996. Examination of the hydroperoxide-induced radical in aspirin-treated prostaglandin H synthase-2. *ASBMB Meeting Abstract.*

41. Pistorius, E. K., and B. Axelrod. 1974. Iron, an essential component of lipoxygenase. *J. Biol. Chem.* **249**:3183–3186.

42. deGroot, J. J. M. C., G. A. Veldink, J. F. G. Vliegenthart, J. Boldingh, R. Wever, and B. F. van Gelder. 1975. Demonstration by EPR spectroscopy of the functional role of iron in soybean lipoxygenase-1. *Biochim. Biophys. Acta* **377**:71–79.

43. Nelson, M. J., S. P. Seitz, and R. A. Cowling. 1990. Enzyme-bound pentadienyl and peroxyl radicals in purple lipoxygenase. *Biochemistry* **29**:6897–6903.

44. Minor, W., J. Steczko, J. T. Bolin, Z. Otwinowski, and B. Axelrod. 1993. Crystallographic determination of the active site iron and its ligands in soybean lipoxygenase L-1. *Biochemistry* **32**:6320–6323.

45. Boyington, J. C., B. J. Gaffney, and L. M. Amzel. 1993. The three-dimensional structure of an arachidonic acid 15-lipoxygenase. *Science* **260**:1482–1486.

46. Cook, H. W., and W. E. M. Lands. 1975. Further studies of the kinetics of oxygenation of arachidonic acid by soybean lipoxygenase. *Can. J. Biochem.* **53**:1220–1231.

47. Egan, R. W., A. N. Tischler. E. M. Baptista, E. A. Ham, D. D. Soderman, and P. H. Gale. 1983. Specific inhibition and oxidative regulation of 5-lipoxygenase. *Adv. Prostaglandin Thromb. Leuk. Res.* **11**:151–157.

48. Seigel, M. I., R. T. McConnell, S. L. Abrahams, N. A. Porter, and P. Cuatrecasas. 1979. Regulation of arachidonate metabolism via lipoxygenase and cyclo-oxygenase by 12-HPETE, the product of human platelet lipoxygenase. *Biochem. Biophys. Res. Commun.* **89**:1273–1280.

49. Yokoyama, C., F. Shinjo, T. Yoshimoto, S. Yamamoto, J. A. Oates, and A. R. Brash. 1986. Arachidonate 12-lipoxygenase purified from porcine leukocytes by immunoaffinity chromatography and its reactivity with hydroperoxyeicosatetraenoic acids. *J. Biol. Chem.* **261**:16714–16721.

50. Slappendel, S., R. Aasa, B. G. Malmstrom, J. Verhagen, G. A. Veldink, and J. F. G. Vliegenthart. 1982. Factors affecting the line-shape of the EPR signal of high-spin Fe(III) in soybean lipoxygenase-1. *Biochim. Biophys. Acta* **708**:259–265.

51. Cheesbrough, T. M., and B. Axelrod. 1983. Determination of the spin state of iron in native and activated soybean lipoxygenase 1 by magnetic susceptibility. *Biochemistry* **22**:3837–3840.

52. Schilstra, M. J., G. A. Veldink, and J. F. G. Vliegenthart. 1994. The dioxygenation rate in lipoxygenase catalysis is determined by the amount of iron(III) lipoxygenase in solution. *Biochemistry* **33**:3974–3979.

53. Nelson, M. J., and S. P. Seitz. 1994. The structure and function of lipoxygenase. *Curr. Opin. Struct. Biol.* **4**:878–884.

54. Corey, E. J., and R. Nagata. 1987. Evidence in favor of an organoiron-mediated pathway for lipoxygenation of fatty acids by soybean lipoxygenase. *J. Am. Chem. Soc.* **109**:8107–8108.

55. Glickman, M. H., J. S. Wiseman, and J. P. Klinman. 1994. Extremely large isotope effects in the soybean lipoxygenase-linoleic acid reaction. *J. Am. Chem. Soc.* **116**:793–794.

56. Hwang, C. C., and C. B. Grissom. 1994. Unusually large deuterium isotope effect in soybean lipoxygenase is not caused by a magnetic isotope effect. *J. Am. Chem. Soc.* **116**:795–796.

57. Wang, Z. X., S. D. Killilea, and D. K. Srivastava. 1993. Kinetic evaluation of substrate-dependent origin of the lag phase in soybean lipoxygenase-1 catalyzed reactions. Biochemistry **32**:1500–1509.

58. Kulmacz, R. J. 1986. Prostaglandin H synthase and hydroperoxides: peroxidase reaction and inactivation kinetics. *Arch. Biochem. Biophys.* **249**:273–285.

59. Dietz, R., W. Nastainczyk, and H. H. Ruf. 1988. Higher oxidation states of prostaglandin H synthase: rapid electronic spectroscopy detected two spectral intermediates during reaction with prostaglandin G2. *Eur. J. Biochem.* **171**:321–328.

60. Kulmacz, R. J., and W. E. M. Lands. 1983. Requirements for hydroperoxide by the cyclooxygenase and peroxidase activities of prostaglandin H synthase. *Prostaglandins* **25**:531–540.

61. Lands, W. E. M., P. J. Marshall, and R. J. Kulmacz. 1985. Hydroperoxide availability in the regulation of the arachidonate cascade, pp. 233–235. *In* S. Yamamoto and O. Hayaishi (eds.), *Advances in Prostaglandins, Thromboxane, and Leukotriene Research,* Vol. 15. Raven Press, New York.

62. Lands, W. E. M. 1988. Peroxide tone in regulating eicosanoid formation, pp. 155–162. *In* A. Sevanian (ed.), *Lipid Peroxidation in Biological Systems.* American Oil Chemical Society, Champaign, IL.

63. Marshall, P. J., R. J. Kulmacz, and W. E. M. Lands. 1984. Lipid hydroperoxides and prostaglandin synthesis, pp. 299–307. *In* W. Bors, M. Saran, and D. Tait (eds.), *Oxygen Radicals in Chemistry and Biology.* deGruyter, Berlin.

64. Reddy, S. T., and H. R. Herschman. 1994. Ligand-induced prostaglandin synthesis requires expression of the TIS10/PGS-2 prostaglandin synthase gene in murine fibroblasts and macrophages. *J. Biol. Chem.* **269**:15473–15480.

65. Aoshima, H., T. Kajiwara, A. Hatanaka, H. Nakatani, and K. Hiromi. 1977. Kinetic study of lipoxygenase-hydroperoxylinoleic acid interaction. *Biochim. Biophys. Acta* **486**:121–126.

66. Schewe, T., S. M. Rapoport, and H. Kuhn. 1986. Enzymology and physiology of reticulocyte lipoxygenase: comparison with other lipoxygenases. *Adv. Enzymol.* **58**:191–272.

67. Yamamoto, S. 1992. Mammalian lipoxygenases: molecular structures and functions. *Biochim. Biophys. Acta* **1128**:117–131.

68. Lands, W. E. M., P. R. LeTellier, L. H. Rome, and J. Y. Vanderhoek. 1973. Inhibition of prostaglandin biosynthesis. *Adv. Biosci.* **9**:15–27.

69. Lands, W. E. M. 1992. Biochemistry and physiology of n-3 fatty acids. *FASEB J.* **6**:2530–2536.

70. Enders, S., R. DeCaterina, E. B. Schmidt, and S. D. Kristensen. 1995. n-3 polyunsaturated fatty acids: update 1995. *Eur. J. Clin. Invest.* **25**:629–638.

71. Kulmacz, R. J., R. B. Pendleton, and W. E. M. Lands. 1994. Interaction between peroxidase and cyclooxygenase activities in prostaglandin-endoperoxide synthase. Interpretation of reaction kinetics. *J. Biol. Chem.* **269**:5527–5536.

72. Laneuville, O., D. K. Breuer, N. Xu, Z. H. Huang, D. A. Gage, J. T. Watson, M. Lagarde, D. L. DeWitt, and W. L. Smith. 1995. Fatty acid substrate specificities of human prostaglandin-endoperoxide H synthase-1 and -2. *J. Biol. Chem.* **270**:19330–19336.

73. Vanderhoek, J., and W. E. M. Lands. 1973. Acetylenic inhibitors of sheep vesicular gland oxygenase. *Biochim. Biophys. Acta* **296**:374–381.

74. Vanderhoek, J. Y., and W. E. M. Lands. 1978. Evidence for H_2O_2 mediating the irreversible action of acetylenic inhibitors of prostaglandin biosynthesis. *Prostagland. Med.* **1**:251–263.

75. Culp, B. R., B. R. Titus, and W. E. M. Lands. 1979. Inhibition of prostaglandin biosynthesis by eicosapentaenoic acid. *Prostagland. Med.* **3**:269–278.

76. Marshall, P. J., M. A. Warso, and W. E. M. Lands. 1985. Selective microdetermination of lipid hydroperoxides. *Anal. Biochem.* **145**:192–199.

77. Miller, J. F., R. J. Kulmacz, and W. E. M. Lands. 1991. Use of a voltage-responsive timer to quantitate hydroperoxide by the cyclooxygenase activation assay. *Anal. Biochem.* **193**:55–60.

78. Marshall, P. J., and W. E. M. Lands. 1986. In vitro formation of activators for prostaglandin synthesis by neutrophils and macrophages from human and guinea pigs. *J. Lab. Clin. Med.* **108**:325–354.

79. Keen, R. R., L. A. Stella, D. P. Flanigan, and W. E. M. Lands. 1990. Differences between arterial and mixed venous levels of plasma hydroperoxides following major thoracic and abdominal operations. *Free Rad. Biol. Med.* **9**:485–494.

80. Keen, R. R., L. A. Stella, D. P. Flanigan, and W. E. M. Lands. 1991. Differential detection of plasma hydroperoxides in sepsis. *Crit. Care Med.* **19**:1114–1119.

81. Lands, W. E. M., and R. R. Keen. 1990. Peroxide tone and its consequences, pp. 657–665. *In* C. C. Reddy, G. A. Hamilton, and K. M. Madyastha (eds.). *Biological Oxidation Systems.* Academic Press, San Diego.

82. Lands, W. E. M. 1981. Actions of antiinflammatory drugs. *Trends Pharmacol. Sci.* **8**:78–80.

83. Rome, L. H., and W. E. M. Lands. 1975. Structural requirements for time-dependent inhibition of prostaglandin biosynthesis by anti-inflammatory drugs. *Proc. Natl. Acad. Sci. USA.* **72**:4863–4865.

84. Lands, W. E. M., H. W. Cook, and L. H. Rome. 1976. Prostaglandin biosynthesis: consequences of oxygenase mechanism upon in vitro assays of drug effectiveness, pp. 7–17. *In* B. Samuelsson and R. Paoletti (eds.), *Advances in Prostaglandin and Thromboxane Research,* Vol. 1. Raven Press, New York.

85. Mancini, J. A., D. Riendeau, J.-P. Falgueyret, P. J. Vickers, and G. P. O'Neill. 1995. Arginine 120 of prostaglandin G/H synthase-1 is required for the inhibition by nonsteroidal anti-inflammatory drugs containing a carboxylic acid moiety. *J. Biol. Chem.* **270**:29372–29377.

86. Bhattacharyya, D. K., M. Lecomte, C. J. Rieke, R. M. Garavito, and W. L. Smith. 1996. Involvement of arginine 120, glutamate 524, and tyrosine 355 in the binding of arachidonate and 2-phenylpropionic acid inhibitors to the cyclooxygenase active site of ovine prostaglandin endoperoxide synthase-1. *J. Biol. Chem.* **271**:2179–2184.

87. Walenga, R. W., S. F. Wall, B. N. Setty, and M. J. Stuart. 1986. Time-dependent inhibition of platelet cyclooxygenase by indomethacin is slowly reversible. *Prostaglandins* **31**:625–637.

88. Callan, O. H., O.-Y. So, and D. C. Swinney. 1996. The kinetic factors that determine the affinity and selectivity for slow binding inhibition of human prostaglandin H synthase 1 and 2 by indomethacin and flurbiprofen. *J. Biol. Chem.* **271**:3548–3554.

89. Stanford, N., G. J. Roth, T. Y. Shen, and P. W. Majerus. 1977. Lack of covalent modification of prostaglandin synthetase (cyclo-oxygenase) by indomethacin. *Prostaglandins* **13**:669–675.

90. Rome, L. H., W. E. M. Lands, G. J. Roth, and P. W. Majerus. 1976. Aspirin as a quantitative acetylating reagent for the fatty acid oxygenase that forms prostaglandins. *Prostaglandins* **11**:23–30.

91. Kulmacz, R. J., and W. E. M. Lands. 1985. Stoichiometry and kinetics of the interaction of prostaglandin H synthase with antiinflammatory agents. *J. Biol. Chem.* **260**:12572–12578.

92. Oullet, M., and M. D. Percival. 1995. Effect of inhibitor time-dependency on selectivity towards cyclooxygenase isoforms. *Biochem. J.* **306**:247–251.

93. Hanel, A. M., and W. E. M. Lands. 1982. Modification of antiinflammatory drug effectiveness by ambient lipid peroxides. *Biochem. Pharmacol.* **31**:3307–3311.

94. Lands, W. E. M., and A. M. Hanel. 1982. Phenolic anti-cyclooxygenase agents and hypotheses of antiinflammatory therapy. *Prostaglandins* **24**:271–278.

95. Vanderhoek, J., and W. E. M. Lands. 1973. The inhibition of the fatty acid oxygenases of sheep vesicular gland by antioxidants. *Biochim. Biophys. Acta* **296**:382–385.

96. Kulmacz, R. J., and W. E. M. Lands. 1985. Quantitative similarities in the several actions of cyanide on prostaglandin H synthase. *Prostaglandins* **29**:175–190.

97. Hsuanyu, Y., and H. B. Dunford. 1992. Prostaglandin H synthase kinetics. The effect of substituted phenols on cyclooxygenase activity and the substituent effect on phenolic peroxidatic activity. *J. Biol. Chem.* **267**:17649–17657.

7

Protein Kinase C as a Sensor for Oxidative Stress in Tumor Promotion and Chemoprevention

Rayudu Gopalakrishna, Zhen-Hai Chen, and Usha Gundimeda

7.1 Introduction

Protein kinase C (PKC) has emerged as a key enzyme that is activated by transmembrane signals such as products of phospholipid hydrolysis and Ca^{2+}.[1-5] PKC represents a family of more than 11 isoenzymes, which are divided broadly into Ca^{2+}-dependent type (α,β,γ) and Ca^{2+}-independent type (δ,ε,ζ, etc.).[1-5] Differences in the subcellular localization as well as the expression of these isoenzymes in various cell types, enable PKC isoenzyme system to respond differentially to a wide variety of cellular stimuli.[1-5] The experimental tumor promoters, phorbol esters, bind to activate PKC, which results in increased membrane association (translocation) of PKC, and ultimately leads to its down-regulation. PKC is also activated and inactivated by oxidant tumor promoters that play an important role in carcinogenesis in human settings.[6-10] The unique structural aspects of PKC make it also a suitable target for the redox-sensitive cancer chemopreventive agents (antioxidants) as well as for growth-inhibiting natural products.[11] This enzyme also has been shown to influence tumor cell properties such as cell motility, adhesion, invasion, and metastasis, and therefore it is involved not only in tumor promotion but also in the later events of carcinogenesis related to tumor progression.[12-15] This is a crucial enzyme currently targeted for the development of cancer chemopreventive agents, antiproliferative, and antimetastatic agents.

Oxidative stress has been implicated in pathophysiology of various diseases[16,17] as well as in tumor promotion[18-23] and cell growth.[24] Sublethal levels of reactive oxygen species could induce signal transduction mechanisms and can bypass normal second messengers to activate cell. This also leads to a development of cellular adoptive responses to oxidative stress. Oxidative stress besides regulating transmembrane signaling influences gene expression through modulating the transcription factors such as activator protein-1 (AP-1), NF-κB, and somatotropin-releasing factor/tissue-coding factor (SRF/TCF).[25-29] In such regulations of transcription factors, mechanisms involving redox modification of critical cysteine residues as well as mechanisms involving protein phosphorylation have been implicated.[30,31] Although modifications of these transcription factors by these two type of mechanisms were well documented in the test tube, in intact cells, the extent to which these two types of mechanisms are regulated is not clearly established. Given the fact that phorbol esters have been used as a tool not only for tumor promotion studies but also for variety of signal transduction studies in diverse fields in biomedical research, the knowledge of PKC redox regulation is useful to understand the relationship between the second messengers (Ca^{2+} and lipid hydrolysis products) and reactive oxygen species.

Although phorbol esters are important experimental tumor promoters, a variety of oxidant-generating agents and inflammatory oxidants have been shown to act as tumor promoters.[18–23] The discovery that phorbol esters bind to activate PKC, and okadaic acid-related agents bind to and inhibit protein phosphatases 1 and 2A supported the importance of protein phosphorylation mechanisms in tumor promotion.[1,32] However, recent studies from our laboratory as well as from others showed that oxidants also can activate serine/threonine and tyrosine protein kinases and can inactivate protein phosphatases and there by can induce an imbalance in protein phosphorylation similar to that elicited by other agents (phorbol esters, okadaic acid) that bind to these receptors.[6–11] The study of the redox modulations of PKC as well as other kinases such as tyrosine protein kinases and mitogen-activated protein (MAP) kinases may help understand how these two types of mechanisms complement each other.

This chapter focuses on three aspects: unique structural aspects of PKC that enable it to function as a sensor for bimodel regulation induced by oxidants, the activation of PKC by structurally diverse oxidant-generating and tumor-promoting agents, and the inactivation of this enzyme by chemopreventive antioxidant (redox) agents. While this chapter focuses on the importance of regulation of PKC by oxidative stress and is generally relevant to various pathophysiological conditions, the emphasis is placed on Ca^{2+}-dependent isoenzymes of PKC as well as on the agents related to tumor promotion and cancer chemoprevention.

7.2 Concept of Oxidants as Tumor Promoters

Reactive oxygen species have been shown to play a major role in mediating the tumor promotion induced by a variety of agents in humans and experimental conditions.[18–23] Phorbol esters induce in a various cell types the generation of oxygen radicals that may modify DNA, proteins, and lipids.[18–23] Furthermore, phorbol esters also have been shown to elevate cellular levels of arachidonic acid and its oxygenation metabolites, which can induce genetic and epigenic events.[33,34] Moreover, phorbol esters as well as the eicosanoids also induce reactive oxygen species by activating inflammatory cells.[35–37] Structurally unrelated oxidants such as organic peroxides (benzoyl peroxide, cumene hydroperoxide, etc.) and *m*-periodate have been shown to be tumor promoters, suggesting that oxidants alone can induce tumor promotion.[18,34,38] Carcinogenesis by heavy metals such as cadmium, chromium, and nickel may be mediated by the generation of reactive oxygen species.[39,40] Oxygen radicals also have been implicated to play a crucial role in pulmonary carcinogenesis induced by asbestos and silica.[41] Furthermore, asbestos has been shown to induce membrane translocation of PKC in hamster tracheal epithelial cells.[42] Although complete carcinogens such as benzo(*a*)pyrene induce the initiation process by covalently binding to DNA after metabolic activation to epoxides, the promotion events may be mediated by H_2O_2 generated as a by-product of its oxidative metabolism.[43-45]

Although oxygen-centered free radicals are produced at some level by all respiring cells, inflammatory cells are particularly effective in generating the entire spectrum of molecular oxygen-derived oxidants.[46] Thus, the activation of the redox metabolism of these cells creates a highly oxidative environment within an organ. The association of chronic inflammation with malignancies has been recognized for centuries.[46] The oxidants generated by macrophages and neutrophils can induce lesions and phenotypic changes characteristic of carcinogens and have been implicated in both tumor promotion as well as in tumor progression.[35-37]

In human settings, cigarette smoke represent a strong source of tumor promoters that can play a crucial role in both tumor promotion and progression.[47,48] Both phases of cigarette smoke, tar and gas, contain free radicals.[47,48] Gas-phase smoke contains high concentrations of nitric oxide, reactive olefins, and dienes; and the tar phase contains polyphenols.[47,48] Nitric oxide can induce nitrosylation of amines, mutations, and cell transformation.[49,50] We have previously reported that nitric oxide can reversibly inactivate PKC by inducing a modification of vicinal thiols present within the catalytic domain of the enzyme.[51] Although nitric oxide by itself is a weak oxidant, it can be converted to strong oxidants such as peroxy nitrite or nitrogen dioxide.[47] Cigarette smoke contains more than 200 semivolatile phenols, but hydroquinone and catechol occur in abundance.[52] Both agents are present in the weakly acidic phenolic fraction of cigarette smoke, which has both cocarcinogenic activity and tumor-promoting activity.[53] The autooxidation of polyphenolic agents generates oxidants such as superoxide, hydrogen peroxide, and hydroxyl radicals as well as semiquinones and quinones.[48,54,55] Due to the hydrophilic nature of polyphenolic agents, it is not clear whether they enter the cell, but they may undergo autooxidation outside the cell.[56] Cigarette smoke has been shown to induce some morphological and biochemical changes such as metaplasia and a decrease in epidermal growth factor receptor in lung epithelial cells somewhat similar to phorbol esters.[57] Catechol also was classified as a chemopreventive agent and can induce phase II enzymes that can alter carcinogen detoxication.[58]

7.3 Chemoprevention by Antioxidant (Redox) Agents

Antioxidant micronutrients are one of the body's primary defenses against oxidants and this protection may, in part, be responsible for their chemopreventive action.[59-64] Carotenoids (β-carotene) by themselves act as antioxidants or bring some of their actions having provitamin A activity.[65,66] Vitamin E is the major lipid-soluble antioxidant found in cell membranes where it protects against lipid peroxidation.[67] Vitamin C is a water-soluble antioxidant. Ellagic acid, a phenolic compound present in fruits, can reduce tumor promotion.[68] Other plant polyphenolic agents such as epigallocatechin gallate and curcumin are inhibitors of carcinogenesis. Suppression of tumorigenesis by organosulfur compounds, diallyl sulfide, and S-allyl-cysteine, present in garlic and onion was reported.[69] These sulfur agents activate of scavenging enzymes and inhibit cytochrome P-450 involved in carcinogen activation.[70] Other sulfur antioxidant agents, such as oltipraz, a derivative of dithiolthione, have also been shown to induce protective enzymes.[71-74] Various chemical forms of selenium have been found to anti-tumor-promoting activities.[75,76]

Conceivably, consistent with the generalization that diverse tumor promoters are oxidants or can generate metabolically oxidants, a variety of structurally unrelated chemopreventive agents are antioxidants. However, these so-called antioxidants are also prooxidants. It is not clear that at the concentrations that each agent required to inhibit the carcinogenesis process whether these agents can act as antioxidant or prooxidants. A prooxidant also can protect the cell from a toxic doses of an oxidant by inducing a global antioxidant adoptive response.

7.4 Unique Structural Aspects of PKC that Makes It a Sensor for Oxidative Stress

Although cell has several targets for oxidants, the structural aspects of some proteins make them more sensitive targets than others. The unique structural aspects of PKC may make it a highly susceptible target for oxidants and a suitable candidate for a bidirectional regulation induced by oxidants.[11] A selective oxidative modification of the regulatory domain results in a Ca^{2+}/lipid-independent activation of the kinase with a loss of phorbol ester binding.[7,8] Alternatively, a selective oxidative modification of the kinase domain results in a generation of an inactivated form of PKC which exhibits only phorbol ester binding.[8] The regulatory domain contains 12 cysteine residues (C1 region) that coordinate the binding of four zinc atoms,[3] and the zinc-thiolate structure is required for binding of phorbol ester and diacylglycerol.[70] The C3 constant region present within the catalytic domain has a putative ATP-binding site and one cysteine residue. The C4 constant region present in the catalytic domain has four to six cysteine residues as well as another putative ATP-binding site. Although these cysteine residues in the primary sequence present apart, studies carried out with periodate, nitric oxide, and selenite from this laboratory suggest that they may be vicinal in nature in the tertiary structure.[11]

Cysteine residues present within the regulatory and catalytic domains may react at a different rate with oxidizing and alkylating agents.[11] Based on the current models of four zinc atoms binding to the regulatory domain, all the cysteine residues in this domain may exist in the zinc-thiolate form, whereas in the catalytic domain cysteine residues may exist as free thiols or thiolates. Since *N*-ethylmaleimide, nitric oxide, chelerythrine, and sanguinarine preferentially modify the catalytic region, it is possible that these thiolates in this region may readily react with alkylating (arylating) agents. As thiolates in the regulatory domain coordinate zinc binding, they may not readily react with alkylating agents. However, anionic oxidants such as hydroxyl radicals, peroxynitrite, superoxide, and hypochlorite at a lower concentrations may preferentially react with positively charged zinc-thiolate followed by the oxidation of thiolates involved in the metal coordination. At higher concentrations, however, these oxidants may also react with cysteine residues present within the catalytic region.

Although the importance of the cysteine residues present within the regulatory domain is recognized,[70] the functional roles for the cysteine residues present within the catalytic domain are not known. Recent studies from our laboratory showed that some redox-active chemopreventive and growth-inhibiting agents such as selenite, curcumin, 4-hydroxy-tamoxifen inactive PKC by oxidizing the vicinal thiols present within the catalytic domain.[11,77,78] Furthermore, chelerythrine, an inhibitor for PKC with a growth-inhibiting potential, reacts covalently with the cysteine residues present within the catalytic domain.[79] Therefore, by having different oxidation susceptible regions in the regulatory and catalytic domains, PKC can respond to both oxidant tumor promoters and to chemopreventive antioxidants (prooxidants) to elicit entirely different response: oxidant tumor promoters activate while chemopreventive agents inactivate the enzyme.

7.5 Activation of PKC by Oxidant Tumor Promoters

7.5.1 Peroxides

Both regulatory and catalytic domains of PKC are susceptible to oxidative modification by H_2O_2.[6] At higher concentrations, H_2O_2 inactivated both phorbol ester binding and kinase activity. At lower concentrations of H_2O_2, a selective modification of the regulatory domain was achieved by protecting the catalytic site with ATP/Mg^{2+}. This modification resulted in a loss of phorbol ester binding but activation of enzyme in the absence of Ca^{2+} and lipids. Although we can not exclude the direct modification of PKC with H_2O_2, the presence of iron accelerated this modification suggesting the involvement of hydroxyl radical in these modifications. In cells treated with H_2O_2, there was a decrease in the activity of Ca^{2+}/lipid-activated form of PKC (proform) eluted from DEAE-cellulose column with 0.1-M NaCl (peak A), while a transient formation of Ca^{2+}-lipid-independent form of PKC eluted with 0.25-M NaCl (peak B) from DEAE-cellulose column was observed. This activation also occurred within a few seconds. There was a down-regulation of this form of enzyme at a later time period. However, H_2O_2 did not induce membrane translocation of PKC in some cell types tested. Although in the test tube, PKC modification required millimolar concentrations of H_2O_2, with cells in a serum-free medium, it required only 100 to 300 μM. When the cells are treated in Hank's balanced salt solution instead of medium, H_2O_2 required to induce PKC modifications was much lower (50 to 100 μM). This is important to take into consideration that PKC is more susceptible to oxidative modification in intact cells than in the test tube.

Activation of PKC by H_2O_2-treated cells was also observed in other studies.[80,81] In some cases, a membrane translocation of PKC was observed,[10,80] while other cases, there was no membrane translocation of PKC.[81] In some cases, a formation of Ca^{2+}/lipid-independent form of PKC was also reported.[81] It is possible that depending on the cell incubation conditions, such as the presence or absence of serum in the medium or balanced salt solution is used, the concentrations of H_2O_2 needed can be varied as well as the changes occurring in PKC in response to H_2O_2. Increase in phosphorylation of myristylated alanine-rich C-kinase substrate was shown in astrocytoma cells treated with H_2O_2, which was blocked by PKC inhibitors.[82] In cells treated with H_2O_2, an activation of MAP kinases, transcriptional factors AP-1 and NF-κB and induction of C-*jun*, C-*fos*, and C-*myc* have been observed.[25–29] In some of these cases a role for oxidatively activated PKC has been implicated. However, it is still not clear whether PKC influences these events directly or if it indirectly triggers these events by influencing the protein kinases that are involved in the downstream. Similarly, the role of protein phosphorylation versus redox modulation is not clear in these processes.

Fatty acid hydroperoxides have been shown to induce a membrane translocation of PKC.[80–83] Arachidonic acid hydroperoxide has been shown to activate directly PKC.[84] However, unsaturated fatty acids also activate PKC at the same concentrations. In our studies, we have found that fatty acid hydroperoxide activated PKC in the test tube in manner similar to unsaturated fatty acids, but did not induce modification of PKC cysteine residues. However, it is possible that in the cells, free radicals formed from the fatty acids can induce PKC modifications. Similar fashion, benzoyl peroxide a known skin tumor promoter alone did not induce oxidative modification of PKC in the test tube. However, in the presence of copper (I) it modified PKC. It is known that copper (I), but not iron, induces the formation of benzoloxyl radical and phenyl radical from benzoyl peroxide.[85] Therefore, it is important to take into consideration not only iron but also copper in tumor-promotion mechanisms.[86]

7.5.2 m-Periodate

Sodium *m*-periodate acts as a tumor promoter and also as a mitogen to some cell types.[38] Periodate has been used to modify protein vicinal thiols in a variety of enzymes.[11] It was suggested that by virtue of having a tetrahedral structure, periodate can bind to vicinal thiols present within the binding site for phosphorylated compound. Therefore, such binding sites may be protected by either phosphate or phosphorylated ligand.

Periodate, at micromolar concentrations, modified the regulatory domain of PKC as determined by the loss of ability to stimulate kinase activity by Ca^{2+}/phospholipid, and also by the loss of phorbol ester binding.[7] This modification resulted in an increase in Ca^{2+}/phospholipid-independent kinase activity (oxidative activation). However, at higher concentrations (> 100 μM) periodate also modified the catalytic domain, resulting in complete inactivation of PKC. The oxidative modification induced by low periodate concentrations (< 0.5 mM) was completely reversed by brief treatment with 2-mM dithiothreitol. In this aspect, the modification induced by periodate was different from that of irreversible modification of PKC induced by H_2O_2. Both reversible and irreversible oxidative activation and inactivation of PKC also was observed in intact cells treated with periodate. Taken together, these results suggest that periodate, by virtue of having a tetrahedral structure, binds to the phosphate-binding regions present within the phosphatidylserine-binding site of the regulatory domain and the ATP-binding site of the catalytic domain, and modifies the vicinal thiols present within these sites. This results in the formation of intramolecular disulfide bridge(s) within the regulatory domain or catalytic domain leading to either reversible activation or inactivation of PKC, respectively.

7.5.3 Polyphenolic Agents—Catechol and Hydroquinone

Catechol and hydroquinone are typical representative compounds in this series. They are classified as both tumor-promoting and chemopreventive agents. Our recent work showed that hydroquinone (50 μM), or catechol (500 μM), can induce a cytosol-to-membrane redistribution of PKC in Lewis lung carcinoma (LL/2) cells.[8] There was a lag period of 5 to 10 min before the onset of PKC membrane translocation, which became maximal at 30 to 45 min after the initiation of treatment. After a 4-h treatment, PKC activity down-regulated in both cytosol and membrane fractions. Catechol required nearly a 10-fold higher concentration to induce membrane translocation of PKC than hydroquinone. This is in agreement with nearly 6- to 10-fold higher rate of autoxidation and increased generation of H_2O_2 for hydroquinone compared to catechol.[54,55] Cigarette smoke condensate prepared from 2RF Kentucky research cigarette also produced similar effects at 50 μ/ml concentrations. Hydroquinone and catechol although structurally unrelated to phorbol 12-myristate 13-acetate (PMA), still induced both cytosol-to-membrane translocation and subsequent down-regulation of PKC. Nevertheless, some differences were noted: PMA induced more pronounced PKC translocation without any of the lag period seen with polyphenolic agents. Furthermore, PMA also induced a rapid down-regulation of PKC within 2 h in this cell type.

In LL/2 lung carcinoma cells treated with all these agents, there was 1.5 to 2.5-fold increase in activity of Ca^{2+}/lipid-independent form, which eluted with 0.25-M NaCl (peak B). Although this type of lipid-independent form of PKC (M-kinase) was previously believed to be caused by calpain-mediated proteolysis of the proenzyme,[21] evidence from our laboratory suggests that part of this Ca^{2+}/ lipid-independent activity represents the oxidatively activated form of PKC.[11]

7.5.4 Complete Carcinogen benzo(a)pyrene

Benzo(*a*)pyrene (BP) is a complete carcinogen having both initiating and promoting activities. The mechanism of initiation by microsomal metabolism to BP-7,8 diol epoxides and alkylation of DNA bases is well established.[43,45] However, the promoting activity of BP may also involve autooxidation to generate reactive oxygen species.[43–45] Previously it was shown that 6-OH derivative of BP by going through diol/dione redoxy cycling generates superoxide and subsequently H_2O_2 and hydroxyl radicals.[43–45]

In our studies, BP (1 to 5 μM) induced a rapid (3 min) and a transient membrane translocation of PKC followed by a weak down-regulation (10 to 25%) of the kinase activity after 15 to 30 min, particularly in the membrane fraction. Unlike phorbol esters, the extent of membrane translocation as well as the down-regulation observed with BP were lower. Compared to polyphenolic agents, the extent of down-regulation induced was also lower. BP also induced an elevation of intracellular Ca^{2+} within 3 min. This Ca^{2+} elevation was proceeded by an increase in release of arachidonic acid during 3 to 60 min treatment with BP. The changes occurring in PKC, Ca^{2+}, and arachidonic acid release all substantially blocked by vitamin E (50 μM) and catalase (2000 U/ml). At this concentration, BP did not alter the activity of purified PKC. Unlike phorbol ester, BP did not support the membrane association of PKC when the isolated membrane and purified PKC were incubated. By incubating LL/2 cells with [³H]BP most of the cell-associated radioactivity (93%) was recovered in the membrane fraction. Taken together these observations suggest that BP by initially partitioning into the membrane, induced an oxidative stress, which in turn induced a transmembrane signal transduction. The mechanism of BP-induced effects on the signal transduction and on PKC resembled that induced by tamoxifen.[78] Although tamoxifen is a reversible inhibitor for PKC in the test tube, as shown before, our recent studies suggested that in intact cells it can induce oxidative regulation of PKC and influence the cellular signal transduction by eliciting oxidative stress in a manner different from its actions in the test tube.[78]

7.6 Differential Response of PKC to Various Generators of Oxidative Stress

Although different agents generate oxidative stress, the nature of the reactive species formed and the compartment in which these are generated could influence the outcome of the cellular response. Various oxidants or oxidant-generating systems induce a variable effects on PKC. Reagent H_2O_2 induced a direct activation of PKC by modifying the enzyme in intact cells. In general, there was no translocation of the enzyme in the cells treated with H_2O_2. In these cells, there was no increase in intracellular Ca^{2+} or arachidonic acid. On the contrary, fatty acid hydroperoxides can induce membrane translocation of PKC but no direct modification of the enzyme.[80,83] Polyphenolic compounds, which are believed to induce generation of superoxide and H_2O_2 did induce membrane translocation of PKC and elevation of intracellular Ca^{2+} and arachidonic acid.[8] However, at a later stage these agents also induced an oxidative activation of PKC probably through generation of H_2O_2. These agents are water soluble and are believed to act primarily in the extracellular medium. However, BP-type agents can partition into the membrane and can induce a membrane translocation and activation of PKC through an indirect mechanism. Therefore, the oxidant-generating system, depending on its nature or its action in a particular cellular compartment, could influence PKC either directly by inducing activation/inactivation or indirectly through the elevation of second messengers such as Ca^{2+} and lipid hydrolysis products. The increased Ca^{2+} as well as oxidants may activate other intracellular Ca^{2+}-dependent enzymes including phospholipases A_2, C, and D to release arachidonic acid and diacylglycerol.[1]

7.7 Inactivation of PKC by Chemopreventive Agents

7.7.1 Selenium Compounds

Experimental studies in animals indicate that selenium supplementation at levels above dietary requirement is capable of protecting tumorigenesis induced by a variety of chemical carcinogens or viruses.[75,76] The form and mode of administration of selenium influence the antitumorigenic properties of this trace element.[87] Selenite was more potent than selenomethionine. It has been shown to inhibit both initiation and promotion phases. The chemopreventive mechanism of action of selenium is not known. At nutritional levels, selenium is essential for the enzyme glutathione peroxidase. However, several experiments have ruled out a function for enhanced level of this enzyme activity in selenium-mediated inhibition of mammary tumorigenesis.[87] It is clear that selenite require activation prior to eliciting its effects. The well-established nonenzymatic activation in the test tube involves its reduction by glutathione into selenide through intermediate formation of selenotrisulfide and selenopersulfide.[85,88,89] This volatile selenide or more stable derivative methylselenide suggested to interact with target proteins containing cysteine-rich regions. However, in intact cells, other cellular thiols including protein may play crucial role in influencing the bioreductive activation of selenite.[90]

Selenium interaction with protein thiols has been suggested to be one of its modes of action. Since PKC has cysteine-rich zinc finger motifs, which are required for phorbol ester binding and also other critical cysteine residues in the catalytic domain, we have determined whether selenium can influence PKC activity.[77] When purified PKC was incubated with low concentrations (0.05 to 5 μM) of selenite, PKC rapidly lost its kinase activity but not the phorbol ester binding (Fig. 7.1). This was reversed by incubation with higher concentrations of dithiothreitol. These studies showed that the vicinal thiols present within the catalytic domain may be involved in the reduction of selenite. Selenocystine also inactivated PKC with an IC_{50} of 6 μM. However, selenate and selenomethionine had no effect on the PKC activity. The treatment of intact breast carcinoma cells with either selenite (0.5 to 10 μM) or selenocystine (10 μM) also resulted in an inactivation of PKC activity. Given the fact that PKC serves as a receptor for not only phorbol esters but also to other structurally unrelated tumor promoters such as oxidants, selenium-mediated inactivation of PKC may have an important role in the chemopreventive action of selenium.

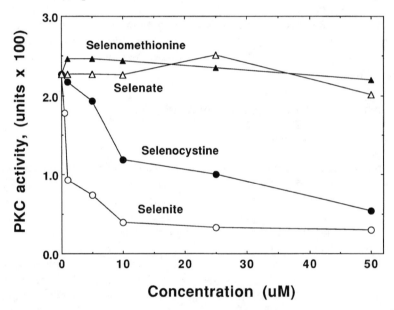

Figure 7.1. Inhibition of purified PKC activity by selenium compounds. Initially from the purified PKC from rat brain (mixture of α, β, and γ isoenzymes). 2-Mercaptoethanol was removed by gel filtration. The indicated concentrations of various selenium compounds were added to the PKC reaction mixture and the kinase activity was determined using histone H1 as the substrate. PKC activity was expressed in units, where one unit of enzyme transfers one nmol of phosphate to histone H1/min at 30°C.

7.7.2 Curcumin

Curcumin is the major yellow pigment of turmeric, a product isolated from the rhizome of the plant *Curcuma longa* Linn. It possesses antitumor promoter, anti-inflammatory, antioxidant properties.[91,92] Curcumin inhibits a variety of phorbol ester-induced effects. It inhibits phorbol ester-induced expression of c-*jun,* c-*fos,* and c-*myc* protooncogenes, NF-κB activation, induction of ornithine decarboxylase (ODC), and induction of xanthine dehydrogenase/oxidase activity.[91–94] Since PKC serves as a receptor for tumor promoters not only phorbol ester, but also for other tumor promoters such as oxidants, it is logical that curcumin may counteract at certain stage in the metabolic events triggered by tumor promoters and PKC. However, the early studies revealed either a lack or a weak inhibition of PKC by curcumin.[92,95]

Curcumin inhibits iron-dependent lipid peroxidation, iron-dependent Fenton reaction, and production of reactive oxygen species by neutrophils.[96] It produces photochemical toxicity as a prooxidant. It forms carbon-centered radicals, and generates singlet oxygen, hydroxyl radicals, and superoxide.[97–99] Curcumin structure is often represented as diketo structure. This was suggested to exist as a keto-enol form based on the its reaction with borate and ferric ions.[100] While the 1,3 ketone role in iron chelation and inhibition of lipid peroxidation without the need for phenolic group was suggested in one study.[101] Certain biological effects of curcumin were shown to be mediated without the need for the phenolic group.[91] In another study, the role of phenolic group in mediating the antioxidant activity curcumin and other related agents was suggested.[102] It could possible that curcumin's broad spectrum properties may be related to phenolic group, metal-binding ketoenol function, and extended conjugation present in the two chromophores.

In our studies, curcumin inhibited purified PKC (Ca^{2+}-dependent isoenzymes) in the absence of thiol agents (IC_{50} = 50 μM).[103] However, in the presence of thiol agents such as dithiothreitol (1 mM) the inhibition was weaker (IC_{50} = 110 μM). Nonetheless, when curcumin was used as its ferric complex, it inhibited PKC strongly (IC_{50} = 3 μM) even in the presence of thiol agents (Fig. 7.2). Furthermore, curcumin inhibited PKC reversibly, whereas curcumin-ferric complex inhibited PKC irreversibly. Other metals tested were found to be in effective in promoting the action of curcumin. Detailed studies carried out with purified PKC suggested that curcumin-ferric complex permanently modified PKC, resulting in loss of both kinase activity and phorbol ester binding. The phenolic group present in curcumin was found to be necessary for mediating this effect.

Curcumin also irreversibly inactivated PKC in intact cells and the sensitivity of various cell types varied. Since curcumin alone inhibits PKC in reversible manner and the observed irreversible inactivation of PKC intact cells as well as the difference in susceptibility of various cells types suggest that other cofactors may be necessary to mediate this effect. It is possible that such an irreversible effect of curcumin on PKC may be mediated by intracellular formation of a curcumin-ferric complex. The difference in sensitivity of various cell types to curcumin-mediated inactivation of PKC may be related to the differences in intracellular accessible iron pool. Since iron plays an important role in oxidations, carcinogenesis, and cell growth, a chelation of iron by curcumin may facilitate

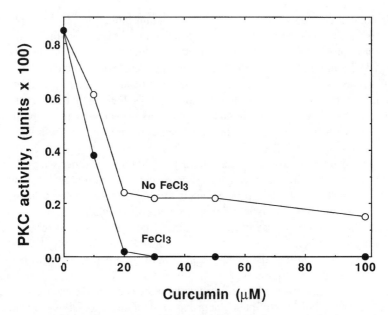

Figure 7.2. Inhibition of purified PKC by curcumin and curcumin/ferric complex. In the absence of thiol agents, curcumin or curcumin/ferric complex was added to purified PKC and then the phosphotransferase activity was determined. The curcumin/ferric complex was prepared by mixing equimolar amounts of curcumin and $FeCl_3$ and the complex was diluted.

its beneficial effects. Taken together these results suggest that curcumin may mediate its cancer chemopreventive effects, at least in part, by chelation of iron and inducing an irreversible inactivation of PKC.

7.7.3 Oltipraz

Oltipraz [4-methyl-5-(2-pyrazinyl)1,2-dithiole-3-thione] is an antischistosomal drug currently under clinical evaluation as a possible chemopreventive agent in humans.[71] It has been proven as an efficient chemopreventive agent in several experimental models of chemical carcinogenesis involving diverse organs.[71–74] Its anti-initiation effects are mediated by its actions on the metabolism and deposition of chemical carcinogens via the induction of electrophile detoxication enzymes.[71–74] Metabolites derived from oltipraz was shown to inhibit the activity of glutathione reductase.[104] It inhibits or inactivates directly the activity of cyto-chrome P-450 and reverse transcriptase.[105,106] A covalent adduct formation of oltipraz with reverse transcriptase has been recently demonstrated.[106] In some models oltipraz exerts postinitiation effects.

Our recent studies showed an inhibition of PKC by oltipraz (IC_{50} = 25 μM) with a mixed type mechanism with respect to ATP and phosphatidylserine.[107]

Both Ca^{2+}/lipid-independent activity of PKC with protamine as the substrate and proteolytically activated enzyme were also inhibited by oltipraz, suggesting that it acts upon the catalytic domain. Direct incubation of PKC with oltipraz in the presence of Ca^{2+} resulted in an inactivation of the enzyme, which was prevented by thiol agents. Addition of oltipraz (5 to 50 µM) to cells produced a transient inactivation of PKC 15 to 30 min after treatment, which returned to control level after 2 h (Fig. 7.3). Oltipraz also inhibited phorbol ester-induced transformation of JB6 and C3H10T½ cells. Thus, in addition to the known anti-initiating actions, oltipraz, may inhibit tumor promotion, mediated at least in part by transient inactivation of PKC.

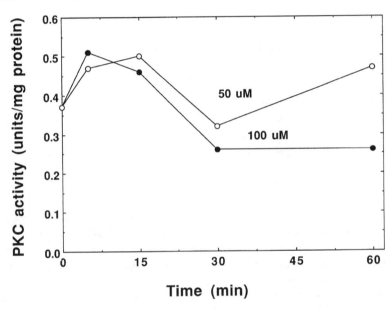

Figure 7.3. Oltipraz-induced reversible inactivation of PKC in B16 melanoma cells. Confluent F10 subline of B16 melanoma cella were treated with indicated concentrations of oltipraz for various periods of time. Then the total PKC activity was extracted with buffer containing detergent. After isolation of PKC with DEAE-cellulose chromatography, the kinase activity was determined. Mercapto agents were not included in the buffers used.

7.8 Paradox of Antioxidant Chemopreventive Agents Acting through an Oxidative Mechanism

Our original hypothesis is that tumor promoters are oxidants and conversely cancer chemopreventive agents are antioxidants to counteract the actions of tumor promoting oxidants. While elucidating the mechanisms of action of chemopreventive agents at the PKC level, however, we have found that these chemopreventive agents also have prooxidant effects. To some extent, they can be distinguished from the oxidant tumor promoters that can modify the regulatory domain of PKC to activate the enzyme directly or can induce the elevation of second messengers. Nevertheless, they may share some common mechanism of action with tumor promoters. This it is not surprising that in early studies in cancer chemoprevention it was realized that the protective agents were often carcinogens themselves and that under appropriate experimental conditions established carcinogens could act as protectors.[58] Indeed, prior treatment with low doses of a carcinogen protected against higher doses of the same carcinogen.[58] Therefore, the sublethal oxidative stress elicited by chemopreventive agents may induce an adoptive response by the cells.

7.9 PKC Oxidations Occurring in the Presence of Glutathione and Other Cellular Antioxidant System

Glutathione is an important cellular thiol that has been implicated in preventing the oxidative stress under certain conditions or depending on the compartment in which the oxidative stress is occurring. Therefore, it is important to understand how PKC oxidative regulation occurs in a cell containing high concentrations of glutathione. Cellular glutathione can decrease modification of DNA by water soluble alkylating agent melphalan. Therefore, pretreating the cells with buthionine sulfoxamine can increase DNA alkylation by this drug. In contrast, in our study, we have noticed that there was no increase in PKC modifications by various oxidants and alkylating agents in cells pretreated with buthionine sulfoxamine (0.1 to 1 mM) for 24 h, where the cellular glutathione levels decreased more than 90%.

The membrane association of PKC may explain this unexpected results. PKC is extractable from cells predominately in the cytosolic (soluble) fraction, and appears to translocate from cytosol to membrane after phorbol esters or other agents are added that induce elevation of phospholipid hydrolysis products.[108] However, in control cells, PKC loosely associated with membrane and dissociates from the membrane during homogenization with a low-ionic strength buffer. When phorbol ester or lipid products bind to PKC, the membrane association of PKC is stabilized and the enzyme is not extracted with a buffer into soluble fraction, but it require detergents to extract from the membrane. A similar type of loose binding to membrane or other subcellular organelles was also observed for other proteins such as calpain and estrogen receptor.[109] It could be possible that cysteine-rich regions in PKC may be facing the inner side of the plasma membrane associated with proteins or lipid, and the cytosolic glutathione may not reach such sites. The externally added lipophilic agents or water soluble agents such as H_2O_2 may directly react with the cysteine residues in PKC without being blocked by cytosolic glutathione.

Cellular catalase can protect PKC from H_2O_2-induced oxidation in the test tube. However, it is important to take into consideration that catalase localized exclusively in peroxisomes in the cell.[110] Therefore, it must be considered that glutathione and various antioxidants systems might have some limitations in preventing the redox changes occurring in PKC.

7.10 PKC Inactivation by Commonly Used PKC Antagonists

The unique oxidation susceptibility of PKC is well appreciated when the mechanism of action of some so-called PKC-specific inhibitors are considered. Calphostin C and hypericin have emerged as relatively specific inhibitors for PKC after screening of various agents. Light is required for inhibition of PKC by these agents. Our recent studies revealed that these agents induced an irreversible inactivation of PKC by inducing a photochemical generation of oxidants presumably singlet oxygen.[111] This inactivation occurred in the test tube in the presence of mercaptoethanol or glutathione. Initially, these drugs bind to a Ca^{2+}-induced hydrophobic site on PKC and then are stimulated by light to produce an oxidant, which then without diffusing into the solution, modifies an adjacent oxidation susceptible site. This "cage" type of oxidative reaction has been described for other enzymes at metal-binding sites.[112] However, these agents can also inactivate PKC in the cells in dark by a mechanism probably involving the redox-cycling of these agents in a manner similar to that described for anthracyline antibiotics.[113]

7.11 Summary and Perspectives

The unique structural aspects of PKC cause it to function as a sensor for oxidative stress. Although the cellular regulation is triggered by the concerted effort of many proteins, the unique structural properties of PKC may enable it to readily respond to reactive oxygen species. Such oxidative modifications of PKC resulting in activation or inactivation may trigger a cellular response. This may involve cascades of protein phosphorylation events further amplified by a series of protein kinases present in the downstream regulations. Such downstream events may also involve events triggered by redox modification of proteins. Therefore, it is important to distinguish the extent to which these two important protein modifying mechanisms, phosphorylations and redox modification of proteins, participate in cellular regulations.

PKC is activated by a variety of structurally unrelated oxidants either directly or indirectly through an induction of transmembrane signals in the cells. Hydrogen peroxide and periodate induce a direct activation/inactivation of PKC by modifying cysteine-rich regions present within the zinc-finger sites of the regulatory domain or in the catalytic domain. Some oxidant tumor promoters (polyphenolic agents) initially induce transmembrane signals such as Ca^{2+} and lipid hydrolysis products, which then induce a membrane translocation (activation) of PKC.

Subsequently, this can facilitate an oxidative modification of PKC, leading to activation followed by inactivation. Some chemopreventive antioxidants (selenite, oltipraz, and curcumin) inactivate PKC by modifying the cysteine-rich region present within the catalytic domain. Conceivably, PKC is a target for not only tumor promoters but also chemopreventive agents. However, PKC differentially regulated by this two classes of agents producing differential responses. The altered kinase activity in response to these redox agents may in turn affect the phosphorylation of cellular proteins in the downstream, which could influence the transformation related expression events or can induce an adoptive response to tumor promoters.

Prior to the discovery of PKC, and protein phosphatases 1 and 2A as receptors for certain tumor promoters and implicated in protein phosphorylation-mediated cellular regulation of tumor promotion, a substantial evidence was accumulated on the importance of oxidants in tumor promotion. These two fields emerged in two different eras of carcinogenesis research and represent two phases of one complex process. This cross talk is one of the poorly understood subjects in tumor promotion research. Since oxidants play key roles in tumor promotion as well as PKC, and protein phosphatases 1 and 2A are emerging as receptors for tumor promoters, the chemopreventive agents may act on this oxidant-influenced phosphorylation system. Further studies in this direction may help us understand the molecular mechanisms involved in cancer prevention and improve chemopreventive strategies.

Acknowledgments

Work from the authors' laboratory was supported in part by grant 93B45 from the American Institute for Cancer Research, and grant CA62146 from the National Cancer Institute.

References

1. Nishizuka, Y. 1992. Intracellular signalling by hydrolysis of phospholipids and activation of protein kinase C. *Science* **258**:607–614.
2. Jaken, S. 1990. Protein kinase C and tumor promoters. Curr. Opin. Cell. Biol. **2**:192–197.
3. Bell, R. M., and D. J. Burns. 1991. Lipid activation of protein kinase C. *J. Biol. Chem.* **266**:4661–4664.
4. Rasmussen, H., C. M. Isales, R. Calle, D. Throcmorton, M. Anderson, J. Gasella Herraiz, and R. McCarthy. 1995. Diacylglycerol production, Ca^{2+} influx, and protein kinase C activation in sustained cellular responses. *Endocr. Rev.* **16**:649–676.
5. Blumberg, P. M. 1991. Complexities of the protein kinase C pathway. *Mol. Carcinog.* **4**:330–344.

6. Gopalakrishna, R., and W. B. Anderson. 1989. Ca²⁺-, and phospholipid-independent activation of protein kinase C by selective oxidative modification of regulatory domain. *Proc. Natl. Acad. Sci. USA* **86**:6758–6762.

7. Gopalakrishna, R., and W. B. Anderson. 1991. Reversible oxidative activation and inactivation of PKC by the mitogen/tumor promoter periodate. *Arch. Biochem. Biophys.* **285**:382–387.

8. Gopalakrishna, R., Z. H. Chen, and U. Gundimeda. 1994. Tobacco smoke tumor promoters, catechol and hydroquinone induce oxidative regulation of protein kinase C and influence invasion and metastasis of lung carcinoma cells. *Proc. Natl. Acad. Sci. USA* **91**:12233–12237.

9. Kass, G. E. N., S. K. Duddy, and S. Orrenius. 1989. Activation of hepatocyte protein kinase C by redox-cycling quinones. *Biochem. J.* **260**:499–507.

10. Larsson, R., and P. Cerutti. 1989. Translocation and enhancement of phosphotransferase activity of protein kinase C following exposure of mouse epidermal cells to oxidants. *Cancer Res.* **49**:5627–5633.

11. Gopalakrishna, R., Z. H. Chen, and U. Gundimeda. 1995. Modification of cysteine-rich regions in protein kinase C induced by oxidant tumor promoters and the enzyme specific inhibitors. *Methods Enzymol.* **252**:134–148.

12. Gopalakrishna, R., and S. H. Barsky. 1988. Tumor promoter-induced membrane-bound protein kinase C regulates hematogenous metastasis. *Proc. Natl. Acad. Sci. USA* **85**:612–616.

13. Korczak, B., C. Whale, and R. S. Kerbel. 1989. Possible involvement of Ca²⁺ mobilization and protein kinase C activation in the induction of spontaneous metastasis by mouse mammary adenocarcinoma. *Cancer Res.* **49**:2597–2560.

14. Bonfil, R. D., S. Momiki, R. Fridman, R. Reddel, C. C. Harris, and A. Klein-Szanto. 1989. Enhancement of the invasive ability of a transformed human bronchial epithelial cell line by 12-O-tetradecanoyl-phorbol-13-acetate and diacyglycerol. *Carcinogenesis* **10**:2335–2338.

15. Liu, B., and K. V. Honn. 1992. Protein kinase C inhibitor calphostin C inhibits B16 melanoma metastasis. *Int. J. Cancer Res.* **52**:147–153.

16. Halliwell, B., and C. E. Cross. 1991. Reactive oxygen species, antioxidants, and aquired immunodeficiency syndrome. *Arch. Intern. Med.* **151**:29–31.

17. Ames, B. N., and R. L. Saul. 1987. Oxidative damage, cancer, and aging. *Ann. Intern. Med.* **107**:536–543.

18. Slaga, T. J., A. J. P. Kleen-Szanto, L. L. Triplet, and L. P. Yotti. 1981. Skin tumor promoting activity of benzoyl peroxide, a widely used free radical generating compound. *Science* **213**:1023–1025.

19. Goldstein, B. D., G. Witz, M. Amoruso, D. S. Stone, and W. Troll. 1981. Stimulation of human polymorphonuclear leukocyte superoxide anion radical production by tumor promoters. *Cancer Lett.* **11**:257–262.

20. Kensler, T. W., and M. A. Trush. 1984. Role of oxygen radicals in tumor promotion. *Environ. Mutagen.* **6**:593–602.

21. Cerutti, P. A. 1985. Prooxidant states and tumor promotion. *Science* **227**:375–381.

22. Borek, C., and W. Troll. 1983. Modifiers of free radicals inhibit in vitro the oncogenic actions of x-rays, bleomycin, and the tumor promoter 12-*O*-tetradecanoylphorbol 13-acetate. *Proc. Natl. Acad. Sci. USA* **80**:1304–1308.

23. Frenkel, K., and K. Chrzan. 1987. Hydrogen peroxide formation and DNA modification by tumor promoter-activated polymorphonuclear leukocytes. *Carcinogenesis* **8**:455–460.

24. Burdon, R. H. 1995. Superoxide and hydrogen peroxide in relation to mammalian cell proliferation. *Free. Rad. Biol. Med.* **18**:775–794.

25. Meyer, M., R. Schreck, J. M. Muller, and P. A. Baeuerle. 1994. Redox control of gene expression by eukaryotic transcription factors NF-κB, AP-1 and SRF/TCF, pp. 217–235. *In* C. Pasquier, C. Auclair, R. Y. Olivier, and L. Packer (eds.), *Oxidative Stress, Cell Activation and Viral Infection.* Birkhauser Verlag, Switzerland.

26. Rao, G. N., B. Lassegue, K. K. Griendling, and W. R. Alexander. 1993. Hydrogen peroxide stimulates transcription of c-*jun* in vascular smooth muscle cells: role of arachidoinic acid. *Oncogene* **8**:2759–2764.

27. Schreck, R. B. 1991. A role for oxygen radicals as second messengers. *Trends Cell Biol.* **1**:39–42.

28. Stauble, B., D. Boscoboinik, A. Tasinato, and A. Azzi. 1994. Modulation of activator protein-1 (AP-1) transcription factor and protein kinase C by hydrogen peroxide and D-α-tocopherol in vascular smooth muscle cells. *Eur. J. Biochem.* **226**:393–402.

29. Amstad, P. A., G. Krupitza, and P. A. Cerutti. 1992. Mechanism of c-*fos* induction by active oxygen. *Cancer Res.* **52**:3952–3960.

30. Abate, C., L. Patel, F. J. Rauscher III, and T. Curran. 1990. Redox regulation of Fos and Jun DNA-binding activity in vitro. *Science* **249**:1157–1161.

31. Matthews, J. R., N. Wakasugi, J. L. Virelizier, J. Yodoi, and R. T. Hay. 1992. Thioredoxin regulates the DNA binding activity of NF-κB by reduction of a disulphide bond involving cysteine 62. *Nucleic Acids Res.* **20**:3821–3830.

32. Cohen, P., C. F. B. Holmes, and Y. Tsukitani. 1990. Okadaic acid: a new probe for the study of cell regulation. *Trans. Biochem. Sci.* **15**:98–102.

33. Fischer, S. M., G. Furstenberger, F. Marks, and T. J. Slaga. 1987. Events associated with mouse skin tumor promoters with respect to arachidonic acid metabolism: a comparison between SENCAR and NMRI mice. *Cancer Res.* **47**:3174–3179.

34. Ohuchi, K., M. Watanabe, K. Yoshizawa, S. Tsurufuji, H. Fujiki, M. Suganuma, T. Sugimura, and L. Levine. 1985. Stimulation of prostaglandin E2 production by 12-O-tetradecanoylphorbol 13-acetate (TPA)-type and non-TPA-type tumor promoters in macrophages and its inhibition by cycloheximide. *Biochem. Biophys. Acta* **834**:42–47.

35. Weitzman, S. A., A. B. Weitberg, E. P. Clark, and T. P. Stossel. 1985. Phagocytes as carcinogens malignant transformation produced by human neutrophils. *Science* **227**:1231–1233.

36. Yamashina, K., B. E. Miller, and G. H. Heppner. 1986. Microphage mediated induction of drug resistant variants in a mouse mammary tumor cell line. *Cancer Res.* **46**:2396–2401.

37. Robertson, F. M., A. J. Beavis, T. M. Oberyszyn, S. M. O'Connell, A. Dokidos, J. D. Laskin, and J. J. Reiners. 1990. Production of hydrogen peroxide by murine epidermal keratinocytes following treatment with the tumor promoter TPA. *Cancer Res.* **50**:6062–6067.

38. Srinivas, L., and N. H. Colburn. 1984. Preferential oxidation of cell surface sialic acid by periodate leads to promotion of transformation in JB6 cells. *Carcinogenesis* **5**:515–519.

39. Zhong, Z., W. Troll, K. L. Koenig, and K. Frenkel. 1990. Carcinogenic sulfides salts of nickel and cadmium induce H_2O_2 formation by human polymorphonuclear leukocytes. *Cancer Res.* **50**:7564–7570.

40. Aiyar, J., H. J. Berkovits, R. Floyd, and K. E. Wetterhahn. 1991. Reaction of chromium (VI) with glutathione or with hydrogen peroxide: identification of reactive intermediates and their role in chromium (VI)-induced DNA damage. *Environ. Health Perspect.* **92**:53–62.

41. Marsh, J. P., and B. T. Mossman. 1991. Role of asbestos and active oxygen species in activation of and expression of ornithine decorboxylase in hamster tracheal epithelial cells. *Cancer Res.* **51**:167–173.

42. Perderiset, M., J. P. Marsh, B. T. Mossman. 1991. Activation of protein kinase C by crocidolite asbestos in hamster tracheal epithelial cells. *Carcinogenesis* **12**:1499–1502.

43. Ashurst, S. W., G. M. Cohen, S. Nesnow, J. Digiovanni, and T. J. Slaga. 1983. Formation of benzo(*a*)pyrene/DNA adducts and their relationship to tumor initiation in mouse epidermis. *Cancer Res.* **43**:1024–1029.

44. Lorentzen, R. J., W. J. Caspary, S. A. Lesko, and P. O. P. Ts'o. 1975. The autoxidation of 6-hydroxybenzopyrene and 6 oxobenzopyrene radical, reactive metabolites of benzo(*a*)pyrene. *Biochemistry* **14**:3970–3977.

45. Frenkel, K., J. M. Donahue, and S. Banerjee. 1988. Benzo(*a*)pyrene-induced oxidative DNA damage: a possible mechanism for promotion by complete carcinogens, pp. 509–524. *In* P. Cerutti et al. (eds.), *Oxy-Radicals in Molecular and Cellular Biology,* Vol. 82. UCLA Symposium. New York: Alan R. Liss.

46. Trush, M. A., and T. W. Kensler. 1991. An overview of the relationship between oxidative stress and chemical carcinogenesis. *Free Rad. Biol. Med.* **10**:201–209.

47. Pryor, W. A., and K. Stone. 1992. Oxidants in cigarette smoke radicals, hydrogen peroxide, peroxynitrate, and peroxynitrite. *Ann. N. Y. Acad. Sci.* **686**:12–27.

48. Nakayama, T., D. F. Church, and W. A. Pryor. 1989. Quantitative analysis of the hydrogen peroxide formed in aqueous cigarette tar extracts. *Free Rad. Biol. Med.* **7**:9–15.

49. Cooney, R. V., A. A. Franke, P. J. Harwood, V. Hatch-Pigott, L. J. Custer, and L. J. Mordan. 1993. γ-Tocopherol detoxification of nitrogen dioxide: superiority to α-tocopherol. *Proc. Natl. Acad. Sci. USA* **90**:1771–1775.

50. Nguyen, T. T., D. Brunson, C. L. Crespi, B. W. Penman, J. S. Wishnok, and S. R. Tannenbaum. 1992. DNA damage and mutation in human cells exposed to nitric oxide. *Proc. Natl. Acad. Sci. USA* **89**:3030–3034.

51. Gopalakrishna, R., Z. Chen, and U. Gundimeda. 1993. Nitric oxide and nitric oxide-generating agents induce a reversible inactivation of PKC and phorbol ester binding. *J. Biol. Chem.* **268**:27180–27185.

52. Hoffmann, D., S. S. Hecht, and E. L. Wynder. 1983. Tumor promoters and cocarcinogens in tobacco carcinogenesis. *Environ. Health Prospect.* **50**:247–257.

53. Hecht, S. S., S. Carmella, H. Mori, and D. Hoffmann. 1981. A study of tobacco carcinogenesis XX. Role of catechol as a major cocarcinogen in the weakly acidic fraction of smoke condensate. *JNCI* **66**:163–171.

54. Leanderson, P., and C. Tagesson. 1990. Cigarette smoke-induced DNA damage: role of hydroquinone and catechol in the formation of the oxidative DNA-adduct, 8-hydroxydeoxyguanosine. *Chem. Biol. Interact.* **75**:71–81.

55. Greenlee, W. F., J. D. Sun, and J. S. Bus. 1981. A proposed mechanism of benzene toxicity formation of reactive intermediates from polyphenol metabolites. *Toxicol. Appl. Pharmacol.* **59**:187–195.

56. Picardo, M., S. Passi, M. Nazzaro-Porro, A. Breathnach, C. Zompetta, A. Faggioni, and P. Riley. 1987. Mechanism of antitumoral activity of catechols in culture. *Biochem. Pharmacol.* **36**:417–425.

57. Willey, J. C., R. C. Grafstrom, C. E. Moser, C. Ozanne, K. Sundqvist, and C. C. Harris. 1987. Biochemical and morphological effects of cigarette smoke condensate and its fractions on normal human bronchial epithelial cells in vitro. *Cancer Res.* **47**:2045–2049.

58. Talalay, P. 1989. Mechanisms of induction of enzymes that protect against chemical carcinogenesis. *Adv. Enzyme Reg.* **28**:237–242.

59. Wattenberg, L. W. 1983. Inhibition of neoplasia by minor dietary constituents. *Cancer Res.* **43** (suppl):2448s–2455s.

60. Slaga, T. J., and W. M. Bracken. 1977. The effects of antioxidants on skin tumor initiation and arylhydrocarbon hydroxylase. *Cancer Res.* **37**:1631–1635.

61. Bertram, J. S., L. N. Kolonel, and F. L. Meyskens, Jr. 1987. Rationale and strategies for chemoprevention of cancer in humans. *Cancer Res.* **47**:3012–3031.

62. Borek, C. 1987. Radiation and chemically induced transformation: free radicals, antioxidants and cancer. *Br. J. Cancer* **55**:74–81.

63. Kennedy, A. R. 1988. Implications for mechanisms of tumor promotion and its inhibition by various agents from studies of *in vitro* transformation, pp. 201–212. *In* R. Langenbach et al. (eds.), *Tumor Promoters: Biological Approaches for Mechanistic Studies and Assay Systems.* (eds) R. Langenbach, E. Elmore, and J. C. Barrett. Raven Press, New York.

64. Boone, C. W., G. J. Kelloff, and W. E. Malone. 1990. Identification of candidate cancer chemopreventive agents and their evaluation in animal and human clinical trails. *Cancer Res.* **50**:2–9.

65. DiMascio, P., S. Kaiser, and H. Sies. 1989. Lycopene as the most efficient biological carotenoid singlet oxygen quencher. *Arch. Biochem. Biophys.* **274**:532–536.

66. Pung, A., J. Rundhaug, C. N. Yoshizawa, and J. S. Bertram. 1988. β-carotene and canthaxanthin inhibit chemically—and physically—induced neoplastic transformation in 10T1/2 cells. *Carcinogenesis* **9**:1533–1539.

67. Perchellet, J. P., M. D. Owen, T. D. Posey, D. K. Orten, and B. A. Scheider. 1985. Inhibitory effects of glutathione level-raising agents and D-α-tocopherol on ornithine decarboxylase induction and mouse skin tumor promotion by 12-O-tetradecanoylphorbol-13-acetate. *Carcinogenesis* **6**:567–573.

68. Gali, H. U., E. M. Perchellet, D. S. Klish, J. M. Johnson, and J. P. Perchellet. 1992. Antitumor-promoting activities of hydrolyzable tannins in mouse skin. *Carcinogenesis* **13**:715–728.

69. Brady, J. F., D. Li, H. Ishizaki, and C. S. Yang. 1988. Effect of diallyl sulfide on rat liver microsomal nitrosamine metabolism and other monooxygenase activities. *Cancer Res.* **48**:5937–5942.

70. Quest, A. F. G., J. Bloomenthal, E. S. G. Bardes, and R. M. Bell. 1992. The regulation domain of protein kinase C coordinates four atoms of Zinc. *J. Biol. Chem.* **267**:10193–10197.

71. Helzisouer, K. J., and T. W. Kensler. 1993. Cancer chemoprotection by oltipraz: experimental and clinical considerations. *Prev. Med.* **22**:783–795.

72. Roebuck, B. D., Y.-L. Liu, A. E. Rogers, J. D. Groopman, and T. W. Kensler. 1991. Protection against aflatoxin B1-induced hepatocarcinogenesis in F344 rats by 5-(2-pyrazinyl)-4-methyl-1,2-dithiole-3-thione (oltipraz): predictive role for short-term molecular dosimetry. *Cancer Res.* **51**:5501–5506.

73. Wattenberg, L. W., and E. Bueding. 1986. Inhibitory effects of 5-(2-pyrazinyl)-4-methyl-1,2-dithiol-3-thione (oltipraz) on carcinogenesis induced by benzo[a] pyrene, diethylnitrosamine and uracil mustard. *Carcinogenesis* **7**:1379–1381.

74. Rao, C. V., K. M. Tokomo, G. Kelloff, and B. S. Reddy. 1991. Inhibition of dietary oltipraz of experimental intestinal carcinogenesis induced by azoxymetahne in male F344 rats. *Carcinogenesis* **12**:1051–1055.

75. Milner, J. A., and C. Y. Hsu. 1981. Inhibitory effects of selenium on the growth of L1210 leukemia cells. *Cancer Res.* **41**:1652–1656.

76. Medina, D., and F. Shepherd. 1981. Selenium-mediated inhibition of 7,12-dimethyl-benzo(a)anthracene-induced mouse mammary tumorigenesis. *Carcinogenesis* **2**:451–455.

77. Gundimeda, U., Z. Chen, and R. Gopalakrishna. 1994. Selenium interacts with cysteine-rich regions of protein kinase C and induces enzyme inactivation: its role in cancer chemoprevention (abstract). *FASEB J.* **8**:3136.

78. Gundimeda, U., Z. Chen, and R. Gopalakrishna. 1995. Tamoxifen modulates protein kinase C via oxidative stress in estrogen receptor-negative breast cancer cells. *J. Biol. Chem.* **271**:13504–13514.

79. Gopalakrishna, R., Z. Chen, K. Yoolee, and U. Gundimeda. 1994. Chelerythrine and sanguinarine induce membrane translocation and irreversible inactivation of

protein kinase C by modifying thiol residues (abstract). *Proc. Am. Assoc. Cancer Res.* **35**:615.

80. Taher, M. M., J. G. Garcia, and V. Natarajan. 1993. Hydroperoixde-induced diacyl-glycerol formation and protein kinase C activation in vascular endothelial cells. *Arch. Biochem. Biophys.* **303**:260–266.

81. Whisler, R. L., M. A. Goyette, I. S. Grants, and Y. G. Newhouse. 1995. Sublethal levels of oxidant stress stimulate multiple serine/threonine kinases and suppress protein phosphatases in jurkat T cells. *Arch. Biochem. Biophys.* **319**:23–35.

82. Brawn, M. K., W. J. Chiou, and K. L. Leach. 1995. Oxidant-induced activation of protein kinase C in UC11MG cells. *Free Rad. Res.* **22**:23–37.

83. Sweetman, L. L., N. Y. Zhang, H. Peterson, R. Gopalakrishna, and A. Sevanian. 1955. Effect of linoleic acid hydroperoxide on endothelial cell calcium homeostasis and phospholipid hydrolysis. *Arch. Biochem. Biophys.* **323**:97–107.

84. O'Brien, C. A., N. E. Ward, I. B. Weinstein, A. W. Bull, and L. J. Marnett. 1988. Activation of rat brain protein kinase C by lipid oxidation products. *Biochem. Biophys. Res. Commun.* **155**:1374–1380.

85. Akman, S. A., T. W. Kensler, J. H. Doroshow, and M. Diadaroglu. 1993. Copper ion-mediated modification of bases in DNA in vitro by benzoyl peroxide. *Carcinogenesis* **14**:1971–1974.

86. Li, Y., and M. A. Trush. 1994. Reactive oxygen-dependent DNA damage resulting from the oxidation of phenolic compounds by a copper-redox cycle mechanism. *Cancer Res.* **354**:1895s–1898s.

87. Ip, C., and H. E. Ganther. 1990. Activity of Methylated forms of selenium in cancer prevention. *Cancer Res.* **50**:1206–1211.

88. Tsen, C. C., and A. L. Tappel. 1958. Catalytic oxidation of glutathione and other sulfhydryl by selenite. **233**:1230–1232.

89. Ganther, H. E. 1968. Selenotrisulfides formation by the reaction of thiols with selenious acid. *Biochemistry* **7**:2898–2905.

90. Frenkel, G. D., and D. Falvey. 1988. Evidence for the involvement of sulfhdryl compounds in the inhibition of cellular DNA synthesis by selenite. *Mol. Pharmacol.* **34**:573–577.

91. Lu, Y. P., R. L. Chang, M. T. Huang, and A. H. Conney. 1993. Inhibitory effect of curcumin on 12-O-tetradecanoylphorbol-13-acetate-induced increase in ornithine decarboxylase mRNA in mouse epidermis. *Carcinogenesis* **14**:293–297.

92. Lu, Y. P., R. L. Chang, Y. R. Lou, M. T. Huang, H. L. Newmark, K. R. Reuhl, and A. H. Conney. 1994. Effect of Curcumin on 12-O-tetradecanoylphorbol-13-acetate and ultraviolet B light-induced expression of c-Jun and c-Fos in JB6 cells and in mouse epidermis. *Carcinogenesis* **15**:2363–2368.

93. Singh, S., and B. B. Aggarwal. 1995. Activation of transcriptional factor NF-κB is suppressed by curcumin (diferuloylmethane). *J. Biol. Chem.* **270**:24995–25000.

94. Lin, J. K., and C. A. Shih. 1994. Inhibitory effect of curcumin on xanthine dehydrogenase/oxidase induced by phorbol-12-myristate-13-acetate in NIH3T3 cells. *Carcinogenesis* **15**:1717–1721.

95. Lui, J. Y., S. J. Lin, and J. K. Lin. 1993. Inhibitory effects of curcumin on protein kinase C activity induced by 12-O-tetradecanoyl-phorbol-13-acetate in NIH 3T3 cells. *Carcinogenesis* **14**:857.

96. Sreejayan and M. N. Rao. 1994. Curcuminoids as potent inhibitors of lipid peroxidation. *J. Pharm. Pharmacol.* **46**:1013–1017.

97. Chignell, C. F., P. Bilski, K. J. Reszka, A. G. Motten, R. H. Sik, and T. A. Dahl. 1994. Spectral and photochemical properties of curcumin. *Photochem. Photobiol.* **59**:295–302.

98. Dahl, T. A., P. Bilski, K. J. Reszka, and C. F. Chignell. 1994. Photocytotoxicity of curcumin. *Photochem.* and *Photobiol.* **59**:290–294.

99. Gorman, A. A., I. Hamblett, V. S. Srinivasan, and P. D. Wood. 1994. Curcumin-derived transients: a pulsed laser and pulse radiolysis study. *Photochem. Photobiol.* **59**:389–398.

100. Srinivasan, K. R. 1953. A Chromatographic study of the curcuminoids in Curcuma Longa, L. *J. Pharm. Pharmacol.* **5**:448–457.

101. Ruby, A. J., G. Kuttan, K. D. Babu, K. N. Rajasekharan, and R. Kuttan. 1995. Antitumor and antioxidant activity of natural curcuminoids. *Cancer Lett.* **94**:79–83.

102. Rajakumar, D. V., and M. N. A. Rao. 1995. Antioxidant properties of phenyl styryl ketones. *Free Rad. Res.* **22**:309–317.

103. Chen, Z., U. Gundimeda, and R. Gopalakrishna. 1996. Curcumin irreversibly inactivates protein kinase C activity and phorbol ester binding: its possible role in cancer chemoprevention. *Proc. Am. Assoc. Cancer Res.* (abstract 1922):282.

104. Moreau, N., T. Martens, M. B. Fleury, and J. P. Leroy. 1990. Metabolism of oltipraz and glutathione reductase inhibition. *Biochem. Pharmacol.* **140**:1299–1305.

105. Langouet, S., B. Coles, F. Morel, L. Becquemont, P. Beaune, P. Guengerich, B. Ketter, and A. Guillouzo. 1995. Inhibition of CYP1A2 and CYP3A4 by oltipraz results in reduction of aflatoxin B1 metabolism in human hepatocytes in primary culture. *Cancer Res.* **55**:5574–5579.

106. Chavan, S. J., W. G. Bornmann, C. Flexner, and H. J. Prochaska. 1995. Inactivation of human immunodeficiency virus 1 reverse transcriptase by oltipraz: evidence for the formation of a stable adduct. *Arch. Biochem. Biophys.* **324**:143–152.

107. Gopalakrishna, R., Z. Chen, U. Gundimeda, and T. Kensler. 1996. Reversible inactivation of protein kinase C by oltipraz through reaction with the catalytic domain: role in inhibition of tumor promotion. *Proc. Am. Assoc. Cancer Res.* (abstract 1817):267.

108. Gopalakrishna, R., S. H. Barsky, P. T. Thomas, and W. B. Anderson. 1986. Factors influencing phorbol diester-induced membrane association of protein kinase C. *J. Biol. Chem.* **261**:16438–16445.

109. Gopalakrishna, R., and S. H. Barsky. 1986. Hydrophobic association of calpains with subcellular organelles—compartmentalization of calpains and the endogenous inhibitor calpastatin in tissues. *J. Biol. Chem.* **261**:13936–13942.

110. Van den Bosch, H., R. B. H. Schutgens, R. J. A. Wanders, and J. M. Tager. 1992. Biochemistry of peroxisomes. *Annu. Rev. Biochem.* **61**:157–197.

111. Gopalakrishna, R., Z. H. Chen, and U. Gundimeda. 1992. Irreversible inactivation of protein kinase C by photosensitive inhibitor calphostin C. *FEBS Lett.* **314**:149–154.

112. Stadtman, E. R., and C. N. Oliver. 1991. Metal catalyzed oxidation of proteins. *J. Biol. Chem.* **266**:2005–2008.

113. Doroshow, J. H. 1986. Role of hydrogen peroxide and hydroxyl radicle in the killing of Ehrlich tumor cells by anticancer quinones. *Proc. Natl. Acad. Sci. USA* **83**:4514–4518.

8

Tyrosine Phosphorylation in Oxidative Stress

Gary L. Schieven

8.1 Introduction

Tyrosine phosphorylation is a central regulatory mechanism for cells to transmit signals from cell surface receptors to the nucleus. Signaling pathways involving tyrosine phosphorylation are conserved in diverse organisms ranging from nematodes and fruit flies to humans.[1,2] Receptors for a wide variety of growth factors and hormones, including epidermal growth factor (EGF), platelet-derived growth factor (PDGF), insulin, and erthropoietin (EPO) induce cellular tyrosine phosphorylation either directly through intrinsic tyrosine kinase activity or by coupling to cytoplasmic tyrosine kinases (see Refs. 3 and 4 for review). Tyrosine kinases thus play key roles in growth and differentiation. Tyrosine kinases are also important in the functioning of the immune system. Lymphocyte antigen receptors are coupled to cytoplasmic tyrosine kinases, as are receptors for the cytokines that coordinate the response of the immune system.

Although tyrosine phosphorylation plays a key role in regulating cellular responses, phosphotyrosine represents only 0.1% of the protein phosphorylation in the cell.[5] In addition to the large number of tyrosine kinases, a variety of phosphotyrosine phosphatases (PTP) also act to regulate these pathways, maintaining the overall level of cellular tyrosine phosphorylation at a low level under normal circumstances. Tyrosine kinases can act as oncogenes if they are overexpressed or if they undergo mutations that increase their activity. For example, overexpression of the HER2 tyrosine kinase plays an important role in human breast and ovarian cancers and is linked to poor prognosis in patients.[6,7]

The tyrosine kinases and tyrosine phosphatases act in complex regulatory networks. The kinases themselves are regulated by tyrosine phosphorylation, and these phosphorylations can either activate or inhibit enzymatic activity depending on the particular enzyme and phosphorylation site involved. The PTP can also be tyrosine phosphorylated, and these phosphorylations may also alter enzymatic

181

activity. Oxidative stress in the form of oxidizing chemicals generated by the organism or the environment or in the form of radiation can alter the activity of these enzymes. Since the enzymes are linked in complex networks where both positive and negative regulation can occur, oxidative stress can activate as well as inhibit tyrosine kinase signal pathways.[8]

Although oxidative stress affects tyrosine phosphorylation and cell function in many cell types, ranging from cells of the immune system[8] to human sperm,[9] this chapter focuses on lymphocytes. B lymphocytes are responsible for antibody production, while T lymphocytes are the primary regulatory cells of the immune system and also have important cytotoxic function. Lymphocytes offer an excellent model system for studying the effects of oxidative stress on tyrosine kinase signal pathways because their responses are primarily regulated by tyrosine phosphorylation signal pathways,[10] while the cells themselves are highly sensitive to oxidative stress.[11,12] Two enzymes that counter oxidative stress, catalase and thioredoxin, act as autocrine growth factors for lymphocytes.[13,14] Lymphocytes are frequently exposed to oxidative stress during the course of their activities in defending against disease. They are exposed to H_2O_2 produced by neutrophils and monocytes at sites of inflammation, and this production of H_2O_2 leads to immunosuppression.[12,15] Oxidative stress occurs in rheumatoid arthritis[16] and increased levels of oxidative stress have been reported in pharmacologically immunosuppressed patients.[17] Lymphocytes are also exposed to oxidative stress as a result of exposure to ionizing and UV radiation during the course of medical therapy. Understanding the impact of oxidative stress on lymphocyte tyrosine phosphorylation thus offers new opportunities in the treatment and understanding of disease. Oxidative stress can, of course, damage cellular components and thus decrease cell responses. Numerous studies have shown that glutathione depletion and other instances of oxidative stress can lead to impaired lymphocyte responses.[18-21] However, recent work has shown that oxidative stress can activate tyrosine phosphorylation signal pathways, and this signaling plays an important role in the cell's responses to oxidative stress.

8.2 Receptor Tyrosine Kinases

Although the wide variety of receptors linked to tyrosine kinases are involved in many complex biological functions, most receptor-linked tyrosine phosphorylation pathways have a conserved set of components. The elucidation of these conserved elements since the identification of Src viral oncogene as the first known tyrosine kinase in 1980[22] has set the stage for recent studies describing the effects of oxidative stress on tyrosine phosphorylation pathways.

Receptors such as the insulin receptor, PDGF receptor, and the EGF receptor induce tyrosine phosphorylation via their intrinsic tyrosine kinase activity. The receptor protein tyrosine kinases (RPTK) contain an extracellular ligand binding domain, a transmembrane domain and a kinase domain. Ligand binding induces dimerization in this type of receptor, either due to the dimeric nature of the ligand, as in the case of PDGF, or via stabilization of receptor dimerization, as in the case of EGF.[3] The intracellular portions of the receptors then phosphorylate each other in *trans* at multiple sites.

This tyrosine phosphorylation of the receptor serves two purposes. First, tyrosine phosphorylation at particular sites within the catalytic domains activates the kinase activity. For example, the insulin receptor tyrosine kinase will remain activated even when insulin dissociates from the receptor, as long as it remains tyrosine phosphorylated.[23] Second, the tyrosine phosphorylation serves as scaffold for recruiting signaling molecules that contain SH2 domains. SH2 domains bind phosphotyrosine in the context of specific amino acid sequences.[24] Multiple signaling molecules containing SH2 domains are recruited to receptor tyrosine kinases in this manner, including Src-family kinases that are essential for mitogenesis induced by PDGF[25] phospholipase C-γ (PLC-γ), the p85 regulatory subunit of phosphatidylinositol 3′ kinase (PI-3 kinase), GTPase activating protein (GAP) that acts as a negative regulator of Ras, and adapter proteins such as Shc.[24,26] Recently, a second phosphotyrosine binding domain known as the PTB (phosphotyrosine binding domain) or PI (phosphotyrosine interaction) domain has been identified in Shc and other molecules, and this domain is believed to have a similar function.[27,28] Recruitment of SH2-containing proteins to the receptor leads them to be tyrosine phosphorylated by the RPTK, which can contribute to activation of enzymatic activity as is the case for PLC-γ. The binding via SH2 interactions can also contribute to enzymatic activation, as is the case for Src kinases (see below) and PI-3 kinase.[29,30]

Another important function of recruiting proteins via SH2 interactions is to localize the proteins to the plasma membrane. The activation of mitogen-activated protein kinase (MAP kinase) via the Ras pathway is the most important example of a pathway utilizing this function (see Ref. 31 for review) and is illustrated relative to antigen receptor signaling in Fig. 8-1. The adapter protein Grb2, which contains one SH2 domain and two SH3 domains, plays a key role in initiating this pathway.[32] Grb2 becomes associated with activated receptors through the binding of its SH2 domain to either the tyrosine phosphorylated receptors directly or via tyrosine phosphorylated adapter proteins such as Shc or p36 (recently identified as Lnk[33]). The SH3 domains of Grb2 bind to proline rich regions of Sos, which acts as a guanine nucleotide exchange factor for Ras. Ras is a small GTP binding protein that is active when it binds GTP and inactive when it binds GDP (see Ref. 32 for review). GAP induces the hydrolysis of GTP to GDP by Ras, serving to inactivate it. Oncogenic Ras, which is involved in many human cancers, is no longer responsive to GAP. Ras is constitutively associated with the inner surface of the plasma membrane by a covalently bound farnesyl lipid group, and Shc binding to RPTK brings Sos to the plasma membrane in proximity to Ras, so Ras can be activated. Activated Ras recruits the serine/threonine kinase

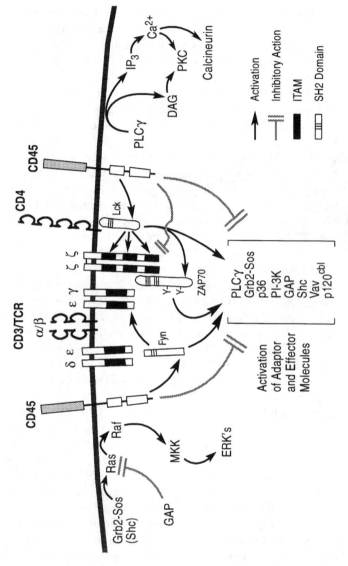

Figure 8.1 T cell receptor signal transduction.

Raf to the membrane, activating Raf. This initiates a kinase cascade of Raf phosphorylating and activating mitogen activated kinase kinase (MKK), which phosphorylates and activates extracellular signal-related kinases (ERKs), also called MAP kinase. The ERKs in turn can phosphorylate a variety of transcription factors, such as Elk1 which regulates c-*fos* expression.[32] Oxidative stress has been reported to induce Ras activation,[34] which in turn would be expected to alter gene expression by this pathway. Note that other pathways, including G-protein-coupled receptors as well as nonreceptor tyrosine kinases and additional kinases such as KSR-1,[1] also contribute to regulation of Ras and Raf responses.

Receptor tyrosine kinase signaling is negatively regulated by PTP, which dephosphorylate and inactivate the receptors and their substrates. The transmembrane PTP known as LAR has recently been found to negatively regulate a variety of receptors.[35] As described below, PTP can be inactivated by oxidation, making them a key site for oxidative stress to act on receptor tyrosine kinase signal pathways.

In addition to tyrosine phosphorylation signal pathways being acted on by oxidative stress, a recent study has demonstrated that PDGF receptor tyrosine kinase signal transduction involves the generation of H_2O_2.[36] Vascular smooth muscle cells were found to produce H_2O_2 in response to PDGF treatment, and blocking accumulation of H_2O_2 in the cells with the antioxidant *N*-acetyl cysteine or the enzyme catalase inhibited downstream PDGF responses such as accumulation of cellular tyrosine phosphorylation, activation of MAP kinase, and DNA synthesis. H_2O_2 inhibition of PTP that would normally oppose signaling by the receptor tyrosine kinase could explain these results.

8.3 Role of Cytoplasmic Tyrosine Kinases in Antigen Receptor Signaling

Many receptors that lack intrinsic tyrosine kinase activity are nonetheless capable of activating tyrosine kinase signal pathways by associating with cytoplasmic tyrosine kinases. Antigen-receptor signaling in lymphocytes is one of the best characterized signal pathways involving nonreceptor tyrosine kinases. The T-cell receptor consists of polymorphic α,β or γ,δ dimers that recognize antigen peptides presented by major histocompatibility complexes (MHC) I or II on antigen presenting cells, plus the invariant CD3 signal transducing chains γ,δ,ε and a dimer of ζ chains or an alternatively spliced η-ζ heterodimer (see Ref. 10 for review). The Src family tyrosine kinases Lck and Fyn are the first tyrosine kinases activated in antigen receptor signaling (Fig. 8.1). Fyn binds directly to the CD3 complex, while Lck is associated with the coreceptors CD4, CD8, and CD2.[10] CD8 and CD4 bind to MHC I and II, respectively, with CD8 being expressed in cytotoxic T cells and CD4 being expressed in T helper cells that provide help in the form of cytokines and cell surface ligands to B cells and monocytes of the immune system to lead to productive responses. Successful T-cell receptor binding of antigen results in the formation of a complex of CD4 or CD8 with the T-cell receptor and activation of Lck and Fyn.[10] This activation leads to the phosphorylation of specific sequences of approximately 26 residues in the CD3 chains known as ITAMS (immunoreceptor tyrosine activation motif) that contain dual tyrosine residues spaced approximately 10 residues apart.[37] Three ITAM sequences are found in the ζ chains. Phosphorylation of the dual tyrosines in ITAM sequences permits binding of the Syk-family tyrosine kinase ZAP-70, called as a zeta-associated protein of 70 kD, by its tandem SH2 domains. Although each SH2 domain alone gives weak binding, the two together give high-avidity binding.[38] Lck then phosphorylates ZAP-70 on tyrosine 493, activating the enzyme.[39] This leads to further phosphorylation of ZAP-70 on additional residues that permit it to participate in the phosphorylation of downstream signaling molecules and the formation of signaling complexes in conjunction with the phosphorylated ITAM sequences of the CD3 and ζ chains much as the receptor tyrosine kinases described above do (Fig. 8-1). One important component of such complexes is PLC-γ, which associates with the 36-kDa adapter protein Lnk.[33,40] Recruitment and activation of PLC-γ results in the hydrolysis of phosphatidylinositol-4,5-bisphosphate to generate diacylglycerol and inositol 1,4,5-trisphosphate (IP_3). Binding of IP_3 to its receptor results in mobilization of Ca^{2+} ions. Ca^{2+} acts with diacylglycerol to activate protein kinase C (PKC) and also acts to activate the calcium-dependent phosphatase calcineurin in an essential step of T-cell activation. Calcineurin is the target of the cyclosporine A/immunophilin complex and this inhibition of calcineurin is the mechanism by which the drug cyclosporine A exerts its potent immunosuppressive effects.[32]

B-cell receptor signaling is similar to T-cell receptor (TCR) signaling, but with differences in the enzymes involved. The first kinases to be activated are the Src-family kinases Lyn, Blk, Fyn, and Lck. These kinases phosphorylate ITAM sequences in Igα and β as well as CD22.[37,41] Syk is directly activated by binding to the phosphorylated ITAM sequences and differs from ZAP-70 in that it does not require additional phosphorylation by other kinases.[42]

PTP, in particular the transmembrane phosphatase CD45, provide both positive and negative regulation of lymphocyte signaling. Expression of CD45 is required for T-cell receptor signal transduction because it is required for the activation of the Src-family kinases Lck and Fyn (see Ref. 10 for review). In fact, all signal transduction involving Src-family kinases may be positively regulated by PTP because of opposing effects of tyrosine phosphorylation at two different sites in these molecules (see Refs. 43 and 44 for review). Src kinases are activated by phosphorylation of a conserved tyrosine in the active site (Y394 in Lck) and are inhibited by phosphorylation of a conserved tyrosine at the C terminus (Y505 in Lck) (Fig. 8.2). Members of the Csk family of tyrosine kinases are responsible for the phosphorylation of the C terminal (see Ref. 44 for review). Tyrosine

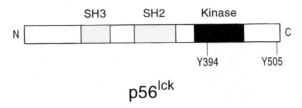

Figure 8.2. The Lck tyrosine kinase

phosphorylation at the C terminus is believed to inhibit the Src family enzymes by binding the SH2 domain.[45] CD45 dephosphorylates Src-family kinases such as Lck and Fyn and thereby permits them to be activated.[46] However, CD45 also negatively regulates T-cell signaling by dephosphorylating the ζ chain[47] and the 36-kDa adapter protein recently identified as Lnk associated with PLC-γ1 and other signaling proteins.[48,49] CD45 can also negatively regulate insulin receptor signaling.[50] PTP1C (also known as HCP), an SH2 domain containing phosphatase, negatively regulates signaling by a variety of receptors, including the B-cell antigen receptor and the erythropoietin receptor.[46] PTP1C, however, is a positive effector for Fas, which induces programmed cell death on binding of the Fas ligand.[51] *Motheaten* mice are defective in PTP1C and suffer from autoimmune disease and severe inflammation that gives rise to their name.

8.4 Activation of Tyrosine Phosphorylation Signal Pathways by Oxidative Stress

H_2O_2 has been employed as a pharmacologic agent to investigate how oxidative stress affects tyrosine phosphorylation signal pathways. H_2O_2 is also naturally generated at sites of inflammation. H_2O_2 is an insulin mimetic agent that increases the tyrosine phosphorylation of the insulin receptor and other cellular proteins, promotes glucose transport and oxidation, and stimulates lipogenesis and glycogen synthase.[52,53] H_2O_2 induces tyrosine phosphorylation in both T and B lymphocytes in patterns that are similar to those observed after antigen receptor stimulation.[54–56] Ca^{2+} signals are also induced within seconds in both T and B cells by H_2O_2 in a dose-dependent manner[54,56] with maximal signaling observed at a concentration of approximately 7.5 mM. As described above, IP_3 binding to Ca^{2+} channels is the basis of the Ca^{2+} signals observed following antigen receptor or growth factor stimulation,[57] and H_2O_2 was found to induce IP_3 production within 10 s of treatment.[56] Both the IP_3 production and the Ca^{2+} signals were blocked by tyrosine kinase inhibitors,[54,56] indicating that the oxidative stress induced by H_2O_2 acted via tyrosine kinase signal pathways similar to those used by the antigen receptors. However, T cells lacking antigen receptor expression still gave Ca^{2+} signals of similar magnitude as normal cells, but with slower kinetics.[55]

These results demonstrate that oxidative stress is able to bypass normal receptor control to activate a key cellular signal pathway.

These similar patterns of tyrosine phosphorylation and Ca^{2+} signaling observed for antigen receptor stimulation and H_2O_2 treatment suggest that both responses involved common signal transducing elements. Examination of cellular tyrosine kinases revealed that the Syk-family kinases Syk and ZAP-70 are highly responsive to H_2O_2 in B and T lymphocytes, respectively, whereas the Src-family kinases Lck, Fyn, Lyn, and Yes are not.[54-56] In Ramos B cells, Syk was rapidly activated by 5 to 10 mM H_2O_2 to levels equal to those observed following antigen receptor stimulation, whereas no activation of the Src-family kinases Lck, Fyn, and Lyn was observed.[54] Subsequent studies have indicated that Syk is also responsive to oxidative stress in other cell types, including neutrophils undergoing a respiratory burst response,[58] and in erythrocytes where Syk may regulate glycolysis.[59] Additional insights into the role of oxidative stress in the regulation of this class of enzymes have been obtained from studies of the T cell line Jurkat, where ZAP-70 is activated within 5 min by 2.5-mM H_2O_2, with maximal response observed at 7.5-mM H_2O_2, a dose response that paralleled the overall cellular tyrosine phosphorylation response.[55] Strikingly, the level of ZAP-70 activation in the cells treated with H_2O_2 was much stronger than was obtained following a maximal stimulation of the CD3/TCR complex with a monoclonal antibody. Interestingly, even in cells defective for CD3/TCR expression, some ZAP-70 could be activated by H_2O_2. Furthermore, H_2O_2 could also induce Ca^{2+} signals in CD3/TCR negative cells, although with slower kinetics than wild-type cells. These results indicate that H_2O_2 is able to activate ZAP-70 and downstream signaling pathways such as Ca^{2+} mobilization by both receptor dependent and independent means. Thus oxidative stress is able to bypass normal receptor control of the Syk family to activate these kinases.

The Syk family kinases are not directly activated by oxidative stress, since direct treatment of immunoprecipitated Syk by H_2O_2 had no effect on enzyme activity, nor did treatment with dithiothreitol (DTT) of Syk immunoprecipitated from cells treated with H_2O_2.[54] These results indicate that oxidative stress must act on other cellular components that regulate the Syk-family kinases.

H_2O_2 is known to inhibit PTP such as CD45,[60] and this inhibition provides an underlying mechanism for the activation of Syk-family kinases by H_2O_2 and the triggering of tyrosine phosphorylation signal pathways in a manner bypassing receptor control. The pattern of activating Syk kinases but not the Src-family kinases by oxidative stress would be expected if inhibition of PTP were responsible. Syk kinases are activated by tyrosine phosphorylation and by the tyrosine phosphorylation of ITAM sequences that activate the Syk kinases by binding to the Syk SH2 domains. The inhibition of PTP can permit these phosphorylations to accumulate in the cell, leading to Syk activation. However, the Syk kinases lack the negative regulatory phosphorylation site that is found at the C terminus of Src-family kinases (Tyr 505 in Lck). Thus, whereas Src kinases may not be readily activated by PTP inhibition because PTP activity is needed to activate them, the Syk kinases do not face this limitation. In cells lacking the phosphotyrosine phosphatase CD45, H_2O_2 gave stronger activation of ZAP-70 and greater phosphorylation of the associated ζ chain, suggesting that CD45 can oppose the action of H_2O_2.[55] Receptor tyrosine kinases may also be activated by such phosphatase inhibition. The insulin receptor is activated both by H_2O_2[52,53] and by the PTP inhibitor vanadate.[53,61]

Phosphotyrosine phosphatases are inactivated by reactive oxygen species because they contain a highly reactive cysteine residue in their active site that is essential for catalytic activity (see Ref. 62 for review). The location of this cysteine in the active site has been determined by X-ray crystallography.[63] This cysteine residue (Cys 215 in PTP1B) has an extremely low pK_a of approximately 4.7 due to its local environment, as compared to a usual pK_a value of 8.3 for cysteine.[64] This makes the cysteine highly reactive as a nucleophile. During the enzymatic reaction, this cysteine residue acts as the phosphoacceptor, and a phosphocysteine intermediate has been detected.[65] The phosphocysteine is hydrolyzed to regenerate the free cysteine. Oxidation of the essential cysteine residue thus blocks the enzymatic activity of the phosphatase. In addition to H_2O_2, nitric oxide and superoxide anion are also produced at sites of inflammation and these two compounds can react to form peroxynitrite. Recently, generation of nitric oxide by macrophages has also been found to inhibit PTP activity.[66]

Compounds that have the potential to selectively react with PTP active sites can increase cellular tyrosine phosphorylation to extraordinarily high levels relative to any known biological stimulation. Phenylarsine oxide, which reacts with vicinal sulfhydryl groups, rapidly induces high levels of tyrosine phosphorylation in a variety of cells.[67,68] Pervanadate, formed from the reaction of hydrogen peroxide and vanadate, induces extremely high levels of tyrosine phosphorylation by combining inhibition of PTP with the activation of tyrosine kinases that include the Src family kinases.[56,69,70]

In addition to inhibiting PTP activity, oxidative stress has been found to induce the expression of the phosphatase identified as CL100 in humans and MAP kinase phosphatase 1 (MKP-1) in the mouse.[71] Both H_2O_2 and UV irradiation induce the expression of this phosphatase. MKP-1/CL100 specifically dephosphorylates the phosphotyrosine and phosphothreonine residues in the MAP kinases or ERKs that are required for activation of these enzymes. MKP-1 also acts on the corresponding residues of the related c-*jun* N-terminal kinase (JNK), which is a member of the stress-activated protein kinase (SAPK) family.[72] JNK activates c-*jun* and thus plays an important role in the activity of the transcription factor AP-1, of which c-*jun* is a component. Induction of expression of the MKP-1 phosphatase by oxidative stress offers a mechanism to modulate signals generated through the ERK and JNK pathways by oxidative stress. The physiological relevance of this induction of MKP-1 by oxidative stress has been demonstrated by the finding that it occurs in antibody induced glomerulonephritis, an acute immune injury of the kidney that involves oxidative stress.[73]

8.5 Ionizing Radiation Induces Altered Gene Expression and Apoptosis via Tyrosine Kinase Signal Pathways

Ionizing radiation generates reactive oxygen species by the radiolysis of bound and solvent water in the cell producing hydroxyl radicals, hydrated electrons, and H_2O_2.[74] The reaction of hydrated electrons with oxygen molecules leads to the production of superoxide anions, which in turn generate H_2O_2 by dismutation.[74] Lymphocytes are an excellent model system for studying the effects of ionizing radiation on cells because of their high degree of sensitivity to radiation. Irradiation of lymphocytes leads to cell cycle arrest and programmed cell death, and these responses form the basis of the use of irradiation followed by bone marrow transplantation in the treatment of leukemia and lymphoma.[75]

Ionizing radiation is well known to induce DNA damage, but recent work has shown that the action of ionizing radiation on signal transduction pathways is responsible not only for the induction of new gene expression, but also for radiation-induced programmed cell death in leukemic cells. Initial evidence for the role of cytoplasmic signal pathways in the regulation of gene expression by ionizing radiation came from studies showing that radiation increases expression of the immediate early genes c-*jun* and *EGRI*[76] via a pathway dependent on radiation-induced activation of PKC.[77] Subsequently, it was found that ionizing radiation in therapeutically relevant doses of 50 to 400 cGy induces cellular tyrosine phosphorylation in leukemic B-cell precursors.[78] This radiation-induced tyrosine phosphorylation was blocked by the tyrosine kinase inhibitors genistein and herbimycin, but it was strongly enhanced by the PTP inhibitor sodium orthovanadate.[78] Kinase assays demonstrated that tyrosine kinases were activated by the ionizing radiation. Further studies have revealed that both Lck[79] and Lyn[80] can be activated by ionizing radiation. The tyrosine kinase inhibitor genistein was found to inhibit radiation-induced apoptosis, whereas the PTP inhibitor sodium orthovanadate, which alone had little effect, greatly increased the radiation-induced death of leukemic B-cell precursors.[78] Tyrosine kinase inhibitors such as herbimycin were found to block the activation of protein kinase C and multiple additional serine kinases,[81] and also blocked the induction of c-*jun*.[82] Tyrosine kinase inhibitors can also block radiation-induced transcription of other immediate early genes such as c-*fos*, c-*myc*, and c-*Ha-ras*.[83] In addition to the Src-family kinases linked to antigen receptor signaling, ionizing radiation can also activate the c-Abl tyrosine kinase, and Abl is required for the radiation induced activation of the stress activated protein kinase JNK.[84]

These results indicate that radiation activates tyrosine kinases leading to a kinase cascade, altered gene expression, and programmed cell death in leukemic B cells, whereas PTP serve to limit the response. The important role for PTP in B cell apoptosis was further elucidated by the finding that a PTP inhibitor with much greater activity on cells than orthovanadate, bis(maltolato)oxovanadium(IV), activates signaling pathways involving Syk and PLC-γ1 leading to programmed cell death in leukemic immature B cells.[85]

Why should activation of these signal pathways lead to programmed cell death? The answer lies in the particular biology of B-cell development. Immature B cells are programmed to die when their antigen receptors bind strongly to self-antigens as part of the process of negative selection, which deletes autoreactive cells that could induce autoimmune disease if they were allowed to mature.[86] This process requires the same signal transduction pathway that in mature B cells would lead to proliferation. The high degree of sensitivity of the leukemic immature B cells to ionizing radiation that permits the use of radiation therapy in the treatment of such diseases is thus due at least in part to the ability of ionizing radiation to bypass normal receptor control to activate the signal pathways that lead to cell death. However, in other cell types such as fibroblasts, the induction of c-*jun* and *Egr-1* expression by ionizing radiation is a protective response that promotes cell survival.[87]

8.6 Ultraviolet Radiation

Although ultraviolet radiation is well known to induce DNA damage by the production of thymidine dimers, recent studies have shown that UV also acts on tyrosine kinase signaling pathways. Lymphocytes are highly sensitive to UVB (302 nm) and UVC (252 nm), whereas the longer wavelength UVA radiation has little effect on lymphocyte biologic responses.[88] Strikingly, both UVB and UVC, but not UVA, irradiation strongly induce tyrosine phosphorylation in T and B lymphocytes.[89] UV irradiation also acts on Ca^{2+} mobilization, but in a cell type specific manner. In normal human peripheral blood lymphocytes, both $CD4^+$ and $CD8^+$ lymphocytes respond to UV irradiation with strong Ca^{2+} signaling, but other leukocytes isolated from blood, including B lymphocytes, are not responsive.[89,90] In the T cells, UV induces Ca^{2+} signaling via tyrosine phosphorylation of PLC-γ1 and associated proteins, and the Ca^{2+} mobilization is blocked by tyrosine kinase inhibitors.[89] UV induces activation of Syk in B cells[54] and ZAP-70 in T cells.[55]

An important clue to the mechanism by which UV acts to induce cellular tyrosine phosphorylation is that UV induction of tyrosine kinase activation and downstream cellular responses requires the surface expression of antigen receptor and CD45, whereas H_2O_2 does not require either to induce tyrosine phosphorylation responses.[55] Specifically, UV does not induce activation of ZAP-70, cellular tyrosine phosphorylation, or Ca^{2+} signaling in T-cell lines lacking surface expression of the TCR either by selection or by the internalization of CD3 induced by monoclonal antibodies.[55] By contrast, H_2O_2 can still induce signals in these cells. UV induces the association of activated ZAP-70 with the ε and ζ chains of the TCR, just as occurs in TCR induced signaling.[55] Similarly, the downstream response of NF-κB activation by UV is dependent on surface expression of the TCR.[55] UV also requires the CD45 expression that is required for receptor induced signaling, whereas H_2O_2 does not.[55] These results indicate that UV acts directly on tyrosine kinase linked receptors to induce signaling, whereas H_2O_2 is able to bypass receptors to act on intracellular molecules.

There are several possible mechanisms by which UV might activate receptors. UV might act directly on receptors or associated proteins to cross-link the receptors or otherwise induce aggregation. UV is known to act on cell membranes by causing lipid peroxidation,[91] and this might also lead to receptor aggregation. UV is able to activate receptors such as the EGF receptor within 30 s,[55] so the mechanism of action must accommodate the rapid activation of diverse protein molecules. Since receptor aggregation is a central activation mechanism of receptor tyrosine kinases, this is a likely pathway for UV.

UV induces the expression of a specific set of genes in what is termed the mammalian UV response.[92–94] This response is quite similar to the set of genes induced by growth factors that signal through receptor tyrosine kinases. In HeLa cells, this response includes the activation of the preexisting transcription factors AP-1 (c-Fos-c-Jun) and TCF/Elk-1 as well as transcriptional activation of the c-*fos* and c-*jun* genes (reviewed in Ref. 95). UV radiation can activate NF-κB in enucleated HeLa cells, demonstrating that DNA damage is not required for this response.[96] The mammalian UV response in HeLa cells requires the activity of Src-family kinases in a signaling cascade requiring active Ras and Raf.[97] In addition, active receptor tyrosine kinases such as the EGF receptor are required.[95] Since receptor tyrosine kinases can activate Src-family kinases via the binding of the SH2 domain of a Src kinase to a tyrosine phosphorylated receptor tyrosine kinase,[26,98] it is likely that Src-family kinase activation by UV is downstream of receptor activation. Thus UV acts through cell surface receptors to alter gene expression via tyrosine phosphorylation signaling pathways in many cell types. This response may promote cell survival by inducing gene expression that aids in the repair of cellular components damaged by irradiation.[97]

8.7 Conclusion

Oxidative stress can damage a variety of cellular components, and in general this leads to loss of function. Some cell types such as lymphocytes are highly sensitive to oxidative stress, and proteins that counter such stress, such as catalase and thioredoxin, can function to promote the growth of these cells. One enzyme class particularly susceptible to oxidative stress is the phosphotyrosine phosphatases, which have roles in both the negative and positive regulation of tyrosine kinase signaling pathways. Acting at least in part by inhibiting phosphotyrosine phosphatase function, oxidative stress can activate tyrosine kinase signaling pathways in a manner which bypasses normal receptor control, leading to altered gene expression. These signals have the potential to aid cell survival or to induce programmed cell death, depending on the specific cell type involved and the extent of the oxidative-stress-induced damage that has occurred.

References

1. Downward, J. 1995. KSR: A novel player in the RAS pathway. *Cell* **83**:831–834.

2. Schlessinger, J. 1993. How receptor tyrosine kinases activate Ras. *Trends Biochem. Sci.* **18**:273–275.

3. Ullrich, A., and J. Schlessinger. 1990. Signal transduction by receptors with tyrosine kinase activity. *Cell* **61**:203–212.

4. Ihle, J. N. 1995. The *Janus* protein tyrosine kinases in hematopoetic cytokine signaling. *Sem. Immunol.* **7**:247–254.

5. Cooper, J. A., B. A. Sefton, and T. Hunter. 1983. Detection and quantitation of phosphotyrosine in proteins. *Methods Enzymol.* **99**:387–402.

6. Slamon, D. J., W. Godolphin, L. A. Jones, J. A. Holt, S. G. Wong, D. E. Keith, W. J. Levin, S. G. Stuart, J. Udove, A. Ullrich, and M. F. Press. 1989. Studies of the HER-2/*neu* proto-oncogene in human breast and ovarian cancer. *Science* **244**:707–712.

7. Pietras, R. J., J. Arboleda, D. M. Reese, N. Wongvipat, M. D. Pegram, L. Ramos, C. M. Gorman, M. G. Parker, M. X. Sliwkowski, and D. J. Slamon. 1995. HER-2 tyrosine kinase pathway targets estrogen receptor and promotes hormone-independent growth in human breast cancer cells. *Oncogene* **10**:2435–2446.

8. Schieven, G. L., and J. A. Ledbetter. 1994. Activation of tyrosine kinase signaling pathways by radiation and oxidative stress. *Trends Endocrinol. Metab.* **5**:383–388.

9. Aitken, R. J., M. Paterson, H. Fisher, D. W. Buckingham, and M. Vanduin. 1996. Redox regulation of tyrosine phosphorylation in human spermatozoa and its role in the control of human sperm function. *J. Cell. Sci.* **108**:2017–2025.

10. Weiss, A., and D. R. Littman. 1994. Signal transduction by lymphocyte antigen receptors. *Cell* **76**:263–274.

11. Ames, B. N, M. K. Shigenaga, and T. M. Hagen. 1993. Oxidants, antioxidants, and the degenerative diseases of aging. *Proc. Natl. Acad. Sci. USA* **90**:7915–7922.

12. El-Hag, A., P. E. Lipsky, M. Bennett, and R. A. Clark. 1986. Immunomodulation by neutrophil myeloperoxidase and hydrogen peroxide: differential susceptibility of human lymphocyte functions. *J. Immunol.* **136**:3420–3426.

13. Sandstrom, P. A., and T. M. Buttke. 1993. Autocrine production of extracellular catalase prevents apoptosis of the human CEM T-cell line in serum-free medium. *Proc. Natl. Acad. Sci. USA* **90**:4708–4712.

14. Yodai, J., and T. Uchiyama. 1993. Diseases associated with HTLV-I: virus, IL-2 receptor dysregulation and redox regulation. *Immunol. Today* **13**:405–411.

15. El-Hag, A., and R. A. Clark. 1987. Immunosuppression by activated human neutrophils. *J. Immunol.* **139**:2406–2413.

16. Mapp, P. I., M. C. Grootveld, and D. R. Blake. 1995. Hypoxia, oxidative stress and rheumatoid arthritis. *Br. Med. Bull.* **51**:419–436.

17. de Lorgeril, M., M. J. Richard, J. Arnaud, P. Boissonnat, J. Guidollet, G. Dureau, S. Renaud, and A. Favier. 1994. Increased production of reactive oxygen species in pharmacologically-immunosuppressed patients. *Chem.-Biol Interact.* **91**:159–164.

18. Duncan, D. D. and D. A. Lawrence. 1989. Differential lymphocyte growth-modifying effects of oxidants: changes in cytosolic Ca^{2+}. *Toxicol. Appl. Pharmacol.* **100**:485–497.

19. Kanner, S. B., T. J. Kavanagh, A. Grossmann, S. L. Hu, J. B. Bolen, P. S. Ravinovitch, and J. A. Ledbetter. 1992. Sulfhydryl oxidation down-regulates T-cell signaling and inhibits tyrosine phosphorylation of phospholipase Cγ1. *Proc. Natl. Acad. Sci. USA* **89**:300–304.

20. Messina, J. P., and D. A. Lawrence. 1989. Cell cycle progression of glutathione-depleted human peripheral blood mononuclear cells is inhibited at S phase. *J. Immunol.* **143**:1974–1981.

21. Kavanagh, T. J., A. Grossmann, E. P. Jaecks, J. C. Jinneman, D. L. Eaton, G. M. Martin, and P. S. Rabinovtich. 1990. Proliferative capacity of human peripheral blood lymphocytes sorted on the basis of glutathione content. *J. Cell. Physiol.* **145**:472–480.

22. Hunter, T., and B. M. Sefton. 1980. Transforming gene product of Rous sarcoma virus phosphorylates tyrosine. *Proc. Natl. Acad. Sci. USA* **77**:1311–1315.

23. Rosen, O. M., R. Herrera, Y. Olowe, L. M. Petruzzelli, and M. H. Cobb. 1983. Phosphorylation activates the insulin receptor tyrosine protein kinase. *Proc. Natl. Acad. Sci. USA* **80**:3237–3240.

24. Songyang, Z., S. E. Shoelson, M. Chaudhuri, G. Gish, T. Pawson, W. G. Haser, F. King, T. Roberts, S. Ratnofsky, R. J. Lechleider, B. G. Neel, R. B. Birge, J. E. Fajardo, M. M. Chou, H. Hanafusa, B. Schaffhausen, and L. C. Cantley. 1994. SH2 domains recognize specific phosphopeptide sequences. *Cell* **72**:767–778.

25. Twamley-Sein, G. M., R. Pepperkok, W. Ansorge, and S. A. Courtneidge. 1993. The Src family tyrosine kinases are required for platelet-derived growth factor-mediated signal transduction in NIH 3T3 cells. *Proc. Natl. Acad. Sci. USA* **90**:7696–7700.

26. Koch, C. A., D. Anderson, M. F. Moran, C. Ellis, and T. Pawson. 1991. SH2 and SH3 domains: elements that control interactions of cytoplasmic signaling proteins. *Science* **252**:668–674.

27. Kavanaugh, W. M., C. W. Turck, and L. T. Williams. 1995. PTB domain binding to signaling proteins through a sequence motif containing phosphotyrosine. *Science* **268**:1177–1179.

28. Bork, P. and B. Margolis. 1995. A phosphotyrosine interaction domain. *Cell* **80**:693–694.

29. Carpenter, C. L., K. R. Auger, M. Chanudhuri, M. Yoakim, B. Schaffhausen, S. Shoelson, and L. C. Cantley. 1993. Phosphoinositide 3-kinase is activated by phosphopeptides that bind to the SH2 domains of the 85-kDa subunit. *J. Biol. Chem.* **268**:9478–9483.

30. Shoelson, S. E., M. Sivaraja, K. P. Wiliams, P. Hu, J. Schlessinger and M. A. Weiss. 1993. Specific phosphopeptide binding regulates a conformational change in the PI 3-kinase SH2 domain associated with enzyme activation. *EMBO J.* **12**:795–802.

31. McCormick, F. 1993. How receptors turn Ras on. *Nature* **363**:15–16.

32. Pastor, M. I., K. Reif, and D. Cantrell. 1995. The regulation and function of p21*ras* during T-cell activation and growth. *Immunol. Today* **16**:159–164.

33. Huang, X., Y. Li, K. Tanaka, K. G. Moore, and J. I. Hayashi. 1995. Cloning and characterization of Lnk, a signal transduction protein that links T-cell receptor activation signal to phospholipase Cγ₁, Grb2, and phosphatidylinositol 3-kinase. *Proc. Natl. Acad. Sci. USA* **92**:11618–11622.

34. Lander, H. M., J. S. Ogiste, K. K. Teng, and A. Novogrodsky. 1995. p21*ras* as a common signaling target of reactive free radicals and cellular redox stress. *J. Biol. Chem.* **270**:21195–21198.

35. Kulas, D. T., B. J. Goldstein, and R. A. Mooney. 1996. The transmembrane protein-tyrosine phosphatase LAR modulates signaling by multiple receptor tyrosine kinases. *J. Biol. Chem.* **271**:748–754.

36. Sundarensan, M., Z. X. Yu, V. J. Ferrans, K. Irani, and T. Finkel. 1995. Requirement for generation of H₂O₂ for platelet-derived growth factor signal transduction. *Science* **270**:296–299.

37. Cambier, J. C. 1995. Antigen and Fc receptor signaling. *J. Immunol.* **155**:3281–3285.

38. Hatada, M. H., X. Lu, E. R. Laird, J. Green, J. P. Morgenstern, M. Lou, C. S. Marr, T. B. Phillips, M. K. Ram, K. Theriault, M. J. Zoller, and J. L. Karas. 1995. Molecular basis for interaction of the protein tyrosine kinase ZAP-70 with the T-cell receptor. *Nature* **377**:32–38.

39. Chan, A. C., M. Dalton, R. Johnson, G. Kong, T. Wang, R. Thoma, and T. Kurosaki. 1995. Activation of ZAP-70 kinase activity by phosphorylation of tyrosine 493 is required for lymphocyte antigen receptor function. *EMBO J.* **14**:2499–2508.

40. Gilliland, L. K., G. L. Schieven, N. Norris, S. B. Kanner, A. Aruffo, and J. A. Ledbetter. 1992. Lymphocyte lineage-restricted tyrosine phosphorylated proteins that bind PLCγ1 SH2 domains. *J. Biol. Chem.* **267**:13610–13616.

41. Law, C. L., S. P. Sidorenko, and E. A. Clark. 1994. Regulation of lymphocyte activation by the cell-surface molecule CD22. *Immunol. Today* **15**:442–449.

42. Kolanus, W., C. Romeo, and B. Seed. 1993. T cell activation by clustered tyrosine kinases. *Cell* **74**:171–183.

43. Bolen, J. B., R. B. Rowley, C. Spana, and A. Y. Tsygankov. 1992. The Src family of tyrosine protein kinases in hemopoietic signal transduction. *FASEB J.* **6**:3403–3409.

44. Chow, L.M.L., and A. Veillette. 1995. The Src and Csk families of tyrosine protein kinases in hemopoietic cells. *Semin. Immunol.* **7**:207–226.

45. Cooper, J. A., and B. Howell. 1993. The when and how of Src regulation. *Cell* **73**:1051–1054.

46. Thomas, M. L. 1995. Positive and negative regulation of leukocyte activation by protein tyrosine phosphatases. *Semin. Immunol.* **7**:279–288.

47. Furukawa, T., M. Itoh, N. X. Krueger, M. Streuli and H. Saito. 1994. Specific interaction of the CD45 protein-tyrosine phosphatase with tyrosine-phosphorylated CD3 zeta chain. *Proc. Natl. Acad. Sci. USA* **91**:10928–10932.

48. Ledbetter, J. A., G. L. Schieven, F. M. Uckun, and J. B. Imboden. 1991. CD45 cross-linking regulates phospholipase C activation and tyrosine phosphorylation of specific substrates in CD3/Ti stimulated cells. *J. Immunol.* **146**:1577–1583.

49. Kanner, S. B., J. P. Deans, and J. A. Ledbetter. 1992. Regulation of CD3-induced phospholipase-Cγ-1 (PLCγ-1) tyrosine phosphorylation by CD4 and CD45 receptors. *Immunology* **75**:441–447.

50. Kulas, D. T., G. G. Freund, and R. A. Mooney. 1996. The transmembrane protein-tyrosine phosphatase CD45 is associated with decreased insulin receptor signaling. *J. Biol. Chem.* **271**:755–760.

51. Su, X., T. Zhou, Z. Wang, P. Yang, R. S. Jope, and J. D. Mountz. 1995. Defective expression of hematopoietic cell protein tyrosine phosphatase (HCP) in lymphoid cells blocks Fas-mediated apoptosis. *Immunity* **2**:353–362.

52. Heffetz, D., and Y. Zick. 1989. H_2O_2 potentiates phosphorylation of novel putative substrates for the insulin receptor kinase in intact Fao cells. *J. Biol. Chem.* **264**:10126–10132.

53. Heffetz, D., I. Bushkin, R. Dror, and Y. Zick. 1990. The insulinomimetic agents H_2O_2 and vanadate stimulate protein tyrosine phosphorylation in intact cells. *J. Biol. Chem.* **265**:2896–2902.

54. Schieven, G. L., J. M. Kirihara, D. L. Burg, R. L. Geahlen, and J. A. Ledbetter. 1993. p72[syk] tyrosine kinase is activated by oxidizing conditions which induce lymphocyte tyrosine phosphorylation and Ca^{2+} signals. *J. Biol. Chem.* **268**:16688–16692.

55. Schieven, G. L., R. S. Mittler, S. G. Nadler, J. M. Kirihara, J. B. Bolen, S. B. Kanner, and J. A. Ledbetter. 1994. ZAP-70 tyrosine kinase, CD45 and T cell receptor involvement in UV and H_2O_2 induced T cell signal transduction. *J. Biol. Chem.* **269**:20718–20726.

56. Schieven, G. L., J. M. Kirihara, D. E. Myers, J. A. Ledbetter, and F. M. Uckun. 1993. Reactive oxygen intermediates activate NF-κB in a tyrosine kinase dependent mechanism and in combination with vanadate activate the p56[lck] and p59[fyn] tyrosine kinases in human lymphocytes. *Blood* **82**:1212–1220.

57. Berridge, M. J. and R. F. Irvine. 1984. Inositol trisphosphate, a novel second messenger in cellular signal transduction. *Nature* **312**:315–321.

58. Brumell, J. H., A. L. Burkhardt, J. B. Bolen, and S. Grinstein. 1996. Endogenous reactive oxygen intermediates activate tyrosine kinases in human neutrophils. *J. Biol. Chem.* **271**:1455–1461.

59. Low, P. S., A. Kiyatkin, Q. Li, and M. L. Harrison. 1995. Control of erythrocyte metabolism by redox-regulated tyrosine phosphatases and kinases. *Protoplasma* **184**:196–202.

60. Hecht, D. and Y. Zick. 1992. Selective inhibition of protein tyrosine phosphatase activities by H_2O_2 and vanadate in vitro. *Biochem. Biophys. Res. Commun.* **188**:773–779.

61. Fantus, I. G., F. Ahmad, and G. Deragon. 1994. Vanadate augments insulin stimulated insulin receptor kinase activity and prolongs insulin action in rat adipocytes: evidence for transduction of amplitude of signaling into duration of response. *Diabetes* **43**:375–383.

62. Barford, D. 1995. Protein phosphatases. *Curr. Opin. Struct. Biol.* **5**:728–734.

63. Barford, D., A. J. Flint, and N. K. Tonks. 1994. Crystal structure of human protein tyrosine phosphatase 1B. *Science* **263**:1397–1404.

64. Zhang, Z. Y., and J. E. Dixon. 1993. Active site labeling of the *Yersinia* protein tyrosine phosphatase: the determination of the pKa of the active site cysteine and the function of the conserved histidine 402. *Biochemistry* **32**:9340–9345.

65. Guan, K. L., and J. E. Dixon. 1993. Evidence of protein-tyrosine-phosphatase catalysis proceeding via a cysteine-phosphate intermediate. *J. Biol. Chem.* **266**:17026–17030.

66. Caselli, A., P. Chiarugi, G. Camici, G. Manao, and G. Ramponi. 1995. In vivo inactivation of phosphotyrosine protein phosphatases by nitric oxide. *FEBS Lett.* **374**:249–252.

67. Frost, S. C., and M. D. Lane. 1985. Evidence for the involvement of vicinal sulfhydryl groups in insulin-activated hexose tranport by 3T3-L1 adipocytes. *J. Biol. Chem.* **260**:2646–2652.

68. Garcia-Morales, P., Y. Minami, E. Luong, R. D. Klausner, and L. E. Samelson. 1990. Tyrosine phosphorylation in T cells is regulated by phosphatase activity: studies with phenylarsine oxide. *Proc. Natl. Acad. Sci. USA* **87**:9255–9259.

69. Secrist, J. P., Burns, L. A., L. Karnitz, G. A. Koretzky, and R. T. Abrahams. 1993. Stimulatory effects of the protein tyrosine phosphatase inhibitor, pervanadate, on T-cell activation events. *J. Biol. Chem.* **268**:5886–5893.

70. O'Shea, J. J., D. W. McVicar, T. L. Bailey, C. Burns, and M. J. Smyth. 1992. Activation of human peripheral blood T lymphocytes by pharmacological induction of protein-tyrosine phosphorylation. *Proc. Natl. Acad. Sci. USA* **89**:10306–10310.

71. Keyse, S. M., and E. A. Emslie. 1992. Oxidative stress and heat shock induce a human gene encoding a protein tyrosine phosphatase. *Nature* **359**:644–647.

72. Liu, Y., M. Gorospe, C. Yang, and N. J. Holbrook. 1995. Role of mitogen-activated protein kinase phosphatase during the cellular response to genotoxic stress. *J. Biol. Chem.* **270**:8377–8380.

73. Feng, L. L., Y. Y. Xia, D. Seiffert, and C. B. Wilson. 1995. Oxidative stress-inducible protein tyrosine phosphatase in glomerulonephritis. *Kidney Int.* **48**:1920–1928.

74. Wardman, P. 1983. Principles of radiation chemistry, p. 51. *In* G.G. Steel and G.E. Adams, (eds.), *The Biological Basis of Radiotherapy*. Elsevier, New York.

75. Hall, E. J., M. Astor, J. Bedford, C. Borek, S. B. Curtis, M. Fry, C. Geard, T. Hei, J. Mitchell, and N. Oleinick. 1988. Basic radiobiology. *Am. J. Clin. Oncol.* **11**:220–252.

76. Sherman, M. L., R. Datta, D. E. Hallahan, R. R. Weichselbaum, and D. W. Kufe. 1990. Ionizing radiation regulates expression of the c-*jun* protooncogene. *Proc. Natl. Acad. Sci. USA* **87**:5663–5666.

77. Hallahan, D. E., V. P. Sukhatme, J. L. Sherman, S. Virudachalam, D. Kufe, and R. R. Weichselbaum. 1991. Protein kinase C mediates X-ray inducibility of nuclear signal transducers EGR1 and JUN. *Proc. Natl. Acad. Sci. USA* **88**:2156–2160.

78. Uckun, F. M., L. Tuel-Ahlgren, C. W. Song, K. Waddick, D. E. Myers, J. Kirihara, J. A. Ledbetter, and G. L. Schieven. 1992. Ionizing radiation stimulates unidentified tyrosine-specific protein kinases in human B-lymphocyte precursors triggering apoptosis and clonogenic cell death. *Proc. Natl. Acad. Sci. USA* **89**:9005–9009.

79. Waddick, K. G., H.P. Chae, L. Tuel-Ahlgren, L. J. Jarvis, I. Dibirdik, D. E. Myers, and F. M. Uckun. 1993. Engagement of the CD19 receptor on human B-lineage leukemia cells activates LCK tyrosine kinase and facilitates radiation-induced apoptosis. *Radiat. Res.* **136**:313–319.

80. Kharbanda, S., Z. M. Yuan, E. Rubin, R. Weichselbaum, and D. Kufe. 1994. Activation of Src-like p56/p53*lyn* tyrosine kinase by ionizing radiation. *J. Biol. Chem.* **269**:20739–20743.

81. Uckun, F. M., G. L. Schieven, L. M. Tuel-Ahlgren, I. Dibirdik, D. E. Myers, J. A. Ledbetter, and C. W. Song. 1993. Tyrosine phosphorylation is a mandatory proximal step in radiation-induced activation of the proten kinase C signaling pathway in human B-lymphocyte precursors. *Proc. Natl. Acad. Sci. USA* **90**:252–256.

82. Chae, H. P. L. J. Jarvis and F. M. Uckun. 1993. Role of tyrosine phosphorylation in radiation-induced activation of c-*jun* protooncogene in human lymphohematopoietic precursor cells. *Cancer Res.* **53**:447–451.

83. Prasad, A. V., N. Mohan, B. Chandrasekar, and M. L. Meltz. 1995. Induction of transcription of "immediate early genes" by low-dose ionizing radiation. *Radiat. Res.* **143**:263–272.

84. Kharbanda, S. R. B. Ren, P. Pandey, T. D. Shafman, S. M. Feller, R. R. Weichselbaum, and D. W. Kufe. 1995. Activation of the c-Abl tyrosine kinase in the stress response to DNA-damaging agents. *Nature* **376**:785–788.

85. Schieven, G. L., A. F. Wahl, S. Myrdal, L. Grosmaire, and J. A. Ledbetter. 1995. Lineage-specific induction of B cell apoptosis and altered signal transduction by the phosphotyrosine phosphatase inhibitor bis(maltolato)oxovanadium(IV). *J. Biol. Chem.* **270**:20824–20831.

86. Cohen, J. J., R. C. Duke, V. A. Fadok, and K. S. Sellins. 1992. Apoptosis and programmed cell death in immunity. *Annu. Rev. Immunol.* **10**:267–293.

87. Hallahan, D. E., E. Dunphy, S. Virudachalam, V. P. Sukhatme, D. W. Kufe, and R. R. Weichselbaum. 1995. c-*jun* and *EGR-1* participate in DNA synthesis and cell survival in response to ionizing radiation exposure. *J. Biol. Chem.* **270**:30303–30309.

88. Kripke, M. L. 1984. Immunological unresponsiveness induced by ultraviolet radiation. *Immunol. Rev.* **80**:87–102.

89. Schieven, G. L., J. M. Kirihara, L. K. Gilliland, F. M. Uckun, and J. A. Ledbetter. 1993. Ultraviolet radiation rapidly induces tyrosine phosphorylation and calcium signaling in lymphocytes. *Mol. Biol. Cell* **4**:523–530.

90. Schieven, G. L., and J. A. Ledbetter. 1993. Ultraviolet radiation induces differential calcium signals in human peripheral blood lymphocyte subsets. *J. Immunother.* **14**:221–225.

91. Black, H. S. 1987. Potential involvement of free radical reactions in ultraviolet light-mediated cutaneous damage. *Photochem. Photobiol.* **46**:213–221.

92. Ronai, Z. A., E. Lambert, and I. B. Weinstein. 1990. Inducible cellular responses to ultraviolet light irradiation and other mediators of DNA damage in mammalian cells. *Cell Biol. Toxicol.* **6**:105–126.

93. Holbrook, N. J., and A. J. Fornace, Jr. 1991. Response to adversity: molecular control of gene activation following genotoxic stress. *New Biol* **3**:825–833.

94. Karin, M., and P. Herrlich. 1989. Cis and transacting genetic elements responsible for induction of specific genes by tumor promoters, serum fctors and stress, pp. 415–440. *In* N.H. Colburn, (ed.), *Genes and Signal Transduction in Multistage Carcinogenesis.* Marcel Dekker, New York.

95. Sachsenmaier, C., A. Radler-Pohl, R. Zinck, A. Nordheim, P. Herrlich, and H. J. Rahmsdorf. 1994. Involvement of growth factor receptors in the mammalian UV response. *Cell* **78**:963–972.

96. Devary, Y., C. Rosette, J. A. Didonato and M. Karin. 1993. NF-κB activation by ultraviolet light not dependent on a nuclear signal. *Science* **261**:1442–1445.

97. Devary, Y., R. A. Gottlieb, T. Smeal, and M. Karin. 1992. The mammalian ultraviolet response is triggered by activation of src tyrosine kinases. *Cell* **71**:1081–1091.

98. Kypta, R. M., Y. Goldberg, E. T. Ulug, and S. A. Courtneidge. 1990. Association between the PDGF receptor and members of the src family of tyrosine kinases. *Cell* **62**:481–492.

9

Reactive Oxygen Intermediates as Signaling Molecules Regulating Leukocyte Activation

Léa Fialkow and Gregory P. Downey

9.1 Introduction

The purpose of the inflammatory process is to combat infection by pathogenic microorganisms. The primary effectors of this response are leukocytes including neutrophils, monocytes, and macrophages. Of necessity, these cells have evolved many properties that facilitate their effective function in inflammation including the ability to move to the site of inflammation (chemotaxis) and to ingest and kill pathogens by release of toxic products including proteolytic enzymes, reactive oxygen intermediates (ROI), and cationic proteins. Leukocytes have also evolved the ability to respond to signals released in an inflammatory milieu such as bacterial products (formyl peptides and lipopolysaccharide), components of the complement and clotting cascades, and soluble factors such as cytokines released by other inflammatory cells. The processes whereby soluble factors activate leukocyte effector functions (chemotaxis, proteolytic enzyme secretion, and the oxidative burst) involve complex and interconnected transmembrane signaling pathways. Many of the components of these pathways have been elucidated including the molecular characterization of membrane receptors, GTP-binding proteins, phospholipases, protein kinases, and phosphatases (reviewed in Refs. 1 and 2). Recent studies have provided evidence that free radicals including reactive oxygen intermediates, traditionally viewed as potent microbicidal agents,[3] may function in the regulation of these signaling pathways. Reactive oxygen intermediates fulfill important prerequisites for intracellular messenger molecules: they are small, diffusible, and ubiquitous molecules that can be synthesized as well as destroyed rapidly (reviewed in Ref. 4). However, because of their toxicity, there might be only a narrow concentration range in which they can function exclusively as second messengers.

The purpose of this chapter is to discuss the evidence that these reactive species participate as physiological signaling molecules in the regulation of intracellular signaling pathways in leukocytes and the implications that this has for regulation of the inflammatory response. While the intent is to focus primarily on neutrophils, where information is not available, we discuss observations in other leukocyte types including monocytes, macrophages, lymphocytes, and cultured leukemic cells and, if necessary, nonhematopoetic cells and attempt to relate this information to conserved pathways that are present in neutrophils.

9.2 Sources of Reactive Oxygen Intermediates in Leukocytes

9.2.1 NADPH Oxidase

Leukocytes possess a membrane-bound multicomponent enzyme complex termed the NADPH oxidase[5-8] that can be activated to produce ROI. This system is responsible for the "respiratory burst" and is separate from the mitochondrial electron transport chain. The term respiratory burst (increased respiration of phagocytosis), was used by Baldridge and Gerard[9] to describe the phenomenon that, during ingestion of microorganisms, phagocytic cells demonstrated a striking increase in oxygen consumption. The oxidase is dormant in resting cells and can be rapidly activated by a variety of soluble mediators and by particulate stimuli that interact with cell-surface receptors. The oxidase exhibits a marked preference for NADPH (a product of the hexose monophosphate shunt) as the electron donor and converts molecular oxygen to its one-electron-reduced product, superoxide (O_2^-), the major end product:

$$NADPH + H^+ + 2O_2 \rightarrow NADP^+ + 2H^+ + 2O_2^-$$

Hydrogen peroxide (H_2O_2) arises from the subsequent dismutation of superoxide:

$$2O_2^- + 2H^+ \rightarrow H_2O_2 + O_2$$

In addition to these two primary ROI, other products are formed including (1) the hydroxyl radical (OH·) and (2) hypochlorous acid (HOCl), which is formed in the presence of a halide such as Cl$^-$ in a reaction catalyzed by the granular enzyme myeloperoxidase.[10] These are highly reactive species that are released into phagosomes and into the extracellular environment where they combine rapidly with nearby proteins and lipids resulting in their oxidation. Reactive species are also released (or leak) into the cytosol, leading to a change in the redox state of the cell.[11] Importantly, in this location they can interact with signaling proteins and modify their function.

In addition to the NADPH oxidase, phagocytes contain other enzyme systems that are capable of generating ROI as side products of electron transfer reactions.

9.2.2 Mitochondrial Electron Transport

ROI, including O_2^- and H_2O_2, are commonly generated as side products of electron transfer reactions that occur during the operation of the mitochondrial electron transport chain (reviewed in Refs. 10 and 12–15). The mechanism of generation of ROI involves "leakage" of electrons from electron carriers that are passed directly to oxygen, reducing it to O_2^-. The importance of these reactions as a source of ROI is attested to by the fact that mitochondria contain their own superoxide dismutase (SOD), an inducible Mn^{2+}-dependent enzyme,[10] for rapid elimination of such reactive species. Although little information exists pertaining directly to leukocytes, based on information in other cell types, the importance of mitochondria as a source of ROI merits further investigation.

9.2.3 Arachidonic Acid Metabolism

The metabolism of arachidonic acid by both cycoloxygenases and lipoxygenases has been shown to generate ROI. Prostaglandin H synthase, which possesses both cyclooxygenase and hydroperoxidase activity, is a key enzyme in the biosynthesis of prostaglandins, prostacyclins, and thromboxanes[16,17] (reviewed in Ref. 18). The cyclooxygenase component incorporates oxygen into arachidonic acid, converting it to hydroperoxy endoperoxide (prostaglandin G_2 [PGG_2]). The hydroperoxidase component reduces hydroperoxides such as PGG_2 to the corresponding alcohol (prostaglandin H_2 [PGH_2]), the precursor of all prostaglandins. Oxidizing equivalents, primarily in the form of O_2^-, are released by the hydroperoxidase activity of this enzyme via side-chain reactions[16,19] and this is dependent on the presence of a suitable reducing substrate, either NADH or NADPH.[17] Metabolism of arachidonic acid by 5-lipoxygenase leads to the formation of leukotrienes. This enzyme converts arachidonic acid to 5-hydroperoxyeicosatetraenoic acid (HPETE), and thence to leukotriene A_4, the precursor of leukotrienes B_4, C_4, and D_4. Like cyclooxygenase, 5-lipoxygenase can produce O_2^- in the presence of either NADH or NADPH.[17]

Agonist stimulation of leukocytes including lymphocytes, neutrophils, and macrophages, results in the formation of increased amounts of arachidonic acid, primarily via activation of phospholipase A_2 on phosphatidylcholine or phophatidylinositol. Arachidonate is then converted to prostaglandins. Stimulated neutrophils produce prostaglandins E and F and leukotrienes such as leukotriene B_4, a proinflammatory lipid capable of activating leukocyte effector functions such as the NADPH oxidase, secretion, and chemotaxis (reviewed by Naccache et al[20]). ROI produced during agonist-induced activation of arachidonate metabolism may regulate other signaling pathways in these cells. There is precedent for this concept in T-lymphocytes where interleukin-2 (IL-2) production and activation of the transcription factor NF-κB can be regulated by ROI production by 5-lipoxygenase.[21]

9.2.4 Cytochrome P-450

Leukocytes express members of the cytochrome P-450 superfamily, ubiquitous enzymes[22,23] that are a source ROI. Members of this superfamily catalyze a series of reactions, the primary purpose of which is to detoxify lipid-soluble drugs and toxic metabolic by-products. These enzymes use high-energy electrons transferred from NADPH to add hydroxyl groups to potentially harmful hydrophobic hydrocarbons dissolved in the lipid bilayer. The result of such reactions is to convert a water-insoluble drug or metabolite, that would otherwise accumulate in cell membranes, to a form sufficiently water soluble to diffuse out of the cell and be excreted in the urine. Both O_2^- and H_2O_2 are released during this catalytic cycle. Human monocytes and lymphocytes have been shown to express aryl hydrocarbon hydroxylase (AHH), a cytochrome P-450-dependent mixed function oxidase that is present in a variety of mammalian tissues including liver and lung[24] and is responsible for the metabolism of certain carcinogenic compounds. Human neutrophils express the leukotriene B_4 ω-hydroxylase, the first member of a new subgroup within the cytochrome P-450 superfamily 4.[25,26] This enzyme catalyzes the ω-oxidation of leukotriene B_4 (LTB_4), a first step in the metabolism of this potent inflammatory mediator. Although no direct evidence exists, it is conceivable that members of the cytochrome P-450 family may be physiologically important sources of ROI in leukocytes.

9.2.5 Xanthine Oxidase

Xanthine oxidase, a cytosolic enzyme that converts xanthine or hypoxanthine to uric acid, is derived from xanthine dehydrogenase by oxidation of essential thiol groups and/or by limited proteolysis. This occurs during the ischemic phase of ischemia/reperfusion injury. Xanthine oxidase uses molecular oxygen as its electron acceptor and generates a mixture of O_2^- and H_2O_2.[27] This system is best described in endothelial cells and has been strongly implicated as a key enzyme contributing to ischemia-reperfusion injury. Xanthine oxidase activity has been demonstrated in murine leukocytes including alveolar and peritoneal macrophages and peritoneal neutrophils.[28] Recent studies have demonstrated that treatment of rat macrophages with interferon-γ results in reduction of xanthine oxidase activity by a mechanism involving nitric oxide production. The authors suggested that nitric oxide-mediated inhibition of xanthine dehydrogenase/xanthine oxidase activity could minimize the potential for tissue injury from xanthine oxidase-generated reactive oxygen species released from macrophages into the inflammatory milieu.[29]

9.2.6 Nitric Oxide Synthase

NO is derived from the amino acid L-arginine by NO synthase and is an important reactive species containing both nitrogen and oxygen. NO synthase isoforms generally fall into two categories: (1) constitutive NO synthases that are dependent on Ca^{2+} calmodulin and (2) inducible NO synthase (iNOS). The former includes neuronal constitutive form (ncNOS), first described in neurons, and endothelial constitutive form (ecNOS), first identified in endothelial cells. Inducible NO synthase is expressed by macrophages and endothelial cells and its activity is independent of $[Ca^{2+}]$ (reviewed by Nathan and Xie[30]). As its name implies, the expression of iNOS is increased by cytokines and other inflammatory stimuli. Production of NO by neutrophils is well documented, but there are important interspecies differences in the amount of NO produced.[31,32] Although information is incomplete, neutrophils express iNOS and likely ncNOS (Dr. Phil Marsden, personal communication, 1996).

NO may affect diverse cellular responses and can have both pro- and anti-inflammatory effects. For example, NO has been shown to inhibit leukocyte adhesion.[33-35] Similar to reactive oxygen species, unregulated NO synthesis may have detrimental effects as has been suggested in sepsis and autoimmune diseases (reviewed by Albina and Reichner[36]). As for ROI, NO may participate in regulation of cellular signaling and examples of these effects include activation of GTP-binding proteins, ion channels, transcription factors, and tyrosine kinases[37-39] (reviewed by Stamler[40]). Importantly, interactions between NO and ROI have been described.[34,41-45] NO has been shown to interact with O_2^- and thus may act as a biologic scavenger or inactivator of ROI.[41,42] Relevant to leukocytes, *in vivo* studies have suggested that the antiadhesive effects of NO are related to its ability to scavenge superoxide.[34] NO and O_2^- may combine to form peroxynitrite ($ONOO^-$), a potentially cytotoxic compound[43,45] (reviewed by Augusto and Radi[46]) that may also function to modify signaling molecules by oxidation and nitrosylation.[47]

9.3 Identity of ROI Produced by Leukocytes

Although there is clear evidence that ROI may function as signaling molecules in a variety of cellular pathways (see below), little is known about the identity of the actual species involved in the regulation of leukocyte signaling. The term ROI encompasses many species including singlet oxygen (O^2), hydrogen peroxide (H_2O_2), the superoxide anion (O_2^-), and the hydroxyl radical (OH·) (reviewed by Halliwell and Gutteridge[48]). The hydroxyl radical is perhaps the most reactive of these species and, by combining rapidly with target molecules in its immediate vicinity, can initiate a free radical chain reaction leading to peroxidation of lipids and proteins. Superoxide is less reactive than OH· and thus may diffuse some distance in the cell before it encounters an appropriate reaction partner. Hydrogen peroxide is even less reactive and therefore longer lived. Additionally, because of its membrane permeability, hydrogen peroxide can be released into the extracellular space from phagocytic cells and diffuse into other cells (a potential paracrine signaling mechanism as discussed below). The various types of ROI can be interconverted by reactions that depend in part on metal ions such as Fe^{2+} and Cu^{2+}, and secondary reactive products can be formed from the primary species. For example, hypochlorous acid (HOCl) is generated from H_2O_2 by the action of the granular enzyme myeloperoxidase (reviewed by Klebanoff).[49] Note that the exact reactive species responsible for any given effect is often difficult to characterize rigorously, in part because of their rapid interconversion.

9.4 Antioxidant Defenses

ROI in sufficient concentration are potent microbicidal agents but can also injure host cells. Thus regulation of the production and intracellular concentration of these highly reactive molecules is of utmost importance to host survival. Accordingly, leukocytes possess many antioxidant defenses including superoxide dismutase, the glutathione peroxidase system, and catalase (reviewed in Refs. 49–54). The intracellular concentration of a given ROI represents a dynamic equilibrium between rates of the production and removal of the species. Note again that the more reactive a species, the more rapidly it will combine with vicinal molecules. Thus the most reactive species will have the shortest half-life and will diffuse only short distances within the cytosol or in the extracellular medium.

9.5 Redox State

The redox state (E_0) of a compound can be defined as the tendency to accept or donate electrons. As ROI are potent oxidizing agents, they can affect the local or general cytosolic balance of oxidation/reduction ("redox state").[55,56] *In vitro*, under defined conditions, this can be readily measured. However, in intact cells with a multitude of pathways that can accept and/or donate electrons, it is much more difficult to define this term. Under physiological conditions, the cellular redox state is characterized by a reducing environment of the cytosol. The major redox "buffer" in the cytosol is glutathione and the vast excess of reduced over oxidized glutathione is largely responsible for the reducing potential of the cytosol.[57] Other redox buffers include NAD/NADH and NADP/NADPH. The ratio of NAD/NADH in leukocytes can change under certain physiological conditions that parallel activation of the NADPH oxidase.[11,58,59] It is presently uncertain whether ROI directly modify (e.g., by directly oxidizing thiol containing amino acid residues) signaling molecules or whether a more general alteration in cytosolic redox state by consumption of redox buffers such as NADH, NADPH, or glutathione by concomitant activation of metabolic pathways influences the components of the signaling pathways.

9.6 Redox Regulation of Signaling Pathways: General

A broad range of evidence obtained from studies in diverse cell types implicates alterations in cellular redox state in the regulation of signaling pathways at many levels including receptor function, enzymatic activity, transcription factor binding, and gene expression. A prime example of redox regulation is found in certain bacteria where the transcriptional regulatory protein, Oxy R, which regulates gene expression in response to oxidative stress, changes DNA-binding activity depending on its redox state: oxidized but not reduced Oxy R is capable of activation of transcription of oxidative-stress-inducible genes *in vitro*.[60] These findings suggest that a distinct conformational change in the Oxy R protein, or a change in its state of oligomerization, is associated with the transition from the reduced (inactive) to the oxidized (active) state and that this conformational change exposes a DNA-binding domain. Redox regulation has been implicated in other pathways including the translational control of ferritin expression (RNA-protein interaction[61]), in metabolic pathways involved in membrane transport,[62] growth inhibition,[63] and hormone-receptor interactions.[64] Changes in the redox potential have also been shown to regulate the activity of several enzymes: the formation of mixed disulfides or intramolecular disulfides can increase or decrease catalytic activity and examples of both are well documented (reviewed in Refs. 65 and 66).

9.7 ROI Regulation of Leukocyte Function

In leukocytes, like other cell types, ROI and alterations in cellular redox state have been shown to influence signaling pathways in diverse ways, including the regulation of enzymatic activity, receptor function, and protein phosphorylation, and they may participate in the signaling cascade triggered by several inflammatory mediators including cytokines and chemotactic agents. A substantial body of evidence implicates these mechanisms in the regulation of protein tyrosine phosphorylation, which is the primary focus of the following discussion.

9.8 Tyrosine Phosphorylation

Tyrosine phosphorylation plays a cardinal role in signaling processes that regulate many key functional responses of leukocytes during acute inflammation. Soluble activating agents including N-formyl-methionyl-leucyl-phenylalanine (fMLP),[67,68] tumor necrosis factor (TNF),[69] granulocyte-macrophage colony-stimulating factor (GM-CSF),[70,71] phorbol myristate acetate (PMA),[68] and platelet-activating factor (PAF)[72] induce tyrosine phosphorylation in neutrophils that parallels activation of effector responses in these cells. For example, activation of the NADPH oxidase parallels increases in protein tyrosine phosphorylation,[73,74] and conversely, tyrosine kinase inhibitors block the respiratory burst.[75-78] Tyrosine phosphorylation has been implicated in the regulation of actin assembly,[79] cell adhesion,[80] migration,[77] and Fc-receptor-mediated phagocytosis.[81,82] In addition, adhesion-dependent potentiation of neutrophil responses, particularly amplification of the respiratory burst has been linked to tyrosine phosphorylation–dependent pathways.[83,84] Importantly, substantial evidence has accumulated that ROI and alterations in cellular redox state regulate tyrosine phosphorylation in leukocytes.

9.9 Effects of Exogenous Oxidants

Exposure of lymphocytes to exogenous H_2O_2 induced increased tyrosine phosphorylation of multiple cellular proteins.[85] In an analogous manner, we observed that exposure of neutrophils to the potent oxidizing agent diamide induced marked increases in tyrosine phosphorylation of multiple cellular proteins, an effect that was abrogated by the antioxidant *N*-acetylcysteine.[86] However, in contrast to lymphocytes, addition of exogenous H_2O_2 to neutrophils resulted in only minimal phosphotyrosine accumulation. As phagocytic cells such as neutrophils have significant levels of antioxidants that might blunt oxidant-induced alterations, the effects of H_2O_2 were reexamined after treatment of cells with aminotriazole, an inhibitor of cellular catalase.[87] Under these conditions, the addition of H_2O_2 resulted in large amounts of tyrosine phosphorylation.[86] Analagous responses were observed in macrophages where the addition of H_2O_2 alone did not elicit increases in tyrosine phosphorylation. However, when cells were exposed to H_2O_2 in the presence of vanadate, a protein tyrosine phosphatase inhibitor, a marked increase in tyrosine phosphorylation of a protein in the range of 50 to 55 kDa was observed.[88] This mechanism is likely to involve the interaction of vanadate with H_2O_2 forming vanadyl hydroperoxide, a more potent tyrosine phosphatase inhibitor than vanadate.[89] Direct addition of vanadyl hydroperoxide has been shown to induce large amounts of tyrosine phosphorylation in neutrophils.[86]

These mechanisms appear to be generalizable to many cells. For example, addition of H_2O_2 to rat adipocytes[90] and hepatoma cells[91] induced phosphorylation of the insulin receptor on serine and tyrosine residues and potentiated the insulin-induced tyrosine phosphorylation of several proteins in these cells.[92,93] Additionally, the combination of hydrogen peroxide and vanadate markedly enhanced cellular tyrosine phosphorylation in rat hepatoma cells.[93]

The studies described above have all utilized direct addition of oxidants to cells. This approach can be criticized as being somewhat unphysiological, especially because of the use of potent and nonphysiological oxidants such as diamide. Studies using more "physiological" oxidants such as H_2O_2, a compound produced in large quantities by leukocytes as well as other cells, may also be subject to criticism as it may be difficult to distinguish a "physiological response" from possible "toxic" effects of this compound. Despite these shortcomings, such studies have provided useful information on the initial characterization of the effects of oxidants because they have allowed investigators to test oxidants of differing chemical structure as well as with different mechanisms of action, which has helped to exclude nonspecific effects of such compounds.

9.10 Role of Endogenously Produced Oxidants: ROI

9.10.1 Use of Permeabilized Cells as a Model System

Permeabilized cells have been used by our group as well as by others as a model to study the effects of endogenously produced ROI on tyrosine phosphorylation. In cells permeabilized by electroporation, activation of the NADPH oxidase by direct addition of GTPγS, a nonhydrolyzable analog of GTP, was found to be sustained, resembling the physiological stimulation elicited by phagocytic stimuli.[74] Importantly, these conditions were associated with marked increases in tyrosine phosphorylation,[86] which was NADPH-dependent and prohibited by diphenylene iodonium (DPI), an inhibitor of the oxidase.[94] Additionally, neutrophils from a patient with chronic granulomatous disease, deficient in the production of ROI, demonstrated no such enhanced tyrosine phosphorylation under identical conditions. These data suggested to us that ROI produced by the NADPH oxidase were in some way responsible for the induction of tyrosine phosphorylation.

Since NO synthase is also both dependent on NADPH and sensitive to DPI,[95] a possible contribution of this enzyme system to the observed enhanced tyrosine phosphorylation was considered. However, additional studies indicated that NO synthase was likely not to be a major source of oxidants that regulated tyrosine phosphorylation in neutrophils.[86] This conclusion was based on the following observations: (1) activation of the oxidase by direct addition of GTPγS (in the presence of NADPH) resulted in minimal NO production and (2) Nω-nitro-L-arginine, a potent inhibitor of NO synthase,[96] did not alter the induction of tyrosine phosphorylation. However, the potential role of nitric oxide synthase in the regulation of tyrosine phosphorylation awaits further investigation.

The use of permeabilized cells as a model system to study the effects of endogenously produced oxidants merits further discussion at this juncture. Electroporated cells maintain a relatively normal cytoarchitecture, remain sensitive to stimulation by a variety of physiological stimuli, and can be equilibrated with buffers mimicking the solute concentrations of the cytosol, such as calcium, Mg^{2+}, and ATP.[97] Additionally, the use of permeabilized cells allows direct control of cytosolic NADPH levels, an essential cofactor of the oxidase. However, it is also important to point out potential shortcomings of the permeabilized cell model. First, many pathways may be activated by a stimulus such as GTPγS making interpretation of results somewhat complex. Second, although the concentration of certain components can be restored by addition of exogenous supplies, it is likely that other compounds such as glutathione leak out therefore altering the level of intracellular antioxidant defenses. Thus the possibility that regulation of protein tyrosine kinases and phosphatases differs in intact and permeabilized cells should be born in mind. However, from the perspective of the inflammatory response, depletion of antioxidant defenses may occur "physiologically" (i.e., in the absence of permeabilization) and thus permeabilization might in fact mimic certain conditions present in an inflammatory milieu.[98]

9.10.2 Study of Receptor-Mediated Agonists in Intact Cells

Agents that activate neutrophils via surface receptors including the formyl peptide fMLP and the complement fragment C5a, induce both activation of the NADPH oxidase and increased protein tyrosine phosphorylation.[67,68] To investigate the role of endogenously produced ROI in such receptor-mediated increases in protein tyrosine phosphorylation, the effects of the formyl peptide fMLP were studied. We observed that both the pattern and kinetics of fMLP-induced tyrosine phosphorylation differed from those induced by addition of GTPγS and NADPH to permeabilized cells.[86] Moreover, the patterns of tyrosine phosphorylation induced by the formyl peptide in intact cells were comparable between normal and chronic granulomatous disease neutrophils. We interpreted these results to mean that the signalling pathway(s) leading to accumulation of phosphotyrosine in response to fMLP differed from those activated by GTPγS in permeabilized cells; only the latter appeared to be dependent on the production of ROI. One potential explanation for these differences was that activation of the NADPH oxidase in response to fMLP was transient and the amount of superoxide produced significantly less (0.5 nmol/min/10^6 cells)[99] when compared to permeabilized cells stimulated with GTPγS (10.3 nmol/min/10^6 cells[100]). However, other potential explanations for the differences between the effects of these stimuli need to be considered. For example, the fMLP receptor of neutrophils has been shown to undergo down-regulation and desensitization and these negative regulatory events may account for the rapid termination of the oxidative burst induced by the formyl peptide.[101] By contrast, activation of permeabilized cells with GTPγS bypasses receptors and results in a large and sustained activation of the oxidative burst with generation of large amounts of ROI. Additionally, as discussed above, potential low-molecular-weight antioxidants such as glutathione might leak out of permeabilized cells lowering the redox "buffering" capacity thus enhancing the effects of agonist-induced alterations.

9.11 Mechanisms of Oxidant/ROI-Induced Tyrosine Phosphorylation in Leukocytes

Tyrosine phosphorylation represents a balance between the activity of protein tyrosine kinases and tyrosine phosphatases. Thus, oxidant-induced tyrosine phosphorylation could conceivably result from activation of tyrosine kinases, inhibition of tyrosine phosphatases or from a combination thereof. There is precedent for both of these mechanisms in the literature, as discussed below.

9.11.1 Regulation of Tyrosine Kinases

Multiple tyrosine kinases are expressed in leukocytes. Lymphocytes are known to express syk (72 kDa); the src-family kinases lck (56 kDa), fyn (59 kDa), hck (56 and 59 kDa), lyn (53 and 56 kDa), fgr (59 kDa), yes (62 kDa), btk (77 kDa), a member of the tec-family of tyrosine kinases, and csk (50 kDa), ZAP-70 (70 kDa), and fak (125 kDa) (reviewed by Mustelin[102]). In lymphocytes, the oxidizing agent diamide induced activation of the tyrosine kinase ltk present in the endoplasmic reticulum.[103] Additional studies in lymphocytes demonstrated that syk is activated by H_2O_2 and suggested that the increased syk kinase activity contributed to the oxidant-induced enhancement of tyrosine phosphorylation observed in these cells.[85]

Neutrophils express syk, hck, lyn, fgr, yes, and btk.[104–119] Recently, studies investigated whether endogenous ROI, generated by the NADPH oxidase, affected the activity of tyrosine kinases in human neutrophils.[109] In these studies, activation of the oxidase was achieved by stimulation of electroporated neutrophils with GTPγS.[74] In accordance with previous observations,[86] GTPγS-induced tyrosine phosphorylation was dependent on the presence of NADPH. Additionally, inclusion of small concentrations of vanadate, a tyrosine phosphatase inhibitor, potentiated the response, suggesting that the stimulatory effect of GTPγS/NADPH was moderated by the presence of active tyrosine phosphatases (see below). Importantly, five kinases were found to be modulated by these conditions as judged by their autophosphorylating activity including syk; btk; and three src-family members lyn, hck, and fgr. ROI production stimulated both the in vitro autophosphorylation of hck, syk, and btk, and their ability to phosphorylate the exogenous substrate enolase. However, in vitro autophosphorylation activities of lyn and fgr were found to be decreased under these conditions. Interestingly, while autophosphorylation of lyn was decreased by ROI, its ability to phosphorylate enolase was markedly increased. The authors ascribed the apparent decrease in lyn autophosphorylation activity to the occupancy of substrate sites by nonradioactive phosphate, which may have occurred prior to immunoprecipitation.

9.11.2 Mechanisms of Tyrosine Kinase Activation

Studies in lymphocytes examining the effects of oxidants on ltk suggested that changes in the redox state of the cell may be one of the mechanisms involved in the oxidant-induced activation of this enzyme. Changing the cellular redox state toward a more oxidizing environment by the addition of either alkylating or thiol-oxidizing agents such as diamide apparently facilitated the formation of disulfide-linked multimers, thereby activating the kinase activity of ltk.[103] In contrast, the mechanism of H_2O_2-induced activation of syk in lymphocytes was not due to direct effects on the kinase but rather to modulation of redox-sensitive regulatory elements somewhere upstream. This conclusion was based on the findings that no enhancement of activity was observed when the isolated kinase was exposed to H_2O_2 *in vitro*.

Studies on the effects of ROI on the activation of tyrosine kinases in neutrophils paralled the findings of *syk* activation in lymphocytes: oxidizing agents were unable to activate *hck in vitro*. Additionally, once *hck* was activated *in situ*, reducing agents failed to inactivate it.[109] In contrast, the same study observed that the antioxidant *N*-acetylcysteine (NAC) effectively attenuated the tyrosine phosphorylation induced by endogenously produced ROI. Taken together, these results suggest that the effects of ROI on certain tyrosine kinases such as *syk* and *hck* are indirect, possibly mediated by induction of alternate post-translational modifications such as protein phosphorylation.

As tyrosine phosphorylation has been shown to regulate the activity of tyrosine kinases, Brumell et al.[109] explored this mechanism related to regulation of neutrophil tyrosine kinases by ROI. They found that production of endogenous ROI by activation of permeabilized neutrophils with GTPγS induced tyrosine phosphorylation of *hck*. Importantly, tyrosine phosphorylation of *hck* paralleled its activation, and dephosphorylation *in vitro* reversed the stimulation, indicating that tyrosine phosphorylation was central to the activation of *hck*. Note that tyrosine phosphorylation is capable of either positively or negatively regulating the activity of tyrosine kinases, depending on the particular tyrosine residue. For example, phosphorylation of tyrosine residues of *lck* can either suppress or activate the enzyme depending on the specific residue: phosphorylation of a specific C terminal residue was found to be inhibitory and dephosphorylation of this residue led to its activation (reviewed by Cooper and Howell[120]). However, phosphorylation of other tyrosine residues of *src* kinases are required for full activity (reviewed in Refs. 120–122). Therefore, by modulation of the level of tyrosine phosphorylation, endogenously produced ROI could conceivably regulate the activity of tyrosine kinases in leukocytes. Additionally, as discussed below, oxidants can regulate the activity of protein tyrosine phosphatases providing an alternate mechanism whereby ROI could regulate tyrosine kinases.

9.11.3 Regulation of Tyrosine Phosphatases

Cellular redox state (oxidation/reduction reactions) and ROI can modulate the activity of protein tyrosine phosphatases, effects which may be of equal importance to those on kinases. Many tyrosine phosphatases are dependent on conserved cysteine groups in their first catalytic domain for full enzymatic activity.[123] As these residues must be in a reduced state for phosphatase activity,[124] oxidants might be expected to inhibit these enzymes. In fact, there is strong evidence that oxidants can regulate the activity of tyrosine phosphatases in leukocytes. For example, studies have demonstrated that phenylarsine oxide, a trivalent arsenical compound that complexes the vicinal sulfhydryl groups of proteins,[125] can regulate tyrosine phosphatase activity in T lymphocytes.[126] This compound is capable of oxidizing cysteine residues and consequently inhibiting phosphatases.[127] Studies in macrophages, leukemic cell lines and neutrophils have demonstrated that the combination of vanadate and H_2O_2, (which yields vanadyl hydroperoxide, as discussed above) greatly potentiated the inhibitory effects on whole cell tyrosine phosphatase activity that was observed with either agents alone.[73,88,89,128,129]

Neutrophils demonstrate significant consitutive tyrosine phosphatase activity that may function in a negative regulatory manner by preventing tyrosine kinases from inducing phosphoprotein accumulation and thus cell activation. This notion is supported by the observations that addition of vanadate, a tyrosine phosphatase inhibitor to neutrophils[74] and to HL-60 cells[89,131] resulted in activation of the respiratory burst. To explore the contribution of specific tyrosine phosphatases to the enhanced tyrosine phosphorylation observed in activated neutrophils, we studied CD45, a tyrosine phosphatase of high abundance in these cells,[132–134] as a paradigm for the regulation of neutrophil tyrosine phosphatases. This phosphatase has been shown to play an important role in the regulation of *src* family tyrosine kinases such as p56[lck] and p69[fyn135,136] by removing inhibitory phosphates from carboxy terminal regulatory tyrosine residues.[135] Addition of GTPγS to permeabilized neutrophils, conditions known to activate the oxidative burst (see above), resulted in inhibition of CD45 activity, as determined by *in vitro* phosphatase assay of CD45 immunoprecipitates.[137] That this inhibition was, at least in part, a consequence of activation of the NADPH oxidase was supported by three observations: (1) GTPγS-induced inhibition of CD45 was NADPH-dependent; (2) pretreatment of cells with diphenylene iodonium, an inhibitor of the oxidase, partially prevented this inhibition; and (3) GTPγS-induced inhibition of CD45 was markedly diminished in neutrophils from two patients with chronic granulomatous disease (CGD).

9.11.4 Mechanism of Tyrosine Phosphatase Inhibition

As described above, CD45 inhibition in permeabilized cells stimulated by addition of GTPγS and NADPH paralleled activation of the NADPH oxidase. Presumably CD45 was inhibited as a consequence of oxidation of the conserved cysteine in the catalytic domain by endogenously produced ROI. Further studies demonstrated that while the inhibition of CD45 could be partially prevented by prior treatment of the cells with antioxidants, direct antioxidant treatment of CD45 immunoprecipitates could not restore activity.[137] However, as demonstrated by others[128] the isolated enzyme was susceptible to direct oxidation and this effect could be reversed *in vitro*. Our interpretation of these results was that while purified CD45 was susceptible to direct oxidation and reduction *in vitro*, part of the inhibitory effects observed in permeabilized cells stimulated with GTPγS/NADPH may have been exerted on regulatory elements in the signaling pathway controlling CD45 activity. Alternatively, it is possible that oxidation of CD45 induced by endogenous reactive species including O_2^-, H_2O_2, and OH· under the conditions present in the cytosol, is not reversible, perhaps due to more extensive alterations in the tertiary structure of the enzyme.

To determine if these mechanisms were important during receptor-mediated activation, we next studied regulation of CD45 in cells exposed to the chemoattractant fMLP, an agent known to both induce tyrosine phosphorylation and activation of the NADPH oxidase. However, these under these conditions, no inhibition of CD45 was noted.[138] While further investigation is necessary to determine whether receptor-mediated agonists modulate the activity of other tyrosine phosphatases, evidence points against a predominant role for endogenous ROI in the regulation of the responses elicited by these agonists.

PMA, a phorbol ester that directly activates protein kinase C, is known to induce a maximal respiratory burst in neutrophils. Early studies demonstrated that exposure of neutrophils to PMA resulted in significant inhibition of whole cell protein tyrosine phosphatase activity.[139,140] More recent studies have focussed on the identity of PMA-sensitive phosphatases in neutrophils. Studies from our laboratory have documented that exposure of neutrophils to PMA resulted in a significant reduction in CD45 activity and suggested that ROI generated under these conditions contributed to inhibition of CD45.[137] In particular, the effects of PMA were greatly attenuated by pretreatment of cells with DPI, an inhibitor of the NADPH oxidase and PMA-induced inhibition of CD45 was much less in neutrophils isolated from CGD patients than from normal donors. However, more prolonged exposure (30 min) of CGD cells to PMA resulted in substantial inhibition of CD45. Two conclusions can be drawn from these results: (1) there is an important role for endogenously degenerated ROI in CD45 regulation in cells stimulated with PMA and (2) the NADPH oxidase is a major source of ROI that participate in this inhibition. However, PMA might also be able to stimulate ROI production by other oxidant-generating systems including the mitochondrial electron transport chain (also partially DPI-sensitive),[141] cytochrome P-450,[142] or other as yet uncharacterized systems known to be present in CGD patients.[143] It is also likely that other mechanisms, such as phosphorylation mediated by activation of PKC, might be involved in PMA-induced inhibition of CD45[144,145] (reviewed by Koretzky[146]). Clearly, this is an area requiring further investigation.

The potential role of oxidants in the regulation of mixed function phosphatases such as MAP Kinase Phosphatase 1 (MKP-1),[147] an enzyme that is capable of dephosphorylating and inactivating mitogen-activated protein kinase (MAP) kinase (see below), remains to be explored. It should be noted that serine-threonine phosphatases are also sensitive to alterations in the cellular redox state.[148] Although less is known about the active site of these phosphatases, the catalytic domain of the serine/threonine phosphatases type 1 (PP1) and type 2A (PP2) have been shown to contain cysteine residues.[149,150] As many phosphatases have cysteine residues in their active site, they appear to be uniquely susceptible to changes in the cellular redox state and alteration by ROI.

In summary, the activity of tyrosine phosphatases can be regulated by exogenous as well as endogenously generated oxidants (i.e., ROI) in leukocytes. A major focus has been on CD45, but other tyrosine phosphatases known to be present in leukocytes including PTP1C,[151] PTP1B, and MEG2[83] may be a target of oxidant modulation.

9.12 Oxidant/ROI Activation of Serine/Threonine Kinases

9.12.1 MAP Kinase

Studies from our laboratory have demonstrated that exogenous oxidants induce tyrosine phosphorylation and activation of MAP kinase, a serine/threonine kinase expressed in human neutrophils.[138] This activation was mediated in part, by activation of MAP kinase (or Erk) kinase (MEK), the putative upstream activator of MAP kinase (reviewed in Refs. 152 and 153a). The mechanisms by which oxidants activate MEK are presently unknown but it is plausible that oxidant-induced activation of MEK might be the result of effects on *ras* and/or *raf* (reviewed by Crews and Erikson[153a]). *Raf* has been shown to bind directly to and phosphorylate MEK, which results in activation of MEK.[153b–155] Alternatively, it is possible that activation of MEK by oxidants could be through activation of a parallel pathway, involving heterotrimeric GTP binding proteins, leading to

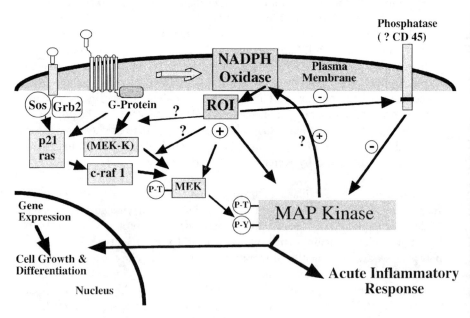

Figure 9.1. Regulation of the MAP kinase signaling pathway by ROI.

activation of a MEK kinase.[156] Oxidants could stimulate MEK via direct activation of GTP-binding proteins by a mechanism involving oxidation of sulfhydryl groups. As CD45 is capable of dephosphorylating and inactivating MAP kinase,[157] oxidant-induced inhibition of CD45 could also contribute to MAP kinase activation (or at least prolongation of activation). Note that receptor-mediated MAP kinase activation can proceed by mechanisms that are independent of production of endogenous ROI because fMLP-induced activation proceeded normally in CGD neutrophils.[138]

9.12.2 Protein Kinase C

Protein kinase C (PKC) has been implicated in a wide variety of cellular responses produced by diverse stimuli including hormones, growth factors, chemotaxins such as fMLP, and tumor promoting phorbol esters. Importantly, the activity of this cysteine-rich enzyme[158] can be regulated by oxidants. Studies in rat hepatocytes demonstrated that generation of oxidants by redox-cycling quinones induced activation of PKC, likely by oxidation of thiol-containing residues such as cysteine.[159] Additionally, *in vitro* studies with purified PKC demonstrated that the activity of this enzyme could be modulated by H_2O_2.[160] As activation of PKC participates in the regulation of many leukocyte responses, the possibility that PKC could be regulated by endogenously produced ROI has potentially important functional implications.

9.13 Oxidant/ROI Activation of Phospholipase A₂

Recent studies have implicated ROI in phorbol-ester mediated phospholipase A_2 activation.[88,129] Phospholipase A_2 (PLA_2) which releases arachidonic acid (AA) from phospholipids, is a key enzyme in leukocyte activation. The basal activity of PLA_2 was rapidly increased after exposure of cells to a variety of soluble and particulate stimuli. In these studies, treatment of macrophages with vanadate plus the phorbol ester PMA led to activation of both PLA_2 and the oxidative burst perhaps by tyrosine phosphorylation–dependent pathways. That these effects were mediated by oxidants was supported by the following observations: (1) exposure of cells to H_2O_2 in combination with vanadate markedly increased PLA_2 activity and (2) inhibition of the NADPH oxidase and ROI production by diphenyleneiodonium (DPI) prevented these effects. Note that PLA_2 may be phosphorylated and activated by MAP kinase[161,162] and that, as discussed above, MAP kinase activation may be modulated by oxidants.

9.14 Potential Functional Implications of ROI/Oxidant-Mediated Effects on Leukocyte Signaling

A plethora of evidence supports an important role for ROI as signaling molecules in diverse cell types in many cellular processes including regulation of receptor function, enzymatic activity, transcription factor binding, and gene expression. In fact, it has been proposed that inflammatory processes may rely on prooxidant conditions.[163] As leukocytes generate prodigious amounts of ROI via activation of the NADPH oxidase, ROI/oxidant regulation of cellular signaling may have particularly important physiological consequences in these cells. The purpose of this section is to speculate on the potential physiological consequences of signaling by ROI in leukocytes.

9.14.1 Modulation of Tyrosine Phosphorylation–Dependent Pathways

We have outlined the evidence implicating ROI in the regulation of tyrosine phosphorylation in leukocytes by a combination of activation of tyrosine kinases and inhibition of tyrosine phosphatases. As tyrosine phosphorylation appears to participate in the regulation of many leukocyte effector functions (see discussion above), it follows that ROI may exert an important influence on leukocyte function via effects on tyrosine phosphorylation.

9.14.2 Modulation of Fc Receptor Function and Phagocytosis

Fc receptors participate in responses vital for host defense including phagocytosis of microbial pathogens, release of lysosomal enzymes (secretion), and activation of the oxidative burst. Neutrophils express two types of Fc receptors, FcγRIIa and FcγRIIIb,[82] which are responsible not only for recognition and binding of opsonized microorganisms but for the initiation of transmembrane signaling pathways leading to internalization (phagocytosis) and intracellular killing of the pathogens. Recent studies have demonstrated that cross-linking of FcγRIIIb can lead to activation of FcγRIIa, which may provide an efficient mechanism for amplification of Fcγ receptor function.[164] Importantly, this response was blocked by inhibitors of reactive oxygen intermediates and by inhibitors of the H_2O_2-myeloperoxidase-chloride system, which generates hypochlorous acid.[164] As serine protease inhibitors also abrogated the response and since oxidants such as HOCl have been shown to inactivate proteinase inhibitors, a potential mechanism of action for the oxidant modulation of Fc signaling might involve inhibition of a protease cascade. This oxidant-dependent modulation of receptor interaction is an example of an autocrine function of ROI.[164] Further evidence that ROI influence neutrophil phagocytosis is provided by the observation that amplification of ingestion during phagocytosis was deficient in neutrophils from patients with chronic granulomatous disease.[165]

9.14.3 Modulation of Tyrosine Kinase Activity

ROI may also participate in signaling events downstream of Fc receptors involved in phagocytosis. For example, certain tyrosine kinases are closely associated with Fc receptors.[81,111,166,167] The tyrosine kinase hck may be involved in the regulation of phagocytosis in macrophages. Activation of *syk* may regulate tyrosine phosphorylation of proteins following cross-linking of either Fcγ RI or II in HL-60 cells,[117] and has been suggested to regulate Fc receptor–mediated phagocytosis in macrophages.[168] Studies in neutrophils have demonstrated that *fgr* coimmunoprecipitates with FcRγII and that its kinase activity is enhanced following crosslinking of FcγRII suggesting that *fgr* is involved in FcγRII-mediated signal transduction pathways.[111] Recent studies in neutrophils suggest that *hck* is associated with FcγRIIIb receptor, and speculate that this interaction modulates the synergistic effects of FcγRII and FcγRIIIb on activation of the oxidase.[169]

9.14.4 Modulation of Tyrosine Phosphatase Activity

Production of ROI may also influence leukocyte effector functions via their affects on tyrosine phosphatases. The regulation and the functional relevance of CD45 has been most extensively investigated in lymphocytes. This phosphatase is crucial in antigen receptor mediated signaling in T and B lymphocytes (reviewed in Refs. 146 and 170–172). Studies in T lymphocytes have demonstrated that depletion of intracellular reduced glutathione levels potentiated TNF-induced tyrosine phosphorylation on stimulation of T cell receptors suggesting that redox changes and ROI produced by leukocytes may potentiate signaling responses elicited by cytokines.[173]

The function of CD45 in phagocytic cells such a neutrophils is presently incompletely understood. CD45 may regulate chemotactic responses induced by LTB4 and C5a[174] and participate in low-affinity FcγR signaling in neutrophils.[175] As ROI are potent inhibitors of tyrosine phosphatases such as CD45, these compounds may influence physiological responses to these chemoattractants. The influence of ROI on other tyrosine phosphatases expressed by neutrophils is an area requiring further investigation.

9.14.5 Actin Polymerization and Cell Motility

The addition of hydrogen peroxide to a murine macrophage cell line was found to initiate profound cytoskeletal changes.[176] The authors speculated that the increase in polymerized actin occurred by a conformational change in actin monomers induced by oxidation of sulfhydryl groups. This would in turn promote monomer release from sequestering proteins such as profilin and the increase in monomeric actin would then drive actin polymerization. It is noteworthy that in lymphocytes[79] and macrophages[81] tyrosine phosphorylation is required for actin assembly. As oxidants can regulate tyrosine phosphorylation, the effects of hydrogen peroxide on the actin cytoskeleton might also be mediated by tyrosine phosphorylation of proteins that regulate actin polymerization.

Cell motility is a complex process that is dependent in part on coordinated alterations in the actin cytoskeleton. There is certainly precedent for the involvement of ROI in regulation of cell motility. Recently, platelet-derived growth factor (PDGF)-induced chemotaxis in vascular smooth muscle cells was found to be mediated by endogenously produced ROI.[177] As neutrophil chemotaxis can also be triggered by PDGF, it is possible that ROI are involved in leukocyte motility. However, as chemotactic agents such as fMLP can induce migration of neutrophils from patients with chronic granulomatous disease, it is unlikely that ROI generated by the NADPH oxidase are exclusive mediators of chemotactic signaling in neutrophils.

9.14.6 Functional Implications of Activation of MAP kinase

The functional relevance of activation of the MAP kinase pathway in neutrophils is presently uncertain. As the kinetics of activation of MAP kinase is too slow to account for the early neutrophil responses such as initiation of activation of the NADPH oxidase, it has been suggested that MAP kinase may be more important in the generation or maintenance of late responses such as degranulation.[178,179] Several proteins are efficient substrates of MAP kinases *in vitro* and these may reflect physiological targets of MAP kinase *in vivo*. For example, phosphorylation of microtubule-associated protein 2 (MAP-2) by MAP kinases may result in alterations of the cytoskeleton and provide a molecular mechanism for the morphological changes induced by hydrogen peroxide in a macrophage cell line.[174] MAP kinases can activate transcription factors such as c-*jun*.[180] While neutrophils do not synthesize substantative amounts of new proteins, these pathways are also active in macrophages, cells in which synthesis of many proteins are regulated at the transcriptional level. Another important MAP kinase substrate is cytosolic phospholipase A_2,[161,181] which, as discussed, may be activated by ROI-induced stimulation of MAP kinase. Activation of this cascade would promote production of arachidonic acid and its metabolites such a prostaglandins and leukotrienes. It is noteworthy that arachidonic acid or products of its metabolism via the 5-lipoxygenase or prostaglandin synthase pathways may be involved in signaling changes in the actin filaments in certain culture cell lines.[182] Thus, oxidant-induced MAP kinase activation could potentially participate in cytoskeletal responses such as actin polymerization[174,182] through activation of PLA_2

9.14.7 ROI and Oxidants in the Regulation of Apoptosis

The process of apoptosis (programmed cell death) is important both in the normal homeostasis of leukocytes and in the removal of these cells during inflammatory responses.[183,184] Oxidative stress can induce apoptosis in several cells.[185–187] Recent studies suggest that tyrosine phosphorylation may participate in the regulation of apoptosis in neutrophils and eosinophils.[188–190] Additionally, protein tyrosine phosphatases,[191] including an isoform of CD45 (CD45 RO) may regulate apoptosis in lymphocytes.[192] Thus it is conceivable that ROI, through their effects on tyrosine phosphatases such as CD45, participate in the regulation of neutrophil apoptosis. ROI such as superoxide might also influence apoptosis via interactions with nitric oxide. Peroxynitrite, an oxidant generated by the interaction of NO and superoxide, has been shown to induce apoptosis in two transformed cell lines, HL-60 and U-937 cells. However, it was ineffective at inducting apoptosis in normal human endothelial and peripheral blood mononuclear cells.[193]

9.14.8 Regulation of Gene Expression by ROI in Leukocytes

Oxidants have been shown to regulate gene expression by effects on transcription factors such as NF-κB,[194] c-*fos*, and c-*jun*.[195-197] The evidence that ROI can activate the transcription factor NF-κB may be directly relevant to leukocyte responses during acute inflammation. NF-κB can be found in a variety of cell types and tissues, including cells of the immune system such as B and T lymphocytes, macrophages, and monocytes. This transcription factor mediates the rapid induction of expression of several genes involved in the acute inflammatory response including cytokines, cytokine receptors, major histocompatibility complex (MHC), antigens, acute-phase proteins and several viral enhancers (reviewed by Bauerle[198]). The implications of NF-κB activation in neutrophils is presently unknown. However, macrophages demonstrate increased NF-κB activity following stimulation with lipopolyssacharide.[199-202] Since macrophages can generate ROI in response to lipopolysaccharide (LPS), it is possible that ROI may participate in LPS-induced NFκB activation in these cells, an effect that would have important consequences for the inflammatory and immune responses.

Cytokines themselves such as TNF and IL-1 can result in NF-κB activation. TNF-mediated NF-κB activation occurs by an effect dependent on the production of ROI.[4,194] ROI have also been shown to mediate TNF-induction of c-*fos* in chondrocytes which may have important consequences in the development of inflammatory diseases such as arthritis.[204] Evidence for a role of mitochondrial-derived ROI in TNF-mediated signaling has been provided by studies in fibrosarcoma cells[205] and TNF cytotoxic effects. Taken together, these observations may provide a basis for a mitochondria-nucleus regulatory pathway in which ROI would participate as physiological signaling molecules.

ROI may also affect gene expression involved in the regulation of other cytokines. Monocytes produce certain cytokines including colony-stimulating factor 1 (CSF-1) and the monocyte chemoattractant protein (MCP-1). Importantly, ROI produced by the NADPH oxidase have been implicated in TNF-α as well as immunoglobulin G–induced expression of the CSF-1 and MCP-1 genes in mesangial cells.[206] The findings in mesangial cells are relevant to glomerular diseases such as immune-mediated glomerular injury. In this model, as there is an influx and activation of monocyte-macrophages, the local generation of MCP-1 and CSF could contribute to the glomerular injury.[206]

9.14.9 Paracrine Effects of ROI

ROI produced by leukocytes may act in a paracrine manner to influence other cells. It is clear from careful stoichiometric studies that the majority of oxidizing equivalents (ROI) produced by the respiratory burst oxidase are released extracellularly.[207] Since ROI are membrane permeant, they may influence intracellular signaling pathways in cells including other leukocytes as well as different cell types present in the inflammatory milieu. The vascular endothelium, by virtue of its location, is a prime target for such oxidant effects. During their migration through the endothelium, neutrophils may release ROI, which may trigger responses in endothelial cells. Indeed, exposure of endothelial cells to exogenous hydrogen peroxide has been shown to activate phospholipase D, an enzyme that catalyzes the hydrolysis of phospholipids.[209] Additionally, studies have demonstrated that ROI induce human endothelial cells to express P-selectin and bind neutrophils.[208] Although not shown directly, neutrophils are a potential source of ROI that could regulate endothelial cell responses and thus endothelial-neutrophil interactions. It has been clearly demonstrated that ROI produced by neutrophils can pass across the plasma membrane through the extracellular space and into the cytosol of adjacent cells such as erythrocytes[59] and tumor cells.[58] Thus, ROI produced by one neutrophil can influence signaling pathways in adjacent cells be they other neutrophils, macrophages, or resident tissue cells such as fibroblasts or endothelial or epithelial cells. Accumulation of activated neutrophils has been implicated in the pathogenesis of acute lung injury, a disorder characterized by inflammatory injury to both the endothelium and epithelium (reviewed by Downey and O'Brodovich[210]). Based on the evidence outlined above, ROI produced by the leukocyte NADPH oxidase may contribute to amplification of inflammatory injury in this disorder.[98] ROI may also influence the reparative phase of inflammation and in particular the production of excess fibrous tissue by promotion of collagen synthesis by fibroblasts (Fig. 9.2)

9.15 Conclusion

In this chapter, we have reviewed the compelling evidence supporting a role for oxidants including ROI in the regulation of diverse signaling pathways in leukocytes. From a physiological perspective, endogenous ROI produced by the NADPH oxidase appear to regulate tyrosine phosphorylation with attendant important effects on effector responses such as phagocytosis. Under pathological circumstances such as inflammatory tissue injury, excess production of ROI may influence vicinal cells such as endothelium or epithelium. Perhaps it is during the initial phase of the inflammatory response that ROI may act as signaling molecules leading to modulation of crucial events including phagocytosis, secretion, gene expression and apoptosis resulting in dysregulation of inflammation.

Acknowledgments

This work was supported by funds from the Medical Research Council of Canada and the Ontario Thoracic Society.

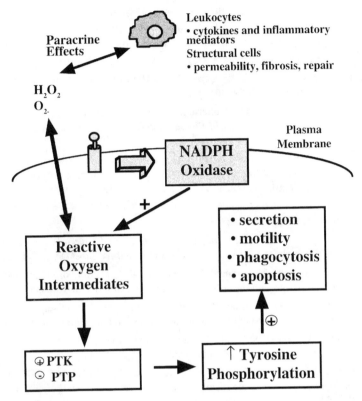

Figure 9.2. ROI and tyrosine phosphorylation: implications for inflammation.

References

1. Sha'afi, R. I., and T.F.P. Molski. 1988. Activation of the neutrophil. *Prog. Allergy* **42**:1–64.

2. Downey, G. P., T. Fukushima, L. Fialkow, and T. K. Waddell. 1995. Intracellular signalling in mechanisms neutrophil priming and activation. *Semin. Cell Biol.* **6**:345–356.

3. Cross, A. R., and O.T.G. Jones. 1989. The molecular mechanism of oxygen reduction by the neutrophil oxidase, pp. 97–111. *In* M. B. Hallet (ed.), *The Neutrophil: Cellular Biochemistry and Physiology.* CRC Press, Boca Raton, FL.

4. Schreck, R., and P. A. Baeuerle. 1991. A role for oxygen radicals as second messengers. *Trends Cell Biol.* **1**:2–3.

5. Rotrosen, D. 1992. The respiratory burst oxidase, pp. 589–601. *In* J. I. Gallin, I. M. Goldstein, and R. Snyderman (eds.), *Inflammation: Basic Principles and Clinical Correlates,* 2nd ed. Raven Press, New York.

6. Bastian, N. T., and J. B. Hibbs, Jr. 1994. Assembly and regulation of NADPH oxidase and nitric oxide synthase. *Curr. Opin. Immunol.* **6**:131–139.

8. Morel, F., J. Doussiere, and P. V. Vignais. 1991. The superoxide-generating oxidase of phagocytic cells. *Eur. J. Biochem.* **201P**:523–546.

9. Balridge, C. W., and R. W. Gerad. 1933. The extra respiration of phagocytosis. *Am. J. Physiol.* **103**:235–236.

10. Halliwell, B., and J.M.C. Gutteridge. 1985. Free radicals as useful species, pp. 246–278. *In Free Radicals in Biology and Medicine*, 2nd ed. Clarendon Press, Oxford.

11. Liang, B. and H. R. Petty. 1992. Imaging neutrophil activation: analysis of the translocation and utilization of NAD(P)H-associated autofluorescence during antibody-dependent target oxidation. *J. Cell. Physiol.* **152**:145–156.

12. Bovaris, A., and B. Chance. 1973. The mitochondrial generation of hydrogen peroxide. *Biochem. J.* **134**:707–716.

13. Turrens, J. F., and A. Bovaris. 1980. Generation of superoxide anions by the NADPH dehydrogenase of bovine heart mitochondria. *Biochem. J.* **191**:421–427.

14. Boveris, A., E. Cadenas, and A.O.M. Stopppani. 1976. Role of ubiquinone in the mitochondrial generation of hydrogen peroxide. *Biochem J.* **156**:435–444.

15. Boveris, A., and E. Cadenas. 1975. Mitochondrial production of superoxide anions and its relationship to antimycin insensitive respiration. *FEBS Lett* **54**:311–314.

16. Egan, R. W., J. Paxton, and F. A. Kiehl, Jr. 1976. Mechanisms for the irreversible self deactivation of prostaglandin synthetase. *J. Biol. Chem.* **251**:7329–7335.

17. Kukreja, R. C., H. A. Kontos, M. L. Hess, and E. F. Ellis. 1986. PGH synthase and lipoxygenase generate superoxide in the presence of NADH or NADPH. *Circ. Res.* **59**:612–619.

18. Taylor, A. A., and S. B. Shappell. 1992. Reactive Oxygen species, neutrophil and endothelial adherence molecules, and lipid-derived inflammatory mediators in muocardial ischemia-reperfusion injury, pp. 65–141. *In*: M. T. Moslen and C. V. Smith (eds.), *Free Radical Mechanisms of Tissue Injury*. CRC Press, Boca Raton, FL.

19. Egan, R. W., P. H. Gale, E. M. Baptista, K. L. Kennicott, W.J.A. VandeHeuvel, R. W. Walker, P. E. Fagerness, and F. A. Kuehl, Jr. 1981. Oxidation reactions by prostaglandin cyclooxygenase-hydroperoxidase. *J. Biol. Chem.* **246**:7352–7361.

20. Naccache, P. H., R. I. Sha'afi, and P. Borgeat. 1989. Mobilization, metabolism, and biological effects of eicosanoids in polymorphonuclear leukocytes, pp. 113–139. *In* M. B. Hallet (ed.), *The Neutrophil: Cellular Biochemistry and Physiology*. CRC Press, Boca Raton, FL.

21. Los, M., J. Schenk, K. Hexel, P. A. Baeuerle, W. Dröge, and K. Schulze-Osthoff. 1995. IL-2 gene expression and NF-kB activation through CD28 requires reactive oxygen production by 5-lipoxygenase. *EMBO J.* **14**:3731–3740.

22. Nebert, D. W., D. R. Nelson, M. J. Coon, R. W. Estabrook, R. Feyereisen, Y. Fujii-Kuriyama, F. J. Gonzalez, F. P. Guengerich, I. C. Gunsalus, E. F. Johnson, J. C. Loper, R. Sato, M. R. Waterman, and D. J. Waxman. 1991. The P450 super-

family: update on new sequences, gene mapping, and recommended momenclature. *DNA Cell Biol.* **10**:1–14.

23. Guengerich, P. 1995. Human cytochrome P450 enzymes, pp. 473–535. *In* R. Ortiz de Montellano (ed.), *Cytochrome P-450: Structure, Mechanism, and Biochemistry,* 2nd ed. New York: Plenum Press.

24. Robie-Suh, K., R. Bodinson, and H. V. Gelboin. 1980. Aryl hydrocarbon hydroxylase is inhibited by antibody to rat liver cytochorme P-450. *Science* **208**:1031–1033.

25. Kikuta, Y., E. Kusunose, K. Endo, S. Yamamoto, K. Sogawa, Y. Fujii-Kuriyama, and M. Kusunose. 1993. A novel form of cytochrome P-450 family 4 in human polymorphonuclear leukocytes: cDNA cloning and expression of leukotriene B$_4$ ω-hydroxylase. *J. Biol. Chem.* **268**:9376–9380.

26. Shak, S., and I. M. Goldstein. 1985. Leukotriene B$_4$ ω-hydroxylase in human polymorphonuclear leukocytes: partial purification and identification as a cytochrome P-450. *J. Clin. Invest.* **76**:1218–1228.

27. McCord, J. M. 1985. Oxygen derived free radicals in post-ischaemic tissue injury. *N. Engl. J. Med.* **312**:159–163.

28. Grum, C. M., T. J. Gross, C. H. Mody, and R. G. Sitrin. 1990. Expression of xanthine oxidase activity by murine leukocytes. *J. Lab. Clin. Med.* **116**:211–218.

29. Rinaldo, J. E., M. Clark, J. Parinello, and V. L. Shepherd. 1994. Nitric oxide inactivates xanthine dehydrogenase and xanthine oxidase in interferon-γ-stimulated macrophages. *Am. J. Respir. Cell. Mol. Biol.* **11**:625–640.

30. Nathan, C., and Q.-W. Xie. 1994. Nitric oxide synthases: roles, tolls, and controls. *Cell* **78**:915–918.

31. Markert, M., B. Carnal, and J. Mauel. 1994. Nitric oxide production by activated human neutrophils. *Biochem. Biophys. Res. Commun.* **199**:1245–1249.

32. McCall, T. B., R.M.J. Palmer, and S. Moncada. 1991. Induction of nitric oxide synthase in rat peritoneal neutrophils and its inhibition by dexamethasone. *Eur. J. Immunol.* **21**:2523–2527.

33. Kubes, P., M. Suzuki, and D. N. Granger. 1991. Nitric oxide: an endogenous modulator of leukocyte adhesion. *Proc. Natl. Acad. Sci. USA* **88**:4651–4655.

34. Gaboury, J., R. C. Woodman, D. N. Granger, P. Reinhardt, and P. Kubes. 1993. Nitric oxide prevents leukocyte adherence: role of superoxide. *Am. J. Physiol.* **34**:H862–H867.

35. Kubes, P., I. Kurose, and D. N. Granger. 1994. NO donors prevent integrin-dependent leukocyte adhesion but not P-selectin-dependent rolling in potischemic venules. *Am. J. Physiol.***36**:H931–H937.

36. Albina, J. E., and J. S. Reichner. 1995. Nitric oxide in inflammation and immunity. *New Horizons* **3**:46–64.

37. Lander, H. M., P. K. Sehajpal, and A. Novogrodsyk. 1993. Nitric oxide signaling: a possible role for G proteins. *Immunology* **151**:7182–7187.

38. Bolotina, V. M., S. Najibi, J. J. Palacino, P. J. Pagano, and R. A. Cohen. 1994. Nitric oxide directly activates calcium-dependent potassium channels in vascular smooth musche. *Nature* **368**:850–853.

39. Gopalakrishna, R., Z. H. Chen, and U. Gundimeda. 1993. Nitric oxide and nitric oxide-generating agents induce a reversible inactivation of protein kinase C activity and phorbol ester binding. *J. Biol. Chem.* **268**:27180–27185.

40. Stamler, J. S. 1994. Redox signaling: nitrosylation and related target interactions of nitric oxide. *Cell* **78**:931–936.

41. Feigl, E. O. 1988. EDRF-a protrective factor? *Nature (London)* **331**:490–491.

42. Rubanyi, G. M., E. H. Ho, E. H. Cantor, W. C. Lumma, and L. H. Botelho. 1991. Cytoprotective function of nitric oxide: inactivation of superoxide radicals produced by human leukocytes. *Biochem. Biophys. Res. Commun.* **181**:1392–1397.

43. Blough, N. V., and O. C. Zafiriou. 1985. Reaction of superoxide with nitric oxide to form peroxonitrite in alkaline aqueous solution. *Inorg. Chem.* **24**:3502–3504.

44. Darley-Usmar, V., H. Wiseman, and B. Halliwell. 1995. Nitric oxide and oxygen radicals: a question of balance. *FEBS Lett.* **369**:131–135.

45. Beckman, J. S., T. W. Bekman, J. Chen, P. M. Marshall, and B. A. Freeman. 1990. Apparent hydroxyl radical production by peroxynitrite: implications for endothelial injury from nitric oxide and superoxide. *Proc. Natl. Acad. Sci. USA* **87**:1620–1624.

46. Augusto, O., and R. Radi. 1995. Peroxynitrite reactivity: free radical generationk thiol oxidation, and biological significance, pp. 83–116. *In* L. Packer and E. Cadenas (eds.), *Biothiols in Health and Disease*. Marcel Dekker, New York.

47. Haddad, I. Y., G. Pataki, P. Hu, C. Galliani, J. S. Beckman, and S. Matalon. 1994. Quantitation of nitrotyrosine levels in lung sections of patients and animals with acute lung injury. *J. Clin. Invest.* **94**:2407–2413.

48. Halliwell, B., and J.M.C. Gutteridge. 1990. Free radicals and metal ions in human disease. *Methods Enzymol.* **186**:1–85.

49. Klebanoff, S. J. 1992. Oxygen metabolites from phagocytes, pp. 541–588. *In* J. I. Gallin, I. M. Goldstein, and R. Snyderman (eds.), *Inflammation: Basic Principles and Clinical Correlates*, 2nd ed. Raven Press, New York.

50. Rest, R. F., and J. K. Spitznagel. 1977. Subcellular distribution of superoxide dismutases in human neutrophils. Influence of myeloperoxidase on the measurement of superoxide dismutase activity. *Biochem. J.* **166**:145–153.

51. Roos, D., R. S. Weening, S. R. Wyss, and H. E. Aebi. 1980. Protection of human neutrophils by endogenous catalase. Studies with cells from catalase-deficient individuals. *J. Clin. Invest.* **65**:1515–1522.

52. Voetman, A. A., and D. Roos. 1980. Endogenous catalase protects human blood phagocytes against oxidative damage by extracellularly generated hydrogen peroxide. *Blood* **56**:846–852.

53a. Spielberg, S. P., L. A. Boxer, J. M. Oliver, J. M. Allen, and J. D. Schulman. 1979. Oxidative damage to neutrophils in glutathione synthetase deficiency. *Br. J. Haematol.* **42**:215–223.

53b. Roos, D., R. S. Weening, A. A. Boetman, M.L.J. Vanschaik, A.A.M. Bot, L. J. Meerhof, and J. A. Loos. 1979. Protection of phagocytic leukocytes by endogenous glutathione. Studies in a family with glutathione reductase deficiency. *Blood* **53**:851–866.

54. Gee, J.B.L., C. L. Vassallo, P. Bell, J. Kaskin, R. E. Basford, and J. B. Field. 1970. Catalase dependent peroxidative metabolism in the alveolar macrophage during phagocytosis. *J. Clin. Invest.* **49**:1280–1287.

55. Michaelis, L. and L. B. Flexner. 1928. Oxidation-reduction systems of biological significance. I. The reduction potential of cysteine: its measurement and significance. *J. Biol. Chem.* **79**:689–722.

56. Guzman-Baron, E. S. 1951. Thiol groups of biological importance. *Adv. Enzymol. Relat. Issues* **11**:201–266.

57. Meister, A. and M. E. Anderson. 1983. Glutathione. [Review]. *Annu. Rev. Biochem.* **52**:711–760.

58. Cao, D., L. A. Boxer, and H. R. Petty. 1993. Deposition of reactive oxygen metabolites onto and within living tumor cells during neutrophil-mediated antibody-dependent cellular cytotoxicity. *J. Cell. Physiol.* **156**:428–436.

59. Maher, R. J., Cao D., L. A. Boxer, and H. R. Petty. 1993. Simultaneous calcium-dependent delivery of neutrophil lactoferrin and reactive oxygen metabolites to erythrocyte targets: evidence supporting granule-dependent triggering of superoxide deposition. *J. Cell. Physiol.* **156**:226–234.

60. Storz, G., L. A. Tartaglia, and B. N. Ames. 1990. Transcriptional regulator of oxidative stress-inducible genes: direct activation by oxidation. *Science* **248**:189–194.

61. Hentze, M. W., T. A. Rouault, J. B. Harford, and R. D. Klausner. 1989. Oxidation-reduction and the molecular mechanism of a regulatory RNA-protein interaction. *Science* **244**:358–359.

62. Ruppersberg, J. P., M. Stocker, O. Pongs, S. H. Heinemann, R. Frank, and M. Koenen. 1991. Regulation of fast inactivation of cloned mammalian $I_k(A)$ channels by cysteine oxidation. *Nature* **352**:711–714.

63. Deiss, L. P., and A. Kimchi. 1991. A genetic toll used to identify thioredoxin as a mediator of a growth inhibitory signal. *Science* **252**:117–120.

64. Grippo, J. F., A. Holmgren, and W. B. Pratt. 1985. Proof that the endogenous, heat-stable glucocorticoid receptor-activating factor is thioredoxin. *J. Biol. Chem.* **260**:93–97.

65. Ziegler, D. M. 1985. Role of reversible oxidation-reduction of enzyme thiols-disulfides in metabolic regulation. *Annu. Rev. Biochem.* **54**:305–329.

66. Gilbert, H. F. 1984. Redox control of enzyme activities by thiol/disulfide exchange. *Methods Enzymol.* **107**:330–351.

67. Huang, C.-K., G. R. Laramee, and J. E. Casnellie. 1988. Chemotactic factor induced tyrosine phosphorylation of membrane associated proteins in rabbit peritoneal neutrophils. *Biochem. Biophys. Res. Commun.* **151**:794–801.

68. Berkow, R. L. and R. W. Dodson. 1990. Tyrosine-specific protein phosphorylation during activation of human neutrophils. *Blood* **75**:2445–2452.

69. Akimaru, K., T. Utsumi, E. F. Sato, J. Klostergaard, M. Inoue, and K. Utsumi. 1992. Role of tyrosyl phosphorylation in neutrophil priming by tumor necrosis

factor-α and granulocyte colony stimulating factor. *Arch. Biochem. Biophys.* **298**:703–709.

70. Gomez-Cambronero, J., M. Yamazaki, F. Metwally, T.F.P. Molsky, V. A. Bonak, C.-K. Huang, E. L. Becker, and R. I. Sha'afi. 1989. Granulocyte-macrophage colony-stimulating factor and human neutrophils: role of guanine nucleotide regulatory proteins. *Proc. Natl. Acad. Sci. USA* **86**:3569–3573.

71. McColl, S. R., J. F. DiPersio, A. C. Caon, P. Ho, and P. H. Naccache. 1991. Involvement of tyrosine kinases in the activation of human peripheral blood neutrophils by granulocyte-macrophage colony-stimulating factor. *Blood* **7**:1842–1852.

72. Gomez-Cambronero, J., E. Wang, G. Johnson, C. K. Huang, and R. I. Sha'afi. 1991. Platelet activating factor induces tyrosine phosphorylation in human neutrophils. *J. Biol. Chem.* **266**:6240–6245.

73. Grinstein, S., W. Furuya, D. J. Lu, and G. B. Mills. 1990. Vanadate stimulates oxygen consumption and tyrosine phosphorylation in electropermeabilized human neutrophils. *J. Biol. Chem.* **265**:318–327.

74. Grinstein, S., and W. Furuya. 1991. Tyrosine phosphorylation and oxygen comsumption induced by G proteins in neutrophils. *Am. J. Physiol. (Cell Physiol.)* **260**:C1019–1027.

75. Gomez-Cambronero, J., C.-K. Huang, V. A. Bonak, E. Wang, J. E. Casnellie, T. Shiraishi, and R. I. Sha'afi. 1989. Tyrosine phosphorylation in human neutrophils. *Biochem. Biophys. Res. Commun.* **162**:1478–1485.

76. Naccache, P. H., C. Gilbert, A. C. Caon, M. Gaudry, C.-K. Huang, V. A. Bonak, K. Umezawa, and S. R. McColl. 1990. Selective inhibition of human neutrophil functional responsiveness by erbstatin, an inhibitor of tyrosine protein kinase. *Blood* **76**:10:2098–2104.

77. Gaudry, M., A. C. Caon, C. Gilbert, S. Lille, and P. H. Naccache. 1992. Evidence for the involvement of tyrosine kinases in the locomotory responses of human neutrophils. *J. Leukocyte Biol.* **51**:103–108.

78. Laudanna, C., F. Rossi, and G. Berton. 1993. Effect of inhibitors of distinct signalling pathways on neutrophil O_2^- generation in response to tumor necrosis factor-α, and antibodies against CD18 and CD11a: evidence for a common and unique pattern of sensitivity to wortmannin and protein tyrosine kinase inhibitors. *Biochem. Biophys. Res. Commun.* **3**:935–940.

79. Melamed, I., G. P. Downey, and C. M. Roifman. 1991. Tyrosine phosphorylation is essential for microfilament assembly in B lymphocytes. *Biochem Biophys. Res. Commun.* **3**:1424–1429.

80. McGregor, P. E., D. K. Agrawal, and J. D. Edwards. 1994. Attenuation of human leukocyte adherence to endothelial cell monolayers by tyrosine kinase inhibitors. *Biochem. Biophys. Res. Commun.* **198**:359–365.

81. Greenberg, S., P. Chang, and S. C. Silverstein. 1993. Tyrosine phosphorylation is required for Fc receptor-mediated phagocytosis in mouse macrophages. *J. Exp. Med.* **177**:529–534.

82. Fukushima, T., S. Grinstein, T. K. Waddell, G. G. Goss, C. K. Chan, M. Woodside, J. Orlowski, and G. P. Downey. 1996. Molecular and pharmacological characterization of Na⁺/H⁺ exchanger in human neutrophils. *J. Cell Biol.* **132**:1037–1052.

83. Fukushima, T., D. Rotin, M. Wheeler, and G. P. Downey. Submitted. Cloning and characterization of protein tyrosine phosphatases expressed in human neutrophils.

84. Waddell, T. K., L. Fialkow, C. K. Chan, T. K. Kishimoto, and G. P. Downey. 1995. Signalling functions of L-selectin: enhancement of tyrosine phosphorylation and activation of MAP kinase. *J. Biol. Chem.* **270**:15403–15411.

85. Schieven, G. L., J. M. Kirihara, D. L. Burg, R. L. Geahlen, and J. A. Ledbetter. 1993. p72^{syk} tyrosine kinase is activated by oxidizing conditions that induce lymphocyte tyrosine phosphorylation and Ca²⁺ signals. *J. Biol. Chem.* **268**:16688–16692.

86. Fialkow, L., C. K. Chan, S. Grinstein, and G. P. Downey. 1993. Regulation of tyrosine phosphorylation in neutrophils by the NADPH oxidase: role of reactive oxygen intermediates. *J. Biol. Chem.* **268**:17131–17137.

87. Margoliash, E., A. Novogrodsky, and A. Schejter. 1960. Irreversible reaction of 3-amino-1:2:4-triazole and related inhibitors with the protein of catalase. *Biochem. J.* **74**:339–348.

88. Zor, U., E. Ferber, P. Gergely, K. Szucs, V. Dombradi, and R. Goldman. 1993. Reactive oxygen species mediate phorbol ester-regulated tyrosine phosphorylation and phospholipase A₂ activation: potentiation by vanadate. *Biochem. J.* **295**:879–888.

89. Trudel, S., M. R. Paquet, and S. Grinstein. 1991. Mechanism of vanadate-induced activation of tyrosine phosphorylation and of the respiratory burst in HL-60 cells: role of reduced oxygen metabolites. *Biochem. J.* **276**:611–619.

90. Hayes, G. R., and D. H. Lockwood. 1987. Role of insulin receptor phophorylation in the insulinomimetic effects of hydrogen peroxide. *Proc. Natl. Acad. Sci. USA* **84**:8115–8119.

91. Koshio, O., Y. Akanuma, and M. Kasuga. 1988. Hydrogen peroxide stimulates tyrosine phosphorylation of the insulin receptor and its tyrosine kinase activity in intact cells. *Biochem. J.* **250**:95–101.

92. Heffetz, D., and Y. Zick. 1989. H₂O₂ potentiates phosphorylation of novel putative substrates for the insulin receptor kinase in intact fao cells. *J. Biol. Chem.* **17**:10126–10132.

93. Heffetz, D., I. Bushkin, R. Dror, and Y. Zick. 1990. The insulinomimetic agents H₂O₂ and vanadate stimulate protein tyrosine phosphorylation in intact cells. *J. Biol. Chem.* **265**:2896–2902.

94. Cross, A. R., and O.T.G. Jones. 1986. The effect of the inhibitor diphenylene iodonium on the superoxide-generating system in neutrophils: specific labelling of a component polypeptide of the oxidase. *Biochem. J.* **237**:111–116.

95. Stuehr, D. J., O. A. Fasehun, N. S. Kwon, S. S. Gross, J. A. Gonzalez, R. Levi, and C. Nathan. 1991. Inhibition of macrophage and endothelial cell nitric oxide synthase by diphenyleneiodonium and its analogs. *FASEB J.* **5**:98–103.

96. Lamas, S., T. Michel, B. M. Brenner, and P. A. Marsden. 1991. Nitric oxide synthesis in endothelial cells: evidence for a pathway inducible by TNF-α. *Am. J. Physiol.* **261**(*Cell Physiol* **30**):C634–C641.

97. Grinstein, S., and W. Furuya. 1988. Receptor-mediated activation of electropermeabilized neutrophils: evidence for a Ca²⁺ and protein kinase C-independent signaling pathway. *J. Biol. Chem.* **263**:1779–1783.

98. Fialkow, L. 1994. Reactive oxygen intermediates as signaling molecules: implications for acute lung injury. Ph.D. Thesis. Institute of Medical Science, University of Toronto, Toronto, Ontario, Canada.

99. Seifert, R., R. Burde, and G. Schultz. 1989. Activation of NADPH oxidase by purine and pyrimidine nucleotides involves G proteins and is potentiated by chemotactic peptides. *Biochem. J.* **259**:813–819.

100. Grinstein, S., and W. Furuya. 1988. Receptor-mediated activation of electropermeabilized neutrophils: evidence for a Ca²⁺ and protein kinase C-independent signaling pathway. *J. Biol. Chem.* **263**:1779–1783.

101. Sklar, L. A., P. A. Hyslop, Z. G. Oades, G. M. Omann, A. J. Jesaitis, R. G. Painter, and C. G. Cochrane. (1985). Signal transduction and ligand-receptor dynamics in the human neutrophil: transient responses and occupacy-response relations at the formyl peptide receptor. *J. Biol. Chem.* **260**:11461–11467.

102. Mustelin, T. 1994. *Src* Family Tyrosine Kinases in Leukocytes, pp. 8–33. R. G. Landes, Austin, TX.

103. Bauskin, A. R., I. Alkalay, and Y. Ben-Neriah. 1991. Redox regulation of a protein tyrosine kinase in the endoplasmic reticulum. *Cell* **66**:1–20.

104. Asahi, M., T. Taniguchi, E. Hashimoto, T. Inazu, H. Maeda, and H. Yamamura. 1993. Activation of protein-tyrosine kinase p72^syk with concanavalin A in polymorphonuclear neutrophils. *J. Biol. Chem.* **268**:23334–23338.

105. Ziegler, S. F., J. D. Marth, D. B. Lewis, and R. M. Perlmutter. 1987. Novel protein-tyrosine kinase gene (*hck*) preferentially expressed in cells of hematopoietic origin. *Mol. Cell. Biol.* **7**:2276–2285.

106. Gutkind, J. S., and K. C. Robbins. 1989. Translocation of the FGR protein-tyrosine kinase as a consequence of neutrophil activation. *Proc. Natl. Acad. Sci. USA* **86**:8783–8787.

108. Yamada, N., Y. Kawakami, H. Kimura, H. Fukamachi, G. Baier, A. Altman, T. Kato, Y. Inagaki, and T. Kawakami. 1993. Structure and expression of novel protein-tyrosine kinases, EMB and EMT, in hematopoietic cells. *Biochem. Biophys. Res. Commun.* **192**:231–240.

109. Brumell, J. H., A. L. Burkhardt, J. B. Bolen, and S. Grinstein. 1996. Endogenous reactive oxygen intermediates activate tyrosine kinases in human neutrophils. *J. Biol. Chem.* **271**:1455–1462.

110. Berton, G., L. Fumagalli, C. Laudanna, and C. Sorio. 1994. β2 integrin-dependent protein tyrosine phosphorylation and activation of the FGR protein tyrosine kinase in human neutrophils. *J. Cell Biol.* **126**:1111–1121.

111. Hamada, F., M. Aoki, T. Akiyama, and K. Toyoshima. 1993. Association of immunoglobulin G Fc receptor II with Src-like protein-tyrosine kinase Fgr in neutrophils. *Proc. Natl. Acad. Sci. USA* **90**:6305–6309.

112. Corey, S., A. Eguinoa, K. Puyanatheall, J. B. Bolen, L. Cantley, F. Mollinedo, T. R. Jackson, P. T. Kawkins, and L. R. Stephens. 1993. Granulocyte macrophage-colony stimulating factor stimulates both association and activation of phosphoinositide 3OH-kinase and *src*-related tyrosine kinase(s) in human myeloid derived cells. *EMBO J* **12**:2681–2690.

113. Gaudry, M., G. F. Barabé, P. E. Poubelle, and P. H. Naccache. 1995. Activation of Lyn is a common element of the stimulation of human neutrophils by soluble and particulate agonists. *Blood* **86**:3567–3574.

114. Ptasznik, A., A. Traynor-Kaplan, and G. M. Bokoch. 1995. G protein-coupled chemoattractant receptors regulate Lyn tyrosine kinase Sch adapter protein signaling complexes. *J. Biol. Chem.* **270**:19969–19973.

115. Skubitz, K. M., K. D. Campbell, K. Ahmed, and A.P.N. Skubitz. 1995. CD66 family members are associated with tyrosine kinase activity in human neutrophils. *J. Immunol.* **155**:5382–5390.

116. Huchcroft, J. E., M. L. Harrison, and R. L. Geahlen. 1992. Association of the 72-kDa protein-tyrosine kinase PTK72 with the B cell antigen receptor. *J. Biol. Chem.* **267**:8613–8619.

117. Agarwal, A., P. Salem, and K. C. Robbins. 1993. Involvement of P72syk, a protein-tyrosine kinase, in Fcγ receptor signaling. *J. Biol. Chem.* **268**:15900–15905.

118. Darby, C., R. L. Geahlen, and A. D. Schreiber. 1994. Stimulation of macrophage FcγRIIIA activated the receptor-associated protein tyrosine kinase syk and induces phosphorylation of multiple proteins including p95vav and p62/GAP-associated protein. *J. Immunol.* **152**:5429–5437.

119. Weiss, A., and D. R. Littman. 1994. Signal transduction by lymphocyte antigen receptors. *Cell* **76**:263–274.

120. Cooper, J. A., and B. Howeel. 1993. The when and how of src regulation. *Cell* **73**:1051–1054.

121. Mustelin, T. 1994. Regulation of *src*-family PTKs, pp. 34–52. *In* T. Mustelin (ed.), *Src* Family Tyrosine Kinases in Leukocytes. R. G. Landes, Austin, TX.

122. Superti-Fuga, G. 1995. Regualtion of the Src protein kinase. *FEBS Lett.* **369**:62–66.

123. Streuli, M., N. X. Krueger, A. Y. Tsai, and H. Saito. 1989. A family of receptor-linked protein tyrosine phosphatases in humans and drosophila. *Proc. Natl. Acad. Sci. USA* **86**:8698–8702.

124. Tonks, N. K., C. D. Diltz, and E. H. Fischer. 1988. Characterization of the major protein-tyrosine-phosphatases of human placenta. *J. Biol. Chem.* **263**:6731–6737.

125. Stocken, L. A., R. H. S. Thompson. 1946. Arsenic derivatives of thiol proteins. *Biochem. J.* **40**:529–535.

126. Garcia-Morales, P., Y. Minami, E. Luong, R. Klausner, and L. E. Samelson. 1992. Tyrosine phosphorylation in T cells is regulated by phopshatase activity: studies with phenylarsine oxide. *Proc. Natl. Acad. Sci. USA* **87**:9255–9259.

127. Balloti, R., S. Tartare, A. Chauvel, J.-C. Scimeca, F. Alengrin, C. Filloux, and E. Van Obberghen. 1991. Phenylarsine oxide stimulates a cytosolic tyrosine kinase activity and glucose transport in mouse fibroblasts. *Exp. Cell Res.* **197**:300–306.

128. Hecht, D., and Y. Zick. 1992. Selective inhibition of protein tyrosine phosphatase activities by H_2O_2 and vanadate *in vitro*. *Biochem. Biophys. Res. Commun.* **188**:773–779.

129. Goldman, R., E. Ferber, and U. Zor. 1992. Reactive oxygen species are second messengers for activation of cellular-phospholipase A_2. *FEBS Lett.* **309**:190–192.

131. Trudel, S., G. P. Downey, and S. Grinstein. 1990. Activation of permeabilized HL60 cells by vanadate: evidence for divergent signalling pathways. *Biochem. J.* **269**:127–131.

132. Pulido, R., V. Alvarez, F. Mollinedo, and F. Sanchez-Madrid. 1992. Biochemical and functional characterization of the leucocyte tyrosine phosphatase CD45 (CD45RO, 180 kD) from human neutrophils. In vivo upregulation of CD45RO plasma membrane expression on patients undergoing haemodialysis. *Clin. Exp. Immunol.* **87**:329–335.

133. Caldwell, C. W., W. P. Patterson, and Y. W. Yesus. 1991. Translocation of CD45RA in neutrophils. *J. Leukocyte Biol.* **49**:317–318.

134. Cui, Y., K. Harvey, L. Akard, J. Jansen, C. Hyghes, R. A. Siddiqui, and D. English. 1994. Regulation of neutrophil responses by phosphotyrosine phosphatase. *J. Immunol.* **152**:5420–5428.

135. Mustelin, T., K. M. Coggeshall, and A. Altman. 1989. Rapid activation of the T-cell tyrosine protein kinase pp56[lck] by the CD45 phosphotyrosine phosphatase. *Proc. Natl. Acad. Sci. USA* **86**:6302–6306.

136. Secrist, J. P., L. A. Burns, L. Karnitz, G. A. Koretzky, and R. T. Abraham. 1993. Stimulatory effects of protein tyrosine phosphatase inhibitor, pervanadate on T-cell activation events. *J. Biol. Chem.* **268**:5886–5893.

137. Fiakow, L., C. K. Chan, and G. P. Downey. Regulation of CD45 in neutrophils by reactive oxygen intermediates. *Immunol.* 1977.

138. Fialkow, L., C. K. Chan, D. Rotin, S. Grinstein, and G. P. Downey.1994. Activation of the mitogen-activated protein kinase signaling pathway in neutrophils: role of oxidants. *J. Biol. Chem.* **269**:31234–31242.

139. Kansha, M., K. Takeshige, and S. Minakami. 1993. Decrease in the phosphotyrosine phosphatase activity in the plasma membrane of human neutrophils on stimulation by phorbol 12-myristate 13-acetate. *Biochim. Biophys. Acta* **1179**:189–196.

140. Kraft, A. S., and R. L. Berkow. 1987. Tyrosine kinase and phosphotyrosine phosphatase activity in human promyelocytic leukemia cells and human polymorphonuclear leokocytes. *Blood* **70**:356–362.

141. Hancock, J. T., and T. G. Jones. 1987. The inhibition by diphenyleneiodonium and its analogues of superoxide generation by macrophages. *Biochem. J.* **242**:103–107.

142. Porter, T. D., and M. J. Coon. 1991. Cytochrome P-450. *J. Biol. Chem.* **266**:13469–13472.

143. Emmendörffer, A., J. Roesler, J. Elsner, E. Raeder, M. L. Lohmann-Matthes, and B. Meier. 1993. Production of oxygen radicals by fibroblasts and neutrophils from a patient with X-linked chronic granulomatous disease. *Eur. J. Haematol.* **51**:223–227.

144. Tonks, N.K, C. D. Diltz, and E. H. Fischer. 1988. Characterization of the major protein-tyrosine-phosphatases of human placenta. *J. Biol. Chem.* **263**:6731–6739.

145. Yamada, A., M. Streuli, H. Saito, D. M. Rothstein, S. F. Scholssman, and C. Morimoto. 1990. Effect of activation of protein kinase C on CD45 isoform expression and CD45 protein tyrosine phosphatase activity in T cells. *Eur. J. Immunol.* **20**:1655–1660.

146. Koretzky, G. 1993. A role of the CD45 tyrosine phosphatase in signal transduction in the immune system. *FASEB J.* **7**:420–426.

147. Sun, H., C. H. Charles, L. F. Lau, and N. K. Tonks. 1993. MKP-1 (3CH134), an immediate early gene product, is a dual specificity phosphatase that dephosphorylates MAP kinase *in vivo. Cell* **75**:487–493.

148. Guy, G. R., J. Cairns, S. B. Ng, and Y. H. Tan. 1993. Inactivation of a redox-sensitive protein phosphatase during the early events of tumor necrosis factor/interleukin-1 signal transduction. *J. Biol. Chem.* **268**:2141–2148.

149. MacKintosh, R. W., K. N. Dalby, D. G. Campbell, P.T.W. Cohen, P. Cohen, and C. MacKintosh. 1995. The cyanobacterial toxin microcystin binds covalently to cysteine-273 on protein phosphatase 1. *FEBS Lett.* **371**:236–240.

150. Shima, H., H. Tohda, S. Aonuma, M. Nakayasu, A. A. DePaoli-Roach, T. Sumigura, and M. Nagao. 1994. Characterization of the PP2Aα gene mutation in okadaic acid-resistant variants of CHO-K1 cells. *Proc. Natl. Acad. Sci. USA* **91**:9267–9271.

151. Brummell, J., C. K. Chan, J. Butler, N. Borregaard, K. Siminovitch, S. Grinstein, and G. P. Downey. Submitted. Regulation of PTP1C during activation of human neutrophils: role of protein kinase C. *J. Biol. Chem.*

152. Ahn, N. G., R. Seger, and E. C. Krebs. 1992. The mitogen-activated protein kinase activator. *Curr. Opin. Cell. Biol.* **4**:992–999.

153a. Crews, C. M., and R. L. Erikson. 1993. Extracellular signals and reversible protein phosphorylation: what to Mek of it all. *Cell* **74**:215–217.

153b. Dent, P., W. Haser, T. A. Haystead, L. A. Vincent, T. M. Roberts, and T. W. Sturgill. 1992. Activation of mitogen-activated protein kinase kinase by v-Raf in NIH 3T3 cells and *in vitro. Science* **257**:1404–1407.

154. Howe, L. R., S. J. Leevers, N. Gomes, S. Nakielny, P. Cohen, and C. J. Marshall. 1992. Activation of the MAP kinase pathway by the protein kinase raf. *Cell* **71**:335–342.

155. Kyriakis, J. M., H. App, X.-F. Zhang, P. Banerjee, D. L. Brautigan, U. R. Rapp, and J. Avruch. 1992. Raf-1 activates MAP kinase-kinase. *Nature* **358**:417–421.

156. Lange-Carter, C. A., C. M. Pleiman, A. M. Gardner, K. J. Blumer, and G. L. Johnson. 1993. A divergence in the MAP kinase regulatory network defined by MEK kinase and Raf. *Science* **260**:315–319.

157. Anderson, N. G., J. L. Maller, N. K. Tonks, and T. W. Sturgill. 1990. Requirement for integration of signals from two distinct phosphorylation pathways for activation of MAP kinase. *Nature* **343**:651–652.

158. Nishizuka, Y. 1988. The molecular heterogeneity of protein kinase C and its implications for cellular regulation. *Nature* **334**:661–665.

159. Kass, G.E.N., S. K. Duddy, and S. Orrenius. 1989. Activation of hepatocyte protein kinase C by redox-cycling quinones. *Biochem. J.* **260**:499–507.

160. Gopalakrishna, R., and W. B. Anderson. 1989. Ca^{2+}- and phospholipid-independent activation of protein kinase C by selective oxidative modification of the regulatory domain. *Proc. Natl. Acad. Sci. USA* **86**:6758–6762.

161. Lin, L.-L., M. Wartmann, A. Y. Lin, J. L. Knopf, A. Seth, and R. J. Davis. 1993. cPLA$_2$ is phosphorylated and activated by MAP kinase. *Cell* **72**:269–278.

162. Durstin, M., S. Durstin, T.F.P. Molski, E. Becker, and R. I. Sha'afi. 1994. Cytolasmic phopholipase A$_2$ translocates to membrane fraction in human neutrophils activated by stimuli that phosphorylate mitogen-activated protein kinase. *Proc. Natl. Acad. Sci. USA* **91**:3142–3146.

163. Schulze-Osthoff, K., M. Los, and P. A. Baeuerle. 1995. Redox signalling by transcription factors NF-kappa B and AP-1 in lymphocytes. *Biochem. Pharmacol.* **50**:735–741.

164. Salmon, J. E., S. S. Millard, N. L. Brogle, and R. P. Kimberly. 1995. Fcγ receptor IIIb enhances FcγIIa function in an oxidant-dependent and allele-sensitive manner. *J. Clin. Invest.* **95**:2877–2885.

165. Gresham, H. D., J. A. McGarr, P. G. Shackelford, and E. J. Brown. 1988. Studies on the molecular mechanisms of human Fc receptor-mediated phagocytosis. Amplification of ingestion is dependent on the generation of reactive oxygen metabolites and is deficient in polymorphonuclear from patients with chronic granulomatous disease. *J. Clin. Invest.* **82**:1192–1201.

166. Connelly, P. A., C. A. Farrell, J. M. Merenda, M. J. Conklyn, and H. J. Showell. 1991. Tyrosine phosphorylation is an early signaling event common to Fc receptor cross-linking in human neutrophils and rat basophilic leukemia cells (RBL-2H3). *Biochem. Biophys. Res. Commun.* **1**:192–201.

167. Scholl P. R., D. Ahern, and R. S. Geha. 1992. Protein tyrosine phosphorylation induced via the IgG receptors FcγRI and FcγRII in the human monocytic cell line THP-1. *J. Immunol.* **149**:1751–1757.

168. Greenberg, S., P. Chang, and S. C. Silverstein. 1994. Tyrosine phosphorylation of the γ subunit of Fcγ receptors, p72^{syk}, and paxillin during Fc receptor-mediated phagocytosis in macrophages. *J. Biol. Chem.* **269**:3897–3902.

169. Zhou, M.-J., D. M. Lublin, D. C. Link, and E. J. Brown. 1995. Distinct tyrosine kinase activation and Triton X-100 insolubility upon FcγRII or FcγRIIIB ligation in human polymorphonuclear leukocytes. *J. Biol. Chem.* **270**:13553–13560.

170. Thomas, M. L. 1994. The regulation of B- and T-lymphocyte activation by the transmembrane protein tyrosine phosphatase CD45. *Curr. Opin. Cell. Biol.* **6**:247–252.

171. Okumura, M., and M. L. Thomas. 1995. Regulation of immune function by protein tyrosine phosphatases. *Curr. Opin. Immunol.* **7**:312–319.

172. Trowbridge, I. S. 1994. CD45: and emerging role as a protein tyrosine phosphatase required for lymphocyte activation and development. *Annu. Rev. Immunol.* **12**:85–116.

173. Staal, F.J.T., Anderson, M. T., Staal, G.E.J., Herzenberg, L. A., Gitler C., and Herzenberg L. N. 1994. Redox regulation of signal transduction: tyrosine phosphorylation and calcium influx. *Proc. Natl. Acad. Sci. USA.* **91**:3619–3622.

174. Harvath, L., J. A. Balke, N. P. Christiansen, A. A. Russell, and K. M. Skubitz. 1991. Selected antibodies to leukocyte common antigen (CD45) inhibit human neutrophil chemotaxis. *J. Immunol.* **146**:949–957.

175. Hoffmeyer, F., K. Witte, U. Gebhardt, and R. E. Schmidt. 1995. The low affinity FCγRIIa and FCγRIIIb on polymorphonuclear neutrophils are differentially regulated by CD45 phosphatase. *J. Immunol.* **155**:4016–4022.

176. Omann, G. M., J. M. Harter, J. M. Burger, and D. B. Hinshaw. 1994. H_2O_2-induced increased in cellular F-actin occur without increased in actin nucleation activity. *Arch. Biochem. Biophys.* **308**:407–412.

177. Sundaresan, M., Z.-X Yu, V. J. Ferrans, K. Irani, and T. Finkel. 1995. Requirement for generation of H_2O_2 for platelet-derived growth factor signal transduction. *Science* **270**:296–299.

178. Grinstein, S., W. Furuya, J. R. Butler, and J. Tseng. 1993. Receptor-mediated activation of multiple serine/threonine kinases in human leukocytes. *J. Biol. Chem.* **268**:20223–20231.

179. Grinstein, S., and W. Furuya. 1992. Chemoattractant-induced tyrosine phosphorylation and activation of microtubule-associated protein kinase in human neutrophils. *J. Biol. Chem.* **25**:18122–18125.

180. Pulverer, B. J., J. M. Kyriakis, J. Avruch, E. Nikolakaki, and J. R. Woodgett. 1991. Phosphorylation of c-*jun* mediated by MAP kinases. *Nature* **353**:670–674.

181. Kramer, R. M. 1993. Structure, function and regulation of mammalian phospholipases A₂. *Adv. Second Messenger Phosphoprotein Res.* **28**:81–89.

182. Peppelenbosch, M. P., L.G.J. Tertoolen, W. J. Hage, and S. W. De Laat. 1993. Epidermal growth factor-induced actin remodeling is regulated by 5-lipoxygenase and cyclooxygenase products. *Cell* **74**:565–575.

183. Haslett, C., S. Savill, M.K.B. Whyte, M. Stern, I. Dransfield, and L. C. Meagher. 1994. Granulocyte apoptosis and the control of inflammation. *Phil. Trans. R. Soc. Lond. B.* **345**:327–333.

184. Savill, J. and C. Haslett. 1995. Granulocyte clearance by apoptosis in the resolution of inflammation. *Semin. Cell Biol.* **6**:385–393.

185. Buttke, T. M., and P. A. Sandstrom. 1994. Oxidative stress as a mediator of apoptosis. *Immunol. Today* **15**:7–10.

186. Lennon, S. V., S. J. Martin, and T. G. Cotter. 1991. Dose-dependent induction of apoptosis in human tumour cell lines by widely diverging stimuli. *Cell Prolif.* **24**:203–214.

187. Greenlund, L. J., T. L. Deckwerth, and E. M. Johnson, Jr. 1995. Superoxide dismutase delays neuronal apoptosis: a role for reactive oxygen species in programmed neuronal death. *Neuron* **14**:303–312.

188. Yao, X. R., and D. W. Scott. 1994. Antisense oligodeoxynecleotides to the blk tyrosine kinase prevent anti-mu-chain-mediated growth inhibition and apoptosis in a B-cell lymphoma. *Proc. Natl. Acad. Sci. USA* **90**:7946–7950.

189. Manabe, A., T. Yi, M. Kumagai, and D. Campana. 1993. Use of strome-supported cultures of leukemic cells to assess antileukemic drugs. Cytotoxicity of interferon alpha in acute lymphoblastic leukemia. *Leukemia* **7**:1990–1995.

190. Yousefi, S., D. R. Grenn, K. Blaser, and H.-U. Simon. 1994. Protein-tyrosine phosphorylation regulates apoptosis in human eosinophils and neutrophils. *Proc. Natl. Acad. Sci. USA* **91**:10868–10872.

191. Schieven, G. L., A. F. Wahl, S. Myrdal, L. Grosmaire, and J. A. Ledbetter. 1995. Lineage-specific induction of B cell apoptosis and altered signal transduction by the phosphotyrosine phosphatase inhibitor bis (maltolato) oxovanadium (IV). *J. Biol. Chem.* **270**:20824–20831.

192. Hanaoka, K., N. Fujita, S. H. Lee, H. Seimiya, M. Naito, and T. Tsuruo. 1995. Involvement of CD45 in adhesion and suppression of apoptosis of mouse malignant T-lymphoma cells. *Cancer Res.* **55**:2186–2190.

193. Lin, K.-T., J.-Y. Xue, M. Nomen, B. Spur, and P. Y.-K. Wong. 1995. Peroxynitrite-induced apoptosis in HL-60 cells. *J. Biol. Chem.* **270**:16487–16490.

194. Schreck, R., P. Rieber, and P. A. Baeuerle. 1991. Reactive oxygen intermediates as apparently widely used messengers in the activation of the NF-κB transcription factor and HIV-1. *EMBO J.* **10**:2247–2258.

195. Crawford, D., I. Zbinden, P. Amstad, and P. Cerutti. 1988. Oxidant stress induces the protooncogenes c-*fos* and c-*myc* in mouse epidermal cells. *Oncogene* **3**:27–32.

196. Nose, K., M. Shibanuma, K. Kikuchi, H. Kageyama, S. Sakiyama, and T. Kuroki. 1991. Transcriptional activation of early-response genes by hydrogen peroxide in a mouse osteoblastic cell line. *Eur. J. Biochem.* **201**:99–106.

197. Xanthoudakis, S., G. Miao, F. Wang, Y.-C.E. Pan, an T. Curran. 1992. Redox activation of Fos-Jun DNA binding activity is mediated by a DNA repair enzyme. *EMBO J.* **11**:3323–3335.

198. Baeuerle, P. A. 1991. The inducible transcrition activator NF-κB: regulation by distinct proteins subunits. *Biochim. Biophys. Acta* **1072**:63–80.

199. Ding, A., S. Hwang, H. M. Lander, and Q. W. Xie. 1995. Macrophages derived from C3H/HeJ (Lpsd) mice respond to bacterial lipopolysaccharide by activating NF-kappa B. *J. Leukocyte Biol.* **57**:174–179.

200. Xie, Q.-W., Y. Kashiwabara, and C. Nathan. 1994. Role of transcription factor NF-κB/Rel in induction of nitric oxide synthase. *J. Biol. Chem.* **269**:4705–4708.

201. Müller, J. M., H.W.L. Ziegler-Heitbrock, and P. A. Baeuerle. 1993. Nuclear factor kappa B, a mediator of lipopolysaccharide effects. *Immunob.* **187**:233–256.

202. Shakhov, A. N., M. A. Collart, P. Vassalli, A. S. Nedospasov, and C. V. Jongeneei. 1990. κB-type enhancers are involved in lipopolysaccharide-mediated transcrip-

tional activation of the tumor necrosis factor α gene in macrophages. *J. Exp. Med.* **171**:35–47.

204. Lo, Y. Y. and T. F. Cruz. 1995. Involvement of reactive oxygen species in cytokine and growth factor induction of c-*fos* expression in chondrocytes. *J. Biol. Chem.* **270**:11727–11730.

205. Schulze-Osthoff, K., R. Beyaert, V. Vandevoorde, G. Haegeman, and W. Fiers. 1993. Depletion of the mitochondrial electron transport abrogates the cytotoxic and gene-inductive effects of TNF. *EMBO J.* **8**:3095–3104.

206. Satriano, J. A., M. Shuldiner, K. Hora, Y. Xing, Z. Shan, and D. Schlondorff. 1993. Oxygen radicals as second messengers for expression of the monocyte chemoattractant protein, JE/MCP-1 and the monocyte colony-stimulating factor, CSF-1, in response to tumor necrosis factor-α and immunoglobulin G. *J. Clin. Invest.* **92**:1564–1571.

207. Root, R. K., J. Metcalf, N. Oshino, and B. Chance. 1975. H_2O_2 release from human granulocytes during phagocytosis. I. Documentation, quantitation, and some regulating factors. *J. Clin. Invest.* **55**:945–955.

208. Patel, K. C., G. A. Zimmerman, S. M. Precott, R. P. McEver, and T. M. McIntyre. 1991. Oxygen radicals induce human endothelial cells to express GMP-140 and bind neutrophils. *J. Cell Biol.* **112**:749–759.

209. Natajaran, V., M. M. Taher, B. Roehm, N. L. Parinandi, H.H.O. Schmid, Z. Kiss, and J.G.N. Garcia. 1993. Activation of endothelial cell phospholipase D by hydrogen peroxide and fatty acid hydroperoxide. *J. Biol. Chem.* 930–937.

210. Downey, G. P., and H. O'Brodovich. In press. Mechanisms of lung injury and repair. *In* L. Tanssig and M. Textbook (eds.), *Textbook of Pediatric Respiratory Medicine*. Mosby–Yearbook, St. Louis, MO.

The Connection of Oxidative Induced Changes in Second Messengers and Transcription Factors with Activation of Gene Expression

10

Reactive Oxygen Intermediates as Primary Signals and Second Messengers in the Activation of Transcription Factors

Klaus Schulze-Osthoff, Manuel Bauer, Markus Vogt, Sebastian Wesselborg, and Patrick A. Baeuerle

10.1 Introduction

Prokaryotic and eukaryotic cells have developed elaborate mechanisms to rapidly respond to changes in their environment by the novel expression of genes. Different forms of cellular stresses constitute primary signals that are transduced via the plasma membrane into the cytoplasm and ultimately stimulate the expression of specific genes in the cell nucleus. Increased production of reactive oxygen intermediates (ROIs), referred to as oxidative stress, was found to be implicated in a variety of distinct cellular stresses, such as heat shock, exposure to ionizing and UV irradiation and environmental pollutants. The products of genes that are newly induced in response to oxidative stress may confer protection against subsequent adversities; repair ROI-mediated damage of cellular components; or serve to signal stress to neighboring cells or cells in other parts of the organism.

Intriguingly, higher multicellular organisms have also evolved enzymes that inducibly synthesize ROIs. In one well-examined case, the reactive compounds are needed as weapons against invading microorganisms. Granulocytes and macrophages have specialized in releasing large amounts of H_2O_2 and superoxide in a respiratory burst reaction. However, many other cell types can also inducibly produce ROIs but in amounts insufficient to kill microorganisms. There is increasing evidence that these small increases in ROI levels fulfill a role as second messengers controlling gene transcription. We and others have proposed that this "signaling burst" serves to regulate the expression of genes whose products serve important functions in the immune response, proliferation control, and differentiation processes during a general pathogenic response.

Recent research has identified several transcription factors that are activated by fluctuations of the intracellular ROI level. In this chapter, we will describe and compare molecular mechanisms that have evolved to translate ROI-mediated signals into altered patterns of gene expression. We especially focus on the redox regulation of NF-κB and activator-protein-1 (AP-1), two higher eukaryotic transcription factors controlled by redox-dependent processes.

10.2 Mechanisms of Gene Transcription

The synthesis of new proteins is most frequently regulated at the transcriptional level by transcription factors binding to regulatory DNA sequences within target genes. In both bacteria and higher organisms, gene expression is controlled by sequence-specific promoter elements that allow the oriented binding of DNA-dependent RNA polymerases. In eukaryotes, the accurate initiation of transcription is regulated by controlled interactions between two different types of DNA-binding proteins.

The first are a group of ubiquitous, general factors required for recognition of the transcriptional start site, the assembly of the preinitiation complex and the subsequent recruitment of RNA polymerase. These basal transcription factors, including TFII-A, -B, -D, -E, -H, and -I interact with RNA polymerase II in the initiation of transcription. Formation of the preinitiation complex is coordinated by binding of the TATA box-binding protein (TBP) and associating factors (TAFs) to the TATA box element.[1]

A second class of transcription factors regulate either tissue-specific or inducible gene expression. These activators recognize site-specific sequences in the upstream promoter region and can dramatically increase or decrease the rate of transcription. Various strategies have evolved to induce gene transcription in response to extracellular signals. One widespread mechanism is the *de novo* synthesis of a transcription factor. This is not a direct event and requires one or more "primary" transcription factors to be activated by post-translational mechanisms. In some cases, inhibitory proteins play a role in controlling activation of transcription. One important example to be discussed is the transcription factor NF-κB. NF-κB resides in the cytoplasm of uninduced cells as an inactive complex with its inhibitory subunit IκB.[2] Exposure to many stimuli rapidly activates NF-κB by dissociation of IκB. This allows translocation of the factor into the nucleus and subsequent DNA binding. A further mechanism of activating transcription factors is binding of accessory proteins. A paradigm is the serum response factor (SRF), which is only active after association on DNA with the ternary complex factor (TCF).[3,4] Perhaps the most common mechanism of controlling the action of preexisting transcription factors is their covalent modification of amino acid residues. Frequently, this involves phosphorylation or dephosphorylation of serine or threonine residues in the DNA binding or transactivation domain of the factor.[5] Examples are activation of the cyclic AMP (cAMP) response binding protein (CREB) by phosphorylation in response to increased cAMP levels and of c-Jun by dephosphorylation in response to activation of protein kinase C. Other post-translational modifications with regulatory potential are reduction-oxidation (redox) reactions. Several transcription factors contain a conserved cysteine residue in their DNA binding region.[6,7] In cell-free systems, oxidation of such cysteine residues has been shown to decrease DNA binding. In the following, we show that oxidative stress-controlled gene expression involves the regulation of transcriptional factors at several distinct levels.

10.3 Genetic Response to Oxidative Stress

When cells are treated with H_2O_2, they become resistant towards subsequent higher amounts of ROIs that would be lethal without pretreatment.[8] This observation indicates that cells can activate an adaptive genetic programme against oxidative stress.[9] The genetic response to oxidative stress has been intensively studied in bacteria.[10–12] In *Escherichia coli,* the synthesis of some 80 proteins is induced on exposure to H_2O_2- or O_2^--generating chemicals. Two transcription factor systems, Oxy R and Sox R/S, have been identified that activate expression of genes whose products are involved in either the protection from oxidative stress or the repair of ROI-mediated damages. Intriguingly, the two transcription factors respond to different ROI species. Oxy R is activated by H_2O_2, while Sox R/S is selectively induced by superoxide anions. Transcription factors which sense changes in dioxygen or ROI concentration have also been found in yeast and plants. In *Saccharomyces cerevisiae,* binding of a heme group, whose synthesis requires oxygen, activates the transcription factor HAP-1, which controls transcription of the catalase and superoxide dismutase gene.[13] In plants, salicylic acid (SA) is known as a mediator for systemic acquired resistance, an equivalent of the vertebrate immune response. Recently, the SA-binding protein has been shown to contain catalase activity, which significantly decreases on ligand binding and thereby results in the accumulation of H_2O_2.[14] Hence, the plant immune response involves generation of ROI messengers by ligand-induced inhibition of an ROI-eliminating enzyme.

In mammalian cells, the mechanisms by which ROIs are sensed or inducibly produced are less well understood, and transcription factors that are exclusively activated by ROIs or that selectively control expression of ROI-protective and -repair enzymes have not yet been identified. The oxidant stress response involves the activation of numerous functionally unrelated genes associated with signal transduction, proliferation, and immunological defense reactions. By analogy to bacteria, it is assumed that the mammalian response to ROI serves a protective function. However, most of the genes induced by oxidative stress can be equally activated by specific physiological signals, such as growth factors or cytokines. This indicates that both groups of signals converge into the same pathway by sharing signaling molecules. Experimental evidence indeed suggests that this is the case. Physiological as well as ROI-triggered signals activate NF-κB and AP-1, two important and widely used transcription factors. As described below, the overlapping effects of both signals may also be explained by the fact that some physiological inducers seem to utilize ROIs as intracellular signaling molecules. In eukaryotes, ROIs may therefore act as second messenger molecules which integrate the diversity of gene-inducing signals into a common genetic response. Hence, ROI-induced gene expression might not be restricted to adverse environmental conditions, but have a more widespread and principal role in cellular metabolism.

10.4 Transcription Factor NF-κB

NF-κB was the first eukaryotic transcription factor shown to respond directly to oxidative stress. The transactivator plays a crucial role in the regulation of numerous genes involved in immune and inflammatory processes.[15-18] A great variety of stimuli can activate NF-κB and initiate transcription of NF-κB-dependent genes. In the following, we describe that the effect of perhaps all activating stimuli relies on the induction of prooxidant conditions.

NF-κB was first identified as a nuclear factor of mature B cells that specifically interacts with a decameric enhancer element (5'-GGGACTTTCC-3') of the immunoglobulin κ light chain gene.[19] Very frequently, NF-κB is composed of the p50 and p65 subunits but at least three other subunits have been reported to participate in dimer formation. In addition to p50 and p65 (now also called RelA), p52, c-Rel, and RelB have been identified.[17,20,21] All share a homologous domain of about 300 amino acids which is required for dimerization, DNA binding, nuclear targeting, and interaction with regulatory proteins. While p50 and p52 may be solely required for DNA binding, p65, c-Rel, and RelB have in addition transactivating activity. It has been shown that combinatorial interactions between the different NF-κB subunits give rise to dimers with distinguishable sequence and transcriptional specificity. In T cells and some other cell types, a transcriptionally inactive p50 homodimer can be found that occupies κB sites for active complexes.[22] This negative regulatory effect of p50 homodimers was suggested to play a physiological role during T cell activation.

A characteristic of NF-κB activation is that it does not require new protein synthesis. In non-stimulated cells, NF-κB is bound to a family of inhibitory proteins, called IκBs, which retain the complex in the cytoplasm and thereby prevent DNA binding.[2,23,24] Activation of NF-κB involves the dissociation of IκB from the NF-κB complex. The best characterized inhibitor is IκB-α, which prevents DNA binding to p65, c-Rel, and RelB. All IκB proteins contain multiple, closely adjacent copies of a characteristic repeat structure of 30 amino acids, called SWI6 ankyrin repeats.[20] Mutational analysis has revealed that these ankyrin repeats are necessary but not fully sufficient to prevent DNA binding.[25] It remains to be established how association of NF-κB subunits with different IκB proteins affects the mechanism of activation. Recent experimental evidence suggests that

IκB-α is apparently not restricted to the cytoplasm, but may also occur within the nucleus.[26] Since IκB decreases the half-life of the NF-κB-DNA complex,[27] and NF-κB upregulates transcription of the IκB-α gene,[28,29] IκB-α may have an additional role as a negative regulator of NF-κB-dependent transcription.

10.5 Activation of NF-κB by ROIs

Potent stimuli activating NF-κB in intact cells are the cytokines tumor necrosis factor (TNF) and interleukin-1 (IL-1), phorbol esters, lipopolysaccharide (LPS), double-stranded RNA, protein synthesis inhibitors, UV and ionizing irradiation, and viral transactivator proteins. Although many of these activators induce a similar pattern of genes as phorbol esters, protein kinase C does not seem to be involved in each of the signaling pathways. A common feature of all inducers, however, is that they are pathogenic or at least proinflammatory stimuli. Interestingly, many, if not all NF-κB-inducing agents lead to an increase in the intracellular concentration of ROIs. In particular, TNF has been reported to induce oxidative stress, which is thought to contribute to its cytotoxic and proinflammatory effects.[30,31] Since TNF is a potent inducer of NF-κB activation, we analyzed whether oxidative stress induced by exposure of cells to H_2O_2 can activate NF-κB. In the human T cell line Jurkat, 30 to 150 µM H_2O_2 rapidly induced NF-κB DNA binding and transactivation activity.[32,33] The activation of NF-κB in response to H_2O_2 was not restricted to lymphoid cells. For instance, in the cervix carcinoma line HeLa, 250 µM H_2O_2 strongly induced NF-κB activation.[34] Another reagent producing peroxides and potently stimulating NF-κB activation was butylperoxide. In contrast, several anthraqinone compounds, such as paraquat, menadione, doxorubicin, or mitomycin, which produce superoxide anions by redox cycling, failed to rapidly induce NF-κB activation.[35] Also the endoperoxide 3,3′-(1,4-naphtylidene)dipropionate (NDPO₂), a substance that releases singlet oxygen, did not cause NF-κB DNA binding. Thus, it appears that H_2O_2 but not other ROIs are the mediators of prooxidant-induced NF-κB activation. Indeed H_2O_2 has the best potential to serve as a messenger. It is the least harmful ROI with the largest stability, diffusion radius and membrane permeability.

Nitric oxide (NO·), which has a well established signaling role in several biological processes, has recently been described to induce NF-κB in peripheral blood lymphocytes.[36] However, our own experiments revealed that neither NO·-generating agents, such as nitroprusside, had an activating effect nor that inhibitors of NO· synthase were capable of suppressing TPA or TNF-induced NF-κB activation.[35] Moreover, several other toxic conditions such as exposure to heavy metals (cadmium, copper salts) or heat shock did not lead to the activation of NF-κB. A recent report, however, shows that the transition metal nickel can induce NF-κB activation as well as expression of the NF-κB target genes Interleukin-6 (IL-6) and Endothelial Cell Adhesion Molecule-1 (ELAM-1) in endothelial cells.[37]

10.6 Effects of Antioxidants on NF-κB Activation

A second line of evidence for the idea that NF-κB activation relies on intracellular ROI formation was provided by the effect of antioxidants. Numerous reports showed that the activation of NF-κB by several stimuli is prevented when cells are pretreated with antioxidative compounds.[35,38] A great variety of compounds act as such inhibitors: thiol compounds (N-acetylcysteine, lipoic acid), dithiocarbamates, thioredoxin, vitamine E and derivatives, iron chelators, salicylate, and other phenolic compounds, such as butylated hydroxyanisol. The NF-κB inhibitors can be divided into two classes. One encompasses scavengers that can directly react with ROIs thereby neutralizing their reactivity. The other class are compounds that interfere indirectly with the production of ROIs. Examples are rotenone, which prevents mitochondrial electron transport[39]; tepoxalin, which inhibits ROI production by cyclooxygenases[40]; or desferal, an iron chelator that prevents hydroxyl radical formation by the Fenton reaction.[35] The NF-κB-inhibitory potential of an antioxidant appears to correlate with its subcellular distribution. The lipophilic vitamine E is more than 100 times less active than its hydrophilic chromane ring,[41] suggesting that cytosolic rather than membrane-bound production of ROIs is relevant for NF-κB activation.

The third line of evidence that H_2O_2 and not other forms of ROIs serve as messenger for NFκB activation comes from molecular genetic experiments.[42] The steady state levels of H_2O_2 and superoxide in the cell are determined by the activities of catalase and superoxide dismutase. This allows to manipulate the cytosolic ROI levels by overexpression of catalase or Cu/Zn-dependent superoxide dismutase (SOD) genes. Catalase overexpression will result in decreased H_2O_2 levels, while SOD overexpression in decreased superoxide but increased H_2O_2 levels. We studied the activation of NF-κB by TNF in mouse keratinocyte cell clones which contain stably integrated extra genes for catalase and SOD resulting in 2.7-fold increased enzymatic activities in comparison to the parental cell clone. Consistent with an important role for H_2O_2 but not superoxide, catalase-overexpressing cells were deficient for NF-κB activation. The catalase inhibitor aminotriazol, however, could restore a normal responsiveness. SOD-overexpressing cells showed a superinduced NF-κB activation, presumably, resulting from an enhanced conversion of superoxide into H_2O_2.

10.7 Inducible ROI Production

Various pathogenic or proinflammatory stimuli, such LPS, viral infection, irradiation, T-cell mitogens, and the cytokines IL-1 and TNF, rapidly lead to NF-κB activation. Therefore the question arises, how this diversity of stimuli commonly induce ROI formation. Since a great variety of agents activate the transcription factor, it is very likely that oxidative stress is elicited by more than one means

Table 10.1.   Mechanisms of ROI-induced NF-κB activation.

NF-κB inducing condition	TNF	anti-CD28	PMA	HIV-Tat
Mechanism of ROI Generation	Mitochondrial respiratory chain	Lipoxygenase	COX-1 (COX-2)	Repression of MnSOD synthesis
Evidence	Pharmacological (Rotenon); Biological (Depletion of the respiratory chain)	Pharmacological, (5-Lipoxygenase inhibition with ICI 230487, MK 886)	Pharmacological (Tepoxalin) COX-1 overexpression	Overexpression, Exogenous addition of HIV-Tat
Reference	38	44	40	45

(Table 10.1). In the case of TNF, a very potent activator of NF-κB, it has been suggested that the respiratory chain of mitochondria is the site of ROI formation, which is important for the cytotoxicity as well as NF-κB activation by TNF.[31,39] In cells lacking a functional mitochondrial respiratory chain as result of drug action or organelle depletion, NF-κB activation was significantly suppressed.

In T cells, a potent NF-κB activation usually requires two signals, one triggering the T cell receptor (TCR) and another one a costimulatory molecule. Recently, it has been observed that triggering CD28, a major costimulatory molecule, causes the rapid decrease of reduced glutathione (GSH) within minutes and enhanced ROI formation, which was maintained for 16 to 24 h after cell stimulation.[43,44] This long-lasting prooxidant stimulus may explain why, in contrast to many other cell types, NF-κB activation in T cells is generally observed for a rather prolonged period. Inhibitor studies have demonstrated that the dioxygenase activity of 5-lipoxygenase was involved in inducing the prooxidant signal observed upon CD28 ligation.[44] In addition to lipoxygenases, other enzymes of the arachidonic acid metabolism may be also involved in NF-κB activation. There are evidences that NF-κB stimulation in response to phorbol esters is mediated by cyclooxygenase-1 (COX-1).[40]

Still another principle of ROI formation has been demonstrated for the Tat protein, a major transactivating factor of human immunodeficiency virus (HIV-1). The potentiation of NF-κB activation by Tat has been attributed to an increased ROI formation by Tat-mediated transcriptional inhibition of the Mn-dependent SOD gene.[45] This situation is reminiscent to the plant system where a redox signal is produced by inhibition of catalase, another ROI-detoxifying enzyme.

10.8 ROI-Induced Signal Transduction

Recently, some steps in the signal transduction pathway leading to NF-κB activation have been elucidated, although the precise mode of how ROIs activate NF-κB is still largely unknown. As described above, NF-κB is mostly present in the cytoplasm as an inactive complex bound to the inhibitory subunit IκB. Activation of cells results in the rapid release of IκB, which allows NF-κB to translocate to the nucleus and to bind to its cognate DNA sequences. It has been demonstrated that NF-κB activation coincides with the rapid proteolytic loss of IκB. The degradation of IκB is absolutely necessary because various protease inhibitors prevent NF-κB activation.[46–48] The degradation of IκB-α is preceeded by phosphorylation on serines 32 and 36.[49] This modification is not sufficient to release IκB from NF-κB but dramatically increases the turnover of IκB presumably by enhancing the conjugation with ubiquitin.[50]

Where in these steps of NF-κB activation do ROIs play a role? Unlike the bacterial peroxide sensor Oxy R that can be activated by oxidation *in vitro,* oxidation of purified NF-κB-IκB does not activate but even causes a loss of DNA-binding activity.[51] A conserved cysteine residue in the DNA-binding domain of NF-κB subunits seems to be responsible for this oxidative inhibition.[7] However, since glutathione (GSH) is highly reduced in the nucleus, the physiological significance of this mechanism remains unclear. A recent study demonstrated that the phosphorylation of IκB-α is blocked by the antioxidant pyrrolidine dithiocarbamate (PDTC).[48] This hinted at the possibility that either an IκB kinase was activated, or that an IκB phosphatase, which constantly counteracts a constitutive IκB phosphorylation, was inactivated. In light of the observation that serines 32 and 36 constitute sites for the constitutive casein kinase type II, the latter scenario appears more likely. It is quite possible that the kinase or phosphatase sensing changes in ROI levels is not directly involved in modifying IκB but is an upstream component of a more complex signaling pathway. Several kinases have been reported to be activated by H_2O_2 and other inducers of oxidative stress, including the mitogen-activated protein (MAP) kinase-related Jun kinases (JNKs) or the tyrosine kinases lymphocyte tyrosine kinase (LTK), epidermal growth factor (EGF) receptor, and Src kinases.[52–57]

Various studies indicated that NF-κB activation involves a cascade of membrane-bound and cytoplasmic protein kinases. For instance, UV irradiation rapidly activates the tyrosine kinase activity of c-Src, an event followed by the activation of the membrane-bound GTP binding protein H-Ras and the Ser/Thr-specific Raf-1 kinase.[57] Overexpression of transdominant negative mutants of these molecules inhibited UV-induced NF-κB activation. Likewise, overexpression of v-Src activated NF-κB in a T-cell line.[58]

In T lymphocytes, treatment with H_2O_2 results in the rapid and extensive tyrosine phosphorylation of multiple proteins. At least two Src kinases, p56^{lck} and p59^{fyn} are activated by oxidizing agents.[55,56] Although it is unclear how these kinases are coupled to the NF-κB pathway, they may provide a useful model for ROI-induced signal transduction. We have recently observed that, similar to NF-κB, p56^{lck} and p59^{fyn} are not directly activated by H_2O_2 in a cell-free system or using recombinant kinases. Strong activation, however, could be seen by addition of oxidized glutathione (GSSG) indicating that, presumably, the formation of a mixed disulfide, a reaction called *S*-thiolation, may be involved in oxidant-induced kinase activation (H. Schenk, W. Dröge, and K. Schulze-Osthoff, unpublished results). Since several inducers of T-cell activation lead to an accumulation of GSSG, *S*-thiolation of a critical cysteine residue may be a physiological mechanism controlling activation of Src-related kinases and other kinases or phosphatases.

10.9 Direct Redox Regulation of Transcription Factors

NF-κB is not the only transcription factor responding to changes in ROI concentrations. As seen with NF-κB, several unrelated factors with different DNA-binding domains contain also cysteine residues whose oxidation state affects DNA binding. Such redox-sensitive cysteines have been found in the basic leucine zipper proteins c-Fos and c-Jun, the helix-loop-helix transcription factor Upstream Stimulating Factor (USF), the zinc-finger protein Egr-1 and other factors, such as c-Myb and p53.[9] These proteins are inactivated by oxidation *in vitro,* and mutation of the critical cysteine residue to a serine residue creates proteins that are resistant toward oxidative inactivation. Since the reduced state of the conserved cysteines is critical for DNA binding, it has been speculated that redox modification of the DNA-binding domain may be a secondary mechanism of controlling transcriptional activity. However, all present data concerning the direct modification of transcription factors by cysteine oxidation rely on cell-free experiments and there are only vague hints supporting a physiological role. A ubiquitous nuclear protein, called Ref-1, has been identified which restores the DNA-binding activity of oxidized Fos and Jun proteins *in vitro.*[6] Ref-1 catalyzes the reduction of thiol groups in these proteins using thioredoxin as a cofactor.

10.10 Transcription FactorAP-1

Heterodimers of members of the Jun and Fos transcription factor family constitute the transcription factor AP-1, which is also tightly regulated by redox processes. In contrast to NF-κB, AP-1 is exclusively localized in the nucleus. AP-1 has mainly been implicated in various processes associated with growth and differentiation events.[59,60] Among the genes known to be controlled by AP-1 are those encoding the genes for metallothionein, collagenase, stromelysin, transforming growth factor β1, interleukin-2, and other cytokine genes. In the past, AP-1 has attracted considerable interest, because it is composed of the c-*jun* and c-*fos* protooncogene product, which have an oncogenic potential. Fos and Jun proteins form homo- (Jun/Jun) or heterodimeric (Jun/Fos) complexes that bind to a transcriptional control element (consensus sequence: 5′-TGANTCA-3′), called the AP-1 site or TPA-responsive element (TRE). Protein interaction occurs through a coiled-coil structure involving basic leucine zipper domains. Several other members of the family, including Jun B, Jun D, and the Fos-related activators Fra-1 and Fra-2 have been identified.[59] Heterodimer formation among the different members of the Fos/Jun family as well as interactions with ATF/CREB members generates a diverse array of protein complexes with overlapping DNA-binding specificities but distinct functional properties.

10.11 Activation of AP-1 by Prooxidants

The activity of AP-1 is controlled by both transcriptional and post-translational mechanisms. Expression of c-Jun and c-Fos synthesis is rapidly and transiently induced by a variety of different extracellular stimuli including mitogens, phorbol esters, differentiation signals, and membrane depolarization. Most important in our context are adverse conditions and stress signals, such as oxygen radicals, UV irradiation, and exposure to heavy metals or DNA-alkylating agents. Intensive past research has revealed that the regulation of AP-1 activity is highly complex and may differ in response to distinct signals. The mechanisms that regulate assembly and functional specificity of different AP-1 complexes are rather poorly understood. Differential expression of family members, interactions with unrelated transcription factors, conformational alterations, and altered DNA-binding specificities of the heterodimers are all likely to play a role.

HeLa cells, in which AP-1 activation has been most intensively studied, contain low amounts of primarily Jun homodimers in their uninduced state. Exposure to UV light and to a lesser extent to H_2O_2 leads to a rapid enhancement of AP-1 DNA binding within a few minutes.[61–63] This immediate induction is not inhibited by cycloheximide and therefore independent of new protein synthesis. The prooxidant-induced activation primarily involves post-translational modifications that are carried out by complex changes in the phosphorylation pattern of c-Jun. The phosphorylation in the DNA-binding domain is decreased resulting in enhanced DNA-binding activity, while phosphorylation of serine-63 and serine-73, located in the transactivation domain, is enhanced.[64,65] Superimposed to these post-translational modifications is the de novo synthesis of AP-1 subunits, which may prolong the expression of target genes. Transcription of both c-jun and c-fos is rapidly induced by exposure to H_2O_2 and other stimuli.[62,63,66]

Although a multitude of different stimuli activates AP-1 and may lead to a similar pattern of induced genes, the used signaling pathway may be different among various inducers. Phorbol ester-treated cells are refractory to restimulation by the same stimulus, but remain fully responsive to UV irradiation and vice versa.[61,62] Comparison of the activation of AP-1 and NF-κB in response to TPA provides an example that one extracellular signal can obviously utilize different signal transduction pathways. As mentioned above, NF-κB activation by TPA is abolished by antioxidants but can be potentiated by H_2O_2.[34] This indicates that NF-κB activation by TPA follows an ROI-dependent pathway. In contrast, H_2O_2

only weakly triggers AP-1 activation and even suppresses activation induced by TPA. This implies that, although TPA induces a prooxidant state in cells, an ROI-dependent pathway is used only for the activation of NF-κB but not of AP-1.

10.12 Antioxidant-Induced AP-1 Activation

Although exposure to ROIs will result in the increased transcription of the c-*fos* and c-*jun* genes, only a very moderate induction of AP-1 DNA-binding and transactivation activity is usually observed.[34,62] Thus, the appearance of mRNAs for Jun and Fos does not necessarily correlate with the activation of AP-1. Exposure of some T-cell lines to H_2O_2 results in virtually no change of AP-1 DNA-binding activity, while NF-κB is strongly activated.[67]

Surprisingly, a strong activation of AP-1 could be observed after treatment of cells with several structurally unrelated antioxidants, including PDTC, *N*-acetyl cysteine, and butylated hydroxyanisole.[34,68] A significant increase in AP-1 DNA-binding and transactivating activity was also detected after transient overexpression or addition of the enzyme thioredoxin, while NF-κB activation was inhibited under the same conditions. The physiological equivalent to the antioxidant treatment may be hypoxia. It has been observed that HeLa cells strongly activate AP-1 in a biphasic response under reduced oxygen pressure.[69,70] In contrast to a previous report,[71] this does not activate NF-κB. We have observed a very strong activation of NF-κB only on subsequent reoxygenation of cell cultures.[70]

The induction of AP-1 by antioxidant treatments requires *de novo* protein synthesis of the c-Jun and c-Fos proteins and is entirely abolished by cycloheximide. This indicates that yet other transcription factors are responsible for the activation of AP-1 in response to cellular redox changes. In the case of the c-*fos* gene, this preexisting factor is Ternary Complex Factor (TCF)/Elk-1; in the case of the c-*jun* gene it is a c-Jun homodimer or a c-Jun/ATF-2 heterodimer. TCF/Elk-1 forms a ternary complex with the serum response factor (SRF) on the serum response element (SRE).[3,4] Phosphorylation of the transactivation domain of TCF/Elk-1 by mitogen-activated kinases (MAPKs or ERKs) induces the transactivating potential of the complex and leads to c-*fos* gene expression. We have found that the SRE is responsible for the activation of c-*fos* expression in response to both antioxidant as well as prooxidant stimulation of cells.[34] Such treatments cause an increased TCF/Elk-1 phosphorylation in the cell nucleus, which is preceded by novel phosphorylation and activation of MAP kinases. TCF/Elk-1 can be considered as a nuclear sensor of the cellular redox imbalance. It is currently unknown which kinases (or phosphatases) in the cascade cause phosphorylation of MAP kinases in response to antioxidants or oxidants.

Preexisting homodimers of c-Jun and heterodimers of c-Jun and ATF-2, which induce the c-*jun* gene in response to prooxidant stimuli and other stresses, are activated by *de novo* phosphorylation of c-Jun in its N-terminal domain.[64] One important kinase is JNK, which activates c-Jun in response to UV light, presumably, depending on redox changes.[52] It is currently unknown what activates c-jun transcription in response to antioxidant stimuli. A candidate factor is NF-Jun.[72] Since strong AP-1 activation is only observed under antioxidant but not under prooxidant conditions, additional posttranscriptional control steps seem to be involved.

10.13 Functional Aspects of Redox-Controlled Transcription Factors

In this chapter, we have described the redox regulation of the transcription factors NF-κB and AP-1 by various forms of oxidative stress. Both factors have a great number of target genes which are all likely to be induced upon changes in the intracellular redox state. Space does not permit us to cover all genes, and the reader is referred to detailed reviews[15-18,73,74] and the following chapters of this volume. In contrast to bacteria, the mammalian response to ROIs is highly complex and involves the induction of genes associated with different cellular processes. These processes include signal transduction (growth factors, growth factor receptors, transcription factors), proliferative events (oncogenes), tissue injury (proteases) and inflammatory processes (cytokines, cytokine receptors, cell adhesion molecules).

NF-κB has a central role in gene control associated with immunological and inflammatory processes. The expression of endothelial cell adhesion molecules is coordinately regulated by NF-κB and it is thus not unexpected to find that the genes coding for ELAM-1, Vascular Cell Adhesion Molecule (VCAM-1), and Intercellular Cell Adhesion Molecule ICAM-1 are induced by H_2O_2.[75,76] This finding might be of relevance for the massive tissue infiltration of leukocytes in inflammatory and oxidative-stress-associated processes. In endothelial and other cell types, various cytokines and growth factors are induced by oxidative stress. Examples are IL-1,[77] IL-2,[67] IL-6,[78] IL-8,[79] TNF,[80] platelet-derived growth factor (PDGF) and Basic Fibroblast Growth Factor (bFGF),[81] colony-stimulating factors,[82] and other proinflammatory gene products. ROI-induced gene expression may be of high relevance for all diseases associated with enhanced formation of ROIs, including septic shock, ischemia/reperfusion, and chronic inflammatory diseases.[83] ROI-inducible cytokines, such as TNF and IL-1, are themselves known to induce ROI formation, suggesting a positive autoregulatory circuit under these pathological conditions. It has recently been found that the transition metal nickel, which is the most important contact allergen, potently activates NF-κB. Expression of NF-κB-dependent target genes, such as IL-6 and endothelial adhesion molecules was completely blocked by the antioxidant PDTC pointing to a potential pharmacological use of antioxidants and other NF-κB inhibitors as anti-inflammatory agents.[37]

Great attention has been paid to the redox control of HIV-1 replication. Proviral expression of HIV-1 is largely controlled by binding of NF-κB to two highly conserved regulatory elements in the HIV enhancer/promoter.[84] HIV-1 expression is strongly inhibited by several antioxidants such as *N*-acetyl-L-cysteine, L-cysteine, and vitamin C.[85–87] In addition, asymptomatic HIV-infected individuals reveal significantly reduced levels of intracellular glutathione and plasma thiols suggesting that a disturbance of the intracellular redox balance might be important in the pathogenesis of AIDS.[88,89]

There is another link between HIV infection and oxidative stress. Several studies have shown that the depletion of CD4-positive T cells, the hallmark of AIDS, is the consequence of a continued apoptotic cell death.[90] Many mediators of oxidative stress lead to apoptosis, while antioxidants protect cells from a variety of toxic insults.[91] The discovery that the protooncogene product Bcl-2, which protects cells after exposure to various apoptotic signals, has antioxidant properties emphasizes the importance of oxidative events.[92] Whereas some physiological inducers of cell death have been proposed to kill cells through the induction of massive ROI generation, other forms of apoptosis, at least in the execution phase of cell death, are redox-independent. Since apoptosis, however, often requires *de novo* protein synthesis, oxidative events may change transcription of certain genes such that the apoptotic pathway is initiated. Intriguingly, transcription factors activated by ROIs, such as AP-1 and NF-κB, have been observed to be also activated in early phases of cell death.[93] It is conceivable that, at least in some cases, Bcl-2 inhibits cells death by interfering with the oxidative activation of transcription factors that control expression of "death genes."

A direct involvement of ROIs has been demonstrated for TNF-induced cell death. It has been suggested that TNF subverts part of the mitochondrial electron transport and redirects it to produce superoxide anions. Consequently, overexpression of Mn-dependent SOD as well as treatment with antioxidants or certain mitochondrial inhibitors protect cells from TNF-induced cytotoxicity.[31,39,94] ROIs are apparently not directly involved in the cytotoxic action of Fas, an important mediator of apoptosis that is structurally related to the TNF receptors.[95] The Fas system has recently been implicated in activation-induced cell death (AICD) during HIV infection.[96] The fact that AICD is associated with the induction of Fas ligand expression and inhibited by neutralizing Fas decoy constructs argues for this assumption. Since antioxidants are able to block AICD but not directly Fas-mediated cytotoxicity, oxidative events are presumably involved in regulation of Fas ligand expression. Indeed, the Fas ligand promoter contains a conserved κB binding site, although its functional importance remains to be established.

There is good evidence to suggest that ROIs and ROI-controlled transcription factors are implicated in the regulation of proliferative events. Although high amounts of ROIs are cytotoxic, sublethal doses of H_2O_2 can be mitogenic for certain cell types. Micromolar amounts of H_2O_2 stimulate T-cell proliferation.[67,97] Various members of the NF-κB family seem to be involved in the control of cell proliferation. V-Rel, p65Δ, Bcl-3, and Lyt-10 were reported to have oncogenic potential.[98-101] A correlation between oxidant stimuli, NF-κB activation and cell proliferation is found in fibroblasts. In 3T3 cells, NF-κB activation has been observed to occur specifically during the G_0/G_1 transition, a time point when various immediate-early genes are induced.[102] The c-*myc* protooncogene is apparently activated by NF-κB on serum stimulation.[103] We have recently observed that 50-μM H_2O_2 strongly increases proliferation of ESb-L T lymphoma cells.[67] This is also accompanied by an increased NF-κB activation and expression of the IL-2 gene. Certainly, reducing conditions may be beneficial for long-term growth in most cell types. Nonetheless, ROIs may be important to initiate certain events associated with growth control. It is possible that a primary prooxidative stimulus initiates the synthesis of glutathione, thioredoxin, or other protective molecules which then shift the intracellular redox state to reducing conditions being more beneficial to DNA synthesis.

It is well established that AP-1 is crucially involved in the control of cell cycle and proliferation. The expression of the protooncogenes c-*fos* and c-*jun* as well as c-*myc* is enhanced by prooxidants.[104] Since AP-1 controls the expression of several proteases such as collagenase, stromelysin, and plasminogen activator, prooxidant conditions are likely to be involved in tumor promotion by supporting spreading of tumor cells in tissues.[105,106]

In conclusion, changes in the cellular redox state toward either prooxidant or antioxidant conditions may exert profound effects on cellular functions. Given our detailed knowledge about the roles and characteristics of transcription factors NF-κB and AP-1, a fine-tuned manipulation of the cellular redox state may therefore provide a potential means and interesting possibility to influence various pathological processes.

References

1. Maldonaldo, E., and D. Reinberg. 1995. News on initiation and elongation of transcription by RNA polymerase II. *Curr. Opin. Cell Biol.* 7:352–361.

2. Baeuerle, P. A., and D. Baltimore. 1988. A specific inhibitor of the NF-κB transcription factor. *Science* 242:540–546.

3. Treisman, R. 1992. The serum response element. *Trends Biochem. Sci.* 17:423–426.

4. Janknecht, R., and A. Nordheim. 1993. Gene regulation by Ets proteins. *Biochim. Biophys. Acta* 1155:346–356.

5. Hunter, T., and M. Karin. 1992. The regulation of transcription by phosphorylation. *Cell* 70:375–387.

6. Xanthoudakis, S., G. Miao, F. Wang, Y.-C.E Pan, and T. Curran. 1992. Redox activation of Fos-Jun DNA binding activity is mediated by a DNA repair enzyme. *EMBO J.* 11:3323–3335.

7. Kumar, S., A. B. Rabson, and C. Gélinas. 1992. The RxxRxRxxC motif conserved in all Rel/κB proteins is essential for the DNA-binding activity and redox regulation of the v-Rel oncoprotein. *Mol. Cell. Biol.* 12:3094–3106.

8. Janssen, Y. M. W., B. Van Houten, P. J. A. Borm, and B. T. Mossman. 1993. Biology of disease. Cell and tissue responses to oxidative damage. *Lab. Invest.* **69**:261–274.

9. Schulze-Osthoff, K., and P. A. Baeuerle. In press. Redox regulation in gene expression. *In* J. M. McCord (eds.), *Oxyradicals in Medical Biology.* JAI Press, Greenwich, CT.

10. Storz, G., L. A. Tartaglia, S. B. Farr, and B. N. Ames. 1990. Bacterial defenses against oxidative stress. *Trends Genet.* **6**:363–368.

11. Demple, B., and C. F. Amabile-Cuevas. 1991. Redox redux: the control of oxidative stress responses. *Cell* **67**:837–839.

12. Pahl, H. L., and P. A. Baeuerle. 1994. Oxygen and the control of gene expression. *Bioessays* **16**:497–502.

13. Zitomer, R. S., and C. V. Lowry. 1992. Regulation of gene expression by oxygen in *Saccharomyces cerevisiae. Microbiol. Rev.* **56**:1–11.

14. Chen, Z., H. Silva, and D. F. Klessing. 1993. Active oxygen species in induction of plant systemic acquired resistance by salicylic acid. *Science* **262**:1883–1886.

15. Baeuerle, P. A. 1991. The inducible transcription activator NF-κB: regulation by distinct protein subunits. *Biochim. Biophys. Acta* **1072**:63–80.

16. Grilli, M., J. J.-S. Chiu, and M. J. Lenardo. 1993. NF-κB and Rel: participants in a multiform transcriptional regulatory system. *Int. Rev. Cytol.* **143**:1–62.

17. Liou, H.-C., and D. Baltimore. 1993. Regulation of the NF-κB/Rel transcription factor and IκB inhibitor system. *Curr. Opin. Cell Biol.* **5**:477–487.

18. Baeuerle, P. A., and T. Henkel. 1994. Function and activation of NF-κB in the immune system. *Annu. Rev. Immunol.* **12**:141–179.

19. Sen, R., and D. Baltimore. 1986. Multiple nuclear factors interact with the immunoglobulin enhancer sequence. *Cell* **46**:705–716.

20. Blank, V., P. Kourilsky, and A. Israël. 1992. NF-κB and related proteins: Rel/dorsal homologies meet ankyrin-like repeats. *Trends Biochem. Sci.* **17**:135–140.

21. Nolan, G. P., and D. Baltimore. 1992. The inhibitory ankyrin and activator Rel proteins. *Curr. Opin. Genet. Dev.* **2**:211–220.

22. Kang, S.-M., A.-C. Tran, M. Grilli, and M. Lenardo. 1992. NF-κB subunit regulation in nontransformed CD4⁺ T lymphocytes. *Science* **256**:1452–1456.

23. Schmitz, M. L., T. Henkel, and P. A. Baeuerle. 1991. Proteins controlling the nuclear uptake of NF-κB, Rel and dorsal. *Trends Cell Biol.* **1**:130–137.

24. Beg, A. A., and A. S. Baldwin, Jr. 1993. The IκB proteins: multifunctional regulators of Rel/NF-κB transcription factors. *Genes Dev.* **7**:2064–2070.

25. Inoue, J. I., L. K. Kerr, D. Rashid, N. Davis, H. R. Bose, and I. M. Verma. 1992. Direct association of pp40/IκB β with rel/NF-κB transcription factors—role of ankyrin repeats in the inhibition of DNA binding activity. *Proc. Natl. Acad. Sci. USA* **89**:4333–4337.

26. Zabel, U., and P. A. Baeuerle. 1990. Purified human IκB can rapidly dissociate the complex of the NF-κB transcription factor with its cognate DNA. *Cell* **61**:255–265.

27. Zabel, U., T. Henkel, M. dos Santos Silva, and P. A. Baeuerle. 1993. Nuclear uptake control of NF-κB by MAD-3, an IκB protein present in the nucleus. *EMBO J.* **12**:201–211.

28. Brown, K., S. Park, T. Kanno, G. Franzoso, and U. Siebenlist. 1993. Mutual regulation of the transcriptional activator NF-κB and its inhibitor, IκB-α. *Proc. Natl. Acad. Sci. USA* **90**:2532–2536.

29. Sun, S.-C., P. A. Ganchi, D. W. Ballard, and W. C. Greene. 1993. NF-κB controls expression of inhibitor IκB-α: evidence for an inducible autoregulatory pathway. *Science* **259**:1912–1915.

30. Yamauchi, N., Y. H. Kuriyama, N. Watanabe, H. Neda, M. Maeda, and Y. Niitsu. 1989. Intracellular hydroxyl radical production induced by recombinant human tumor necrosis factor and its implication in the killing of tumor cells in vitro. *Cancer Res.* **49**:1671–1675.

31. Schulze-Osthoff, K., A. C. Bakker, B. Vanhaesebroeck, R. Beyaert, W. Jacobs, and Fiers. 1992. Cytotoxic activity of tumor necrosis factor is mediated by early damage of mitochondrial functions. Evidences for the involvement of mitochondrial radical generation. *J. Biol. Chem.* **267**:5317–5323.

32. Schreck, R., P. Rieber, P. A. Baeuerle. 1991. Reactive oxygen species intermediates as apparently widely used messengers in the activation of the NF-κB transcription factor and HIV-1. *EMBO J.* **10**:2247–2258.

33. Schreck, R., B. Meier, D. N. Männel, W. Dröge, and P. A. Baeuerle. 1992. Dithiocarbamates as potent inhibitors of NF-κB activation in intact cells. *J. Exp. Med.* **175**:1181–1194.

34. Meyer, M., R. Schreck, and P. A. Baeuerle. 1993. H_2O_2 and antioxidants have opposite effects on activation of NF-κB and AP-1 in intact cells: AP-1 as secondary antioxidant-responsive factor. *EMBO J.* **12**:2005–2015.

35. Schreck, R., K. Albermann, and P. A. Baeuerle. 1992. Nuclear factor κB: an oxidative stress-responsive transcription factor of eukaryotic cells (a review). *Free Rad. Res. Commun.* **17**:221–237.

36. Lander, H. M., P. Sehaipal, D. M. Levine, and A. Novogrodsky. 1993. Activation of human peripheral blood mononuclear cells by nitric oxide generating compounds. *J. Immunol.* **150**:1509–1516.

37. Goebeler, M., J. Roth, E.-B. Bröcker, C. Sorg, and K. Schulze-Osthoff. 1995. Activation of nuclear factor-κB and gene expression in human endothelial cells by the common haptens nickel and cobalt. *J. Immunol.* **155**:2549–2467.

38. Schulze-Osthoff, K., M. Los, and P. A. Baeuerle. 1995. Redox signalling by transcription factors NF-κB and AP-1 in the immune system. *Biochem. Pharmacol.* **50**:735–741.

39. Schulze-Osthoff, K., R. Beyaert, V. Vandevoorde, G. Haegeman, and W. Fiers. 1993. Depletion of the mitochondrial electron transport abrogates the cytotoxic and gene-inductive effects of TNF. *EMBO J.* **12**:3095–3104.

40. Munroe, D. G., E. Y. Wang, J. P. MacIntyre, S. S. C. Tam, D. H. S. Lee, G. R. Taylor, L. Zhou, R. K. Plante, S. M. I. Kazmi, P. A. Baeuerle, and C. Y. Lau. In

press. Novel intracellular signalling function of prostaglandin H synthase-1 in NF-κB activation. *J. Inflamm.*

41. Suzuki, Y. J., and L. Packer. 1993. Inhibition of NF-κB activation by vitamin E and derivatives. *Biochem. Biophys. Res. Commun.* **193**:277–283.

42. Schmidt, K. N., P. Amstad, P. Cerutti, and P. A. Baeuerle. 1995. The roles of hydrogen peroxide and superoxide as messengers in the activation of transcription factor NF-κB. *Chem. Biol.* **2**:13–22.

43. Costello, R., C. Lipcey, M. Algarté, C. Cerdan, P. A. Baeuerle, D. Olive, and J. C. Imbert. 1993. Activation of primary human T lymphocytes through CD2 plus CD28 adhesion molecules induces long-term nuclear expression of NF-κB. *Cell Growth Differ.* **4**:329–339.

44. Los, M., H. Schenk, K. Hexel, P. A. Baeuerle, W. Dröge, and K. Schulze-Osthoff. 1995. IL-2 gene expression and NF-κB activation through CD28 requires reactive oxygen production by 5-lipoxygenase. *EMBO J.* **14**:3731–3740.

45. Westendorp, M., V. Shatrov, K. Schulze-Osthoff, R. Frank, M. Kraft, M. Los, P. H. Krammer, W. Dröge, and V. Lehmann. 1995. HIV-1 Tat potentiates TNF-induced NF-κB activation and cytotoxicity by altering the cellular redox state. *EMBO J.* **14**:546–554.

46. Henkel, T., T. Machleidt, I. Alkalay, M. Krönke, Y. Ben-Neriah, and P. A. Baeuerle. 1993. Rapid proteolysis of IκB-α is necessary for activation of transcription factor NF-κB. *Nature* **365**:182–185.

47. Palombella, V. J., O. J. Rando, A. L. Goldberg, and T. Maniatis. 1994. The ubiquitin-proteasome pathway is required for processing the NF-κB1 precursor protein and the activation of NF-κB. *Cell* **78**:773–785.

48. Traenckner, E. B.-M., S. Wilk, and P. A. Baeuerle. 1994. A proteasome inhibitor prevents activation of NF-κB and stabilizes a newly phosphorylated form of IκB-α that is still bound to NF-κB. *EMBO J.* **13**:5433–5441.

49. Traenckner, E. B.-M., H. L. Pahl, K. N. Schmidt, S. Wilk, and P. A. Baeuerle. 1995. Phosphorylation of human IκB-α on serines 32 and 36 controls IκB-α proteolysis and NF-κB activation in response to diverse stimuli. *EMBO J.* **14**:2876–2883.

50. Traenckner, E. B.-M., and P. A. Baeuerle. 1995. Appearance of ubiquitin-conjugated forms of IκB-α during its phosphorylation-induced degradation. *J. Cell Sci Suppl.* **19**:79–84.

51. Toledano, M. B., and W. J. Leonard. 1991. Modulation of transcription factor NF-κB binding by oxidation-reduction in vitro. *Proc. Natl. Acad. Sci. USA* **88**:4328–4332.

52. Kyriakis, J. M., P. Banerjee, E. Nikolakaki, T. Dai, E. A. Rubie, M. F. Ahmad, J. Avruch, and J. R. Woodgett. 1994. The stress-activated protein kinase subfamily of c-Jun kinases. *Nature* **369**:156–160.

53. Bauskin, A. R., I. Alkalay, and Y. Ben-Neriah. 1991. Redox regulation of a protein tyrosine kinase in the endoplasmic reticulum. *Cell* **66**:685–696.

54. Gamou, S., and N. Shimizu. 1995. Hydrogen peroxide perferentially enhances tyrosine phosphorylation of epidermal growth factor receptor. *FEBS Lett.* **357**:161–164.

55. Schieven, G. L., J. M. Kirihara, D. E. Myers, J. A. Ledbetter, and F. M. Uckun. 1993. Reactive oxygen intermediates activate NF-κB in a tyrosine kinase-dependent mechanism and in combination with vanadate activate the p56[lck] and p59[fyn] tyrosine kinases in lymphocytes. *Blood* **82**:1212–1220.

56. Nakamura, K., T. Hori, N. Sato, K. Sugie, T. Kawakami, and J. Yodoi. 1993. Redox regulation of a src family protein tyrosine kinase p56[lck] in T cells. *Oncogene* **8**:3133–3139.

57. Devary, Y., C. Rosette, J. A. DiDonato, and M. Karin. 1993. NF-κB activation by ultraviolet light not dependent on a nuclear signal. *Science* **261**:1442–1445.

58. Eicher, D. M., T.-H. Tan, R. R. Rice, J. J. O'Shea, and I. C. S. Kennedy. 1994. Expression of v-*src* in T cells correlates with nuclear expression of NF-κB. *J. Immunol.* **152**:2710–2719.

59. Angel, P., and M. Karin. 1991. The role of Jun, Fos, and the AP-1 complex in cell-proliferation and transformation. *Biochem. Biophys. Acta* **1072**:129–157.

60. Karin, K., and T. Smeal. 1992. Control of transcription factors by signal transduction pathways: the beginning of the end. *Trends Biochem. Sci.* **17**:418–422.

61. Büscher, M., H. J. Rahmsdorf, M. Litfin, M. Karin, and P. Herrlich. 1988. Activation of c-*fos* gene by UV and phorbol ester: different signal transduction pathways converge to the same enhancer element. *Oncogene* **3**:301–311.

62. Devary, Y., R. A. Gottlieb, L. F. Laus, and M. Karin. 1991. Rapid and preferential activation of the c-*jun* gene during the mammalian UV response. *Mol. Cell. Biol.* **11**:2804–2811.

63. Nose, K., M. Shibanuma, K. Kikuchi, H. Kageyama, S. Sakiyama, and T. Kuroki. 1991. Transcriptional activation of early-response genes by hydrogen peroxide in a mouse osteoblastic cell line. *Eur. J. Biochem.* **201**:99–106.

64. Binétruy, B., T. Smeal, J. Meek and M. Karin. 1991. Ha-Ras augments c-Jun activity and stimulates phosphorylation of its activation domain. *Nature* **351**:122–127.

65. Boyle, W. J., T. Smeal, L. H. K. Defize, P. Angel, J. R. Woodgett, M. Karin, and T. Hunter. 1991. Activation of protein kinase C decreases phosphorylation of c-*jun* at sites that negatively regulate its DNA-binding activity. *Cell* **64**:573–584.

66. Stein, B., H. J. Rahmsdorf, A. Steffen, M. Litfin, and P. Herrlich. 1989. UV-induced DNA damage is an intermediate step in UV-induced expression of human immunodeficiency virus type 1, collagenase, c-*fos,* and metallothionein. *Mol. Cell. Biol.* **9**:5169–5181.

67. Los, M., W. Dröge, P. A. Baeuerle, and K. Schulze-Osthoff. 1995. Hydrogen peroxide as a potent activator of T-lymphocyte functions. *Eur. J. Immunol.* **25**:159–165.

68. Schenk, H., M. Klein, W. Erdbrügger, W. Dröge, and K. Schulze-Osthoff. 1994. Distinct effects of thioredoxin and other antioxidants on the activation of NF-κB and AP-1. *Proc. Natl. Acad. Sci. USA* **91**:1672–1676.

69. Yao, K. S., S. Xanthoudakis, T. Curran, and P. J. O'Dwyer. 1994. Activation of AP-1 and of nuclear redox factor, Ref-1, in the response of HT colon cancer cells to hypoxia. *Mol. Cell. Biol.* **14**:5997–6003.

70. Rupec, R. A., and P. A. Baeuerle. 1995. The genomic response of tumor cells to hypoxia and reoxygenation: differential activation of transcription factors AP-1 and NF-κB. *Eur. J. Biochem.* **234**:632–640.

71. Koong, A. C., E. Y. Chen, and A. J. Giaccia. 1994. Hypoxia causes the activation of nuclear factor κB through phosphorylation of IκB-α on tyrosine residues. *Cancer Res.* **54**:1425–1430.

72. Brach, M. A., F. Herrmann, Y. Hamada, P. A. Baeuerle, and D. W. Kufe. 1992. Identification of NF-jun, a novel inducible transcription factor that regulates c-jun gene transcription. *EMBO J.* **11**:1479–1486.

73. Herrlich, P., H. Ponta, and H. J. Rahmsdorf. 1992. DNA damage-induced gene expression: signal transduction and relation to growth factor signaling. *Rev. Physiol. Biochem. Pharmacol.* **119**:187–223.

74. Holbrook, N. J., and A. J. Fornace, Jr. 1991. Response to adversity: molecular control of gene activation following genotoxic stress. *New Biol.* **3**:825–833.

75. Lo, S. K., K. Janakidevi, L. Lai, and A. B. Malik. 1993. Hydrogen peroxide-induced increase in endothelial adhesiveness is dependent on ICAM-1 activation. *Am. J. Physiol.* **264** (*Lung Cell. Mol. Physiol.* **8**):L406–L412.

76. Marui, N., M. K. Offermann, R. Swerlick, C. Kunsch, C. A. Rosen, M. Ahmad, R. W. Alexander, and R. M. Medford. 1993. Vascular cell adhesion molecule-1 (VCAM-1) gene transcription and expression are regulated through an antioxidant-sensitive mechanism in human vascular endothelial cells. *J. Clin. Invest.* **92**:1866–1874.

77. Koga, S., S. Ogawa, K. Kuwabara, J. Brett, J. A. Leavy, J. Ryan, Y. Koga, J. Plocinski, W. Benjamin, D. K. Burns, and D. Stern. 1992. Synthesis and release of interleukin 1 by reoxygenated human mononuclear phagocytes. *J. Clin. Invest.* **90**:1007–1015.

78. Brach, M. A., H. J. Gruβ, T. Kaisho, Y. Asano, T. Hirano, and F. Herrmann. 1993. Ionizing radiation induces expression of interleukin 6 by human fibroblasts involving activation of nuclear factor-κB. *J. Biol. Chem.* **268**:8466–8472.

79. Deforge, L. E., A. M. Preston, E. Takeuchi, J. Kenney, L. Boxer, and D. R. Remick. 1993. Regulation of interleukin 8 expression by oxidant stress. *J. Biol. Chem.* **268**:25568–25576.

80. Hallahan, D. E., D. R. Spriggs, M. A. Beckett, D. W. Kufe, and R. R. Weichselbaum 1989. Increased tumor necrosis factor α mRNA after cellular exposure to ionizing radiation. *Proc. Natl. Acad. Sci. USA* **86**:10104–10107.

81. Witte, L., Z. Fuks, A. Haimonovitz-Friedman, I. Vlodavsky, D. S. Goodman, and A. Eldor. 1989. Effects of irradiation on the release of growth factors from cultured bovine, procine and human endothelial cells. *Cancer Res.* **49**:5066–5072.

82. Satriano, J. A., M. Shuldiner, K. Hora, Y. Xing, Z. Shan, and D. Schlondorff. 1993. Oxygen radicals as second messengers for expression of the monocyte chemoattractant protein, JE/MCP-1, and the monocyte colony-stimulating factor, CSF-1, in response to tumor necrosis factor-α and immunoglobulin G. *J. Clin. Invest.* **92**:1564–1571.

83. Halliwell, B., and J. C. M. Gutteridge. 1990. The role of free radicals and catalytic metal ions in human disease: an overview. *Methods Enzymol.* **186**:1–85.

84. Nabel, G., and D. Baltimore, D. 1987. An inducible transcription factor activates expression of human immunodeficiency virus in T cells. *Nature* **326**:711–713.

85. Staal, F. J. T., M. Roederer, L. A. Herzenberg, and L. A. Herzenberg. 1990. Intracellular thiols regulate activation of nuclear factor κB and transcription of human immunodeficiency virus. *Proc. Natl. Acad. Sci. USA* **87**:9943–9947.

86. Roederer, M., F. J. T. Staal, P. A. Raju, S. W. Ela, L. A. Herzenberg, and L. A. Herzenberg. 1990. Cytokine-stimulated human immunodeficiency virus replication is inhibited by N-acetyl-L-cysteine. *Proc. Natl. Acad. Sci. USA* **87**:4884–4888.

87. Mihm, S., J. Ennen, U. Pessara, R. Kurth, and W. Dröge. 1991. Inhibition of HIV-1 replication and NF-κB activity by cysteine and cysteine derivatives. *AIDS* **5**:497–503.

88. Eck, H.-P., H. Gmünder, H. Hartmann, D. Petzoldt, V. Daniel, and W. Dröge. 1989. Low concentrations of acid soluble thiol (cysteine) in the blood plasma of HIV-1 infected patients. *Biol. Chem. Hoppe-Seyler* **370**:101–108.

89. Buhl, R., H. A. Jaffe, K. J. Holroyd, F. B. Wells, A. Mastangeli, C. Saltini, A. M. Cantin, and R. G. Crystal. 1989. Systemic glutathione deficiency in symptom-free HIV-seropositive individuals. *Lancet* **2**:1294–1296.

90. Ameisen, J. C. 1994. Programmed cell death (apoptosis) and cell survival regulation: relevance to AIDS and cancer. *AIDS* **8**:1197–1213.

91. Buttke, T. M., and P. A. Sandstöm. 1994. Oxidative stress as a mediator of apoptosis. *Immunol. Today* **15**:7–10.

92. Hockenbery, D. M., Z. N. Oltvai, X.-M. Yin Yin, C. L. Milliman, and S. J. Korsmeyer. 1993. Bcl-2 functions in an antioxidant pathway to prevent apoptosis. *Cell* **75**:241–251.

93. Grimm, S., M. Bauer, P. A. Baeuerle, and K. Schulze-Osthoff. 1996. Bcl-2 attenuates the activity of NF-κB which is induced upon apoptosis. *J. Cell. Biol.* **134**:13–23.

94. Wong, G. H. W., J. H. Elwell, L. W. Oberley, and D. W. Goeddel. 1989. Manganeous superoxide dismutase is essential for cellular resistance to cytotoxicity of tumor necrosis factor. *Cell* **58**:923–931.

95. Schulze-Osthoff, K., P. H. Krammer, and W. Dröge. 1994. Divergent signaling via APO-1/Fas and the TNF receptor, two homologous molecules involved in physiological cell death. *EMBO J.* **13**:4587–4596.

96. Westendorp, M. O., R. Frank, C. Ochsenbauer, K. Stricker, J. Dhein, H. Walczak, K.-M. Debatin, and P. H. Krammer. 1995. Sensitization of T cells to CD95-mediated apoptosis by HIV-1 Tat and gp120. *Nature* **375**:497–500.

97. Dröge, W., K. Schulze-Osthoff, S. Mihm, D. Galter, H. Schenk, H.-P. Eck, S. Roth, and H. Gmünder. 1994. Functions of glutathione and glutathione disulfide in immunology and immunopathology. *FASEB J.* **14**:1131–1138.

98. Gilmore, T. D. 1991. Malignant transformation by mutant Rel proteins. *Trends Genet.* **7**:318–322.

99. Narayanan, R., J. F. Klement, S. M. Ruben, K. A. Higgins, and C. A. Rosen. 1992. Identification of a naturally occurring transforming variant of the p65 subunit of NF-κB. *Science* **256**:367–370.

100. Neri, A., C. C. Chang, L. Lombardi, M. Salina, P.Corradini, A. T. Maiolo, R. S. Chagnati, and F. R. Dalla. 1991. B cell lymphoma-associated chromosomal translocation involves candidate oncogene *lyt-10,* homologous to NF-κB. *Cell* **67**:1075–1087.

101. Ohno, H., G. Takimoto, and T. W. McKeithan. 1990. The candiate proto-oncogene *bcl-3* is related to genes implicated in cell lineage determination and cell cycle control. *Cell* **60**:991–997.

102. Baldwin, A. S., J. C. Azizkhan, D. E. Jensen, A. A. Beg, and L. R. Coodly. 1991. Induction of NF-κB DNA-binding activity during the G_0 to G_1 transition in mouse fibroblasts. *Mol. Cell. Biol.* **11**:4943–4951.

103. Duyao, M. P., A. J. Buckler, and G. E. Sonnenschein. 1990. Interaction of an NF-κB-like factor with a site upstream of the c-*myc* promoter. *Proc. Natl. Acad. Sci. USA* **87**:4727–4731.

104. Crawford, D., J. Zbinden, P. Amstad, and P.A. Cerutti. 1988. Oxidant stress induces the proto-oncogenes c-*fos* and c-*myc* in mouse epidermal cells. *Oncogene* **3**:27–32.

105. Cerutti, P. A. 1991. Oxidant stress and carcinogenesis. *Eur. J. Clin. Invest.* **21**:1–5.

106. Trush, M. A., and T. W. Kensler. 1991. An overview of the relationship between oxidative stress and chemical carcinogenesis. *Free Rad. Biol. Med.* **10**:201–210.

11

STAT Activation by Oxidative Stress

*Amy R. Simon, Barry L. Fanburg, and
Brent H. Cochran*

11.1 Introduction

Reactive oxygen species (ROS) have been implicated in the pathogenesis of
several human diseases, including adult respiratory distress syndrome, myocardial
infarction, and Parkinson's disease.[1,2] ROS produce a variety of effects on tissues,
including both stimulation and inhibition of cell growth, cell death (both necrotic
and apoptotic), and the expression of novel genes involved in cellular repair
and antioxidant defenses.[3,4] The signal transduction pathways that mediate these
responses are now beginning to be elucidated. Work in both prokaryotes and
eukaryotes has led to the concept that oxidative stress is itself a second messenger
pathway utilized by cells in their vital signaling processes.[5-7] Oxidative stress is
thought to alter gene expression in part by activating latent transcription factors
and several immediate early genes, such as c-*myc* and c-*fos*.[8,9] As noted in prior
chapters, NF-κB and AP-1 are examples of transcription factors involved in
immediate early gene expression that are also activated by ROS.[5,10] Recent work
in our laboratory indicates that the STAT family of transcription factors (signal
transducers and activators of transcription) are activated by ROS.[11,12] This chapter
provides background on the STAT signal transduction pathway and discusses
some newly emerging data about the role of the STAT signal transduction pathway
in mediating the cell's transcriptional response to oxidative stress.

11.2 Background

The STAT proteins were first described as growth factor and cytokine-inducible
DNA binding complexes.[13,14] Currently, there are seven cloned mammalian STAT
family members and very recently the first nonmammalian STATs were cloned.[15-17]
The STATs differ in their tissue distribution and responses to specific ligands.

A list of the growth factors and cytokines that are known to activate various STAT members is provided in Table 11.1.[18,19] The first STATs cloned were found to mediate two distinct modes of interferon (IFN)-regulated gene expression. IFN alpha/beta stimulation activates a transcription factor complex, termed Interferon stimulated gene factor 3 (ISGF3), composed of STAT1, STAT2, and a third non-STAT protein p48 that mediates DNA binding to a regulatory element termed the Interferon-stimulated response element (ISRE) in IFN-α stimulated genes[20,21] (Fig. 11.1). In contrast, IFN-gamma induces a transcription factor complex composed of STAT1 homodimers that bind directly to a different promoter element called the IFN-gamma activated sequence (GAS).[21,22] Both modes of signaling require phosphorylation of STAT members and provide a direct link between cell surface receptors and nuclear regulation of gene expression.

Further cloning efforts have identified five more STAT genes in mammals. STAT3 is interleukin (IL)-6 and growth factor activated in response to tissue injury. STAT4 is activated by IL-12. STAT5a and STAT5b are closely related genes originally identified as being regulated by prolactin. STAT6 is activated by IL-3 and IL-4. Most recently, in Drosophila, D-STAT was identified in two laboratories and found to be involved with early development.[16,17] Given that STATs are so widely utilized in the signal transduction of mammals, it is not surprising that this pathway appears to be conserved during evolution in invertebrates.

Table 11.1. Activation of Jaks and STATs by growth factors and cytokines. The information in the table was complied from various sources including Refs. 41 and 42.

Ligand	Kinase	STAT
IFN-alpha	Tyk2, Jak1	STAT1, STAT2
IFN-gamma	Jak1, Jak2	STAT1
PDGF	?	STAT1,STAT3
EGF	Jak1 ?	STAT1,STAT3
IL-2	Jak1,Jak3	STAT5, STAT3
Il-3	Jak1,Jak2	STAT5, STAT6
IL-4	Jak1,Jak3	STAT6
IL-5	Jak1,Jak2	
IL-6	Jak1,Jak2,Tyk2	STAT1,STAT3
IL-7	Jak1,Jak3	
IL-11	Jak1,Jak2,Tyk2	
IL-12	Jak2,Tyk2	STAT4
G-CSF	Jak1,Jak2	STAT3, Novel
Epo	Jak2	STAT5
GH	Jak2	STAT1,STAT5
Prolactin	Jak2	STAT5
GM-CSF	Jak1,Jak2	STAT5
CNTF	Jak1,Jak2,Tyk2	
LIF	Jak1,Jak2,Tyk2	
Angiotensin II	Jak2, Tyk2	STAT1,2

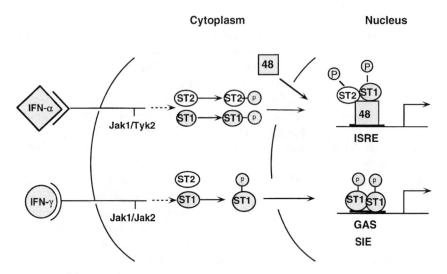

Figure 11.1. Diagram illustrating induction of STATs by interferons.

The structure of the STAT proteins with their conserved domains is illustrated in Figure 11.2. Deletion of the N terminus of STATs prevents STAT activation by impeding tyrosine phosphorylation even though the phosphorylation site is near the C terminus.[15] The amino terminal region contains a series of leucine repeats that is thought to mediate protein-protein interactions. Next is a basic region, which is a novel DNA binding domain, followed by a potential SH3 domain and a functional *src* homology 2 (SH2) domain. The SH3 domain differs in importance sequences from the SH3 consensus and has not yet been shown to be functional. By contrast, the SH2 domain is critical for mediating STAT recruitment to activated receptors and STAT dimerization. The C terminal region contains a tyrosine phosphorylation site critical for STAT activation, and in some cases this is followed by an acidic region that can act as an activation domain. STAT1, STAT3 and STAT5, however, have splicing variants that are missing the terminal amino acids and, hence, are not able to activate transcription. These variants can still bind to DNA and are thought to be naturally occurring dominant negatives.[15] In addition, some STAT members contain a serine in the C terminal region that may require phosphorylation for full transcriptional activation.[23]

All cytokines using the cytokine receptor superfamily signal via one or more

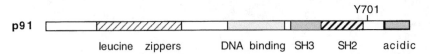

Figure 11.2. Structure of the STAT1 protein. Domains generally conserved among family members are indicated.

STAT proteins. Although these receptors do not have inherent kinase activity they are phosphorylated in response to ligand. This phenomena was subsequently explained by the association of these receptors with a family of tyrosine kinases called the JAK kinases (Janus kinase). JAK kinases are distinguished by having two kinase domains though only one is functional.[24] After ligand binding, the cytokine receptor chain(s) aggregate, enabling the receptor associated JAK kinases to be transphosphorylated and to subsequently phosphorylate the receptor on tyrosine residues (Fig. 11.3). The inactive STATs reside in the cytoplasm and are recruited to the activated receptor where they interact with the phosphorylated tyrosine residues on the receptors via SH2 interactions and subsequently become phosphorylated as well on tyrosine by the JAK kinases.[19] This phosphorylation event mediates homo- and heterodimerization of STAT family members through pTyr-SH2 interactions, resulting in translocation of the proteins to the nucleus and subsequent sequence specific binding to DNA and activation of transcription.[18,25] All STAT proteins, except STAT2, bind to sites in the promoter of STAT

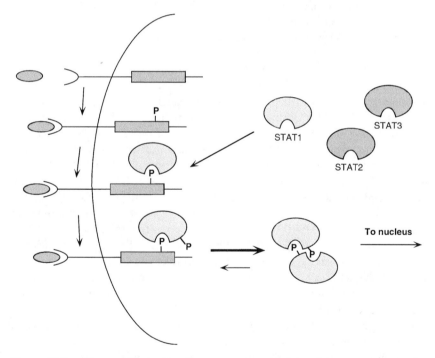

Figure 11.3. Diagram illustrating the proposed mechanism of receptor targeting by STATs. After ligand binding, the receptor becomes phosphorylated on tyrosine residues either by the intrinsic receptor kinase or by an associated kinase such as the JAK kinases. STAT proteins then bind to the receptor via a specific SH2 domain phosphotyrosine interaction. This then leads to tyrosine phosphorylation of the STAT protein by receptor associated kinases and subsequent dimerization with other STAT molecules.

regulated genes that have the approximate consensus TTNCNNAA. Different STATs discriminate in binding specificity to different sites with the largest influence being the length of the spacer region between the near palindromic half-sites.[26] Growth factor receptors with intrinsic tyrosine kinase activity, such as for epidermal growth factor (EGF) and platelet-derived growth factor (PDGF), activate STATs, but do not appear to require JAK kinases for STAT activation.[27] It is currently unclear whether the receptor's tyrosine kinase or another tyrosine kinase, such as the *src* family, or both, are utilized by these receptors for STAT activation.

The specificity of the response to JAK-STAT activation is determined by differential tissue expression of individual STATs and by differential recruitment of STAT members with the various stimuli. For example, STAT5 activates the transcription of whey acidic protein (WAP) in mammary cells, but not in myeloid cells.[28] Likewise, IFN-alpha activates STAT2 but IFN-gamma does not (Fig. 11.1). Specificity in the later case is not controlled by the JAKs, but rather by the receptor's phosphopeptide sequences that recruit different STATs according to their SH2 domain specificity.[29,30]

The biologic roles of the STAT family of transcription factors are only beginning to be understood. A broad array of genes is activated by STATs. STATs are involved in both mitogenic and/or antiproliferative responses. STAT1 appears to be partly responsible for the growth arrest in response to IFN gamma.[31] In contrast, STAT6 knockout mice show impaired B and T cell proliferative responses.[32,33] The STAT pathway may also participate in cellular transformation since induction of STAT3 correlates with transformation by v-*src*.[34] In Drosophila, JAK mutations can result in transformation, and experiments with constitutive cytokine receptor dimerization also cause transformation.[28] Interestingly, Drosophila with a dominant negative D-STAT can suppress proliferation in cells transformed with activated JAKs.[16] Clearly, STATs are required for gene expression mediated by interferon-alpha and -gamma, suggesting an antiviral function. The phenotype of STAT1 knockout mice supports this role. STAT1 deficient animals fail to thrive and are prone to viral disease.[35] STAT1 provides innate immunity, and given the function of many of the other cytokines it is likely that other STATs will play a large role in acquired immunity as well. For instance, STAT6 deficient mice have severely reduced levels of immunoglobulin E (IgE).[32,33] Like NF-κB, the STAT transcription factors are stimulated by cytokines and probably play a role in an organism's stress response. In addition, the STAT factors are known to be involved in the regulation of the immediate early gene c-*fos,* which has also been shown to be responsive to oxidative stress.[8,13,36]

11.3 Response of STAT Pathway to Oxidant Stress

We set out to determine what role, if any, the STAT transcription factors might have in mediating cell signaling in response to oxidative stress. Work in tissue culture in multiple cell lines showed that the STAT transcription factors, in particular STAT1 and STAT3, are activated by ROS. This response appeared to be cell-type-specific. For example, oxidative stress in the form of hydrogen peroxide (H_2O_2) induced STAT activation in Rat-1 fibroblasts, but not in bovine pulmonary artery endothelial cells.[11,12]

In general, fibroblasts show STAT activation in response to H_2O_2.[12] Rat-1 fibroblasts in quiescence have a very low background of STAT activation and are known to release ROS in response to cytokines.[37] When quiescent Rat-1 cells are stimulated with 1-mM H_2O_2 there is activation of STAT DNA binding by electrophoretic mobility shift assay within 5 min (Fig 11.4). The time course of this activation is similar to that seen with growth factors, such as PDGF. STAT

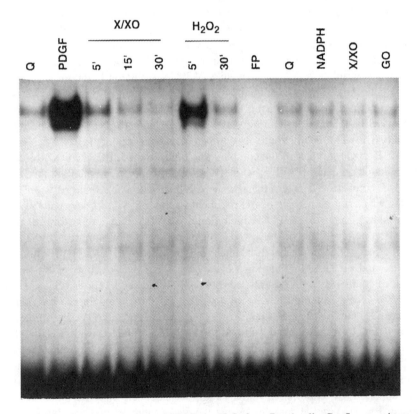

Figure 11.4   Specific activation of STATs by H_2O_2 from Rat-1 cells. Confluent, quiescent Rat-1 cells were treated with either 1-mM H_2O_2 or 0.06 U/ml xanthine oxidase plus 5-µm xanthine for the indicated periods of time and nuclear extracts prepared, as described previously.[12] Extracts were then run on bandshift gels using a high-affinity Sis/PDGF inducible element (SIE) probe and processed for autoradiography. In addition, cells were treated with 2.5-mM NADPH and 7.5 mU/ml glucose oxidase for 15 min.

activation is seen at concentrations of H_2O_2 as low as 100 µM, but is maximal at 1 mM. Work done in JB6 cells has shown that c-*fos* induction is not detected until cells are exposed to 100 to 250 µM H_2O_2.[8] As expected, given the rapid time course of STAT activation after H_2O_2 exposure, preincubation of the cells with cycloheximide to inhibit protein synthesis has no effect on activation (data not shown).

To further understand STAT activation secondary to ROS, we performed similar experiments using other types of ROS besides H_2O_2. Cells treated as long as 1 h with a superoxide stimulus, including xanthine and xanthine oxidase, menadione, NADPH, and paraquat did not exhibit STAT activation (Fig. 11.4 and not shown). Although H_2O_2 is generated via a dismutation reaction from superoxide, we conclude that the levels of H_2O_2 are insufficient to induce STAT activation in these cells. Likewise, nitric oxide donors, such as isosorbide dinitrite, sodium nitroprusside, and nitrosoglutathione all failed to change STAT DNA-binding activity (data not shown). Moreover, STAT activation did not appear to be a generalized cellular stress response since heat shock, heavy metal,, and ultraviolet irradiation did not activate STAT DNA binding (data not shown).

To determine if the DNA binding activity was specific for the SIE, we performed competition experiments using a 100-fold excess DNA from the high-affinity SIE, ISRE, or calcium/cAMP response element (CRE). Only the high-affinity SIE was able to effectively compete for STAT binding (Fig. 11.5). While STAT factors form part of the complex that binds to the ISRE, the DNA binding is mediated by a non-STAT protein. Therefore, it is not surprising that the ISRE did not compete. To determine which STAT members are being activated by ROS, we performed supershift experiments with anti-STAT1 and anti-STAT3 antibodies. From work on PDGF induction of STATs in Rat-1 cells, three distinct STAT complexes can be observed. The upper band is composed of STAT3 homodimers; the middle band of a heterodimer of STAT1 and STAT3; and the lower band of STAT1 homodimers. By comparing the positions of the H_2O_2-induced complexes to those induced by PDGF, it seemed likely that STAT1 and STAT3 were mediating the response to H_2O_2 in fibroblasts. As can be seen in Figure 11.5, antibody made against the C terminal peptide of STAT1 completely shifted the lowermost band, similar to the effect obtained with PDGF. The addition of an anti-STAT3 antibody made against a unique C terminal peptide led to a partial shift of two bands, which are presumably from the top and middle bands, whereas, the addition of the anti-STAT3 antibody almost entirely shifted the upper PDGF band. This finding suggests that there may be other STATs induced by H_2O_2 in the complex or that the STATs may be the same but alternatively spliced or may have different post-translational modifications. Therefore, H_2O_2 induces STAT1 and STAT3 in Rat-1 fibroblasts.

To determine if STAT activation was due to oxidative stress, we used a variety of antioxidants in attempts to block this activation. The antioxidants *N*-acetyl-L-cysteine (NAC), pyrrolidine dithiocarbamate (PDTC), and glutathione (GSH) did

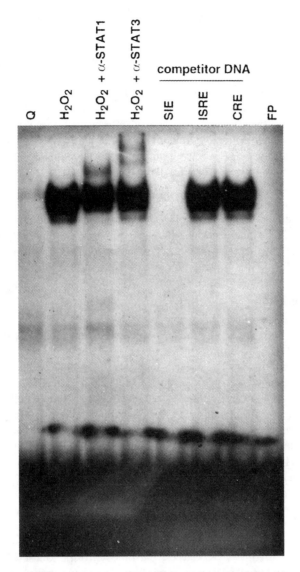

Figure 11.5. Induction of STAT1 and STAT3 by H_2O_2 in Rat-1 fibroblasts. Extracts of quiescent Rat-1 cells or cells treated for 10 min with 1-mM H_2O_2 were assayed using a high affinity SIE probe on band-shift gels. Binding reactions were performed in the presence or absence of the indicated antibodies or competitor DNAs. The anti-STAT1 antibody was made against a peptide with the unique C terminal 37 amino acids of murine STAT1 protein. STAT3 antibody was made against another unique sequence in the C terminal region of human STAT3. Antibodies were incubated with 10 μl (12 μg protein) of nuclear extracts prior to the binding reaction for 30 min on ice at a dilution of 1:10. Subsequently the probe was added bringing the final value to 20 μl. Competitor DNA sequences for the ISRE and CRE were used at a concentration of 100 times greater than that of the ^{32}P labeled probe. FP indicates free probe with no extract added.

not induce STAT activation in quiescent cells. However, cells preincubated with these antioxidants prior to the addition of H_2O_2 failed to exhibit STAT activation.[12] All of these antioxidants can directly scavenge free radicals and PDTC is also thought to have protective effects via its metal chelating properties. In addition, catalase, which converts H_2O_2 to water, also abolished STAT activation (data not shown). The above findings support the concept that H_2O_2-induced oxidative stress is required for STAT activation. However, antioxidants such as catalase and NAC failed to block all STAT activity, as evidenced by the ability of PDGF to still induce STAT DNA binding. This is in distinction to NF-κB whose signaling by all known agonists is abolished in vitro by antioxidants such as NAC.[38] In addition, work on vascular smooth muscle cells indicates that catalase can abolish PDGF signaling through the mitogen-activated protein (MAP) kinase pathway, implying that H_2O_2 is a necessary intermediate in PDGF signal transduction, at least for this pathway.[39]

11.4 Conclusion

The involvement of STATs in most cytokine signaling suggests a role for these transcription factors in an organism's response to stress and inflammation. Diseases such as asthma, where cytokines may act as intermediates for inflammation, are also accompanied by oxidative stress. Cytokine and growth factor stimulation alone have been shown to induce oxidative stress in multiple cell types, hence supporting a second signaling role for ROS. Similar to growth factors and cytokines, we have shown that oxidative stress can activate the STAT pathway in Rat-1 fibroblasts. In particular, STAT1 and STAT3 are activated rapidly in response to H_2O_2. This activation is independent of new protein synthesis and does not appear to be a generalized cell damage response since UV irradiation, heavy metals and heat shock all failed to activate the STATs.

The mechanism of STAT activation by ROS is still unclear. Activation most likely occurs via a cellular tyrosine kinase or inhibition of a cellular phosphatase, given the requirement of tyrosine phosphorylation for STAT DNA binding activity. This could involve JAK kinases, but other tyrosine kinases such as c-*src* may also be important. There is already evidence that some src kinases are activated in response to H_2O_2.[40] Oxidative stress is capable of altering the expression of genes involved in cell growth and repair. Our data show that the STAT pathway also participates in a cell's response to oxidative stress.

References

1. Halliwell, B. 1991. Reactive oxygen species in living systems: source, biochemistry, and role in human disease. *Am. J. Med.* **91**(suppl 3C):14S–22S.
2. Janssen. Y. M. W., B. V. Houten, P. J. A. Borm, and B. T. Mossman. 1993. Biology of disease: cell and tissue responses to oxidative damage. *Lab. Invest.* **69**:261–273.

3. Burdon, R. H. 1995. Superoxide and hydrogen peroxide in relation to mammalian proliferation. *Free Rad. Biol. Med.* **18**:775–794.

4. Simon, A., B. Cochran, and B. Fanburg. In press. Oxidative stress and cell proliferation. *In* D. Massaro and L. Clerch, *Oxygen, Gene Expression and Cellular Function.* Marcel Dekker, New York.

5. Schreck, R., P. Rieber, and P. A. and Baeuerle. 1991. Reactive oxygen intermediates as widely used messengers in the activation of NF-kappa B transcription factor and HIV-1. *EMBO J.* **10**:2247–2258.

6. Thannickal, V., and B. Fanburg. 1995. Activation of a H_2O_2-generating NADH-specific oxidase in human lung fibroblasts by transforming growth factor-Beta 1. *J. Biol. Chem.* **270**:30334–30338.

7. Christman, M. F., R. W. Morgan, F. S. Jacobson, and B. N. Ames. 1985. Positive control of a regulon for defense against oxidative stress and some heat shock proteins in *Salmonella typhorium. Cell* **41**:753–762.

8. Amstad, P. A., G. Krupitza, and P. A. Cerutti. 1992. Mechanism of c-*fos* induction by active oxygen. *Cancer Res.* **52**:3952–60.

9. Shibanuma, M., T. Kuroki, and K. Nose. 1988. Induction of DNA replication and expression of protooncogenes c-myc and c-fos in quiescent Balb/3T3 cells by xanthine/xanthine oxidase. *Oncogene* **3**:17–21.

10. Pahl, H. L., and P. A. Baeuerle. 1994. Oxygen and the control of gene expression. *Bioessays,* **16**:497–502.

11. Simon, A., B. Fanburg, and B. Cochran. 1995. Induction of the STAT pathway by oxidative stress. *J. Cell. Biochem.* (suppl 19C):314.

12. Simon, A., S. Spear, B. Fanburg, and B. Cochran. Activation of STAT transcription factors by oxidative stress. Submitted.

13. Hayes, T. E., A. M. Kitchen, and B. H. Cochran. 1987. Inducible binding of a factor of the c-*fos* regulatory region. *Proc. Natl. Acad. USA Sci.* **84**:1272–1276.

14. Levy, D. E., D. S. Kessler, R. Pine, N. Reich, and J. E. Darnell. 1988. Interferon-induced nuclear factors that bind a shared promoter element correlate with positive and negative transcriptional control. *Genes Dev.* **2**:383–393.

15. Ihle, J. 1996. STATs: signal transducers and activators of transcription. [Minireview]. *Cell* **84**:331–334.

16. Hou, X., M. Melnick, and N. Perrimon. 1996. Marelle acts downstream of the Drosophila Hop/JAK kinase and encodes a protein similar to the mammalian STATs. *Cell* **84**:411–419.

17. Yan, R., S. Small, C. Desplan, and J. Darnell. 1996. Identification of a STAT gene that functions in Drosophila development. *Cell* **84**:421–430.

18. Darnell, J. E., I. M. Kerr, and G. R. Stark, 1994. Jak-STAT pathways and transcriptional activation in response to interferons and other extracellular signaling proteins. [Review]. *Science* **264**:1415–1421.

19. Ihle, J. N., B. A. Witthuhn, F. W. Quelle, K. Yamamoto, W. E. Thierfelder, B. Kreider, and O. Silvennoinen. 1994. Signaling by the cytokine receptor superfamily: JAKs and STATs [Review]. *Trends Biochem. Sci.* **19**:222–227.

20. Fu, X. Y. 1992. A transcription factor with SH2 and SH3 domains is directly activated by an interferon alpha-induced cytoplasmic tyrosine kinase. *Cell* **70**:323–35.

21. Schindler, C., X. Y. Fu, T. Improta, R. Aebersold, and J. E. Darnell. 1992. Proteins of transcription factor ISGF-3: one gene encodes the 91 and 84-kDa ISGF-3 proteins that are activated by interferon alpha. *Proc. Natl. Acad. Sci.* USA **89**:7836–7839.

22. Shuai, K., C. Schindler, V. Prezioso, and J. Darnell. 1992. Activation of transcription by IFN-gamma: tyrosine phosphorylation of a 91kD DNA binding protein. *Science* **258**:1808–1812.

23. David, M., E. R. Petricoin, C. Benjamin, R. Pine, M. J. Weber, and A. C. Larner. 1995. Requirement for MAP kinase (ERK2) activity in interferon alpha- and interferon beta-stimulated gene expression through STAT proteins. *Science* **269**:1721–1723.

24. Wilks, A. F., A. G. Harpur, R. R. Kurban, S. J. Ralph, G. Zurcher, and A. Ziemiecki. 1991. Two novel protein-tyrosine kinases, each with a second phosphotransferase-related catalytic domain, define a new class of protein kinase. *Mol. Cell. Biol.* **11**:2057–2065.

25. Schindler, C., and J. Darnell. 1995. Transcriptional responses to polypeptide ligands: the JAK-STAT pathway. *Annu. Rev. Biochem.* **64**:621–651.

26. Seidel, H. M., L. H. Milocco, P. Lamb, J. J. Darnell, R. B. Stein, and J. Rosen. 1995. Spacing of palindromic half sites as a determinant of selective STAT (signal transducers and activators of transcription) DNA binding and transcriptional activity. *Proc. Natl. Acad. Sci.* USA **92**:3041–3045.

27. Leaman, D. W., S. Pisharody, T. W. Flickinger, M. A. Commane, J. Schlessinger, I. M. Kerr, D. E. Levy, and G. R. Stark. 1996. Roles of JAKs in activation of STATs and stimulation of c-*fos* gene expression by epidermal growth factor. *Mol. Cell. Biol.* **16**:369–375.

28. Ihle, J. 1995. Cytokine receptor signaling. *Nature* **377**:591–594.

29. Stahl, N., T. Farruggella, T. Boulton, Z. Zhong, J. Darnell, and G. Yancopoulos. 1995. Choice of STATs and other substrates specified by modular tyrosine-based motifs in cytokine receptors. *Science* **267**:1349–1353.

30. Heim, M., I. Kerr, G. Stark, and J. Darnell. 1995. Contribution of STAT SH2 groups to specific interferon signaling by the JAK-STAT pathway. *Science* **267**:1347–1349.

31. Chin, Y. E., M. Kitagawa, W.-C. S. Su, Z.-H. You, Y. Iwamoto, and X.-Y. Fu. 1996. Cell growth arrest and induction of cyclin-dependent kinase inhibitor p21 WAF1/CIP1 mediated by STAT1. *Science* **272**:719–722.

32. Takeda, K., T. Tanaka, W. Shi, M. Matsumoto, M. Minami, S. Kashiwamura, K. Nakanishi, N. Yoshida, T. Kishimoto, and S. Akira. 1996. Essential role of Stat6 in IL-4 signalling. *Nature* **380**:627–630.

33. Shimoda, K., J. van Deursen, M. Y. Sangster, S. R. Sarawar, R. T. Carson, R. A. Tripp. C. Chu, F. W. Quelle, T. Nosaka, D. A. Vignali, P. C. Doherty, G. Grosveld, W. E. Paul, and J. N. Ihle. 1996. Lack of IL-4-induced Th2 response and IgE class switching in mice with disrupted Stat6 gene. *Nature* **380**:630–633.

34. Yu, C., D. Meyer, G. Campbell, A. Larner, C. Carter-Su, J. Schwartz, and R. Jove. 1995. Enhanced DNA-binding activity of a STAT 3-related protein in cells transformed by the src oncoprotein. *Science* **269**:81–83.

35. Durbin, J., R. Hakenmiller, M. Simon, and D. Levy. 1996. Targeted disruption of the mouse STAT1 gene results in compromised innate immunity to viral disease. *Cell* **84**:443–450.

36. Crawford, D., I. Zbinden, P. A. Amstad, and P. A. Cerutti. 1988. Oxidant stress induces the protooncogenes c-*fos* and c-*myc* in mouse epidermal cells. *Oncogene* **3**:27–32.

37. Meier, B., H. H. Radeke, S. Selle, M. Younes, H. Seis, K. Resch, and G. G. Habermehl. 1989. Human fibroblasts release active oxygen species in response to interleukin-1 or tumour necrosis factor. *Biochem. J.* **263**:539–545.

38. Schreck, R., B. Meier, D. N. Mannel, W. Droge, and P. A. Baeuerle. 1992. Dithiocarbamates as potent inhibitors of NF-kappa B activation in intact cells. *J. Exp. Med.* **175**:1181–1194.

39. Sundaresan, M., Z. Yu, V. Ferrans, K. Irani, and T. Finkel. 1995. Requirement for generation of H_2O_2 for platelet-derived growth factor signal transduction. *Science* **270**:296–299.

40. Nakamura, K., T. Hori, N. Sato, K. Sugie, T. Kawakami, and J. Yodoi. 1993. Redox regulation of a src family protein tyrosine kinase p 56[lck] in T cells. *Oncogene* **8**:3133–3139.

41. Ihle, J. N., B. A. Witthuhn, F. W. Quelle, K. Yamamoto, W. E. Thierfelder, B. Kreider, and O. Silvennoinen. 1994. Signaling by the cytokine receptor superfamily: JAKs and STATs. [Review]. *Trends Biochem. Sci.* **19**:222–227.

42. Darnell, J. E., I. M. Kerr, and G. R. Stark. 1994. Jak-STAT pathways and transcriptional activation in response to IFNs and other extracellular signaling proteins. [Review]. *Science* **264**:1415–1421.

12

The Antioxidant Response Element
Leonard V. Favreau and Cecil B. Pickett

12.1 Introduction

Oxidative stress results from the production of reactive oxygen intermediates, which can arise from a number of sources including ionizing radiation, inflammation, and electrophilic xenobiotics. Quinones are highly electrophilic compounds and can be very reactive, depending on the degree of reduction. Quinones can undergo either a one-electron reduction by NADPH-cytochrome P-450 reductase to form the semiquinone free radical, a reactive oxygen intermediate, or two-electron reduction by NAD(P)H:quinone oxidoreductase (quinone reductase, DT-diaphorase) to form the hydroquinone, a less reactive species.[1] In the presence of oxygen, the semiquinone autoxidizes to form the original quinone and superoxide anion radical. The repetition of this reaction (oxidation-reduction) can lead to oxidative stress,[2] which can cause cellular damage. One cellular defense mechanism against the toxic and neoplastic effects of quinones is believed to be the enzyme quinone reductase. Rat liver cytosol from 3-methylcholanthrene (3-MC) treated rats containing high levels of quinone reductase was shown to reduce the amount of semiquinone produced *in vitro* using menadione as the quinone acceptor. In contrast, addition of microsomes from phenobarbital-treated rats (which contain high levels of NADPH-cytochrome P-450 reductase) or cytosol containing dicoumerol, a potent quinone reductase inhibitor, resulted in production of the semiquinone.[3] Thus quinone reductase, which produces hydroquinones that can be further metabolized by conjugation and then rapidly eliminated, can function as a cellular protective mechanism against oxidative stress from xenobiotics as well as endogenous compounds.[4]

In the 1980s, it was shown that liver cytosolic quinone reductase activity was elevated after exposure of mice to the phenolic antioxidant butylated hydroxyanisole.[5] Exposure of the mouse hepatoma cell line Hepa 1c1c7 to planar aromatic compounds, azo dyes, and other compounds also lead to an increase in quinone reductase activity.[6] Many of these compounds were quinones or quinonelike compounds that were able to undergo oxidation-reduction.[7] These compounds also lead to an increase in liver cytosolic glutathione S-transferase activity.[8] A survey of these compounds showed that some could induce both cytochrome P-4501A1 and Phase II enzymes, while others could only induce Phase II enzymes. The former acted through the Ah receptor, which when bound with ligand recognizes a specific DNA sequence known as the xenobiotic (or dioxin) response element (XRE or DRE). The latter class of compounds did not induce CYP1A1 and presumably induced Phase II enzymes via a non-Ah receptor mediated pathway. Many of the Phase II enzyme inducers were found to possess a common functional group: most of these inducers (or their metabolites) were electrophilic Michael reaction acceptors.[9] Thus, it was speculated that Phase II enzyme induction was regulated by compounds that all contain (or acquire by metabolism) an electrophilic center.

A number of laboratories have focused on the regulation of four specific genes, i.e., the rat and mouse glutathione S-transferase Ya subunit genes and the rat and human quinone reductase genes. This chapter focuses on the identification and characterization of the *cis*-acting sequences and the *trans*-acting factors that are believed to be involved in the transcriptional regulation of these four genes. Evidence for and against the involvement of Jun/Fos proteins in gene activation through the antioxidant response element (ARE) are reviewed. Finally, a survey of other Phase II genes is presented with evidence supporting the existence of an ARE involved in their transcriptional regulation.

12.2 Rat Glutathione S-Transferase Ya Subunit Gene

12.2.1 Identification of cis-acting sequences

The availability of a cDNA for the glutathione S-transferase Ya subunit allowed for the isolation of genomic clones containing the complete exon and intron sequences for these genes.[10,11] Large segments of the 5'-flanking sequence were also isolated. A fragment containing 1600 nucleotides of the 5'-flanking sequence as well as part of exon I and intron I was fused to the bacterial chloramphenicol acetyltransferase (CAT) gene in the plasmid pSVOCAT.[12] This resulted in a reporter construct under the regulation of the promoter for the rat Ya gene. Transient transfection of this plasmid into hepatoma cells resulted in an elevation of CAT activity when cells were treated with β-naphthoflavone (β-NF). This was the first evidence that *cis*-acting sequences existed that were required for inducible activation of a Phase II gene. A deletion analysis of the flanking region indicated that the *cis*-acting sequences were found between nucleotides −1600 and −700 with respect of the start of transcription of the Ya gene.

Subsequent reports demonstrated that multiple motifs existed in this region.[13,14] A consensus XRE was detected between nucleotides −925 and −875. Consensus binding sites for the hepatocyte specific factors HNF1 and HNF4 were detected between nucleotides −875 and −815 and between nucleotides −790 and −740, respectively. Finally, a region between nucleotides −725 to −650 was identified that contributed to basal level expression and was also responsive to 3-methylcholanthrene and β-NF. The sequence within this region bore no homology to the XRE nor to any other known transcription factor sequence. Deletion mutants made from this region identified a 41-nucleotide fragment with basal level activity as well as inducible activity.[15] Like the XRE, this sequence was three- to fivefold inducible by β-NF and was therefore named the β-NF responsive element.

When Hep G2 cells transfected with CAT constructs containing the β-NF responsive element were subsequently treated with a Phase II enzyme-specific inducer such as tert-butylhydroquinone (t-BHQ), a three- to fivefold stimulation of CAT activity resulted.[15] In parallel experiments, the XRE was shown to be unresponsive to t-BHQ. In mutant mouse hepatoma cell lines that lacked either a functional Ah receptor or a functional CYP1A1, the β-NF responsive element was also found to be responsive to phenolic antioxidants, but not to β-NF. Presumably, compounds like β-NF must first be metabolized to be able to activate the ARE. These results confirmed previous predictions of an Ah receptor–independent mechanism of induction. Therefore, the β-NF responsive element was renamed the antioxidant response element.

12.2.2 Characterization of the ARE

Using deletion analysis and point mutants, the rat Ya ARE enhancer was shown to consist of two similar half-sites (see Table 12.1) arranged in a linear orientation: a distal half-site, 5′-TGGCATTGC-3′, and a proximal half site, 5′-TGA-CAAAGC-3′.[16] As deletions were made in the 5′ end of the 41-nucleotide ARE, about half of the basal level CAT activity was lost when the distal half-site was deleted. However, the proximal half-site was still responsive to β-NF. In contrast, when deletions were made from the 3′ end, all basal and inducible activity was lost when the proximal half-site was deleted. When point mutations were made in the ARE, four nucleotides in the proximal half-site were found to be required for basal and inducible activity, i.e., 5′-<u>TGAC</u>AAAGC-3′. Two additional residues were found to be required for inducible activity but not for basal activity, 5′-TGACAA<u>AG</u>C-3′. Point mutations made upstream or downstream of these residues did not result in altered activity. Thus, based on the results of deletion and point mutants, this sequence was designated the "core sequence."

12.2.3 Identification of trans-Acting Factors

Once the cis-acting sequence had been identified for the Ya gene, nuclear extracts from cultured hepatoma cells were analyzed for ARE-nucleoprotein interactions. Nuclear extracts from Hep G2 cells were found to protect the region containing the ARE from DNase 1 digestion.[17] When extracts from treated cells were used in this assay, no difference in the footprint was detected. Using the electrophoretic mobility shift assay, an ARE-nucleoprotein complex was identified in extracts from Hep G2 cells.[13,14,17] This complex was determined to be specific since it could be competed by an excess of unlabeled oligonucleotide containing the ARE but not by a random oligonucleotide. Using nuclear extracts from Hep G2 cells treated with β-NF or t-BHQ, no new bands appeared and little appreciable difference in the intensity of the complexes could be detected.[14,17]

One major DNA-nucleoprotein complex is observed in Hep G2 cells using a synthetic oligonucleotide containing the 41-nucleotide ARE sequence.[17] This complex was further characterized by methylation interference and protection assays. Extracts from both untreated and treated cells were found to protect guanine residues on the coding strand near the core sequence for the Ya ARE, i.e., 5'-GGTGACAAAGC-3'. Of these three residues, the terminal guanine had previously been shown to be absolutely required for ARE functional activity.[16] However, neither the final guanine on the coding strand nor the two guanines on the noncoding strand in this sequence were protected from methylation. UV cross-linking experiments demonstrated that the ARE-protein complex was composed of two polypeptides of approximate molecular weights of 45 and 28 kDa.

12.3 Mouse Glutathione S-Transferase Ya Subunit Gene

12.3.1 Identification of cis-acting Sequences

Genomic clones for a mouse Ya gene were isolated from a library that was screened with a rat Ya cDNA.[18,19] Initial reports indicated that *cis*-regulatory elements were located in a region between −200 to −1600 base pairs.[20] These elements were shown to be responsive only to t-BHQ in cell lines lacking either a functional Ah receptor or CYP1A1. By deletion analysis, a region between −852 and −693 base pairs was determined to contain the *cis*-acting elements required for basal and inducible activity.[21] Located within this 159 nucleotide region was a 41 nucleotide sequence which differed from the rat Ya ARE in only two nucleotides. This enhancer was named the electrophile responsive element (EpRE) and was shown to be responsive to β-NF, 3-MC, *trans*-4-phenyl-3-butene-2-one, t-BHQ, and TCDD in Hep G2 cells.[21]

12.3.2 Characterization of the ARE

The mouse Ya EpRE was also found to contain two half sites arranged in a linear orientation, a distal half site, 5'-TGACAATGC-3' and a proximal half-site, 5'-TGACAAAGC-3'.[22] As shown for the rat Ya, both half-sites were required for maximal basal level expression. However, in contrast to the rat ARE, both half-sites were required for inducibility by β-NF and t-BHQ. Furthermore, 12-*O*-tetradecamoyl-phorbal-13-acetate (TPA) was also shown to be an inducer of gene expression through the EpRE. Mutation of the T in the sequence 5'-TGACAAAGC-3' completely abolished basal and inducible expression when CAT constructs containing this mutant were transfected into hepatoma cells. Although a single copy of the proximal half-site possessed no activity, the presence of additional copies resulted in a linear increase in basal and inducible activity. Furthermore, if the distal half-site were replaced by a consensus TPA response element (TRE) sequence, a stimulation of basal and inducible expression resulted. Since both half-sites resembled the consensus activator protein-1 (AP-1) binding site, the two half-sites were referred to as AP-1-like binding sites.

12.3.3 Identification of trans-Acting Factors

Initial experiments using a probe containing the mouse glutathione S-transferase Ya EpRE detected a specific complex in nuclear extracts from Hepa 1c1c7 cells.[21] DNase 1 footprinting experiments indicated that less nuclear extract from t-BHQ and 3-MC treated cells was required to protect a region of DNA around the ARE. Furthermore, a binding complex could be detected by EMSA that was more abundant by a factor of 10 in induced cells. This was the first evidence that treatment with inducer compounds lead to an increase in a nuclear binding factor. Subsequent to these studies was the observation that the EpRE contained a binding site for in vitro synthesized Jun/Fos.[22] Furthermore, many of the inducers of the EpRE were found to be inducers of AP-1 in cell-culture lines.[23] Not only could TPA induce AP-1 but β-NF, t-BHQ, 3-MC, and TCDD were also found to stimulate AP-1 binding activity.

12.4 Rat NAD(P)H:QUINONE Oxidoreductase Gene

12.4.1 Identification of cis-Acting Sequences

The synthesis of a cDNA clone for the rat quinone reductase led to the isolation of clones containing the entire genomic sequence.[24,25] Constructs containing the upstream sequence fused to the CAT structural gene were responsive to β-NF in both rat hepatoma (H5-6) and human hepatoblastoma (Hep G2) cell lines.[25] Through a series of deletion mutations of this sequence, a 31-nucleotide motif was identified that mediated almost all of the basal level expression of the full quinone reductase gene promoter.[26] A synthetic oligonucleotide containing this 31-nucleotide sequence was inserted upstream of the rat Ya minimal promoter fused to the CAT structural gene. Cell cultures transiently transfected with this construct produced CAT activity that was fivefold higher than the minimal promoter alone. Treatment of transfected cell cultures with either β-naphthoflavone or t-butylhydroquinone resulted in three- to fivefold stimulation of CAT activity. Since this enhancer was activated by the same inducers as the glutathione S-transferase Ya antioxidant responsive element it was termed an ARE.

12.4.2 Characterization of the ARE

The rat quinone reductase ARE also contains two half-sites, the distal half-site being 5'-GTGACTCTAGA-3' and the proximal half-site being 5'-GTGACTTG GCA-3'.[26] However, instead of the linear arrangement of these two half-sites as seen in the rat and mouse Ya genes, the quinone reductase ARE half-sites are arranged as inverse repeats separated by one nucleotide:

$$5'\text{-TCTAGAGTCAC} \ldots \text{A} \ldots \underline{\text{GTGACTTGGCA-3'}}$$
$$3'\text{-AGAT}\underline{\text{CTCAGTG}} \ldots \text{T} \ldots \text{CACTGAACCGT-5'}$$

Furthermore, a palindromic sequence exists spanning the first six residues in each half-site (underlined).

Deletion analysis of the 31 base nucleotide ARE revealed a complex organization of this enhancer. When the TGAC residues in either half-site were mutated, all basal and inducible activity was lost as assessed by transient transfection into Hep G2 cells.[27] Therefore, both half-sites are required for functional activity. Deleting three of the four adenine residues in the 3' end of the ARE results in a gradual loss of basal level expression without interfering with inducible activity. When the fourth adenine was deleted, there was a dramatic decrease in the ability of the truncated ARE to respond to β-NF. Deletions made in the 5' end of the ARE also resulted in a loss of β-NF responsiveness. Point mutations made in either the distal or proximal half sites resulted in mutant AREs with decreased basal activity. Mutation of five residues in the proximal half-site (5'-GTGACTTG GCA-3') either eliminated or greatly decreased the activity of the ARE. Taken together, these results indicated that both the proximal half-site and at least part of the distal half-site are required for maximal basal and inducible expression.

Since the ARE required the presence of two half-sites to be functional, experiments were designed to determine if the two half-sites were equivalent.[27] Constructs containing an inverse palindromic arrangement of two distal half-sites were without basal or inducible activity when transfected into Hep G2 cells. However, when constructs containing two proximal half-sites arranged in an inverse palindromic orientation were transfected into Hep G2 cells, high basal and inducible CAT expression resulted. Interestingly, the inverse orientation of two proximal half-sites was determined to be much more active than a linear arrangement. These results demonstrated that although both the distal and proximal half-sites were required for functional activity, the proximal half-site was dominant.

12.4.3 Identification of trans-Acting Factors

Nuclear extracts from Hep G2 cells were found to protect a region of DNA within the 31-nucleotide rat quinone reductase ARE.[28] When the ARE was used to probe nuclear extracts from Hep G2 cells, a specific complex was detected that could be completed by an excess of unlabeled quinone reductase ARE and Ya ARE, but not by a random sequence. Using the methylation protection assay, nucleotides within the proximal half-site were found to be involved in the binding of the *trans*-acting factor. Using the methylation protection assay, two guanine residues in the coding strand and two guanine residues in the noncoding strand were found to be involved in complex formation:

5'-GTGACTTGGCA-3'
3'-CACTGAACCGT-5'

Point mutations of the second guanine on the coding strand and the two guanines on the noncoding strand lead to either elimination or a dramatic decrease in functional activity of the ARE. Synthetic oligonucleotides containing two of these point mutants were found to no longer compete with the wild-type ARE, 5'-GTGACTTGGCA-3' and 5'-GTGACTTGGCA-3'. Additional mutants within the proximal half-site (but not in the distal half-site) were also found to no longer compete with the ARE: 5'-GTGACTTGGCA-3' and 5'-GTGACTTGGCA-3'. Taken together, these results indicate that binding of the *trans*-acting factor occurs within the proximal half-site and that sequences in the distal half-site are not required for high-affinity binding. However, functional activity exists only when both half-sites are present.

A survey of cell lines showed that the *trans*-acting factor is not ubiquitous.[28] Using the electrophoretic mobility shift assay, complexes with identical mobility to the complex detected in Hep G2 cells could be detected in rat hepatoma (H4IIEC3) and mouse teratocarcinoma (F9) cell lines. This complex was also detected in rat liver nuclear extracts. However, this complex could not be detected in HeLa or Hepa 1c1c7 cells. The finding of a similar ARE-nucleoprotein complex in rat liver nuclear extracts suggests that results obtained from cell culture lines may be representative of *in vivo* interactions between the ARE found in the rat genome and *trans*-acting factors present in the liver.

Interestingly, both the collagenase TRE and the human quinone reductase ARE

Table 12.1.   ARE Half-Sites.

	Distal	Spacer	Proximal
rNQO₁	GTGACTCTAGA	{A}	GTGACTTGGCA
rNQO₁	GTGACTGCGAT	{A}	GTGACTCAGCA
rGSTP	GTGACTGACTA	{T}	ATGATTCAGCA
rGSTYa	ATGGCATTGCT	{AATG}	GTGACAAAGCA
mGSTYa	ATGACATTGCT	{AATG}	GTGACAAAGCA

All sequences are 5' to 3'. The half-sites for the rat GSTP and the rate and human NQO₁ AREs are inverse repeats with one nucleotide separating the two half-sites. The distal half-site is inverted with respect to the proximal half-site. The half-sites in both Ya genes are direct repeats separated by four nucleotides.

(which contains a TRE) competed in the gel-shift assay for the ARE.[27] However, the ARE-nucleoprotein complex in Hep G2 is not the same as the TRE complex. As shown for the Ya ARE, the ARE does not compete with AP-1/TRE in gel-shift assays when nuclear extracts or *in vitro* synthesized Jun/Fos is used.[17,28] The half-sites in the rat quinone reductase gene contain either one or two deviations from the consensus TRE (5'-TGAC/GTCA-3'), which eliminates the ability to bind to AP-1 with high affinity (see Table 12.1).

12.5 Human NAD(P)H:QUINONE Oxidoreductase Gene

12.5.1 Identification of cis-Acting Sequences

The fourth ARE to be identified was found in the human quinone reductase (NMO1) gene. A cDNA for human NAD(P)H:quinone oxidoreductase$_1$ (NQO$_1$) was generated by screening a cDNA library with a rat NQO$_1$ cDNA.[29] The human cDNA was then used to isolate clones containing the complete genomic sequence for NQO$_1$.[30] A 5'-flanking sequence of 1850 nucleotides was also identified, which was shown to be functional when CAT reported constructs were transiently transfected into human and mouse hepatoma cells. Deletion analysis demonstrated that *cis*-regulatory sequences existed between nucleotides −780 and −365. A survey of this 415 nucleotide sequence revealed a region of significant homology with the rat quinone reductase ARE. Although the rat sequence did not contain any known consensus sequences for transcription factors, the human sequence contained a perfect TRE motif within the ARE (see Table 12.1).

The human ARE was shown to be responsive to the same inducers as the rat gene.[31] When the human ARE was inserted upstream of the thymidine kinase promoter fused to the CAT gene and transfected into human Hep G2, human HeLa, and mouse Hepa 1 cells, an increase in basal level CAT expression was observed. Treatment of transfected cells with β-NF or butylated hydroxyanisole (BHA) resulted in increased CAT activity, indicating that the human ARE was functional in all three cell lines.

12.5.2 Characterization of the ARE

Mutation analysis of the human NQO$_1$ ARE reveals a similar organization to that of the rat NQO$_1$ ARE. Mutating the 5'-TGACTCA-3' residues in the proximal half-site completely eliminated enhancer activity.[31] A third motif with limited sequence homology to the first two half-sites is detected 3' to the ARE. However, this sequence has been shown to play no role in enhancer activity.[32] However, deletion of the distal half-site results in a loss of basal level expression. As seen in the rat quinone reductase ARE, both half-sites are required for functional activity. Like the mouse glutathione S-transferase EpRE and the rat quinone reductase ARE, the presence of one half-site (even containing a perfect TRE) is not sufficient for either maximal basal activity or responsiveness to β-NF.[32]

12.5.3 Identification of trans-Acting Factors

Since the human NQO_1 ARE contains a consensus TRE within the ARE, it was not surprising to observe binding with in vitro synthesized Jun/Fos proteins in the gel-shift assay.[28] When nuclear extracts from either HeLa or mouse hepatoma (Hepa 1) cells were tested, complexes identical to the complex seen with the TRE were formed.[31] Furthermore, supershift experiments employing polyclonal antibodies raised to Jun/Fos proteins have confirmed the presence of Jun B, Jun D and c-Fos in the ARE-nucleoprotein complex generated from Hepa 1 cells.[31,33] A similar complex has been detected in human colon cancer HT-29 cells.[34]

12.6 Is the ARE a TRE?

12.6.1 Evidence for and against AP-1 Binding to the ARE

Both the human NQO_1 ARE and the mouse Ya EpRE have been reported to bind to Jun/Fos.[22,28,31] In contrast, both the rat Ya ARE and the rat quinone reductase ARE have been shown to lack a high-affinity binding site for Jun/Fos.[16,28] In the case of the human NQO_1 ARE, the consensus TRE found within the ARE provides the recognition site for Jun/Fos. In nuclear extracts from uninduced mouse hepatoma Hepa 1 cells, the human ARE-nucleoprotein complex is made up of Jun B, Jun D, and c-Fos.[31,33] However, all three rodent AREs contain sequences that are similar to, but not identical to, the consensus TRE.

The DNA-nucleoprotein complex formed in Hep G2 nuclear extracts with the rat Ya ARE is not competed by the collagenase TRE.[13,17] This suggests that the complex is not composed of Jun/Fos proteins. In vitro synthesized Jun/Fos proteins also do not bind to the ARE under conditions where they do bind to the TRE.[35]

Although the mouse Ya EpRE resembles the rat Ya ARE, it has been shown to bind to in vitro synthesized Jun/Fos.[22] Cotransfection of Jun and Fos expression vectors in mouse F9 cells (which lack AP-1) with a CAT vector under the control of the EpRE resulted in a 100-fold increase in CAT activity. Presumably, the AP-1-like sequences resemble the TRE sufficiently enough to allow for AP-1 binding. The EpRE-nucleoprotein complex in TPA stimulated Hep G2 cells has been shown to contain c-Jun.[36] In support of these results is the observation that many Phase II enzyme inducers also induce AP-1 binding activity and mRNA for several members of the Jun/Fos family.[23,37]

The ability of the mouse EpRE to bind Jun/Fos has recently been reexamined.[38] In this study, the EpRE did not compete with the TRE for AP-1 nor was the major EpRE-nucleoprotein complex in Hep G2 cells composed of Jun/Fos proteins. These results are similar to those previously demonstrated for the rat Ya ARE in Hep G2 cells where the major ARE-nucleoprotein complex was shown to be made up of non-Jun/Fos proteins.[17]

Similar studies have been reported for the rat NQO_1 ARE.[27,28] This ARE does not compete with the TRE for AP-1 in HeLa cells, nor does it compete for *in vitro* synthesized Jun/Fos. The ARE also forms a nucleoprotein complex in hepatoma nuclear extracts, which can be distinguished from the AP-1/TRE complex. A consensus TRE can be made from the rat ARE by changing two nucleotides in the proximal half-site, i.e., 5′-GTGACTGGCA-3′ to 5′-GTGACTCA GCA-3′. This "ARE/TRE" probe can now clearly resolve two complexes in Hep G2 nuclear extracts.[27] Based on gel-shift mobilities and competition profiles one complex was identified as the TRE-nucleoprotein complex and the other as the ARE-nucleoprotein complex. Furthermore, the ARE-nucleoprotein complex has been found in F9 cells that lack detectable levels of AP-1.[27,39] In contrast, little binding to the ARE is detected in HeLa cells that contain a large amount of AP-1 (Fig. 12.1). Therefore, the two nucleotide deviation from the consensus TRE sequence is sufficient to abolish high affinity binding of the rat NQO_1 ARE to Jun/Fos proteins.

12.6.2 Evidence against AP-1 Activation of the ARE

Functionally, a TRE is not an ARE. When the rat Ya ARE is converted to an ARE/TRE by altering two nucleotides in the "core sequence" from 5′-GTGACAA AGC-3′ to 5′-GTGACTCAGC-3′, it retains many of the ARE properties, i.e., basal level expression and responsiveness to TPA and β-NF in Hep G2 cells.[35] However, altering the GC dinucleotide at the 3′ end of the ARE/TRE results in a loss of β-NF responsiveness without eliminating the TPA responsiveness. The GC dinucleotide is also crucial for responsiveness to β-NF in the wild-type ARE. Thus, TPA can transcriptionally regulate gene activity through two pathways, i.e., an AP-1-dependent pathway and an AP-1 independent pathway. In contrast, β-NF functions solely through an AP-1-independent pathway.

Similar conclusions have been reached using the human NQO_1 ARE.[32] However, only complexes containing Jun/Fos proteins have been observed for the human NQO_1 ARE in HeLa and Hepa 1 cells. Interestingly, c-Jun cannot be detected in this complex, possibly because of low c-Jun levels in uninduced cells. However, c-Jun has been detected in the EpRE-nucleoprotein complex formed in TPA stimulated Hep G2 cells.[36] Since Phase II inducers have been shown to increase AP-1 activity in Hep G2 cells, possibly ARE-mediated activation of the NQO_1 gene in this cell line occurs through an increase in AP-1 activity that binds to the consensus TRE sequence. However, the increase in AP-1 activity in Hep G2 cells has recently been shown to be due to increases in c-*jun, jun B, fra1,* and *fra2* expression.[38] Little induction of c-*fos* was observed. These results are significant because Fra1 and Fra2 proteins lack a transcriptional activation function. As *t*-BHQ induces AP-1 binding activity, these complexes will have a decreased ability to transcriptionally activate genes containing an AP-1 motif. Therefore it is plausible that AP-1-dependent and AP-1-independent mechanisms may be operating in different cell lines.

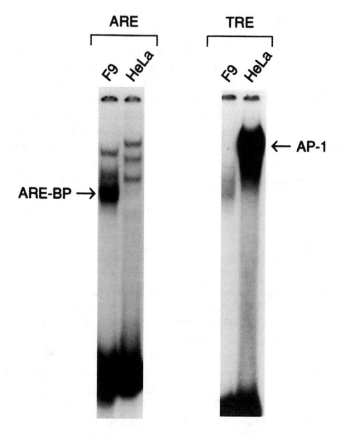

Figure 12.1. Electrophoretic mobility shift analysis of F9 and HeLa nuclear extracts with ARE and TRE probes. ^{32}P-labeled double-stranded oligonucleotides containing ARE or TRE sequences were incubated with nuclear extracts from either F9 or HeLa cells. (See Ref. 27 for details.) The ARE-nucleoprotein complex is detected in F9 cells but *not* in HeLa cells. In contrast, AP-1 is abundant in HeLa cells and *not* detected in F9 cells (see arrows).

12.7 Other Genes Inducible by Phenolic Antioxidants

Several other Phase II enzymes have been shown to be induced by phenolic antioxidants. Epoxide hydrolase is induced in rat and mouse liver by BHA and butylated hydroxytoluene (BHT).[3] A rat aldehyde reductase that metabolizes aflatoxin B_1 is induced by the phenolic antioxidant ethoxyquin.[40] The cytosolic class 3 aldehyde dehydrogenase is induced by phenolic antioxidants in mouse Hepa 1 and human MCF-7 cells.[41] Dihydrodiol dehydrogenase is induced by ethancrynic acid (a Michael acceptor) in human HT29 colon cells.[42] UDP-glucuronosyl transferase is induced by BHA and BHT in the small intestine, kidney, and liver of rats.[43] An ARE-like sequence has been found in the 5'-flanking sequence of a rat UDPGT gene.[44] *t*-BHQ has been shown to induce the γ-glutamylcysteine synthetase heavy subunit in mouse liver and a potential ARE has been identified in this gene as well.[45,46] However, neither of these putative ARE sequences have been shown as yet to be functional. It is highly likely that the existence of AREs will be documented in these and possibly other genes. The genomic sequences for two human alpha class glutathione S-transferases (A1 and A2) have been cloned.[47,48] However, a survey of the first 300 nucleotides of the 5'-flank revealed no sequences with homology to either the ARE or XRE. Analysis of an additional 1000 upstream nucleotides also found no sequences with homology to the ARE.[49]

The rat glutathione transferase P has been shown to be induced by BHA in liver.[50] An enhancer (glutathione transferase P enhancer I [GPEI]) has been identified in the 5'-flanking sequence that resembles the rat and human NQO$_1$ AREs.[51,52] This enhancer is composed of two half-sites that are oriented in the inverse repeat configuration. The GPEI has recently been shown to act as an ARE in human Hep G2 cells.[27] Furthermore, the GPEI competes with the rat NQO$_1$ ARE for the specific nucleoprotein complex found in Hep G2 cells. Transgenic rats containing the CAT gene under the control of the GPEI enhancer have been shown to express high levels of CAT activity in liver preneoplastic nodules.[53,54] Several Phase II enzymes are also increased in these nodules including the glutathione S-transferase Ya subunit and quinone reductase.[55] It will be of interest to determine if the rat Ya and NQO$_1$ AREs are also involved in the regulation of these enzymes in preneoplastic nodules.

12.8 Summary and Future Perspectives

In summary, the AREs for three rodent genes and one human gene are organized as two half-sites that are both required for maximal basal gene expression (Table 12.1). While none of the three rodent genes contain a consensus sequence for AP-1, the human ARE contains a perfect TRE within one half-site.

Of significant importance are the effects of induction of Phase II drug-metabolizing enzymes. Several compounds that provide chemoprotection against the effects of carcinogens and mutagens have been shown to function through activation of the ARE. For example, ellagic acid has been shown to transcriptionally regulate gene expression through the ARE in the rat Ya and NQO$_1$ genes.[56,57] Recently, a compound found in many cruciferous vegetables has been isolated that acts as a potent inducer of quinone reductase activity in Hepa 1c1c7 cells.[58] Suforaphane has also been shown to induce CAT activity through the EpRE in Hep G2 cells.[59] Chemoprevention by naturally occurring or synthetic antioxidants appears to occur through the ARE-dependent induction of Phase II enzymes.

Whether the special orientation of AP-1 and/or AP-1-like sequences leads to binding of Jun/Fos proteins and transcriptional regulation of Phase II genes or whether non-Jun/Fos proteins are involved in the functional binding complex is still unresolved. One potential candidate has recently been identified. The transcription factor Maf (and family members) has been found to recognize a DNA sequence that resembles the ARE core consensus sequence.[60] Maf has also been shown to bind to probes containing ARE sequences.[61] However, it is not known if Maf is present in hepatoma cells or in mouse and rat liver.

Clearly, the most important goal is to identify and characterize the non-AP-1 *trans*-acting factors identified in rat liver and rodent and human cell culture lines. Once this protein (or proteins) has been purified and characterized, it will become possible to determine the precise role it plays in the regulation of Phase II drug-metabolizing genes.

References

1. Iyanagi, T. 1987. On the mechanism of one and two-electron transfer by flavin enzymes. *Chemica Scripta* **27A**:31–36.

2. Farber, J. L., M. E. Kyle, and J. B. Coleman. 1990. Biology of disease: mechanisms of cell injury by active oxygen species. *Lab. Invest.* **62**:670–679.

3. Lind, C., P. Hochstein, and L. Ernster. 1982. DT-diaphorase as a quinone reductase: a cellular control device against semiquinone and superoxide radical formation. *Arch. Biochem. Biophys.* **216**:178–185.

4. Cadenas, E. 1989. Biochemistry of oxygen toxicity. *Annu. Rev. Biochem.* **58**:79–110.

5. Benson, A. M., M. J. Hunkeler, and P. Talalay. 1980. Increase of NAD(P)H:quinone reductase by dietary antioxidants: possible role in protection against carcinogenesis and toxicity. *Proc. Natl. Acad. Sci. USA* **77**:5216–5220.

6. De Long, M. J., H. J. Prochaska, and P. Talalay. 1986. Induction of NAD(P)H:quinone reductase in murine hepatoma cells by phenolic antioxidants, azo dyes, and other chemoprotectors: a model system for the study of anticarcinogens. *Proc. Natl. Acad. Sci. USA* **83**:787–791.

7. Prochaska, H. J., M. J. De Long, and P. Talalay. 1985. On the mechanisms of induction of cancer-preventive enzymes: a unifying proposal. *Proc. Natl. Acad. Sci. USA* **82**:8232–8236.

8. Prochaska, H. J., and P. Talalay. 1988. Regulatory mechanisms of monofunctional and bifunctional anticarcinogenic enzyme inducers in murine liver. *Cancer Res.* **48**:4776–4782.

9. Talalay, P., M. J. De Long, and H. J. Prochaska. 1988. Identification of a common chemical signal regulating the induction of enzymes that protect against chemical carcinogenesis. *Proc. Natl. Acad. Sci. USA* **85**:8261–8265.

10. Pickett, C. B., C. A. Telakowski-Hopkins, G. J.-F. Ding, L. Argenbright, and A. Y. H. Lu. 1984. Rat liver glutathione S-transferases. Complete nucleotide sequence of a glutathione S-transferase mRNA and the regulation of the Ya, Yb, and Yc mRNAs by 3-methylcholanthrene and phenobarbital. *J. Biol. Chem.* **259**:5182–5188.

11. Telakowski-Hopkins, C. A., G. S. Rothkopf, and C. B. Pickett. 1986. Structural analysis of a rat liver glutathione S-transferase Ya gene. *Proc. Natl. Acad. Sci. USA* **83**:9393–9397.

12. Telakowski-Hopkins, C. A., R. G. King, and C. B. Pickett. 1988. Glutathione S-transferase Ya subunit gene: identification of regulatory elements required for basal level and inducible expression. *Proc. Natl. Acad. Sci. USA* **85**:1000–1004.

13. Paulson, K. E., J. E. Darnell, Jr., T. Rushmore, and C. B. Pickett. 1990. Analysis of the upstream elements of the xenobiotic compound-inducible and positionally regulated glutathione S-transferase Ya gene. *Mol. Cell. Biol.* **10**:1841–1852.

14. Rushmore, T. H., R. G. King, K. E. Paulson, and C. B. Pickett. 1990. Regulation of glutathione S-transferase Ya subunit gene expression: identification of a unique xenobiotic-responsive element controlling inducible expression by planar aromatic compounds. *Proc. Natl. Acad. Sci. USA* **87**:3826–3830.

15. Rushmore, T. H., and C. B. Pickett. 1990. Transcriptional regulation of the rat glutathione S-transferase Ya subunit gene: characterization of a xenobiotic-responsive element controlling inducible expression by phenolic antioxidants. *J. Biol. Chem.* **265**:14648–14653.

16. Rushmore, T. H., M. R. Morton, and C. B. Pickett. 1991. The antioxidant responsive element. Activation by oxidative stress and identification of the DNA consensus sequence required for functional activity. *J. Biol. Chem.* **266**:11632–11639.

17. Nguyen, T., and C. B. Pickett. 1992. Regulation of rat glutathione S-transferase Ya subunit gene expression: DNA-protein interaction at the antioxidant responsive element. *J. Biol. Chem.* **267**:13535–13539.

18. Daniel, V., R. Sharon, Y. Tichauer, and S. Sarid. 1987. Mouse glutathione S-transferase Ya subunit: gene structure and sequence. *DNA* **6**:317–324.

19. Daniel, V., Y. Tichauer, and R. Sharon. 1988. 5′-flanking sequence of mouse glutathione S-transferase Ya gene. *Nucleic Acids Res.* **16**:351.

20. Daniel, V., R. Sharon, and A. Bensimon. 1989. Regulatory elements controlling the basal and drug-inducible expression of glutathione S-transferase Ya subunit gene. *DNA* **8**:399–408.

21. Friling, R. S., A. Bensimon, Y. Tichauer, and V. Daniel. 1990. Xenobiotic-inducible expression of murine glutathione S-transferase Ya subunit gene is controlled by an electrophile-responsive element. *Proc. Natl. Acad. Sci. USA* **87**:6258–6262.

22. Friling, R. S., S. Bergelson, and V. Daniel. 1992. Two adjacent AP-1-like binding sites form the electrophile-responsive element of the murine glutathione S-transferase Ya subunit gene. *Proc. Natl. Acad. Sci. USA* **89**:668–672.

23. Daniel, V., S. Bergelson, and R. Pinkus. 1993. Structure and function of glutathione S-transferases: The role of AP-1 transcription factor in the regulation of glutathione S-transferase Ya subunit gene expression by chemical agents. pp. 129–136. *In* K. D. Tew, C. B. Pickett, T. J. Mantle, B. Mannervik, and J. Hayes, eds. CRC Press, Boca Raton, FL.

24. Bayney, R. M., J. A. Rodkey, C. D. Bennet, A. Y. H. Lu, and C. B. Pickett. 1987. Rat liver NAD(P)H:quinone reductase nucleotide sequence analysis of a quinone

reductase cDNA clone and prediction of the amino acid sequence of the corresponding protein. *J. Biol. Chem.* **262**:572–575.

25. Bayney, R. M., M. R. Morton, L. V. Favreau, and C. B. Pickett. 1989. Rat liver NAD(P)H:quinone reductase. Regulation of quinone reductase gene expression by planar aromatic compounds and determination of the exon structure of the quinone reductase gene. *J. Biol. Chem.* **264**:21793–21797.

26. Favreau, L. V., and C. B. Pickett. 1991. Transcriptional regulation of the rat NAD(P)H:quinone reductase gene. Identification of regulatory elements controlling basal level expression and inducible expression by planar aromatic compounds and phenolic antioxidants. *J. Biol. Chem.* **266**:4556–4561.

27. Favreau, L. V., and C. B. Pickett. 1995. The rat quinone reductase antioxidant response element. Identification of the nucleotide sequence required for basal and inducible activity and detection of antioxidant response element-binding proteins in hepatoma and non-hepatoma cell lines. *J. Biol. Chem.* **270**:24468–24474.

28. Favreau, L. V., and C. B. Pickett. 1993. Transcriptional regulation of the rat NAD(P)H:quinone reductase gene. Characterization of a DNA-protein interaction at the antioxidant responsive element and induction by 12-*O*-tetradecanoylphorbol 13-acetate. *J. Biol. Chem.* **268**:19875–19881.

29. Jaiswal, A. K., O. W. McBride, M. Adesnik, and D. W. Nebert. 1988. Human dioxin-inducible cytosolic NAD(P)H:menadione oxidoreductase. cDNA sequence and localization of gene to chromosome 16. *J. Biol. Chem.* **263**:13572–13578.

30. Jaiswal, A. K. 1991. Human NAD(P)H:quinone oxidoreductase (NQO₁) gene structure and induction by dioxin. *Biochemistry* **30**:10647–10653.

31. Li, Y., and A. K. Jaiswal. 1992. Regulation of human NAD(P)H:quinone oxidoreductase gene. Role of AP1 binding site contained within human antioxidant response element. *J. Biol. Chem.* **267**:15097–15104.

32. Xie, T., M. Belinsky, Y. Xu, and A. K. Jaiswal. 1995. ARE- and TRE-mediated regulation of gene expression. Response to xenobiotics and antioxidants. *J. Biol. Chem.* **270**:6894–6900.

33. Li, Y., and A. K. Jaiswal. 1992. Identification of Jun-B as third member in human antioxidant response element-nuclear proteins complex. *Biochem. Biophys. Res. Commun.* **188**:992–996.

34. Yao, K.-S., S. Xanthoudakis, T. Curran, and P. J. O'Dwyer. 1994. Activation of AP-1 and of a nuclear redox factor, Ref-1, in the response of HT29 colon cancer cells to hypoxia. *Mol. Cell. Biol.* **14**:5997–6003.

35. Nguyen, T., T. H. Rushmore, and C. B. Pickett. 1994. Transcriptional regulation of a rat liver glutathione S-transferase Ya subunit gene. Analysis of the antioxidant response element and its activation by the phorbol ester 12-*Q*-tetradecanoylphorbol-13-acetate. *J. Biol. Chem.* **269**:13656–13662.

36. Bergelson, S., and V. Daniel. 1994. Cooperative interaction between Ets an AP-1 transcription factors regulates induction of glutathione S-transferase Ya gene expression. *Biochem. Biophys. Res. Commun.* **200**:290–297.

37. Bergelson, S., R. Pinkus, and V. Daniel. 1994. Induction of AP-1 (Fos/Jun) by chemical agents mediates activation of glutathione S-transferase and quinone reductase gene expression. *Oncogene* **9**:565–571.

38. Yoshioka, K., T. Deng, M. Cavigelli, and M. Karin. 1995. Antitumor promotion by phenolic antioxidants: Inhibition of AP-1 activity through induction of Fra expression. *Proc. Natl. Acad. Sci. USA* **92**:4972–4976.

39. Yang-Yen, H.-F., R. Chiu, and M. Karin. 1990. Elevation of AP1 activity during F9 cell differentiation is due to increased c-*jun* transcription. *New Biol.* **2**:351–361.

40. Ellis, E. M., D. J. Judah, G. E. Neal, and J. D. Hayes. 1993. An ethoxyquin-inducible aldehyde reductase from rat liver that metabolizes aflatoxin B_1 defines a subfamily of aldo-keto reductases. *Proc. Natl. Acad. Sci. USA* **90**:10350–10354.

41. Sreerama, L., G. K. Rekha, and N. E. Sladek. 1995. Phenolic antioxidant-induced overexpression of class-3 aldehyde dehydrogenase and oxazaphosphorine-specific resistance. *Biochem. Pharmacol.* **49**:669–675.

42. Ciaccio, P. J., A. K. Jaiswal, and K. D. Tew. 1994. Regulation of human dihydrodiol dehydrogenase by Michael acceptor xenobiotics. *J. Biol. Chem.* **269**:15558–15562.

43. Kashfi, K., E. K. Yang, J. R. Chowdhury, N. R. Chowdhury, and A. J. Dannenberg. 1994. Regulation of uridine diphosphate glucuronosyltransferase expression by phenolic antioxidants. *Cancer Res.* **54**:5856–5859.

44. Mackenzie, P. I., and L. Rodbourn. 1990. Organization of the rat UDP-glucuronosyltransferase, UDPG*Tr*-2, gene and characterization of its promoter. *J. Biol. Chem.* **265**:11328–11332.

45. Borroz, K. I., T. M. Buetler, and D. L. Eaton. 1994. Modulation of γ-glutamylcysteine synthetase large subunit mRNA expression by butylated hydroxyanisole. *Toxicol. Appl. Pharmacol.* **126**:150–155.

46. Mulcahy, R. T., and J. J. Gipp. 1995. Identification of a putative antioxidant response element in the 5′-flanking region of the human γ-glutamylcysteine synthetase heavy subunit gene. *Biochem. Biophys. Res. Commun.* **209**:227–233.

47. Rozen, F., T. Nguyen, and C. B. Pickett. 1992. Isolation and characterization of a human glutathione S-transferase Ha_1 subunit gene. *Arch. Biochem. Biophys.* **292**:589–593.

48. Röhrdanz, E., T. Nguyen, and C. B. Pickett. 1992. Isolation and characterization of the human glutathione S-transferase A_2 subunit gene. *Arch. Biochem. Biophys.* **298**:747–752.

49. Suzuki, T., S. Smith, and P. G. Board. 1994. Structure and function of the 5′ flanking sequences of the human alpha class glutathione S-transferase genes. *Biochem. Biophys. Res. Commun.* **200**:1665–1671.

50. Satoh, K., A. Kitahara, Y. Soma, Y. Inaba, I. Hatayama, and K. Sato. 1985. Purification, induction, and distribution of placental glutathione transferase: a new marker enzyme for preneoplastic cells in the rat chemical hepatocarcinogenesis. *Proc. Natl. Acad. Sci. USA* **82**:3964–3968.

51. Sakai, M., A. Okuda, and M. Muramatsu. 1988. Multiple regulatory elements and phorbol 12-*O*-tetradecanoate 13-acetate responsiveness of the rat placental glutathione transferase gene. *Proc. Natl. Acad. Sci. USA* **85**:9456–9460.

52. Okuda, A., M. Imagawa, Y. Maeda, M. Sakai, and M. Muramatsu. 1989. Structural and functional analysis of an enhancer GPEI having a phorbol 12-O-tetradecanoate 13-acetate responsive element-like sequence found in the rat glutathione transferase P gene. *J. Biol. Chem.* **264**:16919–16926.

53. Morimura, S., T. Suzuki, S.-I. Hochi, A. Yuki, K. Nomura, T. Kitagawa, I. Nagatsu, M. Imagawa, and M. Muramatsu. 1993. Trans-activation of glutathione transferase P gene during chemical hepatocarcinogenesis of the rat. *Proc. Natl. Acad. Sci. USA* **90**:2065–2068.

54. Suzuki, T., M. Imagawa, M. Hirabayashi, A. Yuki, K. Hisatake, K. Nomura, T. Kitagawa, and M. Muramatsu. 1995. Identification of an enhancer responsible for tumor marker gene expression by means of transgenic rats. *Cancer Res.* **55**:2651–2655.

55. Pickett, C. B., J. B. Williams, A. Y. H. Lu, and R. G. Cameron. 1984. Regulation of glutathione transferase and DT-diaphorase mRNAs in persistent hepatocyte nodules during chemical hepatocarcinogenesis. *Proc. Natl. Acad. Sci. USA* **81**:5091–5095.

56. Barch, D. H., and L. M. Rundhaugen. 1994. Ellagic acid induces NAD(P)H:quinone reductase through activation of the antioxidant responsive element of the rat NAD(-P)H:quinone reductase gene. *Carcinogenesis* **15**:2065–2068.

57. Barch, D. H., L. M. Rundhaugen, and N. S. Pillay. 1995. Ellagic acid induces transcription of the rat glutathione S-transferase-Ya gene. *Carcinogenesis* **16**:665–668.

58. Zhang, Y., P. Talalay, C.-G. Cho, and G. H. Posner. 1992. A major inducer of anticarcinogenic protective enzymes from broccoli: isolation and elucidation of structure. *Proc. Natl. Acad. Sci. USA* **89**:2399–2403.

59. Prestera, T., W. D. Holtzclaw, Y. Zhang, and P. Talalay. 1993. Chemical and molecular regulation of enzymes that detoxify carcinogens. *Proc. Natl. Acad. Sci. USA* **90**:2965–2969.

60. Kataoka, K., M. Noda, and M. Nishizawa. 1994. Maf nuclear oncoprotein recognizes sequences related to an AP-1 site and forms heterodimers with both Fos and Jun. *Mol. Cell. Biol.* **14**:700–712.

61. Kerppola, T. K., and T. Curran. 1994. A conserved region adjacent to the basic domain is required for recognition of an extended DNA binding site by Maf/Nrl family proteins. *Oncogene* **9**:3149–3158.

13

Oxyradicals as Signal Transducers

Roy H. Burdon

13.1 Cell Proliferation Control Proteins

Control of the balance between the life and death of cells is crucial to the proper development of tissues and organs. The proliferation of cells is highly orchestrated with a complexity of molecular checks and balances, but a cell that will not stop dividing presents problems as it can become a tumor. In contrast, a genetically programmed cell death mechanism, *apoptosis,* regulates cell numbers and is a means whereby unwanted or potentially harmful cells are eliminated. Accurate control of cellular life and death involves a plethora of molecular signals and switches, and an essential aspect is how these are activated and terminated at precisely the correct moment.[1]

It is now clear that a wide range of cells generate oxygen-derived free radicals and other reactive oxygen species, which are counterbalanced by antioxidant defence mechanisms. Depending on this redox balance, proliferation can be stimulated or cells can die by apoptotic mechanisms.

In terms of mechanisms that underpin the regulation of cell proliferation, a general outline is beginning to emerge. The fundamental machinery that controls cell division is based on protein kinases. Two family of proteins are important. The first is a family of cyclin-dependent protein kinases (Cdks), which induce "downstream" processes by phosphorylating specific cellular proteins.[2] The second is a family of specialized activating proteins called cyclins that complex with Cdk molecules and control their ability to phosphorylate appropriate target proteins. Two such complexes of proteins regulate the normal cell division cycle, one in late G1 phase, just before the start of DNA replication[3] and the other in G2 phase, just before mitosis.[4]

Proliferation of normal cells is stimulated by growth factors (or cytokines), which interact with specific cell surface receptors that set in motion cascades of intracellular signal or second-messenger systems that in turn, activate specific genes.[5] In some cases, growth factors can interact with plasma membrane receptors with the resultant increase in the formation of diacylglycerol (DG) and inositol trisphoshate (IP$_3$) within cells.[6] This is achieved by virtue of the growth factor receptor communicating a stimulatory signal to the plasma membrane phospholipase C via a group of membrane associated proteins known as G proteins.[7] In turn, IP$_3$ causes the release of calcium ions from the endoplasmic reticulum, and these ions, together with DG, bring about the activation of a specific protein kinase, *protein kinase C*.[8] This important protein kinase can phosphorylate various cellular proteins that are believed to be involved in proliferative control. Besides IP$_3$ and DG, another second messenger of possible significance in cell proliferation is cyclic AMP (cAMP), which can also regulate the activity of protein kinases.[9] cAMP can also directly regulate the transcription of a number of genes.[9] Positive, as well as negative, regulatory DNA sequence elements have been demonstrated in the promoters of cAMP responsive genes. The intracellular level of cAMP is under the influence of a variety of growth factors, which act by binding to specific plasma membrane receptors that subsequently communicate a stimulatory/inhibitory signal to adenyl cyclase through membrane-associated G proteins.

In addition to the mitogenic pathways outlined above, another pathway can also be operational. This pathway involves the direct interaction of certain growth factors (e.g., platelet-derived growth factor) with their cognate plasma membrane receptors with the resultant activation of a protein tyrosine kinase activity, which is an integral element of the plasma membrane receptor molecule itself.[10] Recently a fourth type of signal pathway has been identified, which is utilized by cytokines, that involves a special class of protein tyrosine kinases, the Janus kinases.[11] These are activated by cytokine binding to cognate receptors but are not involved in triggering a complex cascade of reactions. Instead they directly phosphorylate three STAT (signal transducers and activators of transcription) proteins, which then aggregate as a transcription complex on specific genes. Depending on the cell type, two or three of these pathways may be simultaneously operational in stimulating cell growth. These pathways may be entirely independent at the level of membrane signaling and protein kinase activation, partially overlapping at the level of gene expression and protein synthesis, but are believed to converge at the DNA replication phase and mitosis.

The genes that are activated by these intracellular signal or second messenger systems fall into two main categories.[12] *Early response genes* are induced within 15 min of growth factor treatment, whereas *delayed response genes* may require an hour for expression. The best studied early response genes are the *myc, fos,* and *jun* protooncogenes. When overexpressed or hyperactivated by mutation in certain types of cell all of those can cause uncontrolled proliferation.[13] Where *myc* expression is specifically prevented, cells will not divide even in the presence of growth factors.

The transcription of delayed response genes requires the presence of products of early response genes such as Myc[13] and the products of these delayed response genes include Cdks and several cyclins, which are believed to be involved with Cdk proteins in driving cells into the DNA replication phase or mitosis.

Such stimulating signals are suspected to overcome specific inhibitory devices that ensure that cells refrain from proliferating in the absence of positive signals to do so. Such inhibitory devices are encoded by antiproliferative genes that include those originally described as *tumor suppressor genes*. An example is the retinoblastoma gene, which encodes the retinblastoma protein (Rb), which binds to many other proteins, including gene-regulatory proteins depending on its state of phosphorylation.[14] When dephosphorylated it binds to a set of regulatory proteins that favor cell proliferation in such a way as to prevent them acting to facilitate transcription of *fos* and *myc*. In terms of the cell division cycle, growth factors, through the signal pathways described above, somehow relieve the inhibition exerted by Rb by phosphorylating it at a number of sites. This permits the release of the bound regulatory proteins and eventually cells begin to express delayed response genes including those for Cdks and cyclins and commence DNA replication. Rb subsequently becomes dephosphorylated as the cell exits from mitosis.

Another tumor suppressor gene product is p53. Raised levels of p53 protein in a cultured cell will stop its proliferation. The p53 protein binds to DNA and appears to exert its effect, in part at least, by inducing the transcription of another regulatory gene whose protein product of 21 kD binds to complexes of G1 cyclin and Cdk2 protein that normally serve to drive the cell toward the DNA replication phase.[15] Unlike Rb, very little p53 protein is found in most cells under normal conditions.

As already mentioned, the balance between cell population in an organism is controlled between the rates of proliferation, differentiation, and programmed cell death, or apoptosis, of constituent cells.[16] In cancer, for instance, the rate of tumor cell accumulation can be viewed as the rate of cell proliferation minus the rate of cell death.

In many tissues, cells are programmed to kill themselves if they do not receive signals for survival. At least some of the genes that regulate apoptosis have been highly conserved in evolution such as *bcl-2*. Bcl-2 is an acronym for the B-cell lymphoma/leukemia-2 gene. Translocation of this particular gene results in its activation in the majority of follicular non-Hodgkins' B-cell lymphomas. *Bcl-2* can promote cell survival,[17] and the ability of *Bcl-2* protein overproduction to prevent cell death without necessarily affecting cell proliferation suggests *bcl-2* as a new class of oncogene. Gene transfer experiments in hemopoetic cells have shown that *bcl-2* blocks apoptosis induced by c-*myc*.[18] This may account for the aggressive nature of lymphomas and leukemias that contain chromosomal translocation involving both these genes. Because c-*myc* appears to stimulate mitosis as well as apoptosis, simultaneous activation of *bcl-2* can neutralize the apoptotic stimulus of c-*myc*. In certain cell lines, transfection with the protooncogene c-*myc* can lead to "growth factor addiction," that is, cells that would normally undergo cell cycle arrest when growth factors were withdrawn instead went into apoptosis.[19-22] Another report suggests that continuous expression of the protooncogene c-*fos* is a harbinger of apoptosis,[23] whereas others have suggested that p53 is required for apoptosis in certain situations.[24] Most recent findings suggest that p53 coordinates cellular responses to DNA damage. For example, DNA damage results in an increase in the intracellular levels of p53, which can induce apoptosis through transcriptional activation of the gene encoding the protein Bax, which antagonizes the activity of Bcl-2.[25] Alternatively, depending on the extent of DNA damage, DNA repair may be stimulated through p53-stimulated expression of the *gadd 45* gene.[25]

13.2 Superoxide Radicals, Hydrogen Peroxide, and Cell Proliferation

While the mechanisms underlying the regulation of cell proliferation are considerably complex, there is now a growing body of evidence to suggest that superoxide and hydrogen peroxide may play a crucial role in the mechanisms underlying proliferative responses.

Early experiments of ours[26-28] indicated low concentrations of superoxide radicals or hydrogen peroxide (10 nM to 1 μM) to be effective in stimulating the *in vitro* growth of hamster (Fig. 13.1) and rat fibroblasts when added to the culture medium. It is now clear that superoxide and hydrogen peroxide can each stimulate growth and growth responses in a considerable variety of cultured mammalian cell types when added exogenously at low levels.

Besides stimulating the growth of hamster fibroblasts (BHK-21), exogenously added superoxide can also stimulate growth response in human fibroblasts,[29] Balb/3T3 cells,[30,36,37] human amnion cells,[31] mouse epidermal cells (JB6),[32] and human histiocytic leukemia cells (U937).[33] Growth responses are also elicited by hydrogen peroxide in hamster (BHK-21)[26,27] and rat fibroblasts (208F)[28] as well as in Balb/3T3,[36] mouse osteoblastic cells (MC3T3),[35] and primary rabbit kidney epithelial cells. It is notable that these positive responses with low levels of exogenous hydrogen peroxide can be demonstrated in both primary cell cultures and permanent cell lines. Experiments have also been carried out to determine whether or not specific oncogene transformation of cells made them more or less susceptible to the stimulating effects of hydrogen peroxide. We have observed that the growth of rat fibroblasts (208F) transfected with the activated form of *H-ras* from human bladder carcinoma was more notably stimulated by exogenously added hydrogen peroxide than was the growth of their nontransgenic counterparts.[28] In JB6,[32] Balb/3T3,[30,36] and MC3T3 cells, superoxide and hydrogen peroxide stimulated the expression of early growth related genes such as c-*fos* and c-*jun*. Such observations led to suggestions that superoxide and hydrogen peroxide might function as mitogenic stimuli through biochemical processes common to natural growth factors.

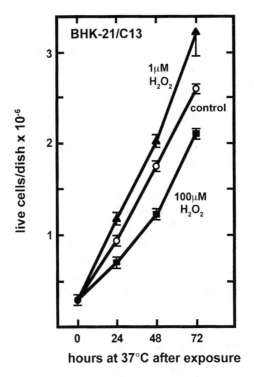

Figure 13.1. The effect of exogenous hydrogen peroxide on the growth of BHK-21/ C13 cells. Triplicate monolayer cultures of BHK cells (0.27×10^6 cells/3.5-cm petri dish) were established in 2-ml Eagle minimal essential medium (MEM) supplemented with 10% calf serum. To some cultures, hydrogen peroxide was added after 24 h and growth was subsequently followed at 37°C. At the times indicated, determinations were made of live cells in the monolayer. Results are expressed as means ± SD ($n = 3$).

In the case of exogenously added superoxide, some of its effects related to growth stimulation are extremely rapid and distinct from those of hydrogen peroxide. For example, intracellular pH is rapidly increased.[31,33] In human amnion cells this occurs within 10 s and is followed by an increase in Ca^{2+} by 20 to 40 s.[31] Such effects, which are not elicited by hydrogen peroxide, appear to be inhibited in the presence of an anion-channel blocker,[31] suggesting that superoxide anions exert their effects intracellularly, possibly after passage through plasma membrane anion channels. Hydrogen peroxide, which can freely permeate cell membranes, however, does not elicit these early changes in pH and Ca^{2+} concentration in human amnion cells. Nevertheless, like superoxide, it does stimulate the expression of the early growth genes mentioned above. At present, such distinctive effects of superoxide compared with hydrogen peroxide are difficult to assess but may have important biological consequences.

Another important feature to emerge from studies on BHK-21 cells with exogenously added hydrogen peroxide is that higher levels such as 100 µM cause growth inhibition (see Fig. 13.1) and, as will be discussed later in this chapter, an increased incidence of apoptosis. Addition of even higher concentrations of hydrogen peroxide to cultures of BHK-21 cells, however, increases cell death due to necrosis, rather than apoptosis. In summary, while low levels of hydrogen peroxide can have positive regulatory effect on mammalian cell growth, increasing concentrations can elicit a "negative" or down-regulatory effect on growth, but a positive signaling role in the propagation of apoptosis.

Although these *in vitro* effects have been observed in the particular cell types mentioned, the generality of these observations *in vitro* and *in vivo* is not yet clear. Other cell types may show distinct differences. Moreover, the overall outcomes of such experimental approaches will always depend on relative concentrations of exogenous superoxide or hydrogen peroxide in relation to the numbers of cells exposed as well as to relative cellular activities of catalase or glutathione redox systems.

While these studies concern the growth of nonimmune cells, there is also evidence implicating reactive oxygen species as signal molecules in lymphocytic activation.[89]

13.3 Cellular Generation of Superoxide Radicals and Hydrogen Peroxide

Despite these observations obtained in cell culture systems to which hydrogen peroxide was added exogenously, a key question relates to their potential biological relevance. In physiological terms, relatively high concentrations of exogenous hydrogen peroxide could conceivably be available at sites of inflammation. Although precise data on such concentrations is not available, it is known that in liver the levels of hydrogen peroxide are in submicromolar range and blood plasma levels of 0.25 to 5 µM have been recorded.[38,39] In an inflamed tissue, cell crowding would cause limited tissue oxygenation, so maximal levels may not be reached. In addition, the duration and frequency of an individual cell's exposure to such external hydrogen peroxide *in vivo* is difficult to assess. For these reasons it is probably important to consider as relevant, effects of hydrogen peroxide over a wide concentration range (nano- to micromolar). Thus, superoxide and hydrogen peroxide generated by immune cells are likely to be physiologically important in contributing to the growth modulation of adjacent noninflammatory cells such as fibroblasts at a site of inflammation.

Whereas both superoxide and hydrogen peroxide are established products of the "respiratory burst" when the plasma membrane NADPH-oxidase of neutrophils and macrophages is activated, it is now becoming clear that superoxide and hydrogen peroxiden are also normally released by a variety of noninflammatory cells (Table 13.1). For instance, primary human skin fibroblasts stimulated

Table 13.1.   Cellular Release of Superoxide Radicals and Hydrogen Peroxide

Released	Cell type	Stimulus	Reference
Superoxide anions	Human fibroblasts	Cytokines/TPA	40
	Human endothelial cells	Cytokines/TPA	42
	Human fat cells	Cytokines/bFGF	46
	Human chondrocytes		49
	BHK-21 cells	None required	50
	Human colonic epithelial cells	TPA/deoxycholate	34
Hydrogen peroxide	Balb/3T3 cells	PDGF/TPA	36
	Rat pancreatic islets	TPA/Ca ionophore	44
	Murine keratinocytes	TPA	43
	Rabbit chondrocytes	Cytokines/bFGF	45
	Human tumor lines	None required	48

TPA, 12-0-tetradecanoylphorbol-13-acetate; PDGF, platelet-derived growth factor; bFGF, basic fibroblast growth factor.

with cytokines, such as interleukin-1 or tumor necrosis factor, release significant levels of superoxide (5nmol/h/10^6 cells). As is the case for neutrophils, the generation of the released superoxide from fibroblasts appears to be catalyzed by a plasma membrane NADPH-oxidase.[40] This conclusion was based on the inhibitory effects of diphenylene iodonium, a known inhibitor of the flavoprotein component of the neutrophil NADPH-oxidase.[41] Endothelial cells also release superoxide,[42] but, as was the case for the primary fibroblasts, this release is greatly stimulated (to 27.6 nmol/h/10^6 cells) by cytokines, in this case interferon and interleukin-1. Superoxide, of course, can dismutate to hydrogen peroxide, and hydrogen peroxide release has also been detected in Balb/3T3 cells,[36] murine epidermal keratinocytes,[43] and pancreatic islets.[44] In each case, release requires the stimulus of a phorbol ester, suggesting the involvement of protein kinase C. In the case of Balb/3T3, cells, platelet-derived growth factor is also stimulatory,[36] and for the rabbit articular chondrocytes, cytokines are required.[45] Cytokines are also necessary for hydrogen peroxide release from human preadipocytes (3T3LI). Intriguingly, however, the basic and acidic forms of fibroblast growth factor have opposing stimulatory and inhibitory effects on the NADPH-dependent generation system in these cells.[46] Bovine articular chondrocytes can also be stimulated to release reactive oxygen species by basic fibroblast growth factor in a reaction mediated by NADPH-oxidase.[47]

In contrast, high levels (up to 0.5nmol/h/10^4 cells) of hydrogen peroxide are *constitutively* released from a wide range of human tumor cells[48] (melanoma, colon carcinoma, pancreatic carcinoma, neuroblastoma, ovarian and breast carcinoma). This hydrogen peroxide, like that described above, is likely to be derived from superoxide released via an NADPH-dependent system, as it is also inhibited by diphenylene iodonium.[48] It may of course be that the apparent *constitutive* release from these tumor cells is due to the release of growth factors from the tumor cells themselves and the interaction of these released factors with their own cell surface receptors. Some tumor cells, on the other hand, are known to have constitutively active growth factor receptors.

In contrast to this cytokine or growth-hormone-regulated generation and release of superoxide and hydrogen peroxide that appears to involve NADPH-oxidase, later studies of ours on hamster (BHK-21) and rat fibroblasts (208F) and HeLa cells however have served to indicate other sources of endogenous superoxide generation.[28,50,51] These include mitochondria and xanthine oxidase. In BHK-21 and HeLa cells the rate of intracellular superoxide generation appears to be negatively regulated by antioxidant serum factors. In L-929 cells (but not BHK-21 or HeLa cells) the mitochondrial generation of superoxide can, however, be positively stimulated by the cytokine, tumor necrosis factor alpha (TNF-α).[52,53] This also appears to be the case in U937 myelomonocytic human cells.[54] Presently, however, it is not possible to decide on the relative quantitative importance of mitochondrial generated superoxide as distinct from that generated via NADPH-oxidase based systems.

13.4 Cellularly Generated Superoxide Radicals and Hydrogen Peroxide as Possible Signals in Relation to Proliferation and Apoptosis

Although exogenously added superoxide or hydrogen peroxide can elicit growth responses in a variety of cultured cell types, an important question is whether these effects have physiological relevance. The findings that a range of nonimmune cells can generate and release superoxide and hydrogen peroxide when stimulated by specific cytokines or growth factors argues for serious consideration of their physiological relevance as potential signal molecules.

Superoxide radicals released or generated by stimulated cells are likely to be dismutated to hydrogen peroxide by superoxide dismutases, which are located both on the outside and inside of a variety of mammalian cell types (Fig. 13.2). Extracellular superoxide dismutase (EC-SOD) is usually found associated with structures on the outside of the plasma membrane, whereas Mn-superoxide dismutase (Mn-SOD) is located within mitochondria and Cu,Zn-superoxide dismutase (Cu,Zn-SOD) in the cytosol. The hydrogen peroxide produced can penetrate cell membranes and therefore in principle could function as an inter- or intracellular signal molecule (Fig. 13.2).

We have examined the cellular levels of hydrogen peroxide in BHK-21 fibroblasts during proliferation[55] and found them to be highly dependent on growth rate. Initially, when growth is low, levels of hydrogen peroxide are high. However, as growth rates increase cellular rates levels of hydrogen peroxide progressively decrease. While cellular levels of hydrogen peroxide may depend on the relative activities of superoxide dismutases, on one hand, they will also be influenced by

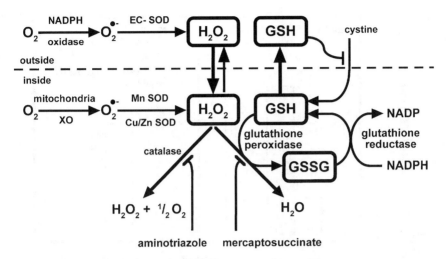

Figure 13.2. Schematic diagram to illustrate the generation and metabolism of superoxide and hydrogen peroxide associated with mammalian cells. XO, xanthine oxidase; EC-SOD, extracellular superoxide dismutase; Mn SOD, manganese superoxide dismutase; Cu/Zn SOD, copper-zinc superoxide dismutase; GSH, reduced glutathione; GSSG, glutathione disulfide.

the activities of catalase and glutathione peroxidase on the other (Fig. 13.2). When these latter activities are examined in growing BHK-21 cells variation with growth is also observed. While catalase activity appears to increase with growth the reverse is true of glutathione peroxidase (Fig. 13.3). In contrast, the activities of both superoxide dismutases and glutathione reductase remain unchanged. Additional experiments with BHK-21 fibroblasts showed that when steps are taken to reduce the growth associated decline in cellular hydrogen peroxide levels by exposure of cells to low molecular weight mimics of superoxide dismutase [Cu-D-isohistidine or CuII(3,5-diisopropylsalicylate)$_2$] or to inhibitors of catalase (aminotriazole) or glutathione peroxidase (mercaptosuccinate) (see Fig. 13.2), proliferation was depressed and apoptosis favored.[55] Thus while kinetic changes in cellular hydrogen peroxide levels are closely related to growth, conditions that favor higher than normal cellular levels appear to promote apoptosis. Since the regulation of cellular hydrogen peroxide is a function of the so-called "antioxidant enzymes" catalase, glutathione peroxidase, and superoxide dismutase, these enzymes now appear to have a hitherto unexpected role as potential regulators of cell proliferation. In fact, they could well be considered as a novel class of therapeutic target molecules. Indeed the superoxide dismutase mimic CuII(diisopropylsalicylate)$_2$ has already been shown to inhibit tumor growth in mouse skin.[56]

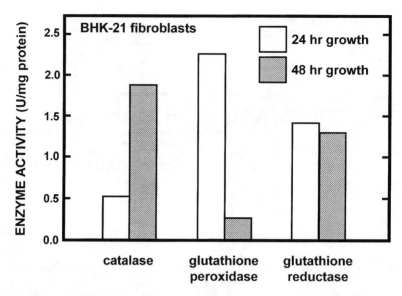

Figure 13.3. The effect of growth on the activity of catalase, glutathione peroxidase, and glutathione reductase in BHK-21/C13 fibroblasts.

13.5 Possible Molecular Mechanisms

13.5.1 Redox Modulation of Cell Proliferation Control Proteins

The observations that superoxide and hydrogen peroxide at low concentration can elicit growth or various growth responses in cultured mammalian cells raises the general question of possible molecular mechanisms. Although there are many possibilities, our appreciation of cell-specific signaling is currently dominated by a simple paradigm whereby this is accomplished by molecules that bind noncovalently to specific receptors through complementarity of shape. In bacteria, such as *Escherichia coli* and *Salmonella typhimurium,* it is becoming clear that specific sets of "defence" genes comprising "regulons" are switched on coordinately, either in response to superoxide and some redox cycling agents (the Sox R regulon), or to hydrogen peroxide (Oxy R regulon).[57–59] On the other hand, to bring about gene responses specific for superoxide, it seems likely that a transcriptional activator protein Sox R senses intracellular stress signals using a redox-active iron sulfur center.[57] In mammalian cells, there is no evidence as yet for proteins similar to Sox R or Oxy R capable of specifically "sensing" superoxide, or hydrogen peroxide, as an initial step in switching on the genes relevant to proliferation responses. This lack of evidence for specific receptor type molecules for superoxide or hydrogen peroxide in animal cells does not imply their absence. On the other hand, mechanisms other than "allosteric" interaction with possible specific receptors have been suggested. Indeed, our own studies with

BHK-21 cells have indicated that growth stimulation can also be achieved with low levels of tertiary butyl hydroperoxide,[27] implying a lack of stringency for hydrogen peroxide.

In relation to the growth promoting effects of low concentrations of hydrogen peroxide, recent studies have shown that following exposure, the early growth response response 1 transcription factor gene (*egr1*) in human HL-525 cells is activated.[60] This appears to involve modification of the serum response factor, which interacts with CArG [GC(A/T)$_6$GG] promoter domains. Low-level hydrogen peroxide exposure in these cells also stimulates c-*jun* activation as well as c-*fos*.[61] In the case of c-*fos* activation, the interaction of the serum response factor and a CArG promoter domain is also required.[62] In mouse epidermal cells studies on c-*fos* induction suggested that protein kinase C is important, and it appears to be activated and translocated following exposure of cells to superoxide.[63,64] Oxidants can activate this important kinase *in vitro*[65,66] (Table 13.2), and it is suggested that this is most likely through the alteration of thiol/disulfide balance of the enzyme. Partially purified protein kinase C from rat brain can also be activated *in vitro* by low concentrations of glutathione disulfise (GSSG) in a glutathione redox buffer.[66] Whereas stress will specifically activate a set of protein kinases that regulate the activity of Jun,[67] a recent observation has been that oxidative stress, including hydrogen peroxide, will activate a human gene encoding a protein-tyrosine phosphatase gene.[68] Moreover, it has been suggested, based on experiments with murine fibroblasts transfected with human epidermal growth factor receptors, that reactive oxygen species and redox processes may contribute to growth factor transduction pathways through alterations in protein-tyrosine kinase and protein tyrosine phosphatase activities.[69] (Table 13.2). Additionally it has been shown that hydrogen peroxide preferentially enhances the tyrosine phosphorylation of the epidermal growth factor itself.[70]

Another cell-signaling target whose activity can be modulated by cellular

Table 13-2. Mammalian Growth Signal Transduction Molecules that are Susceptible to Redox Regulation.

Epidermal growth factor receptor
p21^{ras}
Protein kinase C
Protein tyrosine kinases
Protein tyrosine phosphatases

NF-κB
Fos. Jun heterodimer (AP-1)
Myc
p53
Ets
Rel
Serum response factor (?)

oxidative stress is p21^{ras}. Recent experiments suggest that it too could act as a cellular redox sensor. For example the ability of recombinant p21^{ras} to promote guanine nucleotide exchange can be modified *in vitro* by redox modulators.[71]

Although protein kinases, protein phosphatases, and p21^{ras} pathways play a pivotal role in cellular signal transduction, another group of molecules central to the regulation of signal-induced gene responses are the nuclear transcription factors. Again it may be significant that a large number of recent reports indicate that their activity is highly dependent on their redox status (Table 13.2). For example, the specific DNA-binding *in vitro* of the Fos/Jun heterodimer to the AP-1 promoter site is modulated by reduction-oxidation of a highly conserved cysteine residue in the DNA-binding domain of the proteins.[72] Of potential significance is that reduction of this residue by reducing agents, or by an ubiquitous nuclear redox factor, stimulates AP-1 binding activity *in vitro*, whereas, paradoxically, oxidation of the cysteine has an inhibitory effect on DNA-binding activity.[73] The DNA binding *in vitro* of other transcription factors such as v-Ets,[74] v-Rel,[75] v-Myb,[75] and even p53[76] seems also to be favored by reduction and, in a number of cases, appears particularly to involve the redox state of specific cysteine residues.[76,77] Another transcription factor whose DNA-recognition is redox regulated is NF-κB.[78,79] Mutational analyses indicate that the sulfhydryl group of Cys62 is an important determinant of DNA recognition,[78] again indicating that reduction is likely to favor activity of this protein. In contrast, however, hydrogen peroxide treatment of cells (e.g., T cells) at 100 μM causes "activation" of NF-κB by virtue of causing its release from its inactive cytoplasmic complex with the inhibitory subunit I-κB.[79] That oxidation-reduction is critical in this "activation" of NF-κB from its inactive complex is further supported by the observation that 20-mM N-acetylcysteine (NAC) addition to the cells can counteract the activating effects of hydrogen peroxide.[79]

In other *in vitro* studies it seems that both the activation of NF-κB and the inhibition of its DNA-binding activity may be modulated in intact T cells (Molt-4) by the physiological oxidant glutathione disulfide (GSSG).[80] *In vitro* GSSG inhibits the DNA-binding activity of NF-κB more effectively than that of AP-1, whereas AP-1 is inhibited more effectively by oxidized thioredoxin.[80]

Such observations raise but do not prove the possibility of a novel growth regulatory paradigm superimposed on the established pathways.[81] It is possible to speculate that adjustment of the redox states of proteins involved in these pathways is a prerequisite for optimal functioning. In this context, it is perhaps significant that in the cases mentioned above, where exposure of cells *in vitro* to hydrogen peroxide activates the early growth response I transcription factor gene,[60] or stimulates c-*jun* expression,[61] addition of N-acetylcysteine to the cultured cells again counteracts the effects of hydrogen peroxide. Whereas, as already pointed out, hydrogen peroxide can readily interact directly with transcription factors and so forth to bring about their oxidation, the antagonistic effects of N-acetylcysteine when added to the cultured cells raises the possibility of direct

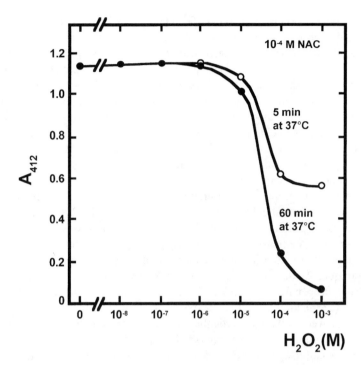

$$H_2O_2(M)$$

Figure 13.4. The ability of various concentrations of hydrogen peroxide to interfere with the reaction of 10^{-4} M *N*-acetylcysteine (NAC) with 5,5'-dithiobis (2-nitrobenzoic acid) (DTNB) 10^{-4} M NAC was allowed to react with various concentrations of hydrogen peroxide at 37°C for 5 or 60 min before being tested for reactive SH group content using the DTNB assay described in reference 126. The formation of 5-thio-2-nitrobenzoate anion (TNB) was monitored at 412 nm.

reversal of the oxidative effect of hydrogen peroxide. As shown in Figure 13.4, *N*-acetyl-L-cysteine can interact directly with hydrogen peroxide. This results in the depletion of cellular levels of hydrogen peroxide.[55] However, the experiments with *N*-acetylcysteine also suggest the possibility that the oxidized-reduced state of the transcription factor can be modulated indirectly through changes in cellular levels of reduced gluthathione (GSH) or glutathione disulphide (GSSG). For example, *N*-acetylcysteine has been reported to bring about elevated GSH levels in cultured cells such as Chinese hamster ovary by promoting cystine uptake and utilization for GSH synthesis,[82] or possibly by directly serving as precursor for the production of cysteine within cells.[83]

13.5.2 Glutathione and Redox Modulation of Control Proteins

In view of the ability of *N*-acetylcysteine to counteract the above growth-related gene responses in cultured cells exposed to oxidants such as hydrogen peroxide, it would be important to consider the relationship between the cellular GSH levels and proliferation responses. An early report suggested that an elevation of intracellular GSH was associated with the mitogenic stimulation of quiescent 3T3 fibroblasts.[84] However, a subsequent study showed that increased levels of GSH in serum-stimulated rat fibroblasts (NRK-49F) are associated not with a response to serum growth factors but with a nutrient repletion.[85] In conditions where this problem was eliminated, it was in fact observed that stimulation of quiescent NRK-49F cells, with epidermal growth factor as mitogen, actually led to a progressive decline in cellular GSH levels. In BHK-21 cells and HeLa cells, we also observe a progressive growth related decline in cellular GSH levels.[86] Indeed when hydrogen peroxide is added to cultures of BHK-21 cells at the low levels required to stimulate proliferation (see Fig. 13.1), while there is actually little change in overall cellular levels of hydrogen peroxide,[55] there is, however, a significant decline in cellular levels of GSH.[86] Such an observation suggests that changes in cellular GSH levels may be more relevant to the successful propagation of growth promotion signals than the actual cellular levels of hydrogen peroxide. As discussed later, procedures that significantly increase cellular levels of hydrogen peroxide have a negative effect on proliferation and can trigger apoptosis (see Section 13.7). Despite these observations there may be considerable variation between cell types, particularly in the case of immune cells. For example, a number of studies have been conducted on the effects of mitogenic stimulation on GSH metabolism in lymphocytes. GSH levels fall abruptly in L1210 cells when placed in culture.[87] The GSH content of splenic T lymphocytes *in vitro* also drops on stimulation with Concanavalin A (Con A),[88] but, in contrast, an increase in GSH is reported in thymocytes after Con A stimulation.[88] In another study, phytohemagglutinin-(PHA)-stimulated EL4 cells (murine lymphoblastoma cells) were found to have less GSH relative to controls, which was in part due to a lowering of the ratio of reduced GSH to the oxidized form, glutathione disulfide (GSSG).[89] Despite possible trivial explanations that relate to culture or nutritional conditions, these differences, although observed *in vitro,* may be biologically important and genuinely reflect differences in cell type. For instance, recently it has become clear that the various functions of cells of the immune system are differentially sensitive to cellular levels of GSH. Interleukin-2 (IL-2) production is inhibited by GSH, whereas other functions, such as IL-2-dependent proliferation, development of $CD8^+$-T cell blasts, and cytotoxic T lymphocyte activity, require high GSH levels.[90] In addition, recent observations relating to redox regulation have shown that decreasing GSH levels in T cells, by using buthionine sulfoximine, abrogates the Ca^{2+} influx normally induced by anti-CD3 antibodies.[91] Moreover, whereas tumor necrosis factor alpha (TNF-α) normally only barely stimulates protein tyrosine phosphorylation in T cells, such GSH depletion creates a situation where TNF-α elicits very significant signal-transduction-related protein tyrosine phosphorylation.[91]

There are a variety of mechanisms that could modulate cellular levels of GSH.[92]

Hydrogen peroxide could most likely influence GSH levels through the activity of glutathione peroxidase and, thus, would generate GSSG and potentially alter the GSH:GSSG equilibrium. It is possible to speculate that some of the growth stimulatory effects of hydrogen peroxide may be propagated through such alterations in cellular GSH levels. Note, however, that the cellular ratio of GSH to GSSG may also reflect the cellular NADP/NADPH ratio,[92] which is determined by a wide variety of NADP- or NADPH-dependent reactions in metabolism. Whatever the reasons, there is evidence to suggest that the GSH:GSSG equilibrium may have a potentially important modulatory role on the activity of certain key signal transduction proteins such as protein kinase C,[66] and a depletion of cellular GSH was recently hypothesized as a component of the signal transduction pathway that regulates the activation of the transcription factor NF-κB.[93] However, with regard to the specific DNA-binding activity of transcription factors such as Ets, Rel, Myc, Fos.Jun (AP-1), p53, and NFκB, it must be remembered, as mentioned already, that the experimental data from *in vitro* experiments paradoxically suggests that maximal DNA binding is favored by *reduction* rather than *oxidation*. It could be speculated that different redox states of signal transduction proteins are appropriate to different cellular requirements for specific gene activation or expression depending on cell type and situation.

13.6 Reactive Oxygen Species as Autocrine Signals in the Redox Regulation of Proliferation

13.6.1 Redox Regulatory Paradigm

Although it is clear that low concentrations of exogenously added superoxide and hydrogen peroxide *in vitro* can elicit growth or growth responses in quite a range of cultured noninflammatory cells, the molecular mechanisms that underlie these effects are not yet clear and may even be distinct in different cell types. Presently there are no data to indicate the existence in mammalian cells of unique highly specific "receptor" or "sensor" molecules for these particular species, as is the case for certain bacteria. This, of course, does not disprove their existence. On the other hand, there are other possibilities. It is evident that hydrogen peroxide, for example, can effect the activation of certain transcription factors that have relevance to growth responses. This might involve the oxidation of specific cysteine residues in such important proteins. In addition, because the redox status of transcription factors, as well as of protein kinases and Ras proteins (e.g., p21^{ras}) appears relevant to their general level of activity (see Table 13.2), a possible speculation is that the overall cellular redox status may be critical. Rather than hydrogen peroxide and superoxide radicals specifically initiating a growth response, by interaction with "cognate" sensors in analogy with growth factor–growth factor receptor interaction, an alternative hypothesis is that they may modulate the efficiency of the *overall* process of growth signal transduction at many sites. This may be through direct oxidation of signal transduction proteins or through their ability to modulate cellular GHS levels to various degrees in different cell types. An attractive and biologically important feature of this hypothesized redox regulatory paradigm is that it can be superimposed on the existant inherent biological specificity that can be achieved through the interaction of specific cytokines, or growth factors, with specific target cell types through cognate cell receptors. Moreover, present evidence suggests that cytokines or growth factors can also specifically regulate the level of superoxide generation in their specific target cells, thus allowing the setting of the appropriate oxidation states for intracellular transmission of their own signal.

This process would thus represent a novel type of "autocrine" coarse control mechanism (see Fig. 13.5). Recent evidence broadly supports this hypothesis. For example, as previously mentioned, treatment of certain cells (e.g., T cells) with hydrogen peroxide causes the activation of NF-κB transcription factor but this can be counteracted over a brief period by N-acetylcysteine.[79] A similar activation of NF-κB can be brought about by exposure of human T cells to TNF-α or interleukin-1. Critically, however, this activation can again be blocked by brief exposure to N-acetylcysteine,[79] implicating reactive oxygen species in the mechanisms that lead to the activation of NF-κB within cells by those cytokines. Although N-acetylcysteine can elevate cellular levels of GSH and thus potentially modulate GSH:GSSG ratios, N-acetylcysteine can also interact directly with hydrogen peroxide itself (Fig. 13.4), thus eliminating a potential "autocrine" signal. Another example is the angiotension II induction of Jun-Fos heterodimer DNA-binding activity and proliferative hypertrophic responses in myogenic cells.[94] Again these responses are abolished by brief exposure to N-acetylcysteine. Similar experiments using N-acetylcysteine suggest the involvement of endogenously generated reactive oxygen species, possibly hydrogen peroxide in the cytokine and growth factor induction of c-*fos* expression in bovine chondrocytes.[47]

Recently, the specific requirement for hydrogen peroxide in the transduction of the platelet-derived growth factor (PDGF) signal in the stimulation of vascular smooth muscle cells has been observed.[95] PDGF transiently increases the cellular concentration of hydrogen peroxide in these cells. This increase could, however, be blunted by N-acetylcysteine or by increasing the intracellular concentration of catalase. Importantly, under these conditions a wide variety of growth associated effects normally induced by PDGF were blocked such as mitogen-activated protein kinase simulation, tyrosine phosphorylation, and DNA synthesis.

13.6.2 Negative versus Positive Growth Regulation

Whereas low levels of hydrogen peroxide, either added exogenously or generated endogenously by cells in response to growth stimulus, can promote the proliferation of a number of cell types, this may not be universal. For example, although hydrogen peroxide acts as a positive signal for quiescent Balb/3T3 fibroblasts,[36]

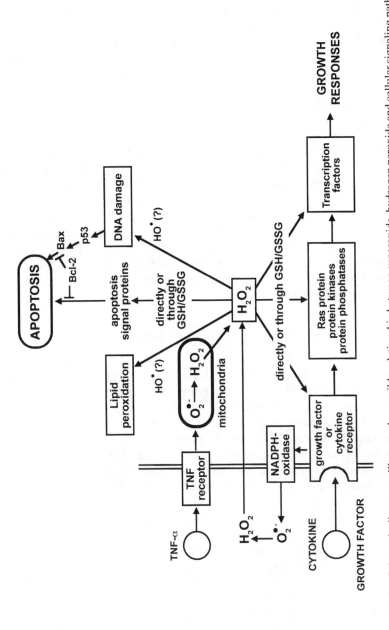

Figure 13.5. Schematic diagram to illustrate the possible relationship between superoxide, hydrogen peroxide and cellular signaling pathways controlling proliferative and apoptotic pathways. TNF-α, tumor necrosis factor alpha; GSH, reduced glutathione; GSSG, glutathione disulfide.

it can act as a negative signal[96] in mouse osteoblastic cells (MC3T3) late in the G_1 phase. Specifically, their studies suggest that hydrogen peroxide served as a "second messenger" for transforming growth factor β1, which inhibits DNA synthesis in these particular cells.[96]

On the other hand, it is clear that in other cells such as BHK-21,[27] 208F,[28] and HeLa,[81] where low levels of hydrogen peroxide were observed to be growth stimulatory, higher levels of exogenous hydrogen peroxide (e.g., 100 µM) have a negative effect on proliferation.

In other experiments where inhibitors of glutathione peroxidase and catalase or mimetics of superoxide dismutase were added to cultures of BHK-21 cells, these caused elevated intracellular levels of hydrogen peroxide, which also resulted in depressed rates of proliferation. Thus in relation to the proposed model of "autocrine" control the ultimate positive or negative control of proliferation induced by endogenously generated reactive oxygen species, or hydrogen peroxide, may well depend on the final cellular levels of those species generated in the response to specific cytokines or growth factors. Such levels would be a function not only of the rates of the generation of those species but also of the rates of their metabolism by the antioxidant enzymes within cells. This would contribute to the setting of the cellular redox balance, which could favor the operation of growth signal transduction proteins in a positive or negative manner, depending on the cell type.

13.7 Reactive Oxygen Species as Signal Transducers in Apoptosis

Another outcome of exposing BHK-21 cells to higher levels of hydrogen peroxide was an increase in the number of cells exhibiting apoptotic properties.[97] Apoptosis is also increased in cells whose intracellular level of hydrogen peroxide was considerably increased by treatment with inhibitors of catalase or glutathione peroxide or with mimetics of superoxide dismutase.[97]

Some recent data implicate oxygen species in cell death by apoptosis. In particular, this view arises from work on a novel oncogene-derived protein Bcl-2, which functions as a repressor of apoptotic cell death in a genetic pathway of cellular suicide common to multicellular animals.[17,98] Bcl-2 has been localized to mitochondria and membranes of the endoplasmic reticulum,[99,100] sites where reactive oxygen species are generated intracellularly.

Although Bcl-2 does not appear to affect the rate of cellular generation of superoxide, gene transfection experiments in lymphocytes indicate that Bcl-2 will protect cells from the lethal effects of H_2O_2 and will suppress oxidative damage to vital cell structures such as lipid membranes.[98,101] Bcl-2 has also been shown to inhibit cell death that follows depletion of glutathione from neural cell lines.[102] It is suggested that Bcl-2 may be protective by virtue of inducing antioxidant enzyme activity in these bacterial cells. Another role for Bcl-2 has been suggested in as much as its homologue in *Caenorhabditis elegans,* Ced9, appears to repress the activity of interleukin-1β converting enzyme (ICE),[103] which is a cysteine protease with specificity for aspartic acid residues.[98] ICE-like molecules are believed to be important in a proteolytic cascade leading to cell death.[104] A suggestion has been that Bcl-2 influences the redox state of cysteine proteases such as ICEs and hence their activities.[98] A complicating factor is, however, that Bcl-2 appears to antagonize the cellular effects of another protein, Bax. While Bcl-2 opposes cell death, Bax promotes death by as yet unknown pathways.[2]

By analogy with the redox mechanisms proposed for growth regulation, however, it is possible to hypothesize that excessively high levels of reactive oxygen species or hydrogen peroxide within cells may result in the cellular redox state, which would favor the operation of apoptosis signaling proteins with the cells rather than proliferative or differentiation pathways. The cytokine TNF-α has been observed to induce apoptosis in neuronal cells[105] and in human U937 cells.[54] While in the former case this outcome can be prevented by over expression of the *bcl-2* gene, in both cases apoptosis can be overcome by use of *N*-acetyl-L-cysteine. Once more this type of observation suggests a mechanism whereby an important component of the cell signaling mechanism is the cytokine-stimulated endogenous generation of reactive oxygen species including hydrogen peroxide. Thus the previously mentioned observation that TNF-α can greatly stimulate the mitochondrial generation of superoxide is of high significance in this context (see Fig. 13.5).

Other mechanisms involving reactive oxygen species including hydrogen peroxide are nevertheless possible. For example, these species are capable of causing damage to cellular DNA by various mechanisms.[106] A response to DNA damage, as mentioned earlier, is the greatly increased expression of the tumor suppressor p53, which can induce apoptosis (see Fig. 13.5) through normal transcriptional activation of the gene encoding the protein Bax, which can antagonize the activity of Bcl-2. Thus as intracellular levels of reactive oxygen species are increased, for example, by (1) exogenous addition of hydrogen peroxide, (2) use of catalase or glutathione peroxidase inhibitors, (3) use of superoxide dismutase mimetics, (4) exposure to redox-cycling drugs, or (4) TNF-α stimulation of mitochondrial superoxide generation, the chances of DNA damage are increased resulting in increased cell levels of p53, and likely apoptosis (see Figure 13.5).

13.8 Free Radical Processes in Membranes as Another Source of Signal Transduction Molecules

Although endogenously generated superoxide radicals and hydrogen peroxide could function as a type of autocrine coarse control system to modulate the redox state of cell signaling molecules either in favor of proliferation or apoptosis, a possible trade-off for this novel signaling mechanism may be peroxidative membrane damage. This would be likely where the cellular generation of superoxide or hydrogen peroxide is excessive, or where cells may be deficient in antioxidant protection.

The phospholipid component of cellular membranes is a highly vulnerable target of oxidative damage due to the susceptibility of its polyunsaturated fatty acid side chains to peroxidation. This can lead to changes in membrane fluidity and permeability characteristics,[107] which can adversely affect cell viability.

Lipid peroxidation may be initiated by any primary free radical that has sufficient reactivity to extract a hydrogen atom from a reactive methylene group of an unsaturated fatty acid.[108] Based on studies with iron chelators,[109,110] it has been argued that an iron-dependent transformation of hydrogen peroxide leads to products capable of initiating lipid peroxidation. These products are as yet unidentified but could include hydroxyl radicals (HO·). Excess superoxide, on the other hand, could interact with protons with membranes with the resultant generation of hydroperoxyl radicals (HO$_2$) which, in principle, could also initiate lipid peroxidation.[108] During lipid peroxidation reactions, cleavage of carbon bonds can result in the formation of alkanals such as malonaldehyde,[111] which can interact with protein thiols, cross-link amino groups of lipids and proteins, and give rise to chromolipids and aggregated proteins. Breakdown of lipid hydroperoxides can also result in alkenals, such as 4-hydroxynonenal,[112] which can modify the activity of membrane-associated enzymes. Cell proteins are also susceptible to attack from radicals intermediates of lipid peroxidation. These may react with proteins closely associated with the peroxidizing lipids. The consequences of such damage may be aggregation and cross-linking, or protein degradation and fragmentation, depending on the nature of the vulnerable protein component, or on the attacking radical species.[113] The outcome of oxidative modification to proteins may be altered enzymic activity and/or altered membrane and cellular function resulting from degradation or cross-linking. Whereas damage to membrane transport proteins, for example, might effect the ionic homeostasis (e.g., Ca^{2+}, Na^+, K^+) of cells, growth signal transduction proteins associated with cell membranes could also be adversely affected. However, the precise role of such damage in contributing to reduced cell proliferation and/or cell death is still the subject of current investigation. Most of the proteins that play key roles in proliferative signal transduction actually function in a membrane environment, or in close association with membranes, and it is well established that the activity of integral membrane proteins is modulated by the lipids of the bilayer. Importantly, protein kinase C has a very specific lipid requirement for its activation in phosphatidyl serine.[114] Another protein whose activity is critical for cell viability and Na^+-K^+ exchange is the sulfhydryl-containing plasma membrane protein Na^+-K^+-ATPase. Recent studies in vitro have shown that when this particular protein is incorporated into liposomes it can be inactivated by radicals produced during lipid peroxidation.[115]

As already mentioned, lipid peroxides can also break down nonenzymically to yield a variety of carbonyls, such as the hydroxyalkenals. These aldehydes, and in particular 4-hydroxynonenal (HNE), can react with thiol and amino groups of nearby proteins affecting several signal transduction activities. These include the enzyme adenyl cyclase[116] and the blockage of c-*myc* gene expression.[117] Perhaps not surprisingly therefore HNE will down-regulate cell proliferation. Although there are no specific reports relating to the effects of HNE in relation to apoptosis, increased lipid membrane lipid peroxidation has been associated with apoptosis. Treatment of thymocytes with hydrogen peroxide/ferrous sulfate mixtures[118] will induce lipid peroxidation and apoptosis. However, the extent of apoptosis can be repressed by the addition of Trolox, a water soluble analogue of vitamin E.[118] Lymphocytes overexpressing *Bcl-2* were protected from oxidative death caused by either hydrogen peroxide or the redox-cycling agent menadione.[101] *Bcl-2* in lymphocytes stimulated into apoptosis appeared not to prevent oxygen radical generation but acted to prevent their damaging effects such as peroxidation of their membrane lipids.[101]

13.9 Are Reactive Oxygen Species Obligatory for Either Proliferation or Apoptosis?

Although one function of *Bcl-2* may be to act as special type of cellular antioxidant, it may be multifunctional and have other primary roles in blocking apoptosis. For example, whether oxygen-derived species are specifically obligatory for apoptosis signaling has recently been called into question. Established inducers of apoptosis in lymphocytes can bring about similar levels of apoptosis at widely differing oxygen tensions,[119] suggesting that oxygen derived species may not always be required for apoptopic signaling. Moreover in lymphocytes and fibroblasts, *Bcl-2* still appears to provide protection from apoptosis in very low oxygen.[120,121] While such apparently contradictory observations suggest that reactive oxygen species, including hydrogen peroxide, are not actually obligatory for apoptosis, they do not rule out the possibility that their presence may simply serve as a type of "coarse control," by virtue of setting an optimal cell redox balance that facilitates the operation of cellular proteins in apoptosis signaling pathways.

In contrast, we have exposed growing BHK-21 cells over an extended period (3 days) to high levels of exogenous catalase, or *N*-acetyl-L-cysteine. While those treatments gradually lead to increased cellular levels of GSH,[86] they can result by direct interaction to severely reduced levels of cellular hydrogen peroxide.[55] Under such conditions of apparent "reductive stress" in BHK-21 cells, increased apoptosis was also observed (unpublished results). Thus paradoxically in BHK-21 cells, the operation of apoptotic signaling pathways appear to be favored by conditions of *either* "oxidative" *or* "reductive" stress (see Fig. 13.6). Whether this situation occurs in any other cell types remains to be investigated although

BHK-fibroblasts

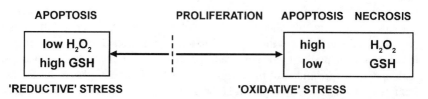

Figure 13.6. Schematic diagram to illustrate a possible relationship of apoptosis in BHK-21 cells with both oxidative and potential reductive stress.

some experiments indicate a similar situation in murine IL-3-dependent pro-B Bo cells.[122] Such observations may contribute much to resolving the dilemma regarding the possible role, or lack of role, of reactive oxygen species in apoptosis. Note, in this context, that the activities of two regulatory molecules, myc and p53, known to be important in apoptotic signal pathways are also sensitive to redox regulation (Table 13.2).

With regard to proliferation, our early experiments pointed to the vital importance of reactive oxygen species in the signaling of proliferative responses. For example, addition of superoxide dismutase or catalase to growing cultures of rodent fibroblasts[50,86] or HeLa[81] cells severely impaired their proliferation. Superoxide dismutase addition was also found by others to inhibit erythroid progenitor cell proliferation and DNA synthesis in these cells.[123–125] When the proliferation responses elicited in smooth muscle cells by PDGF were studied in detail, the generation of hydrogen peroxide turned out to be an obligatory step.[95] Similarly the cellular generation of reactive oxygen species also appears an obligatory component of the angiotensin II induction of proliferation responses in myogenic cells.[94] There is also a requirement for the generation of reactive oxygen species in the processes leading to the induction of c-*fos* expression in rabbit chondrocytes stimulated with tumor necrosis factor α or basic fibroblast growth factor.[45] Surprisingly even in the negative proliferative responses induced in certain cells by Transforming Growth Factor-Beta (TGF-β), hydrogen peroxide appears to play a critical role.[96] In summary present evidence clearly indicates that reactive oxygen species are obligatory for a number of important proliferative responses induced by growth factors or cytokines.

13.10 Overview

It is now clear that cytokines and growth factors can act quite normally to stimulate specifically the generation of superoxide and hydrogen peroxide in their target noninflammatory cells. It is proposed that this process is critical for signaling growth responses as it would allow the setting of a cellular redox balance appropriate for the intracellular transduction of the specific growth signals (Fig. 13.5). In this sense, the cellularly generated superoxide, or hydrogen peroxide, would thus act as novel autocrine second messengers in modulating the redox status of individual components of the signaling pathways such as growth factor receptors, Ras proteins, protein kinases, protein phosphatases, and transcription factors, for optimal activity. On the other hand, excessive generation of superoxide, for example in certain TNF-α-stimulated cells, may be sufficient to propel the cellular redox balance to a more prooxidant state that favors increased operation of apoptotic signaling pathways and peroxidative damage to membrane rather than proliferation (Fig. 13.5). It is likely, however, that different cell types exhibit critical differences in the balance between oxidation states that favor growth as distinct from apoptosis. Indeed paradoxically, apoptosis in BHK fibroblasts is also favored under conditions of apparent "reductive stress."

A final comment concerns the special case of tumor cells. In a number of different types of human tumors, it appears that relatively high levels of cellular generation of hydrogen peroxide are set "constitutively." Additionally, cells harboring mutant protooncogenes appear to be more susceptible to the growth stimulatory effects of low concentrations of hydrogen peroxide.[28] It may be that the advantageous proliferative deregulation that characterizes tumor cells is a function of these features. The latter phenomenon may be a means of providing tumor cells with an unfortunate selective proliferative advantage at sites of chronic inflammation. Despite this, exposure to agents that will considerably raise levels of hydrogen peroxide in tumor cells readily favor apoptosis. Such agents include inhibitors of catalase and glutathione peroxidase as well as low-molecular-weight superoxide dismutase mimetics, and represent a route to a novel class of drugs aimed at promoting tumor cell death by apoptosis. The organometallic superoxide dismutase mimetic CuII(3,5-diisopropylsalicylic acid)$_2$ has already been observed to inhibit tumor development in mouse skin[56] and in our studies rat fibroblasts (208F) transfected with the activated *H-ras* gene seem more readily to become apoptopic when exposed to the mimetic than do their nontransgenic counterparts.[50]

References

1. Hartwell, A. W., and T. A. Weinert. 1989. Checkpoints: controls that ensure the order of cell cycle events. *Science* **246**:614–634.

2. Nigg, E. A. 1993. Targets of cyclin-dependent protein kinases. *Curr. Opin. Cell Biol.* **5**:187–193.

3. Sherr, C. J. 1993. Mammalian GI cyclins. *Cell* **73**:1059–1065.

4. Murray, A. W., M. J. Solomon, and M. W. Kirschner. 1989. The role of cyclin synthesis and degradation in the control of maturation promoting activity. *Nature* **339**:280–286.

5. Cross, M., and T. M. Dexter. 1991. Growth factors in development, transformation and tumorigenesis. *Cell* **64**:271–280.

6. Berridge, M. J. 1984. Inositol trisphosphate and diacycl glycerol as second messengers. *Biochem. J.* **220**:345–360.

7. Friessmuth, M., P. J. Casey, and A. G. Gilman. 1989. G-proteins control diverse pathways of transmembrane signalling. *FASEB J* **3**:2125–2131.

8. Farago, A. and Y. Nishizuka. 1990. Protein kinase C in transmembrane signalling. *FEBS Lett.* **268**:350–354.

9. Dumont, J. E. 1989. The cAMP mediated stimulation of cell proliferation. *Trends Biochem. Sci.* **15**:153–158.

10. Pazin, M. J., and L. T. Williams. 1992. Triggering signalling cascades by receptor tyrosine kinases. *Trends Biochem. Sci.* **17**:374–378.

11. Ihle, J. M., B. A. Witthuhn, F. W. Quelle, K. Yamamoto, W. E. Thierfelder, B. Kreider and O. Silvennoinen. 1994. Signalling by the cytokine receptor superfamily: JAKs and STATs. *Trends Biochem. Sci.* **19**:222–227.

12. Naeve, G. S., A. Sharma, and A. S. Lee. 1991. Temporal events regulating the early phases of the mammalian cell cycle. *Curr. Opin. Cell Biol.* **3**:261–268.

13. Evan, G. I., and T. D. Littlewood. 1993. The role of c-*myc* in cell growth. *Curr. Opin. Genet. Dev.* **3**:44–49.

14. Cobrink, D., S. F. Dowdy, P. W. Hinds, S. Mittnacht, and R. A. Weinberg. 1992. The retinoblastoma protein and the regulation of cell cycling. *Trends Biochem. Sci.* **17**:312–315.

15. Chernova, O. B., M. V. Chernov, M. L. Agarwal, W. R. Taylor, and G. R. Stark. 1995. The role of p53 in regulating genomic stability when DNA and RNA synthesis are inhibited. *Trends Biochem. Sci.* **20**:431–434.

16. Collins, M. K. L., and A. L. Rivas. 1993. The control of apoptosis in mammalian cells. *Trends Biochem. Sci.* **18**:307–309.

17. Rees, J. C. 1994. Bcl-2 and the regulation of programmed cell death. *J. Cell Biol.* **124**:1–6.

18. Vaux, D., S. Cory, and J. Adams. 1988. Bcl-2 gene promotes heamopoetic cell survival and co-operates with c-*myc* to immortalize pre-B cells. *Nature* **335**:440–442.

19. Askew, D., R. Ashmun, B. Simmons, and J. Cleveland. 1992. Constitutive c-*myc* expression in an IL-3 dependent myeloid cell line suppressed cell cycles arrest and accelerates apoptosis. *Oncogene* **6**:1915–1922.

20. Evan, G. I., A. H. Wyllie, C. S. Gilbert, T. D. Littlewood, H. Land, M. Brooks, C. M. Waters, L. Z. Penn, and D. C. Hancock. 1992. Induction of apoptosis in fibroblasts by c-*myc* protein. *Cell* **69**:119–128.

21. Shi, Y., R. Bisonette, J. Glynn, L. Guilbert, T. Cotter, and D. Green. 1992. Rok for c-*myc* in activation-induced apoptotic death in T-cell hybridomas. *Science* **257**:212–214.

22. Bisonette, R., F. Echeverri, A. Mahboudi, and D. Green. 1992. Apoptotic cell death induced by c-*myc* is inhibited by *bcl-2*. *Nature* **359**:552–554.

23. Smeye, R., M. Vendrell, M. Hayward, S. Barker, G. Mison, K. Schilling, L. Robertson, T. Curran, and J. Morgan. 1993. Continuous c-*fos* expression precedes programmed cell death *in vivo*. *Nature* **363**:168–169.

24. Lowe, S., E. Schmitt, S. Smith, B. Osborne, and T. Jacks. 1993. p53 is required for radiation induced apoptosis in mouse thymocytes. *Nature* **363**:847–849.

25. Enoch, T., and C. Norbury. 1995. Cellular responses to DNA damage: cell cycle checkpoints, apoptosis and the roles of p53 and ATM. *Trends Biochem. Sci.* **20**:426–430.

26. Burdon, R. H., and C. Rice-Evans. 1989. Free radicals and the regulation of mammalian cell proliferation. *Free Rad. Res. Commun.* **6**:345–358.

27. Burdon, R. H., V. Gill, and C. Rice-Evans. 1990. Cell proliferation and oxidative stress. *Free Rad. Res. Commun.* **7**:149–159.

28. Burdon, R. H., V. Gill, and C. Rice-Evans. 1990. Oxidative stress and tumour cell proliferation. *Free Rad. Res. Commun.* **11**:65–76.

29. Murrell, G. A. C., M. J. O. Francis, and L. Bromley. 1990. Modulation of fibroblast proliferation by oxygen free radicals. *Biochem. J.* **265**:659–665.

30. Shibanuma, M., T. Kuroki, and M. Nose. 1988. Induction of DNA replication and expression of protooncogenes c-*myc* and c-*fos* in quiescent Balb/3T3 cells by xanthine-xanthine oxidase. *Oncogene* **3**:17–21.

31. Ikebuchi, Y., K. Masumoto, K. Tasaka, K. Kuike, K. Kashahara, A. Miyake, and O. Tanizawa. 1991. Superoxide anion increases intracellular pH, intracellular free calcium and arachidonate release in human amnion cells. *J. Biol. Chem.* **266**:13233–13237.

32. Crawford, D., I. Zbinden, P. Amstad, and P. Cerutti. 1989. Oxidant stress induces the protooncogenes c-*fos* and c-*myc* in mouse epidermal cells. *Oncogene* **3**:27–32.

33. Shibanuma, M., T. Kuroki, and K. Nose. 1988. Superoxide as a signal for increase in intracellular pH. *J. Cell. Physiol.* **136**:379–383.

34. Craven, P. A., J. Pfanstiel, and F. R. DeRobertis. 1986. Role of reactive oxygen in bile salt stimulation of colonic epithelial proliferation. *J. Clin. Invest.* **77**:850–859.

35. Nose, K., M. Shibanuma, K. Kikuchi, H. Kazeyana, S. Sakiyama, and T. Kuroki. 1991. Transcriptional activities of early response genes in a mouse osteoblastic cell line. *Eur. J. Biochem.* **201**:99–106.

36. Shibanuma, M., T. Kuroki, and J. K. Nose. 1990. Stimulation by hydrogen peroxide of DNA synthesis competence family gene expression and phosphorylation of a specific protein in quiescent Balb/3T3 cells. *Oncogene* **3**:27–32.

37. Stirpe, F., T. Higgins, P. L. Tazzori, and E. Rozengurt. 1991. Stimulation by xanthine oxidase of 3T3 Swiss fibroblasts and human lymphocytes. *Exp. Cell Res.* **192**:635–638.

38. Hannigan, B. M., S. Ranjibar, and L. Crombie. 1994. The effect of reactive oxygen species (ROS) on human T and lymphoid cells, pp. 59–63. In C. Pasquier, R. Y.

Olivier, C. Auclair and L. Packer, (eds.), *Oxidative Stress, Cell Activation and Viral Infection.* Birkhauser Verlag, Basel.

39. Test, T., and S. J. Weiss. 1984. Quantitative and temporal characteristics of the extracellular H_2O_2 pool generated by human neutrophils. *J. Biol. Chem.* **259**:399–405.

40. Meier, B., H. H. Radeke, S. Selle, M. Younes, H. Seis, K. Resch, and G. G. Habermehl. 1989. Human fibroblasts release active oxygen species in response to interleukin-1 or tumour necrosis factor-α. *Biochem. J.* **263**:539–545.

41. Meier, B., A. R. Cross, J. T. Hancock, F. Kamp, and O. T. G. Jones. 1991. Identification of a superoxide generating NADPH-oxidase system in human fibroblasts. *Biochem. J.* **275**:241–245.

42. Matsubara, T., and M. Ziff. 1986. Increased superoxide anion release from endothelial cells in response to cytokines. *J. Immunol.* **137**:3295–3304.

43. Robertson, F. M., A. J. Beavis, T. M. Oberszyn, S. M. O'Connell, A. Dokidos, D. L. Laskin, J. D. Laskin, and J. J. Reines. 1990. Production of hydrogen peroxide by murine epidermal keratinocytes following treatment with the tumour promoter 12-*O*-tetradecanoyl-13-acetate. *Cancer Res.* **50**:6062–6067.

44. Takasu, N., M. Komatsu, T. Aizawa, and T. Yamada. 1988. Hydrogen peroxide generation in whole rat pancreatic islets: synergistic regulation by cytoplasmic free calcium and protein kinase C. *Biochem. Biophys. Res. Commun.* **155**:569–575.

45. Tiku, M. L., J. B. Liesch, and F. M. Robertson. 1990. Production of hydrogen peroxide by rabbit articular chondrocytes: enhancement by cytokines. *J. Immunol.* **145**:690–697.

46. Kather, H., and H. I. Kreiger-Bauer. 1992. A stimulus sensitive NADPH-oxidase present in human fat cell plasma membrane defines c novel pathway of signal transduction. *Biol. Chem. Hoppe-Seyler* **373**:746.

47. Lo, Y. Y. C., and T. F. Cruz. 1995. Involvement of reactive oxygen species in cytokine and growth factor induction of c-*fos* expression in chondrocytes. *J. Biol. Chem.* **270**:11727–11730.

48. Szatrowski, T. P., and C. F. Nathan. 1991. Production of large amounts of hydrogen peroxide by human tumour cells. *Cancer Res.* **51**:794–798.

49. Herontin, Y., G. Debydupont, G. Deby, M. Debruyn, M. Lamy, and P. Franchimont. 1993. Production of active oxygen species by isolated chondrocytes. *Bri. J. Rheumatol.* **32**:562–567.

50. Burdon, R. H. 1992. Cell proliferation and oxidative stress: basis for anticancer drugs. *Proc. R. Soc. Edinburgh* **99B**:169–176.

51. Burdon, R. H., V. Gill, and C. Rice-Evans. 1993. Reduction of a tetrazolium salt and superoxide generation in human tumour cells (HeLa). *Free Rad. Res. Commun.* **18**:369–388.

52. Hennet, T., C. Richter, and C. Peterhans. 1993. Tumour necrosis factor-α induces superoxide generation in mitochondria of L-929 cells. *Biochem. J.* **289**:587–592.

53. Goosens, V., J. Grooten, K. DeVos, and W. Fiers. 1995. Direct evidence for tumour necrosis factor-induced mitichondrial reactive oxygen intermediates and their involvement in cytotoxicity. *Proc. Natl. Acad. Sci. USA* **92**:8115–8119.

54. Cossarizza, A., C. Francheschi, D. Monti, S. Salvioli, E. Bellesia, R. Rivabena, L. Biondo, G. Rainaldi, A. Tinari, and W. Malorni. 1995. Protective effect of N-acetylcysteine in tumour necrosis factor-α-induced apoptosis in U937 cells: the role of the mitochondria. *Exp. Cell Res.* **220**:232–240.

55. Burdon, R. H., D. Alliangana, and V. Gill. 1995. Hydrogen peroxide and the proliferation of BHK-21 cells. *Free Rad. Res.* **23**:471–486.

56. Kensler, T. W., D. M. Bush, and W. J. Kozumbo. 1983. Inhibition of tumour promotion by a biomimetic superoxide dismutase. *Science* **221**:75–77.

57. Hildago, E., and B. Demple. 1994. An iron-sulphur centre essential for transcription activation by redox sensing Sox R protein. *EMBO J.* **13**:138–146.

58. Stortz, G., L. A. Tartaglia, and B. Ames. 1990. Transcriptional regulator of oxidative stress inducible genes—direct activation by oxidation. *Science* **248**:189–194.

59. Stortz, G., and L. A. Tartaglia. 1992. OxyR: a regulator of antioxidant genes. *J. Nutr.* **122**:627–630.

60. Datta, R., N. Taneja, V. P. Sukhatma, S. A. Qreshi, R. Weichselbaum, and D. W. Kufe. 1993. Reactive oxygen intermediates target CC(A/T)₆ GG sequences to mediate activation of early growth response transcription factor gene by ionising radiation. *Proc. Natl. Acad. Sci. USA* **90**:2419–2422.

61. Datta, R., D. E. Hallahan, S. M. Kharbanda, E. Rubin, M. K. Sherman, E. Huberman, R. R. Weichselbaum, and D. W. Kufe. 1992. Involvement of reactive oxygen species in the induction of c-*jun* gene transcription by ionising radiation. *Biochemistry* **31**:8300–8306.

62. Treisman, R. H. 1980. Identification of a protein-binding site that mediates transcriptional response of the c-*fos* gene to serum factors. *Cell* **46**:567–574.

63. Larsson, R., and P. Cerutti. 1989. Translocation and enhancement of phosphotransferase activity of protein kinase C following exposure of mouse epidermal cells to oxidants. *Cancer Res.* **49**:5627–5632.

64. Amstad, P. A., G. Krupitza, and P. Cerutti. 1992. Mechanism of c-*fos* induction by active oxygen. *Cancer Res.* **52**:3952–3960.

65. Gopalakrishna, R., and W. B. Anderson. 1989. Ca²⁺ and phospholipid independent activation of protein kinase C by selective oxidative modification of the regulating domain. *Proc. Natl. Acad. Sci. USA* **86**:6758–6762.

66. Kass, G. E. N., S. K. Duddy, and S. Orrenius. 1989. Activation of protein kinase C by redox-cycling quinones. *Biochem. J.* **260**:499–507.

67. Kyriakis, J. M., P. Banerjee, E. Nickolakaki, T. Dai, E. A. Rubie, M. F. Ahmad, J. Avruch, and J. R. Woodgett. 1994. The stress-activated protein kinase subfamily of c-*jun* kinases. *Nature* **369**:156–160.

68. Keyse, S. M., and E. A. Emslie. 1992. Oxidative stress and heat shock induce a human gene encoding a protein tyrosine phosphatase. *Nature* **359**:644–647.

69. Stern, A. 1994. Oxidative stress and growth factor-mediated signal transduction, pp. 35–42. *In* C. Pasquier, R. Y. Olivier, C. Auclair, and L. Packer (eds.), *Oxidative Stress, Cell Activation and Viral Infection*. Birkhauser Verlag, Basel.

70. Gamou, S. and N. Shimizu. 1995. Hydrogen peroxide preferentially enhances the tyrosine phosphorylation of epidermal growth factor receptor. *FEBS Lett.* **357**:161–164.

71. Lander, H. M., J. S. Ogiste, K. K. Teng, and A. Novogrodsky, 1995. p21^{ras} as a common signalling target of reactive free radicals and cellular redox stress. *J. Biol. Chem.* **270**:21195–21198.

72. Abate, C., L. Patel, F. J. Rauscher, and T. Curran. 1990. Redox regulation of Fos and Jun DNA-binding activity *in vitro. Science* **249**:1157–1161.

73. Xanthoudakis, S., G. Miao, F. Wang, Y.-C. E. Pan, and T. Curran. 1992. Redox activation of Fos-Jun DNA-binding activity is mediated by a DNA repair enzyme. *EMBO J.* **11**:3323–3335.

74. Wasylyk, C. and B. Wasylyk. 1993. Oncogenic conversion of Ets affects redox regulation *in vivo* and *in vitro. Nucleic Acids Res.* **21**:523–529.

75. Kumar, S., A. B. Rabson, and C. Gelmas. 1992. The RxxRxRxx C motif conserved in all Rel/kappaB proteins is essential for the DNA-binding activity and redox regulation of the v-Rel oncoprotein. *Mol. Cell. Biol.* **12**:3094–3106.

76. Guehmann, S., G. Vorbrueggen, F. Kalkbrenner, and K. Moelling. 1992. Reduction of a concerved Cys is essential for Myb DNA-binding. *Nucleic Acids Res.* **20**:2279–2286.

77. Hainaut, P., and J. Milner. 1993. Redox modulation of p53 conformation and sequence specific DNA-binding *in-vitro. Cancer Res.* **53**:4469–4473.

78. Matthews, J. R., W. Kaszudska, G. Turcatti, T. N. C. Wells, and R. T. Hay. 1993. Role of cysteine 62 in DNA recognition of NF-κB. *Nucleic Acids Res.* **21**:1727–1734.

79. Schreck, R., P. Rieber, and P. A. Baeuerle. 1991. Reaction oxygen intermediates as apparently widely used messengers in the activation of NF-κB transcription factor and HIV-1. *EMBO J.* **10**:2247–2258.

80. Galter, D., S. Mihm, and W. Droge. 1994. Distinct effects of glutathione disulphide on the nuclear transcription factors κB and activator protein-1. *Eur. J. Biochem.* **221**:639–648.

81. Burdon, R. H., and V. Gill. 1993. Cellularly generated active oxygen species and Hela cell proliferation. *Free Rad. Res. Commun.* **19**:203–213.

82. Issels, R. D., A. Nagele, K.-G. Eckert, and W. Willmans. 1988. Promotion of cysteine uptake and its utilisation for glutathione biosynthesis induced by cysteamine and N-acetylcysteine. *Biochem. Pharmacol.* **36**:127–131.

83. Burgunder, M., A. Varriale, and B. H. Lanterberg. 1988. Effect of N-acetylcysteine on plasma cysteine and glutathione following paracetamol administration. *Eur. J. Clin. Pharmacol.* **36**:881–886.

84. Shaw, J. P., and I. M. Chou. 1986. Elevation of intracellular glutathione content associated with mitogenic stimulation of quiescent fibroblasts. *J. Cell. Physiol.* **129**:193–198.

85. Kang, Y.-J., and M. D. Enger. 1991. Increased glutathione levels in quiescent serum stimulated NRK-49 cells are associated not with a response of growth but with nutrient repletion. *J. Cell. Physiol.* **148**:197–201.

86. Burdon, R. H., D. Alliangana, and V. Gill. 1994. Endogenously generated active oxygen species and cellular glutathione levels in relation to BHK-21 cell proliferation. *Free Rad. Res.* **21**:121–133.

87. Ishii, T., I. Hishinuma, S. Bannai, and Y. Sugita. 1981. Mechanism for growth promotion of mouse lymphoma L1210 cells *in vitro* by feeder layers or 2-mercaptoethanol. *J. Cell. Physiol.* **104**:215–223.

88. Lancombe, P., L. Kraus, M. Fay, and J.-J. Pocidalo. 1987. Glutathione status of rat thymocytes and splenocytes during early events of their ConA proliferation responses. *Biochimie* **69**:37–44.

89. Fidelus, R. K. 1988. The generation of oxygen radicals: a positive signal for lymphocyte activation. *Cell. Immunol.* **113**:175–182.

90. Droge, W., H.-P. Eck, and S. Mihm. 1992. HIV-induced cysteine deficiency of T-cell dysfunction: a rationale for treatment with N-acetylcysteine. *Immunol. Today* **13**:211–214.

91. Staal, F. J. J., M. T. Anderseon, G. E. S. Staal, L. A. Hertzenberg, C. Gitler, and L. A. Hertzenberg. 1994. Redox regulation of signal transduction phosphorylation and calcium influx. *Proc. Natl. Acad. Sci. USA* **91**:3619–3622.

92. Meister, A., 1988. Metabolism and function of glutathione, pp. 367–474. *In* D. Dolphin, R. Poulson, and O. Avramovic, (eds.), *Glutathione: Chemical, Biochemical and Medical Aspects.* John Wiley, New York.

93. Staal, F. J. J., M. Roederer, and L. A. Hertzenberg. 1990. Intracellular thiols regulate the activation of nuclear factor κB and transcription of human immunodeficiency virus. *Proc. Natl. Acad. Sci. USA* **87**:9943–9949.

94. Puri, P. L., M. L. Avantaggiata, V. L. Burgio, P. Chirillo, D. Collepordo, C. Natoli, C. Balsano, and M. Levrero. 1995. Reactive oxygen intermediates mediate angiotensin II-induced c-Jun·c-Fos heterodimer DNA binding activity and proliferative hypertrophic responses in myogenic cells. *J. Biol. Chem.* **270**:22129–22134.

95. Sundaresan, M., Z.-X. Yu, V. J. Ferrans, K. Irani, and T. Finkel. 1995. Requirement for generation of H_2O_2 for platelet-derived growth factor signal transduction. *Science* **270**:296–299.

96. Shibanuma, M., S. Arata, M. Murata, and K. Nose. 1995. Activation of DNA synthesis and expression of JE gene by catalase in mouse osteoblastic cells: possible involvement of hydrogen peroxide in negative growth regulation. *Exp. Cell Res.* **218**:132–136.

97. Burdon, R. H., V. Gill, and D. Alliangana. Accepted. Hydrogen peroxide in relation to proliferation and apopotosis in BHK-21 hamster fibroblasts. *Free Rad. Res.* **24**:81–93.

98. Korsmeyer, S. J. 1995. Regulation of cell death. *Trends Genet.* **11**:101–105.

99. Krajewski, S., S. Tanaka, S. Takayama, M. J. Schibler, W. Fenton, and J. C. Reed. 1993. Investigation of the subcellular distribution of the Bcl-2 oncoprotein: residence in the nuclear envelope, endoplasmic reticulum and outer michondrial membranes. *Cancer Res.* **53**:4701–4714.

100. Nguyen, M., D. G. Millar, V. W. Yong, S. J. Korsmeyer, and G. C. Shore. 1993. Targeting of Bcl-2 to the mitochondrial membrane by a COOH-terminal signal anchor sequence. *J. Biol. Chem.* **286**:25264–25268.

101. Hockenbery, D. M., Z. N. Oltavi, X.-M. Ying, C. L. Milliman, and S. J. Korsmeyer. 1993. Bcl-2 functions in an anti-oxidant pathway to prevent apoptosis. *Cell* **75**:241–251.

102. Kane, D. J., T. A. Sarafian, R. Anton, H. Hahn, E. B. Gralla, J. S. Valentine, T. Ord, and D. E. Bredesen. 1993. Bcl-2 inhibition of neural death: decreased generation of reactive oxygen species. *Science* **262**:1274–1277.

103. Yuan, J., S. Shaham, S. Ledoux, H. M. Ellis, and H. R. Horvitz. 1993. The *C. elegans* cell death gene *ced-3* encode a protein similar to mammalian interleukin-1β-converting enzyme. *Cell* **75**:641–651.

104. Lazebnik, Y. A., S. H. Kaufman, S. Desnoyners, G. G. Poirer, and W. C. Earnshaw. 1994. Cleavage of poly (ADP-ribose) polymerase by a proteinase with properties like ICE. *Nature* **371**:346–341.

105. Talley, A. K., S. Dewhurst, S. W. Perry, S. C. Dollard, S. Gummuluru, S. M. Fine, D. New, L. G. Epstein, H. E. Gendelman, and H. A. Gelbard. 1995. Tumour necrosis factor alpha-induced apoptosis in human neuronal cells: protection by the antioxidant N-acetylcysteine and the genes *bcl2* and *crmA*. *Mol. Cell. Biol.* **15**:2359–2366.

106. Dizdaraglu, M. 1993. Chemistry of free radical damage to DNA and nucleoproteins, pp. 19–39. *In* B. Halliwell and O. I. Aruoma (eds.), *DNA and Free Radicals*. Ellis Horwood, New York, London.

107. Urano, S., K. Yano, and M. Matsuo. 1988. Membrane-stabilizing affect of vitamin E: effect of α-tocopherol and its model compounds on the fluidity of lecethin liposomes. *Biochem. Biophys. Res. Commun.* **150**:469–475.

108. Rice-Evans, C., and K. R. Bruckdorfer. 1992. Free radicals, lipoproteins and cardiovascular dysfunction. *Mol. Asp. Med.* **13**:1–111.

109. Melho-Filho, A. C., M. E. Hoffman, and R. Meneghini. 1984. Cell killing and DNA damage by hydrogen peroxide are mediated by intracellular iron. *Biochem. J.* **218**:273–275.

110. Janero, D. R., D. Hreniok, and H. M. Sharif. 1991. Hydrogen peroxide induced oxidative stress to the mammalian heat-muscle cell (cardiomyocyte): lethal peroxidation membrane injury. *J. Cell. Physiol.* **149**:347–364.

111. Tappel, A. L., and C. J. Dillard. 1981. *In vivo* lipid peroxidation: measurement via exhaled pentane and protection by vitamin E. *Fed. Proc.* **40**:174–178.

112. Esterbauer, H. 1985. Lipid peroxidation product formation, chemical properties and biological activation, pp. 29–47. *In* G. Poli, K. Cheesman, M. V. Dianzani, T. Slater (eds.), *Free Radicals in Liver Injury*. IRL Press, Oxford.

113. Wolff, S. P., A. Garner, and R. P. Dean. 1986. Free radicals, lipids and protein degradation. *Trends Biochem. Sci.* **11**:27–31.

114. Bell, R. M., and D. J. Burns. 1990. Lipid activation of protein kinase C. *J. Biol. Chem.* **266**:4661–4664.

115. Thomas, C. E., and D. J. Reed. 1990. Radical-induced inactivation of kidney Na+, K+/-ATPase: sensitivity to membrane lipid peroxidation and protective effect of Vitamin E. *Arch. Biochem. Biophys.* **281**:96–105.

116. Paradisi, L., C. Panagini, M. Parola, G. Barrera, and M. U. Dianzani. 1985. Effect of 4-hydroxynonenal on adenyl cyclase and 5'-nucleotidase activity in rat liver plasma membranes. *Chem. Biol. Interact.* **53**:209–221.

117. Fazio, V. M., G. Barrera, S. Martinotti, M. G. Farace, B. Giglioni, L. Frati, V. Manzari, and M. U. Dianzani. 1992. 4-hydroxynonemal, a product of cellular lipid peroxidation, which modulates c-*myc* and globin gene expression in K-562 erythroleukaemic cells. *Cancer Res.* **52**:4866–4871.

118. Forrest, V. J., Y.-H. Kang, D. E. McClain, D. H. Robinson, and N. Ramakrishnan. 1994. Oxidative stress-induced apoptosis prevented by Trolox. *Free Rad. Biol. Med.* **16**:675–684.

119. Muschel, R. J., E. J. Bernhard, L. Garza, W. G. McKenna, and C. J. Koch. 1995. Induction of apoptosis at different oxygen tensions—evidence that oxygen radicals do not mediate apoptosis signalling. *Cancer Res.* **55**:995–998.

120. Jacobson, M. D., and M. C. Raff. 1995. Programmed cell death and Bcl-2 protection in very low oxygen. *Nature* **374**:814–816.

121. Shimizu, S., Y. Eguchi, H. Kosaka, W. Kamiike, H. Matsuda, and Y. Tsujimoto. 1995. Prevention of hypoxia-induced cell death by Bcl-2 and Bcl-xL. *Nature* **374**:811–813.

122. Garland, J. M., A. Halestrap, and J. Knight. Accepted. Bcl-2 inhibits apoptosis in haemopoetic cells induced by IL-3 removal or ATP depletion without changes in energy metabolism or free radical production. Evidence that bcl-2 promotes a state of metabolic interaction. *J. Biol. Chem.*

123. Pluthero, F., M. Shreeve, and D. Eskinazi. 1990. Purification of an inhibitor of erythroid progenitor cell cycling and antagonist to interleukin 3 from mouse marrow cell supernatants and its indentification as cytosolic superoxide dismutase. *J. Cell Biol.* **111**:1217–1223.

124. Pluthero, F., M. Shreeve, H. Eskinazi, H. Gaag, and A. Axelrad. 1991. Superoxide dismutase specifically inhibits erythroid cell DNA synthesis and proliferation. *Growth Factors* **4**:397–304.

125. Pluthero, F., and A. Axelrad. 1990. Superoxide dismutase as an inhibitor of erythroid progenitor cell cycling. *Ann. N.Y. Acad. Sci.* **1**:222–232.

SECTION III
Gene Expression in Response to Oxidative Stress

14

Modulation by Oxidants and Antioxidants of Signal Transduction and Smooth Muscle Cell Proliferation

*Angelo Azzi, Daniel Boscoboinik, Orazio Cantoni,
Agata Fazzio, Dominique Marilley, Valerie O'Donnell,
Nesrin Kartal Özer, Stefan Spycher,
Shirin Tabataba-Vakili, and Andrea Tasinato*

Arteriosclerosis, restenosis, and hypertension as well as a number of other vascular diseases are characterized by vascular smooth muscle cell proliferation.[1-4] Smooth muscle cell growth is regulated by specific factors released from blood cells,[1,5] by the vessel wall cells[3,6] and by natural compounds, such as tocopherols. Active oxygen species (i.e., O_2^-, H_2O_2, and OH·) may act as vascular smooth muscle growth factors by inducing proto-oncogene expression and DNA synthesis,[7,8] and antioxidants can retard atherosclerosis and foam cell formation as shown in animal model,[9,10] and human studies.[11-14] Tocopherols, well-established natural antioxidants, may in some cases stimulate cell proliferation by preventing the formation of growth-inhibitory lipid peroxides.[15-19] In addition to having antioxidant properties, RRR-α-tocopherol is a cell growth inhibitor, this effect being unrelated with its radical scavenger properties.[20,21] Parallel to the inhibition of cell proliferation, RRR-α-tocopherol has been shown to inhibit protein kinase C activity in smooth muscle cells.[20,21] The mechanism of this inhibition appears to be related with a diminished phosphorylation of protein kinase C-α.

One of the principal elements of the information transfer chain of events initiated by extracellular signals, including hormones, neurotransmitters, and growth factors,[22] is protein kinase C. It is part of a family of proteins, similar in size, structure and mechanism of activation. Activation of protein kinase C by *sn*-1,2-diacylglycerol or phorbol esters may result in physiological proliferation or tumor-growth-promoting activity, respectively. The activation of protein kinase C by oxidants[7,8,23-27] suggests that this enzyme family may play a central role in the RRR-α-tocopherol and oxidant regulation of cell proliferation. Antioxidants and tocopherols also cause activation or inhibition of cellular functions, indirectly or directly, through the modulation of other signal transduction proteins.[28-35]

14.1 Effect of Oxidants on Cell Signal Transduction

14.1.1 Phospholipases A₂

Phospholipases A_2 ($cPLA_2$) are important effectors in transmembrane signaling. The cytosolic protein ($cPLA_2$) requires physiological levels of Ca^{2+} and is responsive to G-protein activation.[36,37] In its turn, it is the substrate for protein kinase C and mitogen-activated protein kinase (MAPK).[38] Tyrosine kinases can also mediate the activation of $cPLA_2$.[39] Such activation is indirect and involves phosphorylation on serine residues[40] produced by a tyrosine-kinase-dependent serine kinase, e.g., protein kinase C or MAPK.

H_2O_2 stimulates the release of arachidonic acid in the extracellular milieu of vascular smooth muscle cells, an effect that is prevented by the PLA_2 inhibitor mepacrine.[8] H_2O_2 induces phosphorylation of the $cPLA_2$ via protein kinase C and MAP kinase.[41]

14.1.2 Protein Kinase C

Reactive oxygen species affect protein kinase C (for recent reviews, see Refs. 22 and 42), in a complex manner, with results, not infrequently of contrasting nature, on signal transduction, cell regulation and proliferation. It can initially be activated and subsequently inactivated by increasing oxidative modification by H_2O_2, suggesting an effective on/off signal mechanism able to influence cellular events.[43] Thus, oxidants may be able to bypass normal transmembrane signaling systems to directly activate pathways involved in cellular regulation. In hepatocytes, the production of quinone-generated active oxygen species, induces a several-fold increase in protein kinase C activity which is reversed by thiol-reducing agents.[44] Nitric oxide, an important cellular regulator, is able to reversibly inactivate protein kinase C either purified or in melanoma cells.[45]

H_2O_2 activates protein kinase C in endothelial cells,[46] C-6 glioma cells,[43] Jurkat T cells,[47] and in vascular A7r5 smooth muscle cells.[48,49] These cells contain the α, δ, ε, and ζ protein kinase C isoforms; the first three isoforms, unlike the latter, can be down-regulated by prolonged exposure to phorbol esters. H_2O_2-stimulation of protein kinase C is resistant to down-regulation, showing that H_2O_2 activates not only protein kinase C α, δ, ε, μ, but also ζ. The modulation of protein kinase C activity by H_2O_2 might play a critical role in regulating DNA synthesis in growth-arrested smooth muscle cells.[7,49,50] One, two, or all the down-regulatable isoforms of protein kinase C (α,δ,ε) negatively control the progression from G_0 through the S phase of A7r5 cells treated with H_2O_2.[51,52] On the other hand, the ζ-isoform, critical for mitogenic signal transduction[53] and activated by phosphatidylinositol 3,4,5-trisphosphate,[54] is the most likely candidate in mediating H_2O_2-induced DNA synthesis. More recently, it has been reported that the expression of protein kinase C-ζ is markedly diminished in apoptotic cells thus indicating a potential role for this isoform in preventing apoptosis.[55]

In summary, H_2O_2 produces strong alterations at the level of the protein kinase C signal transduction pathway in smooth muscle cells. Activation seems to occur at the level of the conventional and novel as well as of atypical isoform(s). These events certainly play a role in the activation of the DNA synthesis machinery, although it is still unclear whether they can also regulate the lethal response.

14.1.3 Tyrosine Kinases

Normal cellular processes such as cell duplication, embryonic development and synaptic transmission are based on a correct and coordinated function of receptor- (for a review see Ref. 56) and non-receptor-tyrosine kinases (for reviews see Refs. 57 and 58).

Oxidants produced by redox cycling of naphthoquinones stimulate an adenosine-insensitive phosphatidylinositol kinase in rat liver membranes through tyrosine phosphorylation of the enzyme.[59] Exogenous oxidants or endogenous superoxide production induce in human neutrophils tyrosine phosphorylation of several proteins possibly via an oxidation-sensitive tyrosine kinases and/or phosphatases.[60] Quinones markedly increase (possibly via a free-radical-mediated mechanism) protein tyrosine phosphorylation in rat liver plasma membranes and particulate fractions. This reaction is inhibited by superoxide dismutase, catalase, and desferrioxamine.[61] These data support the concept of an oxidant-mediated increase in tyrosine protein phosphorylation as an early event in the signal transduction cascade of growth factor receptors, leading to increase of cell proliferation.[62,63]

In T cells, H_2O_2 and diamide induce phosphorylation of p56^{lck}, an Src-related protein-tyrosine kinase, both at Tyr-394 (autophosphorylation site) and Tyr-505 (negative regulatory site).[64] Signal transduction in T cells involves the activation of phospholipase C-γ1 by tyrosine phosphorylation through an oxidation-sensitive intermediate between surface receptors and tyrosine kinases, perhaps including the interaction between CD4 and pp56^{lck}.[65] Leukocyte tyrosine kinase is an unusual membrane protein lacking an extracellular domain that resembles a receptor tyrosine kinase. It has a short, glycosylated, and cysteine-rich N-terminal domain, and it appears to function via a ligand-independent reaction. Its gene also produces a putative receptor tyrosine kinase for an unknown ligand.[66] Its *in vivo* catalytic activity is markedly enhanced by alkylating and thiol-oxidizing agents.[67]

14.1.4 Protein Phosphatases

Tumor necrosis factor/interleukin-1 (TNF/IL-1)-initiated signal transduction is inactivated by an SH-dependent protein phosphatase and inhibitors of protein phosphatases and H_2O_2 can mimic the early effects of TNF/IL-1 on cells.[68] Heat shock in human skin cells has been shown to induce a mRNA coding for a protein that after expression and purification shows phosphatase activity.[69] Keyse's group[70] has found that the expression of the human CL100 gene (homologue of mouse 3CH134) is increased up to 40-fold in fibroblasts exposed to oxidative/heat stress and growth factors. CL100 is a phosphatase that can react with phosphorylated serines, threonines, or tyrosines. MAP kinase is one of its physiological substrates. CL100 is a protein phosphatase responsible for modulating the deactivation of nuclear MAP kinase. This further involves MAP kinases in the cellular response to stress.[71] Treatment of Jurkat T cells with low H_2O_2 concentrations results in a remarkable inhibition of both protein tyrosine and protein-phosphatases 2A.[47] Similar results on tyrosine phosphatases were obtained in HER 14 cells.[72]

In summary, H_2O_2 not only activates protein kinase C and tyrosine kinases but may also exert effects via inhibition of protein phosphatases.

14.1.5 Mitogen-Activated Protein Kinases

Activation of many tyrosine kinases receptors results in stimulation of MAP kinases.[73–75] Via a cascade of phosphorylations initiated by a MAP kinase kinase kinase, followed by the MAP kinase kinase (a tyrosine-threonine kinase), two MAP kinases ERK1 and ERK2 (extracellular regulated kinases) are activated. Both tyrosine and serine residues are phosphorylated as result of the activation, which depends as well on specific phosphatases (as CL100) belonging to the family of dual specificity Tyr/Thr protein phosphatases.[71,76–78]

Hydrogen peroxide activates the phosphorylation of ERK1 and ERK2.[41,79] In Jurkat T cells, H_2O_2 mimics the activation of MAP kinase promoted by either anti-CD3 antibodies or the mitogen phytohemoagglutinin.[47] This reaction is dependent on protein kinase C and tyrosine kinase phosphorylation. H_2O_2 was also found to inhibit both tyrosine and protein phosphatases.[47] The phosphorylation of ERK1 and ERK2 promoted by H_2O_2 in smooth muscle cells is markedly reduced by the tyrosine kinase inhibitor tyrphostin A23. Complete suppression of MAP kinase activity is achieved by inhibition of both protein kinase C and tyrosine kinases.[48]

The ultraviolet response of mammalian cells is also mediated by reactive oxygen species.[80–82] JNK1, a relative of the MAP kinase group, is activated by dual phosphorylation at Thr and Tyr during the UV response. It binds to the c-Jun transactivation domain where it phosphorylates Ser-63 and Ser-73.[33,83] This is followed by a rapid AP-1 and NF-κB-mediated increase in gene expression.

In summary, oxidants or UV irradiation activate the MAP kinase cascade by stimulating the tyrosine kinase and protein kinase C pathways. These events are similar to those observed following exposure to mitogenic stimuli, although at present it is still unclear whether MAP kinases stimulated by H_2O_2 participate in the complex series of reactions which are ultimately responsible for cell division or, more generally, are part of a stress response taking place in oxidatively injured cells.

14.1.6 Transcription Factors

Activator proteins, such as NF-κB and AP-1 control inducible gene transcription in eukaryotes. Cell treatment with phorbol esters, lipopolysaccharides, interleukin-1 and tumor necrosis factor-α results in NF-κB activation. Protein kinase C and other kinases phosphorylate the inhibitory subunit I-κB, causing its release from a cytoplasmic complex of subunits p65 and p50; this results in their migration to the nucleus and their DNA binding.[84–89] In intact cells, phosphorylation of I-κB is not sufficient for NF-κB activation,[88] and reactive oxygen intermediates as well as an active protease are also necessary.[30,90,91,92,93]

The transcription factor AP-1 is a dimeric complex made of two proteins belonging to the Fos and Jun families.[94–97] It is activated through a phosphorylation cascade involving receptor-linked tyrosine kinases, protein kinase C, and the MAP kinases. AP-1 binding occurs to a DNA specific sequence and depends on the reduction by Ref-1 (nuclear redox factor) of a single conserved cysteine residue located in the DNA-binding domain of the protein.[98] The DNA specific sequences required for DNA binding of AP-1 are named TRE (phorbol ester responsive element) or ARE (antioxidant responsive element). Antioxidants likewise favor transcriptional induction of c-*jun* and c-*fos* genes.[99] On the other hand, an oxidant, hydrogen peroxide, can also induce c-*jun* mRNA via PLA_2 catalyzed release of arachidonic acid in growth-arrested rat aortic smooth muscle cells.[25]

The ARE has been identified in the 5'-flanking region of the rat glutathione S-transferase Ya subunit gene and the NAD(P)H:quinone reductase gene.[100–102] The sequence, 5'-puGTGACNNNGC-3' 3'-pyCACTGNNNCG-5', where N is any nucleotide, represents the core sequence of the ARE required for transcriptional activation. ARE is responsive to H_2O_2 and phenolic antioxidants that undergo redox cycling. These latter data suggest that the ARE is responsive to reactive oxygen species and thus may represent part of a signal transduction pathway that allows eukaryotic cells to sense and respond to oxidative stress.[100–103]

14.1.7 Gene Expression

A large number of genes are induced in cells undergoing oxidative stress,[104–108] which suggests that exposure to reactive oxygen species could have major consequences on normal cell growth and differentiation. In rat vascular smooth muscle cells, 100-μM H_2O_2 produce up to a four-fold reversible overexpression of aldose reductase mRNA in a dose- (10 to 100 μM) and time (7 to 12 h)-dependent way. A 2.5-fold increase in aldose reductase protein and almost a doubling of enzyme activity can be observed. Since aldose reductase metabolizes several aldehyde compounds, including 4-hydroxynonenal, a toxic peroxidation product, induction of this enzyme by H_2O_2 may represent a novel cellular defense mechanism against oxidative stress (Spycher, S.; Tabataba, S.; O'Donnell, V.; Palomba, L.; and Azzi, A. ([1997, FASEB] in press).

14.1.8 Cell Death and Proliferation

Hydrogen peroxide and other reactive oxygen species are important mediators of cell injury. H_2O_2 reacts with divalent iron, leading to the production of the highly reactive—and thus toxic—hydroxyl radical (OH·), which can damage virtually all the biomolecules present in the cell. It is therefore a common opinion that all those sites in which transition metals are bound might represent a potential target for OH·, for the simple reason that in these sites the OH· is generated and in these sites it will produce its damage. H_2O_2 can produce DNA single strand breaks,[109,110,111] activation of the enzyme poly(ADP-ribose)polymerase followed by a decline in NAD^+ and ATP levels[110,112–114] oxidation and depletion of glutathione[115] disruption of intracellular calcium homeostasis,[116] peroxidation of membrane lipids, mitochondrial damage, and necrotic cell death. Hydrogen peroxide can also induce apoptotic death,[117–120] an event that is not biochemically isolated from the alternative mode of necrotic cell death.[48,109]

As summarized above, a number of studies have shown that small concentrations of reactive oxygen species have profound effects on cell signaling and, in particular, it has been suggested that these events triggered by H_2O_2 might stimulate cell proliferation. Vascular smooth muscle cell proliferation is a significant pathological occurrence in a number of vascular diseases, i.e. atherosclerosis, hypertension, endothelial injury response, and restenosis.[2,121–124] Changes in the cellular redox state may be important in the control of cell growth and a role for oxidative stress in atherogenesis has been also proposed.[125,126] Reactive oxygen species have been reported in situations such as ischemia, thrombosis, and angioplasty[127–131] that are associated with vascular smooth muscle proliferation and accelerated atherosclerosis. Oxidants also regulate nuclear factors and protein phosphorylation.*

Lipid-soluble antioxidants, in cholesterol-fed rabbits, reduce intima proliferation and macrophage accumulation showing the participation of reactive oxygen species in the intimal response to injury.[137,138,139] Low-density lipoproteins, possibly containing oxidized moieties,[123] stimulate human smooth muscle cell growth in culture.[21,140,141] This suggests another possible mode of action of antioxidants as therapeutical agents as exemplified by carvedilol, an antioxidant drug and an inhibitor of artery vascular smooth muscle cell proliferation *in vitro* and of neointimal growth following angioplasty by 84%,[142] *in vivo*.

14.2 Effect of Antioxidants on Cell Signal Transduction

14.2.1 Effect of Tocopherols on Protein Kinase C Activity

RRR-α-Tocopherol inhibits protein kinase C activity, a signaling element that can regulate cell proliferation in some but not all cells. The inhibition is not due to

---

*References 8, 29, 30, 67, 93, 132–136.

a direct binding of RRR-α-tocopherol to protein kinase C and depends on the presence of RRR-α-tocopherol early during the cell cycle.[22] In the late G_1 phase, cells become refractory to RRR-α-tocopherol inhibition. Inhibition of activity is not caused by a decrease in the protein kinase C level. The inhibition of protein kinase C by RRR-α-tocopherol, the lack of inhibition by RRR-β-tocopherol and the prevention by RRR-β-tocopherol[143] indicate that the mechanism involved is not related to the radical-scavenging properties of these two molecules, which are essentially equal.[144] Based on the analysis of several other tocopherols and similar compounds, a ligand-interaction type mechanism has been proposed for the inhibition of protein kinase C and of cell proliferation by RRR-α-toco-pherol.[145,146] The α-protein kinase C immunoprecipitated with specific antibodies from RRR-α-tocopherol-treated cells is less phosphorylated and less active. The isoenzyme protein kinase C-ζ,-μ, or -δ are not affected. (Tasinato et al., unpublished). RRR-α-tocopherol may prevent the permissive post-translational phosphorylation of the enzyme.[147,148] Alternatively the RRR-α-tocopherol inhibition may be caused by a protein kinase C phosphatase which dephosphorylates and inactivates the enzyme.[148,149] The inhibition of protein kinase C by physiological concentrations of RRR-α-tocopherol is associated with inhibition of smooth muscle cell proliferation. Such a correlation may account for the *in vivo* quiescence of smooth muscle cells that multiply only under stress conditions.[1,150] A local, dietary or oxidative diminution of RRR-α-tocopherol consequent to an oxidative stress[151,152] would end up in cell activation[153] and growth. *In vivo* evidence using cholesterol-fed rabbits shows a protein kinase C increase in activity and expression by a high-cholesterol diet as well as protection by a tocopherol (Ozer et al., unpublished).

14.2.2 Effect of RRR-α-Tocopherol on MAP Kinases

MAP kinase activation by oxidants was adequately discussed previously. The regulation of these proteins by antioxidants, however is less known. In serum-deprived cells, RRR-α-tocopherol produces an increased phosphorylation of myelin basic protein (Tasinato et al., in preparation). This activation, however, is lower than that produced by 100-nM phorbol myristate acetate (PMA). When both compounds are added together a diminution on MAP kinase activity produced by PMA is observed. Calphostin C abolishes the RRR-α-tocopherol-induced MAP kinase activity, an indication of protein kinase C involvement. Whether the effects of RRR-α-tocopherol are due to its antioxidant properties is not clear at present.

14.2.3 Effect of RRR-α-Tocopherol on Inducible Gene Expression

Inhibition of the tumor necrosis factor-α–induced NF-κB activation by 2,2,5,7,8-pentamethyl-6-hydroxychromane, a vitamin E analogue, is visible in Jurkat T cells.[154]

RRR-α-tocopherol activates activator protein-1 (AP-1), while RRR-β-tocopherol, under the same conditions, has no effect. However, when AP-1 is activated by

PMA, RRR-α-tocopherol acts as a strong inhibitor. In this case also, RRR-β-tocopherol had no effect. In both situations, RRR-β-tocopherol protects against the effects of RRR-α-tocopherol. The effects of tocopherols are cycloheximide-insensitive and protein kinase C-independent, suggesting that the regulatory effects observed with RRR-α-tocopherol occur at the level of AP-1 activation.[50]

An effect similar to that of RRR-α-tocopherol has been described for phenolic antioxidants such as *tert*-butylhydroquinone.[155] AP-1 DNA binding activity is induced by this phenolic antioxidant, but the induction of AP-1 transcriptional activity by the tumor promoter PMA is inhibited by this compound. Treatment with *tert*-butylhydroquinone causes the appearance of AP-1 complexes containing high levels of Fra rather than Fos proteins. Fra proteins heterodimerize with Jun proteins to form stable AP-1 complexes. However, Fra-containing AP-1 complexes have low transactivation potential. Furthermore, Fra-1 represses AP-1 activity induced by either PMA or expression of c-Jun and c-Fos. It has therefore been concluded that inhibitory AP-1 complexes composed of Jun/Fra heterodimers, induced by *tert*-butylhydroquinone, antagonize the transcriptional effects of PMA which are mediated by Jun/Fos heterodimers.[155]

Similarly to phenolic antioxidants, RRR-α-tocopherol is also able to activate AP-1 binding and to modulate transcription of reporter genes containing TPA-responsive elements (S. Marilley and A. Fazzio, 1997 submitted). However, while the effects of phenolic antioxidants are mostly at the transcriptional level, it appears that the activation of AP-1 DNA binding by RRR-α-tocopherol occurs in a cycloheximide-insensitive process.

14.2.4 Effect of Tocopherols on Cell Proliferation

In cells, possibly when their growth is inhibited by lipid peroxidation, antioxidants (and therefore also RRR-α-tocopherol) may restimulate proliferation by removing the inhibitory lipid peroxide.[15–17,156,157] RRR-α-Tocopherol also has a direct effect as cell growth inhibitor and this effect does not appear to be mediated by its reduction-oxidation properties.[20,21,158–161] In fact, RRR-β-tocopherol, an equally potent antioxidant,[144] is not capable of inhibiting cell proliferation. The inhibition by RRR-α-tocopherol of cell proliferation is cell type specific and depends on the mitogen responsible for stimulating growth.

RRR-α-tocopherol (50 μM) inhibits cell growth by approximately 50%. However, RRR-β-tocopherol, an analogue of RRR-α-tocopherol lacking a methyl group in position 7 of the chromanol ring, does not show any inhibition of cell proliferation. Neither the amount of RRR-α- and RRR-β-tocopherol present in the cells after 24 h incubation nor their kinetics of cell uptake are significantly different. Furthermore, no significant competition for uptake between RRR-α-tocopherol and RRR-β-tocopherol was visible. All this indicates that the lack of inhibition by RRR-β-tocopherol is not due to its diminished uptake.[20,50] Furthermore, as with protein kinase C and of AP-1 activation, RRR-β-tocopherol prevents the growth inhibition induced by RRR-α-tocopherol. The effect of RRR-α-tocopherol depends on the type of mitogen utilized to stimulate cell proliferation. The inhibition of cell proliferation produced by 50-μM RRR-α-tocopherol is maximal using platelet-derived growth factor (PDGF), endothelin, or low-density lipoprotein (LDL) as mitogens. Other stimulants such as lysophosphatidic acid, bombesin, and fetal calf serum (FCS) are less effective. When streptolysin-O-permeabilized smooth muscle cells are employed and a peptide substrate for protein kinase C is introduced, RRR-α-tocopherol inhibits protein kinase C activity, whereas RRR-β-tocopherol is much less effective.[20] Although the effect of RRR-α-tocopherol depends on the cell type, all the smooth muscle cell lines tested are inhibited by RRR-α-tocopherol, including several human primary cultures. The molecular basis of RRR-α-tocopherol cell specificity is not yet clear. It could be based on a different signaling pathway used for proliferation in the various cell types. It may be that RRR-α-tocopherol transport and metabolism differ, depending on the cell type. Finally, it is conceivable that RRR-α-tocopherol binding proteins related to RRR-α-tocopherol inhibition are present in some cells and not in others.

Acknowledgments

The technical help of Mrs. Maria Feher and the support by the Swiss National Science Foundation and by F. Hoffmann-La Roche AG for the studies reported in this chapter are gratefully acknowledged.

References

1. Raines, E. W., and R. Ross. 1993. Smooth muscle cells and the pathogenesis of the lesions of atherosclerosis. *Br. Heart. J.* **69**:S30–S37.

2. Badimon, J. J., V. Fuster, J. H. Chesebro, and L. Badimon. 1993. Coronary atherosclerosis: a multifactorial disease. *Circulation* **87**:113–116.

3. Fuster, V., L. Badimon, J. J. Badimon, and J. H. Chesebro. 1992. The pathogenesis of coronary artery disease and the acute coronary syndromes (2). *N. Engl. J. Med.* **326**:310–318.

4. Fuster, V., L. Badimon, J. J. Badimon, and J. H. Chesebro. 1992. The pathogenesis of coronary artery disease and the acute coronary syndromes (1). *N. Engl. J. Med.* **326**:242–250.

5. Schwartz, S. M., R. L. Heimark, and M. W. Majesky. 1990. Developmental mechanisms underlying pathology of arteries. *Physiol. Rev.* **70**:1177–1209.

6. Ross, R., and L. Agius. 1992. The process of atherogenesis—cellular and molecular interaction: from experimental animal models to humans. *Diabetologia* **35**:S34–S40.

7. Rao, G. N., and B. C. Berk. 1992. Active oxygen species stimulate vascular smooth muscle cell growth and proto-oncogene expression. *Circ. Res.* **70**:593–599.

8. Rao, G. N., B. Lassegue, K. K. Griendling, R. W. Alexander, and B. C. Berk. 1993. Hydrogen peroxide-induced c-fos expression is mediated by arachidonic acid release: role of protein kinase C. *Nucleic Acids Res.* **21**:1259–1263.

9. Kita, T., Y. Nagano, M. Yokode, K. Ishii, N. Kume, A. Ooshima, H. Yoshida, and C. Kawai. 1987. Probucol prevents the progression of atherosclerosis in Watanabe heritable hyperlipidemic rabbit, an animal model for familial hypercholesterolemia. *Proc. Natl. Acad. Sci. USA* **84**:5928–5931.

10. Bjorkhem, I., A. Henriksson Freyschuss, O. Breuer, U. Diczfalusy, L. Berglund, and P. Henriksson. 1991. The antioxidant butylated hydroxytoluene protects against atherosclerosis. *Arterioscler. Thromb.* **11**:15–22.

11. Jialal, I., and C. J. Fuller. 1993. Oxidized LDL and antioxidants. *Clin. Cardiol.* **16**:I6–I9.

12. Gey, K. F. 1990. The antioxidant hypothesis of cardiovascular disease: epidemiology and mechanisms. *Biochem. Soc. Trans.* **18**:1041–1045.

13. Rimm, E. B., M. J. Stampfer, A. Ascherio, E. Giovannucci, G. A. Colditz, and W. C. Willett. 1993. Vitamin E consumption and the risk of coronary heart disease in men (see comments). *N. Engl. J. Med.* **328**:1450–1456.

14. Stampfer, M. J., C. H. Hennekens, J. E. Manson, G. A. Colditz, B. Rosner, and W. C. Willett. 1993. Vitamin E consumption and the risk of coronary disease in women (see comments). *N. Engl. J. Med.* **328**:1444–1449.

15. Morisaki, N., K. Yokote, and Y. Saito. 1992. Atherosclerosis from a viewpoint of arterial wall cell function: relation to vitamin E. *J. Nutr. Sci. Vitaminol.* **Spec. No**:196–199.

16. Lindsey, J. A., H. F. Zhang, H. Kaseki, N. Morisaki, T. Sato, and D. G. Cornwell. 1985. Fatty acid metabolism and cell proliferation. VII. Antioxidant effects of tocopherols and their quinones. *Lipids* **20**:151–157.

17. Gavino, V. C., J. S. Miller, S. O. Ikharebha, G. E. Milo, and D. G. Cornwell. 1981. Effect of polyunsaturated fatty acids and antioxidants on lipid peroxidation in tissue cultures. *J. Lipid Res.* **22**:763–769.

18. Kuzuya, M., M. Naito, C. Funaki, T. Hayashi, K. Yamada, K. Asai, and F. Kuzuya. 1991. Antioxidants stimulate endothelial cell proliferation in culture. *Artery* **18**:115–124.

19. Morisaki, N., J. M. Stitts, L. Bartels Tomei, G. E. Milo, R. V. Panganamala, and D. G. Cornwell. 1982. Dipyridamole: an antioxidant that promotes the proliferation of aorta smooth muscle cells. *Artery* **11**:88–107.

20. Chatelain, E., D. O. Boscoboinik, G. M. Bartoli, V. E. Kagan, F. K. Gey, L. Packer, and A. Azzi. 1993. Inhibition of smooth muscle cell proliferation and protein kinase C activity by tocopherols and tocotrienols. *Biochim. Biophys. Acta* **1176**:83–89.

21. Ozer, N. K., P. Palozza, D. Boscoboinik, and A. Azzi. 1993. d-alpha-Tocopherol inhibits low density lipoprotein induced proliferation and protein kinase C activity in vascular smooth muscle cells. *FEBS Lett.* **322**:307–310.

22. Azzi, A., D. Boscoboinik, and C. Hensey. 1992. The protein kinase C family. *Eur. J. Biochem.* **208**:547–557.

23. Chakraborti, S., and J. R. Michael. 1993. Role of protein kinase C in oxidant—mediated activation of phospholipase A₂ in rabbit pulmonary arterial smooth muscle cells. *Mol. Cell. Biochem.* **122**:9–15.

24. Chakraborti, S., S. K. Batabyal, G. Dutta, and J. R. Michael. 1992. Role of serine esterase in hydrogen peroxide-mediated activation of phospholipase A₂ in rabbit pulmonary arterial smooth muscle cells. *Ind. J. Biochem. Biophys.* **29**:477–481.

25. Rao, G. N., B. Lassègue, K. K. Griendling, and R. W. Alexander. 1993. Hydrogen peroxide stimulates transcription of c-jun in vascular smooth muscle cells: role of arachidonic acid. *Oncogene* **8**:2759–2764.

26. Gopalakrishna, R., and W. B. Anderson. 1991. Reversible oxidative activation and inactivation of protein kinase C by the mitogen/tumor promoter periodate. *Arch. Biochem. Biophys.* **285**:382–387.

27. Palumbo, E. J., J. D. Sweatt, S. J. Chen, and E. Klann. 1992. Oxidation-induced persistent activation of protein kinase C in hippocampal homogenates. *Biochem. Biophys. Res. Commun.* **187**:1439–1445.

28. Azzi, A., D. Boscoboinik, and N. K. Ozer. 1994. Vascular smooth muscle cells: regulation and deregulation by reactive oxygen species, pp. 423–445. *In* C. K. Sen, L. Packer, and O. Hänninen (eds.), *Exercise and Oxygen Toxicity.* Elsevier Science Publishers B. V., Amsterdam.

29. Abate, C., L. Patel, F. J. Rauscher, and T. Curran. 1990. Redox regulation of fos and jun DNA-binding activity in vitro. *Science* **249**:1157–1161.

30. Toledano, M. B., and W. J. Leonard. 1991. Modulation of transcription factor NF-kappa B binding activity by oxidation-reduction in vitro. *Proc. Natl. Acad. Sci. USA* **88**:4328–4332.

31. Bannister, A. J., A. Cook, and T. Kouzarides. 1991. In vitro DNA binding activity of Fos/Jun and BZLF1 but not C/EBP is affected by redox changes. *Oncogene* **6**:1243–1250.

32. Frame, M. C., N. M. Wilkie, A. J. Darling, A. Chudleigh, A. Pintzas, J. C. Lang, and D. A. Gillespie. 1991. Regulation of AP-1/DNA complex formation in vitro. *Oncogene* **6**:205–209.

33. Derijard, B., M. Hibi, I. H. Wu, T. Barrett, B. Su, T. Deng, M. Karin, and R. J. Davis. 1994. JNK1: a protein kinase stimulated by UV light and Ha-Ras that binds and phosphorylates the c-Jun activation domain. *Cell* **76**:1025–1037.

34. Engelberg, D., C. Klein, H. Martinetto, K. Struhl, and M. Karin. 1994. The UV response involving the Ras signaling pathway and AP-1 transcription factors is conserved between yeast and mammals. *Cell* **77**:381–390.

35. Minden, A., A. Lin, T. Smeal, B. Dérijard, M. Cobb, R. Davis, and M. Karin. 1994. c-Jun N-terminal phosphorylation correlates with activation of the JNK subgroup but not the ERK subgroup of mitogen-activated protein kinases. *Mol. Cell. Biol.* **14**:6683–6688.

36. Clark, J. D., L. L. Lin, R. W. Kriz, C. S. Ramesha, L. A. Sultzman, A. Y. Lin, N. Milona, and J. L. Knopf. 1991. A novel arachidonic acid-selective cytosolic PLA₂

contains a Ca(2+)-dependent translocation domain with homology to PKC and GAP. *Cell* **65**:1043–1051.

37. Burch, R. M., A. Luini, and J. Axelrod. 1986. Phospholipase A$_2$ and phospholipase C are activated by distinct GTP-binding proteins in response to alpha 1-adrenergic stimulation in FRTL5 thyroid cells. *Proc. Natl. Acad. Sci. USA* **83**:7201–7205.

38. Lin, L. L., M. Wartmann, A. Y. Lin, J. L. Knopf, A. Seth, and J. R. Davis. 1993. cPLA$_2$ is phosphorylated and activated by MAP kinase. *Cell* **72**:269–278.

39. Glaser, K. B., A. Sung, J. Bauer, and B. M. Weichman. 1993. Regulation of eicosanoid biosynthesis in the macrophage. Involvement of protein tyrosine phosphorylation and modulation by selective protein tyrosine kinase inhibitors. *Biochem. Pharmacol.* **45**:711–721.

40. Gupta, S. K., E. Diez, L. E. Heasley, S. Osawa, and G. L. Johnson. 1990. A G protein mutant that inhibits thrombin and purinergic receptor activation of phospholipase A$_2$. *Science* **249**:662–666.

41. Rao, G. N., M. S. Runge, and R. W. Alexander. Hydrogen peroxide activation of cytosolic phospholipase A$_2$ in vascular smooth muscle cells. *Biochim. Biophys. Acta* **1265**:67–72.

42. Mahoney, C. W., and K. P. Huang. 1994. Molecular and catalytic properties of protein kinase C, pp. 16–63. *In* J. Kuo (ed.), *Protein Kinase C*. Oxford University Press, New York.

43. Gopalakrishna, R., and W. B. Anderson. 1989. Ca2+- and phospholipid-independent activation of protein kinase C by selective oxidative modification of the regulatory domain. *Proc. Natl. Acad. Sci. USA* **86**:6758–6762.

44. Kass, G. E., S. K. Duddy, and S. Orrenius. 1989. Activation of hepatocyte protein kinase C by redox-cycling quinones. *Biochem. J.* **260**:499–507.

45. Gopalakrishna, R., Z. H. Chen, and U. Gundimeda. 1993. Nitric oxide and nitric oxide-generating agents induce a reversible inactivation of protein kinase C activity and phorbol ester binding. *J. Biol. Chem.* **268**:27180–27185.

46. Siflinger Birnboim, A., M. S. Goligorsky, P. J. Del Vecchio, and A. B. Malik. 1992. Activation of protein kinase C pathway contributes to hydrogen peroxide-induced increase in endothelial permeability (see comments). *Lab. Invest.* **67**:24–30.

47. Whisler, R. L., M. A. Goyette, I. S. Grants, and Y. G. Newhouse. 1995. Sublethal levels of oxidant stress stimulate multiple serine/threonine kinases and suppress protein phosphatases in Jurkat T cells. *Arch. Biochem. Biophys.* **319**:23–35.

48. Cantoni, O., D. Boscoboinik, M. Fiorani, B. Stäuble, and A. Azzi. 1996. The phosphorylation state of MAP-kinases modulates the cytotoxic response of smooth muscle cells to hydrogen peroxide. *FEBS Lett.* **389**:285–288.

49. Fiorani, M., O. Cantoni, A. Tasinato, D. Boscoboinik, and A. Azzi. 1995. Hydrogen peroxide- and fetal bovine serum-induced dna synthesis in vascular smooth muscle cells: positive and negative regulation by protein kinase C isoforms. *Biochim. Biophys. Acta* **1269**:98–104.

50. Stauble, B., D. Boscoboinik, A. Tasinato, and M. Azzi. 1994. Modulation of activator protein-1 (AP-1) transcription factor and protein kinase C by hydrogen

peroxide and D-α-tocopherol in vascular smooth muscle cells. *Eur. J. Biochem.* **226**:393–402.

51. Sasaguri, T., C. Kosaka, M. Hirata, J. Masuda, K. Shimokado, M. Fujishima, and J. Ogata. 1993. Protein kinase C-mediated inhibition of vascular smooth muscle cell proliferation: the isoforms that may mediate G1/S inhibition. *Exp. Cell. Res.* **208**:311–320.

52. Sasaguri, T., C. Kosaka, K. Zen, J. Masuda, K. Shimokado, and J. Ogata. 1995. Protein kinase C isoforms that may mediate G1/S inhibition in cultured vascular smooth muscle cells. *Ann. N.Y. Acad. Sci.* **748**:590–591.

53. Berra, E., M. T. Diaz Meco, J. Dominguez, M. M. Municio, L. Sanz, J. Lozano, R. S. Chapkin, and J. Moscat. 1993. Protein kinase C zeta isoform is critical for mitogenic signal transduction. *Cell* **74**:555–563.

54. Nakanishi, H., K. A. Brewer, and J. H. Exton. 1993. Activation of the zeta isozyme of protein kinase C by phosphatidylinositol 3,4,5-trisphosphate. *J. Biol. Chem.* **268**:13–16.

55. Pongracz, J., P. Clark, J. P. Neoptolemos, and J. M. Lord. 1995. Expression of protein kinase C isoenzymes in colorectal cancer tissue and their differential activation by different bile acids. *Int. J. Cancer* **61**:35–39.

56. Cadena, D. L., and G. N. Gill. 1992. Receptor tyrosine kinases. *FASEB J.* **6**:2332–2337.

57. Toyoshima, K., Y. Yamanashi, K. Inoue, K. Semba, T. Yamamoto, and T. Akiyama. 1992. Protein tyrosine kinases belonging to the src family. *CIBA Found. Symp.* **164**:240–248.

58. Glenney, J. R., Jr. 1992. Tyrosine-phosphorylated proteins: mediators of signal transduction from the tyrosine kinases. *Biochim. Biophys. Acta* **1134**:113–127.

59. Chen, Y. X., D. C. Yang, A. B. Brown, Y. Jeng, A. Tatoyan, and T. M. Chan. 1990. Activation of a membrane-associated phosphatidylinositol kinase through tyrosine-protein phosphorylation by naphthoquinones and orthovanadate. *Arch. Biochem. Biophys.* **283**:184–192.

60. Fialkow, L., C. K. Chan, S. Grinstein, and G. P. Downey. 1993. Regulation of tyrosine phosphorylation in neutrophils by the NADPH oxidase. Role of reactive oxygen intermediates. *J. Biol. Chem.* **268**:17131–17137.

61. Chan, T. M., E. Chen, A. Tatoyan, N. S. Shargill, M. Pleta, and P. Hochstein. 1986. Stimulation of tyrosine-specific protein phosphorylation in the rat liver plasma membrane by oxygen radicals. *Biochem. Biophys. Res. Commun.* **139**:439–445.

62. Chen, Y., and T. M. Chan. 1993. Orthovanadate and 2,3-dimethoxy-1,4-naphthoquinone augment growth factor-induced cell proliferation and c-*fos* gene expression in 3T3-L1 cells. *Arch. Biochem. Biophys.* **305**:9–16.

63. Zor, U., E. Ferber, P. Gergely, K. Szucs, V. Dombradi, and R. Goldman. 1993. Reactive oxygen species mediate phorbol ester-regulated tyrosine phosphorylation and phospholipase A_2 activation: potentiation by vanadate. *Biochem. J.* **295**:879–888.

64. Nakamura, K., T. Hori, N. Sato, K. Sugie, T. Kawakami, and J. Yodoi. 1993. Redox regulation of a src family protein tyrosine kinase p56lck in T cells. *Oncogene* **8**:3133–3139.

65. Kanner, S. B., T. J. Kavanagh, A. Grossmann, S. L. Hu, J. B. Bolen, P. S. Rabinovitch, and J. A. Ledbetter. 1992. Sulfhydryl oxidation down-regulates T-cell signaling and inhibits tyrosine phosphorylation of phospholipase C gamma 1. *Proc. Natl. Acad. Sci. USA* **89**:300–304.

66. Toyoshima, H., H. Kozutsumi, Y. Maru, K. Hagiwara, A. Furuya, H. Mioh, N. Hanai, F. Takaku, Y. Yazaki and H. Hirai. 1993. Differently spliced cDNAs of human leukocyte tyrosine kinase receptor tyrosine kinase predict receptor proteins with and without a tyrosine kinase domain and a soluble receptor protein. *Proc. Natl. Acad. Sci. USA* **90**:5404–5408.

67. Bauskin, A. R., I. Alkalay, and Y. Ben Neriah. 1991. Redox regulation of a protein tyrosine kinase in the endoplasmic reticulum. *Cell* **66**:685–696.

68. Guy, G. R., J. Cairns, S. B. Ng, and Y. H. Tan. 1993. Inactivation of a redox-sensitive protein phosphatase during the early events of tumor necrosis factor/interleukin-1 signal transduction. *J. Biol. Chem.* **268**:2141–2148.

69. Keyse, S. M., and E. A. Emslie. 1992. Oxidative stress and heat shock induce a human gene encoding a protein-tyrosine phosphatase. *Nature* **359**:644–647.

70. Alessi, D. R., C. Smythe, and S. M. Keyse. 1993. The human CL100 gene encodes a Tyr/Thr-protein phosphatase which potently and specifically inactivates MAP kinase and suppresses its activation by oncogenic ras in Xenopus oocyte extracts. *Oncogene* **8**:2015–2020.

71. Keyse, S. M. 1995. An emerging family of dual specificity MAP kinase phosphatases. *Biochim. Biophys. Acta* **1265**:152–160.

72. Sullivan, S. G., D. T. Chiu, M. Errasfa, J. M. Wang, J. S. Qi, and A. Stern. 1994. Effects of H_2O_2 on protein tyrosine phosphatase activity in HER14 cells. *Free Rad. Biol. Med.* **16**:399–403.

73. Seger, R., and E. G. Krebs. 1995. The MAPK signaling cascade. *FASEB J.* **9**:726–735.

74. Egan, S. E., and R. A. Weinberg. 1993. The pathway to signal achievement (news). *Nature* **365**:781–783.

75. Blumer, K. J., and G. L. Johnson. 1994. Diversity in function and regulation of MAP kinase pathways. *Trends Biochem. Sci.* **19**:236–240.

76. Cobb, M. H., and E. J. Goldsmith. 1995. How MAP kinases are regulated. *J. Biol. Chem.* **270**:14843–14846.

77. Clarke, P. R. 1994. Signal transduction. Switching off MAP kinases. *Curr. Biol.* **4**:647–650.

78. Duff, J. L., B. P. Monia, and B. C. Berk. 1995. Mitogen-activated protein (MAP) kinase is regulated by the MAP kinase phosphatase (MKP-1) in vascular smooth muscle cells. Effect of antinomycin D and antisense oligonucleotides. *J. Biol. Chem.* **270**:7161–7166.

79. Stevenson, M. A., S. S. Pollock, C. N. Coleman, and S. K. Calderwood. 1994. X-irradiation, phorbol esters, and H_2O_2 stimulate mitogen-activated protein kinase activity in NIH-3T3 cells through the formation of reactive oxygen intermediates. *Cancer Res.* **54**:12–15.

80. Reid, T. M., and L. A. Loeb. 1993. Tandem double CC → TT mutations are produced by reactive oxygen species. *Proc. Natl. Acad. Sci. USA* **90**:3904–3907.

81. Hruza, L. L., and A. P. Pentland. 1993. Mechanisms of UV-induced inflammation. *J. Invest. Dermatol.* **100**:35S–41S.

82. Trenam, C. W., D. R. Blake, and C. J. Morris. 1992. Skin inflammation: reactive oxygen species and the role of iron. *J. Invest. Dermatol.* **99**:675–682.

83. Hibi, M., A. Lin, T. Smeal, A. Minden, and M. Karin. 1993. Identification of an oncoprotein- and UV-responsive protein kinase that binds and potentiates the c-Jun activation domain. *Genes Dev.* **7**:2135–2148.

84. Baeuerle, P. A., and D. Baltimore. 1989. A 65-kappaD subunit of active NF-kappaB is required for inhibition of NF-kappaB by I kappaB. *Genes Dev.* **3**:1689–1698.

85. Baeuerle, P. A., and D. Baltimore. 1988. I kappa B: a specific inhibitor of the NF-kappa B transcription factor. *Science* **242**:540–546.

86. Baeuerle, P. A., and D. Baltimore. 1988. Activation of DNA-binding activity in an apparently cytoplasmic precursor of the NF-kappa B transcription factor. *Cell* **53**:211–217.

87. Baeuerle, P. A., M. Lenardo, J. W. Pierce, and D. Baltimore. 1988. Phorbol-ester-induced activation of the NF-kappa B transcription factor involves dissociation of an apparently cytoplasmic NF-kappa B/inhibitor complex. *Cold Spring Harbor Symp. Quant. Biol.* **53**(2):789–798.

88. Henkel, T., T. Machleidt, I. Alkalay, M. Kronke, Y. Ben Neriah, and P. Baeuerle. 1993. Rapid proteolysis of I kappa B-alpha is necessary for activation of transcription factor NF-kappa B. *Nature* **365**:182–185.

89. Li, S., and J. M. Sedivy. 1993. Raf-1 protein kinase activates the NF-kappa B transcription factor by dissociating the cytoplasmic NF-kappa B-I kappa B complex. *Proc. Natl. Acad. Sci. USA* **90**:9247–9251.

90. Schreck, R., P. Rieber, and P. A. Baeuerle. 1991. Reactive oxygen intermediates as apparently widely used messengers in the activation of the NF-kappa B transcription factor and HIV-1. *EMBO J.* **10**:2247–2258.

91. Schreck, R., K. Albermann, and P. A. Baeuerle. 1992. Nuclear factor kappa B: an oxidative stress-responsive transcription factor of eukaryotic cells (a review). *Free Rad. Res. Commun.* **17**:221–237.

92. Schreck, R., H. Zorbas, G. L. Winnacker, and P. A. Baeuerle. 1990. The NF-kappa B transcription factor induces DNA bending which is modulated by its 65-kD subunit. *Nucleic Acids Res.* **18**:6497–6502.

93. Toledano, M. B., D. Ghosh, F. Trinh, and W. J. Leonard. 1993. N-terminal DNA-binding domains contribute to differential DNA-binding specificities of NF-kappa B p50 and p65. *Mol. Cell. Biol.* **13**:852–860.

94. Lin, A., T. Smeal, B. Binetruy, T. Deng, J. C. Chambard, and M. Karin. 1993. Control of AP-1 activity by signal transduction cascades. *Second Messenger Phosphoprotein Res.* **28**:255–260.

95. Angel, P., and M. Karin. 1991. The role of Jun, Fos and the AP-1 complex in cell-proliferation and transformation. *Biochim. Biophys. Acta* **1072**:129–157.

96. Skroch, P., C. Buchman, and M. Karin. 1993. Regulation of human and yeast metallothionein gene transcription by heavy metal ions. *Prog. Clin. Biol. Res.* **380**:113–128.

97. Chiu, R., M. Imagawa, R. J. Imbra, J. R. Bockoven, and M. Karin. 1987. Multiple-*cis*- and *trans*-acting elements mediate the transcriptional response to phorbol esters. *Nature* **329**:648–651.

98. Okuno, H., A. Akahori, H. Sato, S. Xanthoudakis, T. Curran, and H. Iba. 1993. Escape from redox regulation enhances the transforming activity of Fos. *Oncogene* **8**:695–701.

99. Meyer, M., R. Schreck, and P. A. Baeuerle. 1993. H_2O_2 and antioxidants have opposite effects on activation of NF-kappa B and AP-1 in intact cells. AP-1 as secondary antioxidant-responsive factor. *EMBO J* **12**:2005–2015.

100. Rushmore, T. H., R. G. King, K. E. Paulson, and C. B. Pickett. 1990. Regulation of glutathione S-transferase Ya subunit gene expression: identification of a unique xenobiotic-responsive element controlling inducible expression by planar aromatic compounds. *Proc. Natl. Acad. Sci. USA* **87**:3826–3830.

101. Friling, R. S., A. Bensimon, Y. Tichauer, and V. Daniel. 1990. Xenobiotic-inducible expression of murine glutathione S-transferase Ya subunit gene is controlled by an electrophile-responsive element. *Proc. Natl. Acad. Sci. USA* **87**:6258–6262.

102. Li, Y., and A. K. Jaiswal. 1992. Regulation of human NAD(P)H:quinone oxidoreductase gene. Role of AP1 binding site contained within human antioxidant response element. *J. Biol. Chem.* **267**:15097–15104.

103. Pinkus, R., S. Bergelson, and V. Daniel. 1993. Phenobarbital induction of AP-1 binding activity mediates activation of glutathione S-transferase and quinone reductase gene expression. *Biochem. J.* **290**:637–640.

104. Storz, G., and L. A. Tartaglia. 1992. OxyR: a regulator of antioxidant genes. *J. Nutr.* **122**:627–630.

105. Wiese, A. G., R. E. Pacifici, and K. J. Davies. 1995. Transient adaptation of oxidative stress in mammalian cells. *Arch. Biochem. Biophys.* **318**:231–240.

106. Keyse, S. M., and E. A. Emslie. 1992. Oxidative stress and heat shock induce a human gene encoding a protein-tyrosine phosphatase. *Nature* **359**:644–647.

107. Rao, G. N., B. Lassègue, K. K. Griendling, and R. W. Alexander. 1993. Hydrogen peroxide stimulates transcription of c-*jun* in vascular smooth muscle cells: role of arachidonic acid. *Oncogene* **8**:2759–2764.

108. Nose, K., M. Shibanuma, K. Kikuchi, H. Kageyama, S. Sakiyama, and T. Kuroki. 1991. Transcriptional activation of early-response genes by hydrogen peroxide in a mouse osteoblastic cell line. *Eur. J. Biochem.* **201**:99–106.

109. Palomba, L., P. Sestili, F. Cattabeni, A. Azzi, and O. Cantoni. (1996). Prevention of necrosis and activation of apoptosis in oxidatively injured human myeloid leukemia U937 cells. *FEBS Lett.* **390**:91–94.

110. Cantoni, O., F. Cattabeni, V. Stocchi, R. E. Meyn, P. Cerutti, and D. Murray. 1989. Hydrogen peroxide insult in cultured mammalian cells: relationships between DNA single-strand breakage, poly(ADP-ribose) metabolism and cell killing. *Biochim. Biophys. Acta* **1014**:1–7.

111. Mello Filho, A. C., M. E. Hoffmann, and R. Meneghini. 1984. Cell killing and DNA damage by hydrogen peroxide are mediated by intracellular iron. *Biochem. J.* **218**:273–275.

112. Schraufstatter, I. U., P. A. Hyslop, D. B. Hinshaw, R. G. Spragg, L. A. Sklar, and C. G. Cochrane. 1986. Hydrogen peroxide-induced injury of cells and its prevention by inhibitors of poly(ADP-ribose) polymerase. *Proc. Natl. Acad. Sci. USA* **83**:4908–4912.

113. Junod, A. F., L. Jornot, and H. Petersen. 1989. Differential effects of hyperoxia and hydrogen peroxide on DNA damage, polyadenosine diphosphate-ribose polymerase activity, and nicotinamide adenine dinucleotide and adenosine triphosphate contents in cultured endothelial cells and fibroblasts. *J. Cell. Physiol.* **140**:177–185.

114. Kirkland, J. B. 1991. Lipid peroxidation, protein thiol oxidation and DNA damage in hydrogen peroxide-induced injury to endothelial cells: role of activation of poly(ADP-ribose)polymerase. *Biochim. Biophys. Acta* **1092**:319–325.

115. Schraufstatter, I. U., D. B. Hinshaw, P. A. Hyslop, R. G. Spragg, and C. G. Cochrane. 1986. Oxidant injury of cells. DNA strand-breaks activate polyadenosine diphosphate-ribose polymerase and lead to depletion of nicotinamide adenine dinucleotide. *J. Clin. Invest.* **77**:1312–1320.

116. Hyslop, P. A., D. B. Hinshaw, I. U. Schraufstatter, L. A. Sklar, R. G. Spragg, and C. G. Cochrane. 1986. Intracellular calcium homeostasis during hydrogen peroxide injury to cultured P388D1 cells. *J. Cell Physiol.* **129**:356–366.

117. Lennon, S. V., J. J. Martin, and T. G. Cotter. 1991. Dose-dependent induction of apoptosis in human tumour cell lines by widely diverging stimuli. *Cell Prolif.* **24**:203–214.

118. Lennon, S. V., S. A. Kilfeather, M. B. Hallett, A. K. Campbell, and T. G. Cotter. 1992. Elevations in cytosolic free Ca^{2+} are not required to trigger apoptosis in human leukaemia cells. *Clin. Exp. Immunol.* **87**:465–471.

119. Ueda, N., and S. V. Shah. 1992. Endonuclease-induced DNA damage and cell death in oxidant injury to renal tubular epithelial cells. *J. Clin. Invest.* **90**:2593–2597.

120. Nosseri, C., S. Coppola, and L. Ghibelli. 1994. Possible involvement of poly(ADP-ribosyl) polymerase in triggering stress-induced apoptosis. *Exp. Cell Res.* **212**:367–373.

121. Weissberg, P. L. 1991. Mechanisms of vascular smooth muscle cell proliferation. *Ann. Acad. Med. Singapore* **20**:38–42.

122. Schwartz, S. M., G. R. Campbell, and J. H. Campbell. 1986. Replication of smooth muscle cells in vascular disease. *Circ. Res.* **58**:427–444.

123. Ross, R. 1993. The pathogenesis of atherosclerosis: a perspective for the 1990s. *Nature* **362**:801–809.

124. Schwartz, C. J., A. J. Valente, E. A. Sprague, J. L. Kelley, and R. M. Nerem. 1991. The pathogenesis of atherosclerosis: an overview. *Clin. Cardiol.* **14**:I1–16.

125. Fridovich, I. 1978. The biology of oxygen radicals. *Science* **201**:875–880.

126. Halliwell, B. 1989. Free radicals, reactive oxygen species and human disease: a critical evaluation with special reference to atherosclerosis. *Br. J. Exp. Pathol.* **70**:737–757.

127. Redl, H., H. Gasser, G. Schlag, and I. Marzi. 1993. Involvement of oxygen radicals in shock related cell injury. *Br. Med. Bull.* **49**:556–565.

128. Ambrosio, G., J. L. Zweier, C. Duilio, P. Kuppusamy, G. Santoro, P. P. Elia, J. Tritto, P. Cirillo, M. Condorelli, and M. Chiariello. 1993. Evidence that mitochondrial respiration is a source of potentially toxic oxygen free radicals in intact rabbit hearts subjected to ischemia and reflow. *J. Biol. Chem.* **268**:18532–18541.

129. Horton, J. W., and P. B. Walker. 1993. Oxygen, radicals, lipid peroxidation, and permeability changes after intestinal ischemia and reperfusion. *J. Appl. Physiol.* **74**:1515–1520.

130. Granger, D. N., P. R. Kvietys, and M. A. Perry. 1993. Leukocyte—endothelial cell adhesion induced by ischemia and reperfusion. *Can. J. Physiol. Pharmacol.* **71**:67–75.

131. Kontos, H. A., E. P. Wei, J. T. Povlishock, and C. W. Christman. 1984. Oxygen radicals mediate the cerebral arteriolar dilation from arachidonate and bradykinin in cats. *Circ. Res.* **55**:295–303.

132. Malviya, A. N., and P. Anglard. 1986. Modulation of cytosolic protein kinase C activity by ferricyanide: priming event seems transmembrane redox signalling. A study on transformed C3H/10T1/2 cells in culture. *FEBS Lett.* **200**:265–270.

133. Xanthoudakis, S., and T. Curran. 1992. Identification and characterization of Ref-1, a nuclear protein that facilitates AP-1 DNA-binding activity. *EMBO J* **11**:653–665.

134. Staal, F. J., M. Roederer, and L. A. Herzenberg. 1990. Intracellular thiols regulate activation of nuclear factor kappa B and transcription of human immunodeficiency virus. *Proc. Natl. Acad. Sci. USA* **87**:9943–9947.

135. Dartsch, P. C., R. Voisard, and E. Betz. 1990. In vitro growth characteristics of human atherosclerotic plaque cells: comparison of cells from primary stenosing and restenosing lesions of peripheral and coronary arteries. *Res. Exp. Med.* **190**:77–87.

136. Dartsch, P. C., R. Voisard, G. Bauriedel, B. Hofling, and E. Betz. 1990. Growth characteristics and cytoskeletal organization of cultured smooth muscle cells from human primary stenosing and restenosing lesions. *Arteriosclerosis* **10**:62–75.

137. van Poppel, G., A. Kardinaal, H. Princen, and F. J. Kok. 1994. Antioxidants and coronary heart disease. *Ann. Med.* **26**:429–434.

138. Lafont, A. M., Y. C. Chai, J. F. Cornhill, P. L. Whitlow, P. H. Howe, and G. M. Chisolm. 1995. Effect of alpha-tocopherol on restenosis after angioplasty in a model of experimental atherosclerosis. *J. Clin. Invest.* **95**:1018–1025.

139. Haramaki, N., L. Packer, H. Assadnazari, and G. Zimmer. 1993. Cardiac recovery during post-ischemic reperfusion is improved by combination of vitamin E with dihydrolipoic acid. *Biochem. Biophys. Res. Commun.* **196**:1101–1107.

140. Grainger, D. J., H. L. Kirschenlohr, J. C. Metcalfe, P. L. Weissberg, D. P. Wade, and R. M. Lawn. 1993. Proliferation of human smooth muscle cells promoted by lipoprotein(a). *Science* **260**:1655–1658.

141. Buhler, F. R., V. A. Tkachuk, A. W. Hahn, and T. J. Resink. 1991. Low- and high-density lipoproteins as hormonal regulators of platelet, vascular endothelial and smooth muscle cell interactions: relevance to hypertension. *J. Hypertens. Suppl.* **9**:S28–S36.

142. Ohlstein, E. H., S. A. Douglas, C. P. Sung, T. L. Yue, C. Louden, A. Arleth, G. Poste, R. R. Ruffolo, Jr., and G. Z. Feuerstein. 1993. Carvedilol, a cardiovascular drug, prevents vascular smooth muscle cell proliferation, migration, and neointimal formation following vascular injury. *Proc. Natl. Acad. Sci. USA* **90**:6189–6193.

143. Tasinato, A., D. Boscoboinik, G. M. Bartoli, P. Maroni, and A. Azzi. 1995. *d*-α-tocopherol inhibition of vascular smooth muscle cell proliferation occurs at physiological concentrations, correlates with protein kinase C inhibition, and is independent of its antioxidant properties. *Proc. Natl. Acad. Sci. USA* **92**:12190–12194.

144. Pryor, A. W., J. A. Cornicelli, L. J. Devall, B. Tait, B. K. Trivedi, D. T. Witiak, and M. Wu. 1993. A rapid screening test to determine the antioxidant potencies of natural and synthetic antioxidants. *J. Org. Chem.* **58**:3521–3532.

145. Azzi, A., D. Boscoboinik, D. Marilley, N. K. Özer, B. Stäuble, and A. Tasinato. 1995. Vitamin E: A sensor and an information transducer of the cell oxidation state. *Am. J. Clin. Nutr.* **62**:1337S–1346S.

146. Azzi, A., O. Cantoni, N. Ozer, D. Boscoboinik, and S. Spycher. 1996. The role of hydrogen peroxide and RRR-α-tocopherol in smooth muscle cell proliferation. *Cell Death Differen.* **3**:79–90.

147. Pears, C., S. Stabel, S. Cazaubon, and P. J. Parker. 1992. Studies on the phosphorylation of protein kinase C-alpha. *Biochem. J.* **283**:515–518.

148. Newton, A. C. 1995. Protein kinase C: structure, function, and regulation. *J. Biol. Chem.* **270**:28495–28498.

149. Clarke, P. R., S. R. Siddhanti, P. Cohen, and P. J. Blackshear. 1993. Okadaic acid-sensitive protein phosphatases dephosphorylate MARKS, a major protein kinase C substrate. *FEBS Lett.* **336**:37–42.

150. Clowes, A. W., and S. M. Schwartz. 1985. Significance of quiescent smooth muscle migration in the injured rat carotid artery. *Circ. Res.* **56**:139–145.

151. Traber, M. G., L. L. Rudel, G. W. Burton, L. Hughes, K. U. Ingold, and H. J. Kayden. 1990. Nascent VLDL from liver perfusions of cynomolgus monkeys are preferentially enriched in RRR- compared with SRR-alpha-tocopherol: studies using deuterated tocopherols. *J. Lipid Res.* **31**:687–694.

152. Burton, G. W., and K. U. Ingold. 1989. Vitamin E as an in vitro and in vivo antioxidant. *Ann. N.Y. Acad. Sci.* **570**:7–22.

153. Kanno, T., T. Utsumi, H. Kobuchi, Y. Takehara, J. Akiyama, T. Yoshioka, A. A. Horton, and K. Utsumi. 1995. Inhibition of stimulus-specific neutrophil superoxide generation by alpha-tocopherol. *Free Rad. Res.* **22**:431–440.

154. Suzuki, Y. J., and L. Packer. 1993. Inhibition of NF-kappa B activation by vitamin E derivatives. *Biochem. Biophys. Res. Commun.* **193**:277–283.

155. Yoshioka, K., T. Deng, M. Cavigelli, and M. Karin. 1995. Antitumor promotion by phenolic antioxidants: inhibition of AP-1 activity through induction of Fra expression. *Proc. Natl. Acad. Sci. USA* **92**:4972–4976.

156. Morisaki, N., J. A. Lindsey, J. M. Stitts, H. Zhang, and D. G. Cornwell. 1984. Fatty acid metabolism and cell proliferation. V. Evaluation of pathways for the generation of lipid peroxides. *Lipids* **19**:381–394.

157. Hennekens, C. H., and J. M. Gaziano. 1993. Antioxidants and heart disease: epidemiology and clinical evidence. *Clin. Cardiol.* **16**:10–13.

158. Boscoboinik, D., E. Chatelain, G. M. Bargoli, and A. Azzi. 1992. Molecular basis of alpha-tocopherol inhibition of smooth muscle cell proliferation in vitro, pp. 164–177. *In* I. Emerit and B. Chance (eds.). *Free Radicals and Aging.* Birkhäuser Verlag, Basel, Switzerland.

159. Boscoboinik, D., N. K. Özer, U. Moser, and A. Azzi. 1995. Tocopherols and 6-hydroxy-chroman-2-carbonitrile derivatives inhibit vascular smooth muscle cell proliferation by a nonantioxidant mechanism. *Arch. Biochem. Biophys.* **318**:241–246.

160. Boscoboinik, D., A. Szewczyk, and A. Azzi. 1991. Alpha-tocopherol (vitamin E) regulates vascular smooth muscle cell proliferation and protein kinase C activity. *Arch. Biochem. Biophys.* **286**:264–269.

161. Boscoboinik, D., A. Szewczyk, C. Hensey, and A. Azzi. 1991. Inhibition of cell proliferation by alpha-tocopherol. Role of protein kinase C. *J. Biol. Chem.* **266**:6188–6194.

15

The Role of Heme Oxygenase-1 in the Mammalian Stress Response: Molecular Aspects of Regulation and Function

Stefan W. Ryter and Rex M. Tyrrell

15.1 Introduction

Heme oxygenase-1 (HO-1; EC 1:14.99.3), first described as a distinct enzymatic activity by Tenhunen et al.,[1] serves a vital physiologic function in regulating, by catabolism, the intracellular concentration of heme in mammalian tissues.[2] In particular, HO-1 metabolizes the hemoglobin heme released during the phagocytic destruction of senescent erythrocytes in the reticuloendothelial system.[3] In addition to this normal physiologic role, heme oxygenase-1 belongs to a broad classification of proteins termed "shock" or "stress" proteins, which constitute the mammalian "stress" response. The up-regulation of most "stress" proteins, including HO-1, results from transcriptional activation (*de novo* synthesis of RNA) in response to an extracellular stimulation, usually associated with cellular "injury."[4] The resulting increase in HO-1 protein synthesis responds to a wide spectrum of chemical and physical agents, which include the substrate heme, oxidants, photosensitizers, heavy metals, organic solvents, sulfhydryl reagents, hormones, cytokines, and bacterial endotoxins as well as physical stresses such as heat, hypoxia, and ultraviolet A (UVA, 320 to 380 nm) radiation.[5-12] The large variety of agents that are able to modulate HO-1 expression complicates an important question in toxicology: What is the function of an increased cellular capacity for heme degradation following stress? A potential antioxidant function for HO-1 was revealed following the discovery that the bile pigments biliverdin and bilirubin, the end products of heme metabolism, have *in vitro* and serum antioxidant properties.[13] HO-1 also provides a mechanism to reduce the concentration of cellular "free" heme, a potential prooxidant, while facilitating the transfer of the heme-iron to potentially less catalytic forms.[14] The induction of ferritin, which sequesters intracellular chelatable iron, may be a late step in such a pathway.[15-17] Finally, carbon monoxide, an inevitable catabolite of heme oxygenase reactions, has been implicated as a neural second-messenger molecule.[18] The effects of CO on signal transduction and gene expression are poorly understood. The HO-1 system provides a powerful model for gene regulation studies designed to answer the following question: By what mechanism(s) do diverse extracellular stimulants trigger a common cellular response that leads to the transcription of a specific gene?

This chapter focuses on HO-1 as an inducible stress protein in mammalian cells with an emphasis on its potential function in cellular defense mechanisms. The enzymology and physiology of heme oxygenase and various aspects of heme metabolism have been reviewed elsewhere,[2,5-9] but are considered here in the context of understanding the role of HO-1 in the stress response. Recent advances in the molecular regulation of HO-1 gene expression as a function of stress is also described.

15.2 Mammalian Stress Proteins

When mammalian cells are subjected to stress, they respond by alterations in cellular protein synthesis, characterized by the transient induction of distinct protein species above basal levels, against an overall attenuation of cellular protein and mRNA synthesis. The stress condition may take the form of chemical agents (xenobiotics, drugs, transition metals), physical agents (heat, cold, ultraviolet and ionizing radiation), and nutrient deprivation (glucose starvation, hypoxia).[4,19,20] A 32 to 34 kDa mammalian stress protein was initially characterized by its induction on sodium dodecylsulfate–polyacrylamide gel electrophoresis (SDS-PAGE) gels after sodium arsenite, UVA radiation, sulfhydryl reagents, hydrogen peroxide, tumor promoters, and heavy metal stress.[21-23] The identity of the protein was later established as the enzyme, HO-1.[12,24,25]

There is a strong degree of overlap between the specific stress protein responses. Table 15.1 illustrates that certain known inducers of HO-1 may also induce other proteins associated with cellular stress, including the heat shock proteins (HSPs), the glucose-regulated proteins (GRPs), metallothionein (MT), and the growth arrest and DNA damage inducible proteins (GADD).

15.2.1 Overlap of the HO-1 Response with HSPs and GRPs

The classical example of a stress protein response is the induction of the heat shock protein family (HSPs) following cellular hyperthermia treatment (42 to 45°C) or cellular stress from amino acid analogs, ethanol, heavy metals, and sulfhydryl reagents. The mammalian HSPs consist of proteins belonging to the following molecular weight classes: M_r: 20 to 30 kDa, 70 to 73 kDa, 90 kDa, and 104 to 110 kDa.[4,19] The rat heme oxygenase-1 protein (HO-1) has been classified as a heat shock protein (HSP-32) since it is induced in response to thermal stress (42°C), and the 5′ regulatory region of its gene contains consensus

Table 15.1.   Overlap of Mammalian Stress Protein Responses to Various Inducing Agents.

	HO-1	HSPs	GRPs	MT	GADD	References
Heat	+ᵃ/−	+	−			10, 21, 26–29
Heme	+			+ᵇ		30–35
Hypoxia	+	−	+	+	+ᶜ	11, 36–38
Heavy metals	+	+	+	+		21, 23, 27, 30, 39–45
Sodium arsenite	+	+		+		12, 23, 43, 46–49
UVA (320 to 380 nm)	+	+ᵈ				12, 22
UVB/C (< 320 nm)	+/−ᵉ	−	−	+	+	48, 50, 51
H₂O₂	+	+/−ᶠ		+	+	22, 48, 52–55
TPA	+	−	−	+	+/−ᵍ	21, 50, 51, 56
Cytokines	+ʰ			+		57–65
Photosensitizers	+	+	+		+	66–68
Aminoacid analogs		+	+			45
Ca ionophores	−	+	+		+ᶜ	38, 45, 69, 70
Glucose starvation	−	−	+		+ᶜ	38, 70
Alkylating agents	+			+	+	21, 50, 51

Table notation: (+) or (−) designates that references have been cited that refer to the general induction or lack of induction of the protein family by a specific agent. HO-1, heme oxygenase-1; HSPs, heat shock proteins; GRPs, glucose regulated proteins; MT, metallothionein; GADD, growth arrest and DNA damage inducible proteins; UVA, ultraviolet A radiation (320 to 380 nm); UVB-UVC, ultraviolet radiation (< 320 nm); TPA, 12-O-tetradecanoyl-phorbol-13-acetate.

ᵃ In rodent systems only.

ᵇ Preferentially as a heme-hemopexin complex.

ᶜ gadd 145, 153.

ᵈ 102-kDa HSP only (R. M. Tyrell, 1988).

ᵉ HO-1 weakly induced by UVB and UVC in human fibroblasts.

ᶠ Conflicting reports on HSP translation induced by H₂O₂.

ᵍ gadd 153 only, weakly TPA responsive.

ʰ Includes interleukins IL-1, IL-6, IL-11; transforming growth factor B (TGF-B); tumor necrosis factor alpha (TNF-α); and bacterial endotoxin.

heat shock elements (HSEs) similar to those originally discovered in the promoter regions of other heat shock genes (i.e., HSP70).[10,26,71,72] The molecular aspects of the transcriptional regulation of the rat HO-1 gene by heat are discussed in Section 15.7.2.2. HO-1 mRNA and protein accumulate to a high degree after whole-body hyperthermia (42°C) in various rat organs, including the liver, heart, and kidney[73,74] and in the glioma and neural cells of the brain.[72,75] The increase in the rate of transcription of the HO-1 gene by heat shock has been demonstrated in cultured rat glioma cells, reaching a maximum at 3 h following a 42°C incubation.[26] Heat shock weakly induces HO-1 in several murine cell lines (Balb/c 3T3, Hepa).[21] Heat is ineffective as an inducer of HO-1 in most human cell lines studied (HeLa, glioma, macrophages, HL-60, HepG2 hepatoma),[24,31] and thus cannot be classified as an HSP in humans, with the exception of Hep3B hepatoma cells,[28] and erythroleukemic YN-1-O-A.[76] The induction of human HO-1 may, however, overlap the expression of the other HSPs in response to stress from heavy metals and sodium arsenite.[42,43,47]

A distinct stress protein family, GRPs (M_r: 75, 78 to 80, and 94 to 100 kDa) share sequence homology with the HSPs, but do not share homology with HO-1, and do not respond to hyperthermia.[4] Expression of the GRPs occurs in response to glucose starvation, hypoxia, sulfhydryl reductants, perturbation of calcium homeostasis, inhibition of protein glycosylation, and expression of misfolded proteins.[70,77-79] The GRPs and HSPs are coexpressed in human cells in response to heavy metals or amino acid analogs.[45] The expression of GRPs has been demonstrated to overlap the expression of HO-1 following cellular hypoxic stress.[11,36]

15.2.2 Overlap of HO-1 Response with M Induction

Metallothioneins (MTs) are proteins that are expressed following heavy metal and oxidant stress. MT contains multiple free cysteine thiol groups available for metal chelation, and is thought to represent a cellular mechanism for the detoxification of metals.[80] Heavy metals can potentially affect the regulation of both HO-1 and MT. For example, cadmium and zinc induce both HO-1 and MT mRNA levels in HeLa cells.[44] Sodium arsenite or diethylmaleate, oxidative stress from H_2O_2 and synthetic metalloporphyrins, are other treatment conditions that may induce both MT and HO-1.[23,42,49,54,82] Metallothionein is strongly induced by ultraviolet radiation in the UVC range.[50] In contrast, HO-1 is preferentially induced by fluences in the UVA range (320 to 380 nm), and only weakly induced by fluences in the UVB or UVC ranges (<320 nm).[48]

15.2.3 Overlap of HO-1 Response with DNA Damage Inducible Responses

DNA damaging agents induce a large number of genes.[51] Among these, the growth arrest and DNA-damage inducible genes (*gadd 153, 45, 34, 33, 7*) are defined by their induction in response to agents that chemically modify DNA bases such as the alkylating agent methyl methane sulfonate (MMS), UVC radiation, or cytostatic agents such as H_2O_2 and hypoxia treatment.[51] Thus the HO-1 response may partially overlap the GADD response, in particular when the HO-1 inducing agent has a potential for producing DNA damage. For example, the alkylating agent MMS has been shown to induce both HO-1[21] and *gadd* (*153/45*).[83] With respect to UV radiation, fluences in the UVA range, which are associated with oxidative membrane damage produce a strong induction of HO-1 in mammalian cells, whereas fluences in the UVB/UVC range, which produce DNA photodamage, induce GADD proteins[51] but have relatively little effect on HO-1.[48,84] There is apparently no link between DNA damage and HO-1 induction by UV. Finally, hypoxic stress produces a strong overlap between HO-1, the GADD proteins 153 and 45, and GRP78, but has little effect on HSP induction[11,36,38] (see section 15.5.2).

15.2.4 Stress Proteins and Cellular Tolerance

The correlation between the expression of HSPs and the development of thermotolerance (cellular adaptation to heat stress),[85,86] or cross resistance to chemotherapeutic agents,[87] has been extensively studied. Following hyperthermia, HSPs bind thermally denatured protein in the cytosol, mitochondria, and nucleus, acting effectively as a detergent to prevent the aggregation of insoluble protein.[88] Likewise, the GRPs are thought to solubilize denatured, misfolded, or improperly glycosylated proteins in the endoplasmic reticulum (ER), preventing their secretion or aggregation, thus serving as a general protective mechanism following the disruption of ER protein-processing functions.[88] Expression of the GRPs has been implicated in the development of transient adaptation to subsequent chemical stress from oxidants and chemotherapeutic agents such as adriamycin.[89,90] While HO-1 has no known protein binding activity, its induction could represent a cellular protective response following the denaturation of proteins that contain heme as a prosthetic group (see Section 15.7.2.2). Relatively little evidence exists which directly links HO-1 expression with transient adaptation to the chemical agents that induce it. However, Chinese hamster ovary (CHO) cells exposed to CdCl2 a powerful HO-1 inducer, subsequently do become more resistant to cytotoxicity induced by H_2O_2.[91] This phenomenon was ascribed to MT induction. HO-1 expression has been correlated to sodium arsenite tolerance in sodium arsenite resistant human cell lines.[92] HO-1 expression has been implicated in protection against UVA induced membrane damage.[16] Recently, overexpression of HO-1 in rabbit endothelial cells was correlated with protection against cell lysis by exogenous heme treatment.[93]

15.3 Enzymology of Heme Catabolism

HO-1, the rate limiting enzyme in heme degradation, occurs in the microsomal fraction of all mammalian tissues tested.[7] Human HO-1 displays a molecular mass of 32 kDa on SDS-PAGE gels, and has been purified to homogeneity from the microsomes of human liver.[94] HO-1 (32 to 34 kDa) has also been purified from rat liver, porcine and bovine spleen, and unicellular red algae.[95-98] Expression of HO-1 cDNA in bacteria and monkey kidney cells has revealed that the expressed protein spontaneously associates with cellular membranes, a property that was assigned to the hydrophobic domain in the C terminus of the HO-1 protein.[99,100]

A distinct isozyme of heme oxygenase, heme oxygenase-2 (HO-2), has been isolated from rat liver, spleen, brain, and testes.[101-104] HO-1 and HO-2 proteins differ in molecular weight, amino acid composition, thermostability, immunoreactivity, and reaction kinetics.[7] Furthermore, HO-2 is normally refractory to induction by chemicals that induce HO-1 activity.[101] HO-1 and HO-2 share less than 50% amino acid homology, yet they share a conserved domain of about 24 amino acid residues, which is thought to represent a conserved catalytic heme-binding domain.[105] HO-1 and HO-2 represent products of two distinct genes,[106,107] which are located in separate chromatin regions (HO-1: 22q12; HO-2: 16p13.3).[108,109]

The heme oxygenase reaction cycle with its proposed intermediates has been extensively reviewed. For every mole of heme oxidized, HO-1 or HO-2 activity consumes 3 mol of molecular oxygen, to produce 1 mol each of biliverdin IXα,, carbon monoxide (CO), and iron (Fe-II).[1,110,111] Additionally, NADPH and NADPH:cytochrome P-450 reductase are required for HO activity.[112] Recently, spectroscopic analysis of the structure of the heme-heme oxygenase complex has suggested that the reactive oxidizing intermediate in the reaction center of HO-1 is likely a "peroxo" form of heme-iron rather than the oxo-ferryl species proposed for cytochrome P-450 catalysis.[113–115] The prefered substrate for HO activity is heme b (Fe-protoporphyrin IX). The K_m values for hematin are 0.24 and 0.40 μM for HO-1 and HO-2, respectively.[2] The substrate specificity of HO enzymes toward other forms of heme and synthetic metalloporphyrins has been reviewed elsewhere.[2] Despite the similarities, which originally implied cytochrome P-450 as the terminal oxidase in heme metabolism,[3,110] HO proteins are not members of this family.[1,9,116–118] In mammalian tissues, heme metabolism is completed by cytosolic NAD(P)H biliverdin reductase (EC 1.3.1.24), which divalently reduces biliverdin to produce bilirubin IXα.[80] Biliverdin reductase activity has been detected in various rat and bovine tissues[2] including liver, kidney, brain, and testes.

15.4 Regulation of Heme Oxygenase by Its Substrate, Heme

Heme oxygenase serves a vital physiological function by regulating the intracellular concentrations of its substrate heme, in all mammalian tissues. Heme (Fe-protoporphyrin IX) induces heme oxygenase activity, the rate-limiting step in its catabolism, exemplifying substrate mediated enzyme induction that depends on *de novo* mRNA and protein synthesis.[6,30,31,119,120] Heme negatively regulates the synthesis of δ-aminolevulinic acid synthase (ALAS), the rate limiting step in its biosynthetic pathway.[6,121] The effects of heme loading on heme metabolic activities has been extensively studied *in vivo*.[2] Injections of free heme into rats produces reciprocal oscillations of hepatic HO and ALAS activity.[121] These effects can be mimicked by other treatments that effectively increase plasma or hepatic heme concentrations.[3,122–124] The regulation of HO by heme has been observed in many cultured cell lines, including mouse hepatoma, rat glioma, human glioma, and pig alveolar macrophages.[26,30,32,119] The transcriptional regulation of the HO-1 gene by heme has been demonstrated in mouse hepatoma cells, with an apparent maximum rate of 2 h post-treatment[30] and in primary chick embryo hepatocytes, with an apparent maximum at 5 h post-treatment.[125] In the latter system, the induction of HO-1 by heme was associated with an acquired refractoriness to a subsequent induction of the gene by heme (administered at the peak of transcriptional activity).[125] The mechanism for the aquired refractoriness is unknown.

The potent stimulatory effects of heme on HO activity led to theories that a flux in intracellular free heme might serve as a regulatory signal with respect to the induction of HO by nonheme xenobiotics.[1,126] Certain hormones, chemicals, and endocrine factors mimic the effects of heme on HO and ALAS activities, and cause transient increases in free hepatic heme pools in parallel with these phenomenon. Such agents include carbon disulfide, bacterial endotoxin, insulin, glucagon, epinephrine, cyclic AMP, and interferon-inducing agents, as well as the physiological states of starvation and hypoglycemia.[62,126–129] The transient increases in intracellular heme were postulated to arise from impaired hemoprotein assembly or increased hemoprotein turnover.[127–129] In these cases, the free heme pools were estimated by the heme saturation of tryptophane 2,3 dioxygenase, a hemoprotein with variable heme content. The reciprocal regulation of HO and cytochrome P-450 levels in the liver in response to endotoxin shock led to the postulate that increased cytochrome P-450 turnover could be a source of the increase in free heme levels, which in turn would stimulate HO expression.[126] This theory was questioned by Kikuchi and Yoshida[5] since they observed that tissue cytochrome P-450 content did not correlate with the response, and that the kinetics of free heme accumulation were not consistent with a cause and effect relationship with HO-expression. Finally, many other agents that potently induce hepatic HO-1 do not cause a measurable increase in free hepatic heme (i.e., bromobenzene, heavy metals, and sulfhydryl reagents)[5] or cytochrome P-450 degradation.[117] Thus, while heme potently regulates HO expression, it is not a universal intermediate in the regulation of HO by all inducing agents.

15.5 Induction of Heme Oxygenase by Chemical and Physical Stress

The multiple exogenous chemical and physical agents which induce HO-1 can be grouped into four broad categories: (1) agents that increase the prooxidant status of the cell by directly or indirectly generating reactive oxygen intermediates (ROI), (2) agents that directly complex intracellular glutathione (GSH), (3) endogenously produced autocrine substances such as cytokines and agents that may modulate their production such as bacterial endotoxins, and (4) cellular stress produced by hyperthermia (in rodent systems). The molecular aspects of HO-1 regulation by these agents will be discussed in Section 15.7.

15.5.1 Oxidative Stress

HO-1 induction has been identified as a general inducible response to cellular "oxidative" stress generated by many systems, including UVA radiation, photo-sensitizers, hydrogen peroxide (H_2O_2), redox-cycling compounds, aqueous-phase cigarette smoke, and endogenous oxidant production during erythrophago-cytosis.[22,66,130–132]

15.5.1.1 Ultraviolet A (320 to 380 nm) Radiation and Hydrogen Peroxide

UVA radiation (320 to 380 nm) has been shown to generate an oxidative stress to cultured cells.[133,134] Both irradiation in the UVA range and treatment with hydrogen peroxide sharply increase the rate of transcription of the HO-1 gene, and consequently the steady-state levels of the corresponding mRNA in human skin fibroblasts.[48,135] The induction of HO-1 by UVA and H_2O_2 can be inhibited by iron chelators such as desferrioxamine, and o-phenanthroline, suggesting a role for iron-catalyzed reactions in the manifestation of the response to these agents.[52] The induction of HO-1 by UVA may involve singlet molecular oxygen, since it can be potentiated in deuterium oxide and inhibited by compounds that quench singlet oxygen (sodium azide).[136] Scavengers of hydroxyl radical, however, have no effect on the UVA activation of the HO-1 gene. Singlet molecular oxygen has recently been detected by infrared spectroscopy as a reaction product of an *in vitro* Haber Weiss reaction between superoxide (O_2^-) and H_2O_2.[137] While the ultimate reactive oxidizing agents produced by H_2O_2 and UVA are still a matter of controversy, it remains possible that H_2O_2 and UVA may act via similar reactive oxygen intermediates, and share a common mechanism leading to HO-1 induction.

The depletion of cellular glutathione by buthionine (S,R) sulfoximine (BSO), which inhibits γ-glutamyl-cysteinyl-synthetase (γ-GCS), the rate limiting step in the de novo synthesis of glutathione,[138] strongly sensitizes human fibroblasts to the toxic effects of UVA and UVB (290 to 320 nm) radiation, and hydrogen peroxide treatment.[139,140] While BSO treatment alone moderately increased HO-1 mRNA accumulation in human skin fibroblasts, it strongly potentiated the inducing effects of H_2O_2 and UVA, and lowered the UVA fluence threshold necessary to induce HO-1 mRNA steady state levels.[141] High fluences of UVA alone (and to a lesser extent, H_2O_2 alone) produce a significant depletion of intracellular GSH.[142] However, the induction of HO-1 mRNA by these agents is stronger than that expected from GSH depletion alone. These observations indicate that the depletion of GSH levels in combination with an increase in oxidative stress to the cell is more potent with respect to HO-1 induction, than the inhibition of GSH synthesis alone.

Recently, it has been shown that UVA pretreatment of human skin fibroblasts produces a refractoriness to HO-1 induction by a second UVA dose given 15 h later, which lasts up to 48 h following the initial irradiation.[143]

15.5.1.2 Photooxidative Stress

Photosensitizers are compounds that generate reactive oxygen intermediates as a consequence of absorption of light at wavelengths defined by the optical absorption spectrum of the chemical. Photosensitizer-mediated oxidative stress is thought to involve singlet molecular oxygen generated by an energy transfer reaction between an excited photosensitizer molecule and molecular oxygen (type II photochemical mechanism).[68] The photosensitizer photofrin II is a mixture of ether and/or ester linked dimers of hematoporphyrin, a partially purified form of hematoporphyrin derivative (HpD), and is currently in phase III clinical trials for the photodynamic therapy (PDT) of cancer.[144] Long-term incubation (16 h) of Chinese hamster fibroblasts with photofrin II, followed by irradiation with visible red light, induced HO-1 mRNA, and protein accumulation.[66] Dark incubation with photofrin II also produced a lesser but significant induction. Similar results were observed as a result of the treatment of human skin fibroblasts with HpD and visible red light.[67] Photosensitization with the xanthene dye rose bengal (but not incubation with rose bengal under dark conditions), also increased HO-1 mRNA and protein levels. Since irradiation of rose bengal produces singlet oxygen, these experiments imply that oxidative stress generated from structurally dissimilar (nonporphyrin) photosensitizers, is also capable of eliciting the response.[66] Zn-phthalocyanine (Zn-PC), which is a poor singlet oxygen generator, produced a strong induction of HO-1 mRNA accumulation following dark incubation, which was not further increased by light treatment.[67] These results, taken together, suggest that oxidant production by photosensitizers is capable of inducing HO-1. However, photosensitizer dark incubation may contribute a major portion of the response when the photosensitizer bears a metalloporphyrinlike structure, and this probably involves a distinct mechanism of induction.

15.1.3 Redox Active Quinones

Quinones such as menadione (1,4 naphthoquinone) produce superoxide anion radical and hydrogen peroxide as a consequence of intracellular redox cycling. Menadione may also diminish cellular GSH levels directly by complexation, and indirectly, through the generation of reactive oxygen intermediates.[145] The induction of HO-1 by menadione treatment has been observed in human skin fibroblasts and human breast cancer cells.[12,130] In the later case, no correlation was established between HO-1 induction and menadione-induced oxidative DNA strand breakage.[130] Currently, few studies exist on menadione induction of HO-1, however, quinonoid compounds, which differ in redox cycling and GSH conjugating properties, may be useful tools for studying the relationships between oxidative stress, modulation of glutathione levels, and induction of HO-1.

15.5.1.4 Cigarette Smoke

Cigarette smoke fractions (CS) has been shown to represent an oxidative stress to cultured cells, and can cause lipid peroxidation, protein thiol oxidation, and DNA single-strand breakage. Aqueous-phase cigarette smoke causes transient induction of HO-1 mRNA transcription and protein levels in Swiss 3T3 cells.[131] In contrast to observations with UVA radiation, inhibitors of the Fenton reaction, such as the metal chelator o-phenanthroline or the peroxide degrading enzyme catalase failed to block HO-1 induction by CS, whereas they protected against DNA strand breakage caused by CS. The reactive oxidants that are present in cigarette smoke may therefore be capable of eliciting the HO-1 response independently of endogenous iron. The HO-1 induction by CS could be blocked by cysteine, and was accompanied by a depletion of GSH levels.

15.5.1.5 Erythrophagocytosis

Macrophages or monocytes engaged in erythrophagocytosis display high levels of HO expression above those levels normally present in resting macrophages.[132] This induction is thought to represent a response to substrate loading from high levels of erythrocyte hemoglobin. A portion of the HO-1 induction under these conditions has been postulated to occur as a consequence of oxygen radical production during the macrophage respiratory burst. Flavonoid antioxidants inhibit the synthesis of HO-1 during erythrophagocytosis, but do not inhibit the process of erythrophagocytosis itself.[146] Protein kinase C (PKC) inhibitors, however, do not inhibit HO-1 synthesis under these conditions, indicating that the effect of the flavonoids was due to antioxidant properties rather than their PKC inhibition properties. Furthermore the thiolprotector WR-1065 also inhibits HO-1 synthesis during erythrophagocytosis,[147] These studies, taken together, suggest that HO-1, which is inducible by exogenous oxidative stress, may also represent a response to increased endogenous production of reactive oxygen intermediates from activated macrophages, and that GSH depletion may be involved in this process.

15.5.2 Hypoxia

Oxygen deprivation or hypoxia induces a set of stress proteins in CHO cells termed oxygen regulated proteins (ORPs, M_r: 260, 150, 100, 80, 33).[36] Three of these proteins are thought to be identical to the mammalian stress proteins, GRP-94, GRP-78, and the 32-kDa HO-1 (by immunoblot analysis).[11] Hypoxia has been shown to induce the expression of other genes involved in drug detoxification, including NADPH:quinone oxidoreductase and γ-glutamyl-cysteinyl-synthetase, as well as two of the growth arrest and DNA damage inducible (GADD) genes (*gadd45* and *gadd 153*).[38,148]

Hypoxia-mediated induction of HO-1 shares common features with HO-1 induction by oxidative stress, in that it is accompanied by a depletion of cellular GSH content. These observations taken together implicate that redox-sensitive mechanisms could be involved in the induction of HO-1 and other genes by hypoxia.[11] The induction of stress proteins, including HO-1, by hypoxia may be of significance in tumor tissue, which contains hypoxic domains that have been implicated in the development of tumor resistance to anticancer treatment modalities.[11,149]

15.5.3 Induction of MO-1 by Heavy metals and related compounds

15.5.3.1 Heavy Metals

Transition metals are able to modulate various aspects of heme metabolism (For reviews see Refs. 2, 6, 39, and 150). An increase in HO enzymatic activity by inorganic heavy metals in cultured cells was first demonstrated by $CoCl_2$ treatment of chicken embryo hepatocytes.[151] The increased synthesis of a 32-kDa stress protein (later shown to be HO-1), as a consequence of treatment with inorganic heavy metals has been observed in many cell culture systems. These include the $CdCl_2$ treatment of Hela, and HL60 cells, mouse peritoneal macrophages, rat hepatoma,[24,27,46] CHO cells,[152] human and murine melanoma cells,[23] as well as the treatment of mouse embryo fibroblasts with $CdCl_2$, $CuSO_4$, or K_2CrO_4.[21] An increase in HO-1 mRNA steady-state levels has been documented following the $CdCl_2$ treatment of human skin fibroblasts,[12] mouse hepatoma cells,[30] and human hepatoma cell lines (HepG2 Hep3B).[28,42] Nuclear run-off analyses have confirmed that the induction of HO-1 *de novo* protein synthesis and mRNA levels by heavy metals is regulated at the transcriptional level. An increase in the HO-1 transcription rate occurs in mouse hepatoma cells following $CdCl_2$ treatment, and in human hepatocytes following $SnCl_2$ treatment, as early as 1 h following stimulation.[30,42] The transcriptional regulation of HO-1 by $CoCl_2$ has also been demonstrated in rat liver and kidney, with similar kinetics.[153,154]

Heavy metals form complexes with thiol compounds such as cysteine and GSH; and exert a biphasic effect on reduced glutathione levels, manifested as an initial decrease followed by a rebound increase to elevated levels. The complexation of metals with cysteine or GSH blocks the hepatic HO induction produced in rats by the metals alone.[39] The prior administration of diethylmaleate potentiates HO induction by metals *in vivo* and in cultured rat hepatocytes.[39] Furthermore, the induction of rat hepatic HO activity by $CoCl_2$ or Co-protoporphyrin-IX (Co-PPIX) is accompanied by a depletion of hepatic GSH levels.[155] The *in vivo* and *in vitro* thiol-binding properties of heavy metals suggest that the mechanism for the heavy metal induction of HO-1 may share a common pathway with other thiol-reactive substances.

15.5.3.2 Metalloporphyrins

The regulatory effects of certain metals on heme metabolism can be modulated by chelation with protoporphyrin IX (PPIX). For example, Co-PP, Sn-PP, and Zn-PP competitively inhibit HO activity in reconstituted microsomal systems.[150] Co-PP, however, potently induces *in vivo* (hepatic HO enzyme activity), whereas Sn-PP and Zn-PP are inhibitors of *in vivo* enzyme activity.[150,156] Co-PP and Zn-PP (but not Sn-PP) induce HO-1 mRNA transcription in hepatoma cells.[35,42] Co-PP induces HO-1 mRNA steady state levels in rat liver and kidney.[153] While inorganic cobalt (CoCl$_2$) is a potent inducer of HO-1 *in vivo* (rat liver) it fails to induce HO-1 in cultured human hepatocytes, unless administered as Co-PP, or in combination with PPIX or δ-aminolevulinic acid. These observations indicate that hepatic chelation of Co by PPIX may be a prerequisite for the effects of CoCl$_2$ on HO-1 regulation.[42,155]

15.5.3.3 Proposed Mechanisms of HO-1 Activation by Metals

In conclusion, heavy metals may induce HO-1 transcription by several different mechanisms:

1. When the metal serves as a ferrochelatase substrate (Fe, Cu, Co, Zn, Cd) disposing it to chelation by PPIX, the mechanism of HO-1 induction may share a common pathway with that of other metalloporphyrins.[42] The difference in induction kinetics between inorganic metals and corresponding metalloporphyrins argues that two different induction mechanisms may be involved.

2. As a consequence of forming thiol complexes with cellular GSH, transiently lowering cellular sulfhydryl buffering capacity and thus altering cellular redox potential, heavy metals may share a similar mechanism of HO-1 induction with xenobiotics which conjugate GSH or oxidants that deplete the compound.

3. The possibility that metals directly modulate HO-1 gene transcription by forming thiol complexes with nuclear transcription factors cannot be ruled out. However, the mechanism of HO-1 induction by metals is now known to be distinct from that of the metallothionein gene.[157]

4. Finally, DNA sequence elements potentially recognized by AP-1-like factors and CAAT-enhancer binding factors (C/EBP), have recently been implicated in the regulation of HO-1 by CdCl$_2$[158] (see Section 15.7.3).

15.5.4 Thiol Reactive Substances

The induction of HO-1 by thiol reactive substances, including sodium arsenite, diethylmaleate, iodoacetamide, and menadione, has been observed in many cell culture systems. The conjugation of cellular glutathione (nonprotein thiols) by such agents has been correlated with the induction of the 32-kDa HO-1 protein and may be an obligatory step in the induction pathway. The increased synthesis of a 32-kDa protein by sodium arsenite has been documented in mouse peritoneal macrophages, mouse and avian embryo fibroblasts, mouse and human melanoma cells, and cultured human skin fibroblasts.[21-23,46,47] The 32-kDa protein induced by sodium arsenite in human fibroblasts was first directly demonstrated to be HO-1 by molecular cloning and the sequence analysis of its corresponding cDNA.[12] The identity of the protein was also confirmed as HO-1 by Western immunoblot analysis, and chemical and enzymatic proteolytic digestion.[12,24] Sodium arsenite also increases steady state HO-1 mRNA levels in several human cell lines including human skin fibroblasts and lymphoblastoid cell lines, but does not significantly change the levels in primary human keratinocyte lines, which contain high levels of constitutive HO-2 activity.[159] However, the response was also observed in many other cultured placental and nonplacental mammalian cell lines, including fibroblasts from the monkey, mouse, hamster, and opposum.[48]

Diethylmaleate (DEM) is an electrophillic compound which conjugates glutathione in a glutathione S-transferase (GST)-catalyzed reaction. Depletion of cellular glutathione (GSH) content occurs as a result of efflux of the mixed disulfide (GSSR) from the cell. Induction of a 32-kDa protein as a consequence of glutathione depletion (>80%) by DEM was observed in fetal rat fibroblasts (SDS-PAGE analysis) and CHO cells (2-D protein analysis, Western immunoblot analysis).[152,160,161]

The sulfhydryl oxidant diamide, which oxidizes GSH to GSSG, induces high-molecular-weight HSPs, but does not induce HO-1 in systems where HO-1 is not an HSP.[152] Intraperitoneal injection of diamide was reported to induce HO activity in rat liver.[155] An important difference between diamide and DEM is that oxidized glutathione (GSSG) produced by diamide may be reduced by NADPH:glutathione reductase, whereas mixed disulfides (GSSR) produced by DEM, may not be enzymatically reduced, and are excreted from the cell. The thiol reactive substance *N*-ethylmaleimide has little effect on HO-1 induction, presumably because of a preferential reactivity for protein thiols rather than glutathione.[152]

Glutathione depletion by BSO treatment weakly induced HO-1 protein in CHO cells and in mouse peritoneal macrophages, when assayed by two-dimensional electrophoresis and immunoblot analysis, respectively, and has been reported ineffective at inducing 32-kDa protein synthesis when assayed on one-dimensional SDS-PAGE gels.[160,161] Thus the evidence in these rodent cell systems is more consistent with the model that the conjugation of cellular GSH, rather than its depletion, represents a critical regulatory event in the induction of HO-1 by thiol reactive substances. However, GSH depletion by BSO leads to a significant (fivefold) increase in basal levels of HO-1 mRNA in cultured human fibroblasts,[141] so that depletion rather than conjugation appears to be involved in this case.

The list of thiol reactive substances which induce HO-1 includes those xenobiotic agents which undergo microsomal biotransformation by cytochrome (P-450/ P-448) catalyzed monooxygenation reactions, to form electrophillic intermediates that may complex glutathione directly, or as a consequence of GST-catalyzed reactions.[2,6] Such xenobiotics include the halogenated hydrocarbons, polychlorinated biphenyls, and carbon disulfide (CS_2).[129] The cytostatic arachidonic acid metabolite Δ12-prostaglandin J2, which induces HO in porcine aortic endothelial cells, also forms a glutathione conjugate *in vitro* and *in vivo*.[162] The aldehyde 4-hydroxynonenal (4-HNE), which is formed as a by-product of membrane lipid peroxidation, also reacts with GSH in a GST catalyzed reaction.[163] Thus 4-HNE production may be a link between GSH depletion and membrane damaging agents. 4-HNE induces a 32-kDa protein in rat hepatoma cells[164] and human fibroblasts.[165]

15.5.4.1 Redox Regulation of Gene Expression

The mechanisms by which alterations in cellular free thiol status (conjugation or oxidation) produce, or are associated with, the transcriptional induction of HO-1 remain unknown. However, it has been predicted that certain components of signal transduction pathways regulating HO-1 gene transcription may be "redox" sensitive.[141] Potential molecular targets for redox regulation include protein kinases, phosphoprotein phosphatases, or nuclear transcription factors. As an example of such a mechanism, it has been shown that the inhibition of a redox-sensitive protein phosphatase is the mechanism for the increase in HSP27 phosphorylation in human fibroblasts by the thiol reactive substances, H_2O_2, interleukin-1, tumor necrosis factor alpha (TNF-α), all of which also induce HO-1.[166] A peroxide and UVA inducible MAP kinase specific protein tyrosine phosphatase (CL100) has recently been cloned from human skin cells.[167] Hydrogen peroxide has been shown to induce the membrane translocation and activity of protein kinase C, which has been postulated to contain redox-sensitive sulfhydryl groups.[168] Thus the inhibition or induction of protein phosphatases and/or kinases may influence the pathways by which oxidants alter the phosphorylation and activation state of critical factors involved in HO-1 regulation. Redox-sensitive mechanisms have been implicated in the regulation of *in vitro* DNA binding activity of mammalian nuclear transcription factor families, including AP-1[169,170] and c-Rel,[171] although the relevance of these mechanisms *in vivo* is a matter of controversy.

15.5.5 Endocrine Factors and Bacterial Endotoxins

The hepatic acute phase response (APR) is a systemic response to injury or infection in the liver characterized by physiological changes that include the induced or repressed synthesis of a number of hepatic proteins.[172] The hepatic APR is strongly induced by bacterial endotoxins (lipopolysaccharide [LPS]), which stimulates cytokine production from monocytes and macrophages. The physiological changes characteristic of the APR are regulated by cytokines, including interleukin-6 (IL-6), interleukin-1 (IL-1), and TNF-α. HO-1 has been identified as a positively regulated APR protein in humans, rats, and mice. The intraperitoneal injection of bacterial endotoxin (LPS) induces HO activity in rat peritoneal macrophages, and in hepatic parenchyma and sinusoidal cells.[62]

The intraperitoneal injection of LPS, TNF-α, and IL-1 induces hepatic HO enzymatic activity and, in the latter case, HO-1 mRNA transcription in mice.[57] The induction of hepatic HO activity by IL-1 and LPS was blocked by dexamethasone, an illustration of the antagonistic effects of glucocorticoids on cytokine mediated responses.[58] Furthermore, the induction of hepatic HO-1 mRNA levels by LPS was potentiated by glutathione depletion and diminished by the antioxidant *N*-acetyl cysteine, suggesting an influence of cellular redox status in the induction mechanism.[63]

In human Hep3B hepatoma cells, IL-6 treatment slightly induces HO-1 and strongly induces haptoglobin mRNA levels.[114] Interleukin 11 (IL-11) produces similar effects in human HepG2 cells.[60] Transforming growth factor-B (TGF-B) induces HO-1 mRNA and protein levels in cultured human retinal epithelial cells.[61] In conclusion, the role of HO-1 as a stress protein is not limited to stress from exogenous chemical and physical agents, but may also be important during systemic stress caused by injury or infection.

15.5.6 Tumor Promoters

The tumor promoting mitogen 12-*O*-tetradecanoyl phorbol-13-acetate (TPA) has pleiotropic biological effects that include the induction of growth-related phenomenon, such as protein kinase C activation, stimulation of DNA synthesis and mitosis, and the activation of c-*fos* and c-*jun* transcription.[173] TPA and the structurally related compound phorbol-12,13-didecanoate induce 32-kDa protein synthesis[21,56] and HO-1 mRNA accumulation in murine embryo fibroblasts.[25] The diacylglycerol analog 1-oleoyl-2-acetyl-glycerol (OAG) mimics the effects of TPA, a finding consistent with a role for protein kinase C activation in HO-1 activation by such agents.[21] The increased synthesis of the 32-kDa protein was also observed in murine embryo fibroblasts treated with structurally dissimilar tumor promoters, including indole alkaloid and polyacetates.[174] A correlation was observed between the strength of the tumor promoter and its ability to induce the 32-kDa protein.[56,174] TPA treatment induced HO-1 mRNA levels in human (THP-1) and mouse (M1) myelomonocytic cell lines in parallel with their differentiation to macrophages by this agent.[175,176]

15.6 Molecular Genetics of the Heme Oxygenases

15.6.1 Heme Oxygenase Genomic Clones

The human, mouse and rat HO-1 genes share a similar structure in that they are divided into five exons and four introns.[71,157,177] The human gene is longer (14

Table 15.2. Comparison of Heme Oxygenase 1 (HO-1) cDNA Clones and Their Predicted Polypeptides.

Species	Source	Length	Protein	Homologies	Reference
Rat	Spleen	1.53 kb	289 aa, 33 kDa	80% human, 93% mouse	100
Human	Macrophage skin fibroblasts	1.47 kb	288 aa, 32.8 kDa	80% (rat)	31 12
Chicken	Liver	1.25 kb	296 aa, 33.5 kDa	62% (rat)	178
Mouse	Fibroblasts	1.51 kb	289 (281) aa, 32.9 (31.9) kDa	93% (rat), 80% (human)	25
Pig	Spleen	1.55 kb	288 aa, 33 kDa	91% human, 88% rat, 90% mouse, 78% chick	179

kb) than the rat and mouse genes (7 kb). Fluorescence *in situ* hybridization studies have localized the human HO-1 gene to chromosome 22q12.[108,109]

15.6.2 HO-1 cDNA Clones

HO-1 cDNA has been cloned from several species including human, rat, mouse, chicken, and pig (Table 15.2).

A 1.5-kb rat cDNA for HO-1 was originally cloned from a rat spleen expression library using antiheme oxygenase monoclonal antibodies.[100] A human HO-1 cDNA coding for a 32-kDa protein was later cloned from hemin treated human macrophages using the rat cDNA clone as a probe.[31] A cDNA encoding the major sodium arsenite inducible stress protein (p32) was cloned from a human fibroblast cDNA library by the mRNA hybrid selection technique and found to be identical to human HO-1.[12] Murine cDNA corresponding to the low-molecular-weight stress protein was also cloned from fibroblasts by differential screening on the basis of sodium arsenite inducibility.[25] The murine cDNA of 1.5 kb was identified as the murine homolog of rat HO-1 on the basis of high sequence homology with the rat clone (93%), as well as sodium arsenite and cadmium inducibility of its corresponding mRNA. A porcine HO-1 cDNA clone has also been isolated.[179]

15.7 Molecular Regulation of Heme Oxygenase Transcription

The induction of HO-1 protein synthesis by chemical and physical agents is regulated at the level of transcription for all inducing agents so far tested ($CdCl_2$, heme,[30] UVA radiation,[135] sodium arsenite,[135] CoPPIX,[42] cigarette smoke,[131] and hyperthermia[26]). The molecular mechanisms that regulate basal and inducible HO-1 transcription have been partially resolved by sequence analysis and functional studies of the rat,[71,180–182] mouse,[157,158,183,184] and human[177,185,186] HO-1 5′ flanking sequences.

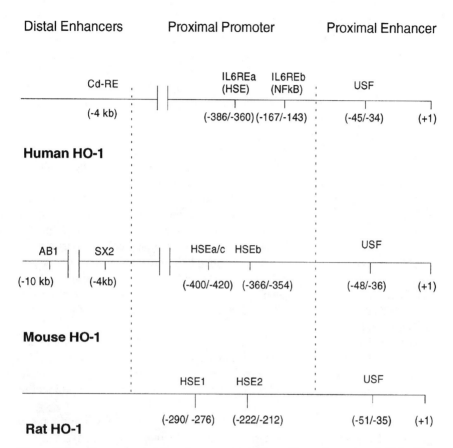

Figure 15.1. Comparison of the rat, mouse, and human Heme Oxygenase-1 5' flanking sequences. USF, upstream stimulatory factor binding site; HSE, heat shock element; IL6RE, Interleukin 6 responsive element, NFkB, Nuclear factor kB binding cite; Cd-RE, Cadmium responsive element; SX2, −4 kb enhancer region; AB1, −10 kb enhancer region.

For the purpose of this chapter, the HO-1 5' flanking sequence has been divided into three regions (Fig. 15.1) (the proximal enhancer region (0/−50 bp), the proximal promoter (−50 bp/−1 kb), and the distal enhancers (−4 kb, −10 kb regions). Initial studies of HO-1 regulation focused on sequence elements of the promoter, which are located within 1 kb upstream of the transcriptional start site. Recent advances have identified distal elements important for basal and inducible HO-1 expression.

15.7.1 Proximal Enhancer Region

A proximal promoter element involved in basal regulation of the rat HO-1 gene was identified upstream of the TATA box at position (−51/−35).[181] This sequence element (CCACCACGTGACTCGAG) has strong homology to the adenovirus 2 major late promoter (AD2MLP) and contains within it a core recognition sequence (CACGTGAC) for upstream stimulatory factor (USF). Another sequence element (TGACTCG), which overlaps the USF box, imperfectly resembles the AP-1 consensus sequence TGA(G/C)TCA, a binding site for the activator protein-1 (AP-1:Fos/Jun) family of transcription factors. The *in vivo* significance of this imperfect AP-1 site is not clear. The AD2MLP homologous element containing USF and AP-1-like sites also occurs in the proximal promoter regions of the mouse (−48/−36) and human (−45/−34) HO-1 genes.[157,185]

In the rat HO-1 promoter, the USF box binds a factor from rat glioma nuclear extracts termed heme oxygenase transcription factor-1 (HOTF-1) thought to represent a rat homolog of the USF.[181] Deletion analysis has shown that the (−51/−40) region of the rat HO-1 promoter is required for accurate *in vitro* transcription of the HO-1 gene, which is stimulated by purified HOTF/USF nuclear protein fraction. The (−45/−34) region of the human HO-1 promoter also constitutively binds a nuclear factor from a partially purified USF fraction that can be visualized by *in vitro* footprinting assays.[185] The USF was identified as a transcription factor that regulates the adenovirus major late promoter, however, this factor also exists in uninfected human cells, and participates in the basal regulation of acute phase response genes.[187]

The irradiation of human fibroblasts with ultraviolet A, under conditions that induce HO-1, altered the USF-DNA complex as evidenced by DNase-1 footprinting and by changes in the protein binding properties of a HO-1 proximal promoter sequence containing the CACGTGAC motif. UVA caused an increased mobility of protein complexed with USF oligonucleotides in gel mobility shift assays.[188] However, subsequent studies from this laboratory failed to show a difference in the methylation interference pattern generated by extracts from irradiated or non-irradiated cells. Furthermore, there is evidence that the changes in the mobility shift pattern may have resulted from a release of specific proteases by UVA radiation, resulting in partial proteolysis of the USF complex.[189] Since such a process has not been shown to occur *in vivo*, we believe it is doubtful that modulation of the USF complex is involved in UVA inducible HO-1 transcription.

15.7.1.1 Putative Myc:Max Recognition Sequence

The (−48/−36) region element within the murine HO-1 promoter contains a potential binding site for a cMyc:Max complex, since it contains a region of sequence identity with the core Myc:Max recognition sequence (CACGTG) (−42/−47).[157] However, when taken in the surrounding nucleotide context, the HO-1 sequence (−38/−50) (CACCACGTGACC) differs from the ideal Myc/Max (PurACCACGTGGTPyr) and Max:Max (Pur ANCACGTGNTPyr) consensus recognition sequences that have recently been determined by *in vitro* oligonucleotide selection assays. Specifically, the presence of an A nucleotide immediately

3′ of the core element disfavored the *in vitro* binding and *in vivo* transactivation by Myc.[190] This recent evidence renders it unlikely that Myc:Max recognizes the HO-1 promoter.

15.7.2 Proximal Promoter Region

The nucleotide sequences of the human, mouse, and rat proximal promoter regions contain several additional *cis*-elements associated with basal gene expression, including C/EBP, SP-1, AP-2, and AP-4.[71,157,177,191,192] The three promoters are quite different with respect to the configuration of these elements and this lack of conservation make it unlikely that they are of critical functional significance. Functional analysis of the human HO-1 proximal promoter region (−1.2 kb) linked to a bacterial chloramphenicol acetyl transferase (CAT) reporter in transient transfection assays revealed that this sequence did not respond to chemical induction (H_2O_2, sodium arsenite, and TPA) or to UVA radiation.[186] However, it was clear from this study that distal enhancer sequences were necessary for HO-1 regulation. This conclusion was substantiated by the studies of Alam[158] and Alam et al.,[184] who found that distal sequences occuring at distances of approximately −4 kb and −10 kb upstream of the transcriptional start site are necessary for the induction of the gene by heme and heavy metals.

15.7.2.1 Metallothionein-like Metal Responsive Elements

The proximal 5′ flanking sequences of the rodent HO-1 genes contain metal responsive elements (MREs) similar to those originally identified in the human metallothionein IIa promoter, and this led to the suggestion that the regulation of HO-1 by heavy metals could occur by a mechanism identical to the regulation of metallothionein genes.[71] In the mouse metallothionein (IA and IIA) genes, metal induction is mediated by an array of MREs with the overall consensus sequence (TGCPuCNC), which vary in their affinities for metal responsive transcription factors (MBF-1, MTF-1).[81]

The mouse HO-1 flanking sequence contains two such elements, MREa (TGCA CTC) at (−697/−690) and MREb (GGAGAGCA) at (−733/−726) in inverted orientation.[157] The rat HO-1 promoter also contains a putative MRE (GGGTGCTG CACTC) at position (−569/−581).[71] Interestingly, the human HO-1 gene proximal promoter (0/−1 kb) does not contain metallothionein-like metal responsive elements.[177] Recent evidence suggests that the MRE-like elements in the mouse HO-1 promoter may be nonfunctional. Under $CdCl_2$ and $ZnCl_2$ treatment conditions, which activated the metallothionein promoter, the proximal promoter (0/−1 kb) region linked to a CAT reporter did not respond to induction by $CdCl_2$ in transient transfection assays. Furthermore, multiple copies of MT-1 MREs confer metal responsiveness to a MT-1 minimal promoter, whereas multiple copies of the mouse HO-1 MREa do not.[157] The primary region mediating metal responsiveness of human and mouse HO-1 promoters has recently been localized to the −4 kb enhancer region, which does not contain metallothionein-like MRE sequences[158] (see Section 15.7.3).

15.7.2.2 Heat Shock Element

The HO-1 genes contain heat shock elements. HSEs were originally identified in the promoters of heat shock protein (hsp) genes. The HSEs consist of inverted repeats of the sequence element 5'NGAAN3'. The rat HO-1 gene can be induced by cellular hyperthermia and contains two HSEs at positions HSE1 (−290/−276) and HSE2 (−222/−212).[180] The HSE-1 contains three inverted repeats of the core element 5'NGAAN3' in alternating orientation, while the HSE2 consists of only two inverted repeats. The rat HO-1 HSE-1 confers heat inducibility but not hemin inducibility when fused to the *Escherichia coli* xanthine guanine phosphoribosyl transferase (*gpt*) gene fusion constructs and transfected into mouse amelanotic melanoma and rat glioma cells.[26] While the rat HSE-1 can mediate heat induction of HO-1 in the absence of HSE2, but not vice versa, both elements are required for maximal induction of the rat HO-1 gene. Furthermore, both HSE1 and HSE2 bound a heat inducible nuclear factor from rat glioma cells. The heat-inducible factor could be a rat homolog of the heat shock factor (HSF), but this has not been shown directly. HSE2 also formed a faster migrating complex with a nuclear factor from both heat-treated and normal cells, which has been designated as the heme oxygenase promoter binding protein (HOBP). The HOBP was distinguished from the constitutive heat shock element binding activity (CHBA) which was identified originally in nonshocked HeLa cells.[180]

The mouse HO-1 promoter also contains three putative HSEs identified by sequence analysis but these have not been characterized for protein binding properties or *in vivo* function.[157]

In contrast to the rat gene, the human HO-1 gene, which is noninducible by hyperthermia in most tissues (except Hep3B hepatoma cells), contains an HSE (−383/−366) which failed to confer heat inducibility to *gpt* fusion constructs in the same transfection system as used to test the rat HSEs.[26] Furthermore, unlike the rat, the human HO-1 promoter does not contain a second HSE in the HSE2 configuration. However, the putative human HO HSE does bind to an unidentified heat-inducible factor in nuclear extracts from heat treated Hep3B cells.[193] Since the heat-inducible binding activity was blocked by actinomycin D and cyclohexi-mide, it is distinct from human HSF, which regulates the heat inducible transcription of the hsp 70 gene, by a post-translational (cycloheximide insensitive) mechanism.

15.7.2.3 κB Elements

The human HO-1 promoter also contains a putative NF-κB binding site at position (−156/−166).[191,192] A fragment containing this region was protected from DNase-1 digestion by purified p50 subunit of NF-κB in *in vitro* footprinting assays. A second NF-κB-like sequence overlaps the putative HSE at position (−379/−370).[191] While NF-κB has been recently defined as a transcription factor whose nuclear accumulation and DNA binding activity responds to induction by oxidants such as H_2O_2,[194] the role of the proximal κB element in mediating oxidant induction of the heme oxygenase gene has not been established. Interestingly, the *in vitro* binding activity of NF-κB to synthetic κB oligonucleotides was induced by heme, the substrate of HO-1.[192]

15.7.2.4 Interleukin Responsive Elements

The human HO-1 promoter contains two interleukin-6 responsive elements (IL6-REa) at position (−386/−360), which overlaps the putative HSE element, and (IL6-REb) at position (−167/−143) which overlaps the putative NF-κB binding site.[59,191,193] The IL6-REb bound a nuclear factor from Hep3B hepatoma cells whose binding was not modulated by IL-6 treatment.[59] It should be noted, however, that any hexanucleotide will occur by chance every 1000 bp on average.

The IL6-RE hexanucleotide motif (CTGGGA) occurs in the promoters of other acute phase protein genes including rat and human B-fibrinogen, and rat α-2-macroglobulin, and represents a putative binding site for the acute phase response factor (APRF), which mediates the transcription of these genes in response to IL-6.[195]

15.7.2.5 Tissue-Specific TPA Responsive Elements

A *cis*-acting DNA element that mediates the tissue specific induction of HO-1 in myelomonocytic cell lines in parallel with their differentiation by TPA, occurs in the human HO-1 promoter at position (−156/−147).[176] The macrophage-specific TPA responsive element (MTE) consists of a 10 bp palindrome (GTCAT ATGAT), which contains within it an E-box core element (CANNTG), a binding site for basic helix-loop-helix transcription factors (bHLH). This element was shown to bind proteins from THP-1 nuclear extracts in *in vitro* gel-shift assays and DNase-1 footprinting assays.

15.7.2.6 The δ-12 Prostaglandin J2 Responsive Element

The induction of the rat HO-1 gene by δ-12 prostaglandin J2 (δ12 PGJ2) was recently shown to be mediated by a *cis* element occuring at position (−660/−690) of the 5′ flanking region. This element contains an E-box motif and confers position independent enhancer activity to an SV40 minimal promoter, in response to δ12 PGJ2.[196]

15.7.3 Distal (−4 kb) Enhancer Region

Deletion mutagenesis analysis of the mouse HO-1 5′ flanking sequence (−1/−7 kb) linked to a (CAT) reporter gene revealed a basal level enhancer element at

−4 kb, which was localized to a 268 bp fragment designated as (SX2).[183] The SX2 element functions in a position independent manner to elevate basal activity (2- to 10-fold) and mediate TPA induction in the presence of the proximal HO-1 5′ flanking region (0/−1 kb). The SX2 fragment, when fused to an SV40 minimal promoter and CAT reporter, confers transcriptional activation to the construct, in response to TPA treatment, or in response to the simultaneous overexpression of c-Fos and c-Jun. The TPA responsiveness of the SX2-CAT construct can only be detected in cell lines with a low constitutive AP-1 activity. The SX2 fragment contains four sites which bind protein from rat c6 glioma nuclear extracts, two of which (TGAGTCA) (AP-1a) and (TGTGTCA) (AP-1b), resemble the AP-1 consensus TGA(G/C)TCA. Purified Jun/Jun homodimer bound both elements, although AP-1a, which matches exactly the AP-1 consensus, had a higher binding affinity than AP-1b. Upstream of the AP-1 sites, two repeats of the sequence TGAGGAAAT represent binding sites for the CCAAT enhancer binding protein family (C/EBP).[158] The −4 kb distal enhancer region SX2 was shown to mediate the $CdCl_2$ and lipopolysaccharide (LPS) induction of the mouse HO-1 gene. The two distal AP-1 sites were essential for the response.[158,197]

In human cells, Takeda et al. have also localized the Cd response to a region between −4 and −4.5 kb of the heme oxygenase-1 5′ flanking region in transient transfection assays using a luciferase reporter gene.[182] The relatively weak effect of Cd in this assay system supports the conclusions of Alam et al. that stable integration of the fusion gene is required for optimal Cd induction.[157] Takeda et al. have further localized the Cd responsive element to a 10 bp sequence TGCTAGATTT within the −4 kb enhancer region immediately upstream from the first distal AP-1 site. These results differ from the results of Alam et al. in that neither AP-1-like sequence elements present in the human −4 kb distal enhancer region, within sequences highly homologous to the mouse −4 kb enhancer region, were included in the putative Cd responsive region.

In unpublished studies from this laboratory, we have observed that the −4 kb region of the human HO-1 promoter contains a constitutive DNase-1 hypersensitive site consistent with a putative regulatory sequence.[198]

15.7.4 Distal −10 kb Enhancer

A second distal enhancer region responsible for basal activity and chemical (cadmium, heme, HgCl, $ZnCl_2$, and hydrogen peroxide) induction of the mouse HO-1 gene has recently been identified. The sequence element designated "AB1" occurs 10 kb upstream of the transcriptional start site and contains three tandem repeats each containing two core elements (A, B) resembling AP-4 and AP-1 consensus sequences, respectively.[184] The B elements (AP-1-like) (GCTGAGTCA NGG) were necessary for the induction response and bound protein from Hepa nuclear extracts.

15.7.5 Conclusions

Nucleotide sequence analysis of HO-1 promoters has revealed a host of putative recognition sequences for known transcription factors. The existence of such elements does not necessarily imply *in vivo* function. Furthermore, the demonstration of nuclear factor binding activity to such elements *in vitro* does not prove that such an interaction occurs *in vivo*. It is clear that the proximal promoter region, which contains many of these putative protein binding sites, is not sufficient to mediate the transcriptional induction of HO-1 genes by many inducing agents, unless accompanied by enhancer sequences which lie several kilobases upstream. The mechanism by which the distal enhancer regions cooperate with the proximal promoter elements and their binding factors to initiate HO-1 gene transcription remain unknown.

15.8 Role of Heme Oxygenases in the Stress Response

15.8.1 Antioxidant Properties of the Bile Pigments

In vitro studies have shown that bilirubin, the end product of the heme catabolic pathway initiated by heme oxygenases, may function as a physiological antioxidant in human blood plasma and bile.[14] Bilirubin formed in tissues is excreted into blood serum where it circulates as a 1:1 complex with serum albumin.[199,200]

In vitro studies show that albumin-bound bilirubin prevents the peroxidation of albumin-bound fatty acids and the oxidative destruction of serum albumin.[201] Bilirubin and biliverdin efficiently prevent the autoxidation of linoleic acid under physiological oxygen concentrations.[13] Conjugated bilirubin prevents the peroxidation of phosphatydyl choline in multilamellar liposomes, inhibits the luminol-enhanced chemoluminescence of activated human macrophages, and scavenges hypochlorous acid (HOCl).[202,203]

Recent studies demonstrate that bilirubin may function as a coantioxidant with low-density lipoprotein (LDL)-associated α-tocopherol in inhibiting the oxidation of the lipid components of purified serum LDL.[204] Conjugated bilirubin and biliverdin act synergistically with α-tocopherol in preventing peroxidation of phosphatydyl choline liposomes, and inhibit consumption of α-tocopherol, possibly by reducing α-tocopheroxyl radical.[205]

Increased production of biliverdin and bilirubin as a consequence HO activity has been proposed to have beneficial antioxidant properties in serum with respect to a whole organism subjected to oxidative stress, particularly from dietary sources.[202] The activity of HO in organs involved in the breakdown of heme from disintegrating erythrocytes probably contributes to such a mechanism. It remains unclear whether bilirubin, which is a lipid-soluble cytotoxic waste product, would have any beneficial antioxidant function at the cellular level prior to its efflux into blood serum.

The *in vitro* antioxiant properties of bilirubin are unlikely to explain the oxidant mediated induction of HO-1 in cells not specialized in heme breakdown such as human skin fibroblasts. It also remains unclear whether the intracellular concentration of biliverdin and bilirubin produced from HO-1 and biliverdin reductase activity would significantly contribute to the existing antioxidant capacity of the cell, given the levels of the other antioxidants already present, that include ascorbate, β-carotene, α-tocopherol, and millimolar concentrations of glutathione (3 to 5 mM).[206] Despite the lack of evidence supporting the role of bilirubin as an intracellular antioxidant, this theory has been widely cited in the recent literature as an "antioxidant" function for HO-1 at the cellular level.

15.8.2 Relationship between HO and Intracellular Iron Metabolism

Heme oxygenase affects intracellular iron metabolism by releasing the iron bound to heme, transiently returning it to a low-molecular-weight "free" or chelatable iron pool. The prooxidant effects of iron have been demonstrated *in vivo,* and these include the promotion of membrane lipid peroxidation, the oxidative modification of DNA,[207,208] and the sensitization of endothelial cells to the cytotoxic and membrane damaging effects of H_2O_2, menadione, or exposure to activated phagocytes.[209]

The sequestration of intacellular iron by the iron storage protein ferritin is thought to limit its potential catalysis of oxidative reactions. Ferritin (M_r:450 kDa) consists of 24 subunits of two types (heavy chain, H-F, $M_r \sim 21$ kDa; and light chain L-F, $M_r \sim 19$ kDa) which sequester iron in a crystalline core. The sequestered iron (approximately 4,500 Fe^{3+} ions per ferritin molecule) is maintained in an oxidized (Fe^{3+}) state by the ferritin H-chain, which possesses a ferroxidase activity.[210] A direct relationship exists between the amount of intracellular chelatable iron and the synthesis of ferritin, which is regulated by a posttranscriptional mechanism.[211]

Since HO-1 activity releases iron from heme, a relationship exists between HO-1 activity and the synthesis of ferritin.[17] The treatment of rat fibroblasts with heme induced HO activity and subsequently increased ferritin synthesis. Oxidative stress generated by the irradiation of human skin fibroblasts with UVA resulted in an increase in HO-1 activity which was followed by a doubling of ferritin protein levels 24 to 48 h postirradiation. These increases in ferritin by heme or UVA treatments were inhibited by the iron chelator desferrioxamine or by subsequent treatment with the metalloporphyrin inhibitor of HO activity (Sn-PPIX). Taken together, these observations support the theory that free iron, including iron released from heme by the action of HO activity, regulates ferritin synthesis.[15,17]

Recently, the potential roles of HO and ferritin in cellular defense have been examined. The treatment of endothelial cells with heme or methemoglobin sensitizes the cells to H_2O_2, menadione, or activated phagocytes after short-term incubation, whereas the prolonged exposure to heme or methemoglobin (16 h) renders these cells resistant to such oxidative stress treatments. This protection against lethal damage by extended heme treatment was observed in association with induction of HO-1 and ferritin synthesis.[212,213] Irradiation of human skin fibroblasts with UVA confers transient resistance to oxidative membrane damage induced by a subsequent (24 h) UVA exposure, and this has been associated with the induction of HO activity and ferritin synthesis. Antisense oligonucleotides specific for the transcriptional start site of HO-1 mRNA abolish the cytoprotective effect of UVA pretreatment, and inhibit both the induction of HO-1 and ferritin synthesis. The iron chelator desferrioxamine also abolishes the cytoprotection, and inhibits ferritin synthesis while having no effect on HO-1 expression.[16] Taken together, these studies suggest that ferritin has a cytoprotective role following its induction by heme or UVA pretreatment, and that HO-1 represents an intermediate in this process by releasing "regulatory" iron from heme.

Paradoxically, it has also been proposed that ferritin could be a potential source of catalytic iron during oxidative stress conditions.[214] In conclusion, further studies designed to monitor the intracellular flow of iron between hemoprotein, free heme pools, low-molecular-weight iron pools, and ferritin pools will increase the understanding of the putative pro- and antioxidant roles of ferritin and HO-1, during *in vivo* oxidative stress conditions.

15.8.2.1 Antioxidant Implications of the Relationship between HO Activity and Hemoprotein Turnover

HO-1 degrades intracellular free heme including that derived from the turnover of cellular hemoprotein, and thus may limit the potential involvement of heme in catalyzing nonspecific oxidative reactions. Protein-bound heme serves as a vital component in cellular respiration, in both oxygen transport and electron transport functions, and catalyzes many biological oxidation reactions, serving as catalytic cofactor in a number of monooxygenase and dioxygenase enzymes. These enzyme functions depend on the heme-iron and its reversible oxidation state.[6] Free heme, however, can catalyze potentially deleterious reactions, including its own nonspecific oxidation in the presence of reducing agents, and may propagate lipid peroxidation in liposomal membranes in the presence of organic hydroperoxides.[215,216] The prooxidant properties of heme have been demonstrated in cultured cells.[212]

The protein architecture of hemoproteins such as cytochrome P-450 positions the heme and a bound substrate in such a way that an activated oxygen molecule bound to the heme-iron can stereospecifically attack the bound substrate.[217] The disruption of hemoprotein tertiary structure may allow for heme-iron in high-oxidation states to react with molecules in the surrounding solution rather than a specific substrate, in essence converting a caged reactive species to one freely reactive. Damage to protein structure, including fragmentation, cross-linking, and changes in physical properties, such as net charge, hydrophobicity, solubility, electrophoretic mobility, and susceptibility to proteolytic degradation, have been described.[218,219] Thus, it can be postulated that during oxidative stress conditions, the accumulation of damaged hemoproteins may have deletereous effects to the cell, including the possibility of Fenton catalysis by exposed or released heme prosthetic groups. It has been shown that microsomal cytochrome P-450 isozymes become substrates for HO activity *in vitro* only after partial proteolytic degradation to the cytochrome P-420 form.[220] Likewise, the heme in cytochrome *c* is not accessible for degradation by HO. However, a proteolytic fragment of cytochrome *c*, heme *c*-undecapeptide, serves as a HO substrate.[221] It can be postulated that HO may serve a protective cellular function by degrading the heme moieties of hemoproteins that have sustained oxidative damage, and as a consequence, proteolytic degradation; thus preventing the iron in exposed heme from catalyzing deleterious free radical reactions.

15.8.3 Role of Carbon Monoxide, a Heme Oxygenase Reaction Product, in Cellular and Neuronal Signal Transduction

Carbon monoxide (CO), which originates from the oxidation and release of the heme α-methene bridge carbon during HO mediated heme degradation, is generally considered a toxic waste product of the HO-1 reaction.[222] At high concentrations CO is an asphyxiating gas that interferes with oxygen transport by forming tight binding complexes with the heme-iron of hemoglobin and myoglobin.[223] CO can form complexes with the heme-iron of guanylate cyclase, a hemoprotein that converts GTP into the soluble second-messenger cyclic GMP (cGMP). The binding of CO to G-cyclase results in an activation of this enzyme *in vitro*.[224] The binding and induction of G-cyclase clearly contributes to the *in vivo* function of another gas, nitric oxide (NO), which has been identified as a vasodilator (endothelial-derived relaxing factor), a neurotransmitter, and a cytotoxic component of the oxidative burst.[225] Furthermore, nitric oxide synthases (NOS) have been discovered which synthesize NO *de novo* from arginine.[226,227] Given that CO has some properties in common with NO (i.e., a low-molecular-weight gas that binds guanylate cyclase) it has been suggested that CO may have a physiological function in minute quantities as a soluble neural second messenger.[228] In analogy with the NO/NOS system, it has been postulated that heme oxygenase-2 (HO-2), which exists at high levels in the brain, might function as a physiological regulator of CO production in certain sections of this organ. The CO produced as a result of heme metabolism is postulated to regulate signal transduction cascades responsive to the production of cGMP by guanylate cyclase.

Evidence in support of this theory includes *in situ* hybridization studies of brain slices, which demonstrate that the distribution of HO-2, the noninducible isozyme of HO, closely matches that of NADPH cytochrome P-450 reductase and guanylate cyclase.[18] The induction of guanylate cyclase in cultured olfactory neurons by olfactory stimulants can be blocked by metalloporphyrin inhibitors of HO such as ZnPP, but not inhibitors of NOS.[18] Since HO-2 exists at high levels in the brain, and is not inducible by agents that induce its isozyme HO-1, it is not clear how the CO output from HO-2 would be regulated according to need. This problem is compounded by the fact that CO, unlike NO, is a relatively stable compound. There is recent evidence that shows that corticosteroids induce HO-2 mRNA transcription in the brain, while having a negative regulatory effect on NOS mRNA transcription.[229] Physiologic pathways that cause fluctuations in substrate availability (i.e., free heme levels) would also influence the rate of CO production in the brain. It has also been proposed that endogenous metalloporphyrins may regulate HO-2 activity, and thus CO production in the brain.[230,231]

Carbon monoxide generated from HO activity has also been suggested to serve as a retrograde messenger in long-term potentiation (LTP), a physiological process involved in memory.[231–233] For example, ZnPPIX inhibits tetanic stimulation of LTP in voltage-clamped rodent pyramidal cells, and in the dentate gyrus of whole rats.[234,235] The theory that CO from HO activity is involved in regulating neurophysiological phenomenon such as LTP remains controversial. Transgenic knockout mice lacking HO-2 were shown to have normal hippocampal LTP, which was inhibitable by ZnPPIX.[236] Further complications with the theory reside in an assumed specificity of the metalloporphyrins as heme oxygenase inhibitors in the brain. For example, certain metalloprophyrins inhibit HO-activity (Sn-mesoporphyrin, Zn-deuteroporphyrin-bis glycol) but have no effect on hippocampal LTP.[237] Additionally, certain metalloporphyrins that inhibit HO activity are also able to inhibit NOS activity (ZnPPIX, chromium mesoporphyrin, CrMP), and have photosensitizer properties (SnPPIX), which may lead to the generation of reactive oxygen intermediates ($^1O_2$, O_2^-).[237,238] In conclusion, pleiotropic effects of metalloporphyrins must be considered when using them to establish proof of new physiological roles for HO in tissue culture systems.

Since CO is an expected by-product of heme oxygenase activity (HO-1 or HO-2) in many tissues that contain it, it can also be postulated that CO may have regulatory effects in other tissues as well, as a consequence of either constitutive HO-2 or inducible HO-1 activity. A recent study demonstrates the elevation of cGMP levels in rat heart tissue from 1 to 6 h following hyperthermia (42°C), in parallel with the transcriptional induction of HO-1.[74] Furthermore, ZnPPIX, but not NOS inhibitors, inhibit smooth muscle relaxation in the opposum internal anal sphincter produced by nonadrenergic noncholinergic (NANC) nerve stimulation.[239] In isolated perfused rat liver, ZnPPIX diminished CO levels detectable in the effluent, and increased the perfusion pressure under the constant flow conditions. These effects were reversed by the addition of CO or cGMP analogues in the perfusate.[240] These provocative pieces of evidence encourage further examination of not only CO/cGMP signal transduction cascades and their possible regulation by heme oxygenases, but also the search for physiological targets potentially influenced by such a pathway.

15.9 Conclusion

This chapter has emphasized the role of HO-1 as a major component of mammalian stress inducible responses. The transcriptional induction of HO-1 occurs in response to a wide variety of chemical and physical cellular insults and overlaps other mammalian stress-inducible responses.

While recent progress has been achieved in understanding the transcriptional regulation of the HO-1 gene following cellular heavy metal stress, the mechanisms that regulate HO-1 as a consequence of oxidative stress, thiol-reactive substances, and associated changes in cellular glutathione status remain unresolved. Likewise the molecular mechanisms that regulate the transcriptional induction of the HO-1 gene following increased substrate availability (heme) are still unknown.

While theoretical models have been discussed, the function of the inducible HO-1 response remains an enigma. Little evidence exists that directly links HO-1 induction to cytoprotection against oxidative stress. Recent progress, however, has implicated HO-1 as a mechanism important in regulating the flow of intracellular iron, and that ferritin synthesized as a consequence of HO activity may have a cytoprotective function during oxidative stress. Finally, carbon monoxide, a by-product of HO activity, has been implicated as a second-messenger molecule. Further work in this field will no doubt contribute to a greater understanding of how genes are regulated as a consequence of environmental stimuli, and provide insight into potential mechanisms contributing to cellular resistance during clinical anticancer therapies.

Acknowledgments

Work from this laboratory described herein has been supported by grants from the Association for International Cancer Research (U.K.), the Swiss National Science Foundation (31-30880-91 and 3100-040720.94/1), the Swiss League Against Cancer of Central Switzerland, and the Neuchateloise League Against Cancer. S. Ryter was supported by a grant from the Emma Muschamp Foundation. We wish to thank Egil Kvam for critical reading of this manuscript, and Alexander Noel for reference databases. We would also like to thank Charles J. Gomer of Childrens Hospital Los Angeles, in whose laboratory S. Ryter began his studies in stress protein responses to cellular photosensitization.

References

1. Tenhunen, R., H. Marver, and R. Schmid. 1969. Microsomal heme oxygenase, characterization of the enzyme. *J. Biol. Chem.* **244**:6388–6394.
2. Maines, M. D. 1992. *Heme Oxygenase: Clinical Applications and Functions.* CRC Press, Boca Raton, FL.

3. Tenhunen, R., H. Marver, and R. Schmid. 1970. The enzymatic catabolism of hemoglobin: stimulation of microsomal heme oxygenase by hemin. *J. Lab. Clin. Med.* **75**:410–421.

4. Welch, W. 1990. The mammalian stress response: cell physiology and biochemistry of the stress response, pp. 223–278. *In* R. Morimoto, A. Tissiers, and C. Georgopoulous (eds.), *Stress Proteins in Biology and Medicine.* Cold Spring Harbor Laboratory Press, Cold Spring Harbor, NY.

5. Kikuchi, G., and T. Yoshida. 1983. Function and induction of the microsomal heme oxygenase. *Mol. Cell. Biochem.* **53/54**:163–183.

6. Maines, M. D. 1984. New developments in the regulation of heme metabolism and their implications. *Crit. Rev. Toxicol.* **12**:241–314.

7. Maines, M. D. 1988. Heme oxygenase: function, multiplicity, regulatory mechanisms, and clinical applications. *FASEB J.* **2**:2557–2568.

8. Abraham, N., J. Lin, M. Schwartzman, R. Levere, and S. Shibahara. 1988. The physiological significance of heme oxygenase. *Int. J. Biochem.* **20**:543–558.

9. Schacter, B. 1988. Heme catabolism by heme oxygenase, physiology, regulation, and mechanism of action. *Semin. Hematol.* **25**:349–369.

10. Shibahara, S. 1988. Regulation of heme oxygenase gene expression. *Semin. Hematol.* **25**:370–376.

11. Murphy, B. J., K. R. Laderoute, S. M. Short, and R. M. Sutherland. 1991. The identification of heme oxygenase as a major hypoxic stress protein in chinese hamster ovary cells. *Br. J. Cancer* **64**:69–73.

12. Keyse, S. M., and R. M. Tyrrell. 1989. Heme oxygenase is the major 32-kDa stress protein induced in human skin fibroblasts by UVA radiation, hydrogen peroxide, and sodium arsenite. *Proc. Natl. Acad. Sci. USA* **86**:99–103.

13. Stocker, R., Y. Yamamoto, A. McDonagh, A. Glazer, and B. Ames. 1987. Bilirubin is an antioxidant of possible physiological importance. *Science* **235**:1043–1045.

14. Stocker, R. 1990. Induction of haem oxygenase as a defence against oxidative stress. *Free Rad. Res. Commun.* **9**:101–112.

15. Vile, G. F., and R. M. Tyrrell. 1993. Oxidative stress resulting from ultraviolet A irradiation of human skin fibroblasts leads to a heme oxygenase-dependent increase in ferritin. *J. Biol. Chem.* **268**:14678–14681.

16. Vile, G. F., S. Basu-Modak, C. Waltner, and R. M. Tyrrell. 1994. Heme oxygenase 1 mediates an adaptive response to oxidative stress in human skin fibroblasts. *Proc. Natl. Acad. Sci. USA* **91**:2607–2610.

17. Eisenstein, R. S., D. Garcia-Mayol, W. Pettingel, and H. N. Munro. 1991. Regulation of ferritin and heme oxygenase synthesis in rat fibroblasts by different forms of iron. *Proc. Natl. Acad. Sci. USA* **88**:688–692.

18. Verma, A., D. J. Hirsch, C. E. Glatt, G. V. Ronnett, and S. H. Snyder. 1993. Carbon monoxide: a putative neural messenger. *Science* **259**:381–384.

19. Subjeck, J. R., and T. Shyy. 1986. Stress protein systems of mammalian cells. *Am. J. Physiol.* **250**:C1–C17.

20. Schlessinger, M., M. Ashburner, and A. Tissieres. 1982. *Heat Shock: From Bacteria to Man.* Cold Spring Harbor Laboratory Press, Cold Spring Harbor, NY.

21. Hiwasa, T., and Sakiyama S. 1986. Increase in the synthesis of a Mr 32,000 protein in BALB/c 3T3 cells after treatment with tumor promoters, chemical carcinogens, metal salts and heat shock. *Cancer Res.* **46**:2474–2481.

22. Keyse, S. M., and R. M. Tyrrell. 1987. Both near ultraviolet radiation and the oxidizing agent hydrogen peroxide induce a 32-kDa stress protein in normal human skin fibroblasts. *J. Biol. Chem.* **262**:14821–14825.

23. Caltabiano, M. M., T. P. Koestler, G. Poste, and R. G. Grieg. 1986. Induction of 32- and 34-kDa stress proteins by sodium arsenite, heavy metals and thiol-reactive agents. *J. Biol. Chem.* **261**:13381–13386.

24. Taketani, S., H. Kohono, T. Yoshinaga, and R. Tokunaga. 1989. The human 32 kDa stress protein induced by exposure to arsenite and cadmium ions is heme oxygenase. *FEBS Lett.* **245**:173–176.

25. Kageyama, H., T. Hiwasa, K. Tokunaga, and S. Sakiyama. 1988. Isolation and characterization of a complementary DNA clone for a Mr 32,000 protein which is induced with tumor promoters in BALB/c 3T3 cells. *Cancer Res.* **48**:4795–4798.

26. Shibahara, S., R. Muller, and H. Taguchi. 1987. Transcriptional control of rat heme oxygenase by heat shock. *J. Biol. Chem.* **262**:12889–12892.

27. Taketani, S., H. Kohno, T. Yoshinaga, and R. Tokunaga. 1988. Induction of heme oxygenase in rat hepatoma cells by exposure to heavy metals and hyperthermia. *Biochem. Int.* **17**:665–672.

28. Mitani, K., H. Fujita, S. Sassa, and A. Kappas. 1990. Activation of heme oxygenase and heat shock protein 70 genes by stress in human hepatoma cells. *Biochem. Biophys. Res. Commun.* **166**:1429–1434.

29. Wu, B., C. Hunt, and R. Morimoto. 1985. Structure and expression of the human gene encoding major heat shock protein HSP70. *Mol. Cell. Biol.* **5**:330–341.

30. Alam, J., S. Shibahara, and A. Smith. 1989. Transcriptional activation of the heme oxygenase gene by heme and cadmium in mouse hepatoma cells. *J. Biol. Chem.* **264**:6371–6375.

31. Yoshida, T., P. Biro, T. Cohen, R. M. Müller, and S. Shibahara. 1988. Human heme oxygenase cDNA and induction of its mRNA by hemin. *Eur. J. Biochem.* **171**:457–461.

32. Shibahara, S., T. Yoshida, and G. Kikuchi. 1978. Induction of heme oxygenase by hemin in cultured pig alveolar macrophages. *Arch. Biochem. Biophys.* **188**:243–250.

33. Alam, J., and A. Smith. 1989. Receptor mediated transport of heme by hemopexin regulates gene expression in mammalian cells. *J. Biol. Chem.* **264**:17637–17640.

34. Alam, J., and A. Smith. 1992. Heme-hemopexin-mediated induction of metallothionein gene expression. *J. Biol. Chem.* **267**:16379–16384.

35. Smith, A., J. Alam, P. V. Escriba, and W. T. Morgan. 1993. Regulation of heme oxygenase and metallothionein gene expression by the heme analogs, cobalt-, and tin-protoporphyrin. *J. Biol. Chem.* **268**:7365–7371.

36. Sutherland, R., J. Freyer, W. Mueller-Klieser, R. Wilson, C. Heacock, J. Sciandra, B. Sordat. 1986. Cellular Growth and Metabolic Adaptations to Nutrient Stress Environments in Tumor Microregions. *Int. J. Radiat. Oncol. Biol. Phys.* **12**:611–615.

37. Murphy, B. J., K. R. Laderoute, R. J. Chin, and R. M. Sutherland. 1994. Metallothionein IIA is up-regulated by hypoxia in human A431 squamous carcinoma cells. *Cancer Res.* **54**:5808–5810.

38. Price, B., and S. Calderwood. 1992. Gadd 145 and gadd 153 messenger RNA levels are increased during hypoxia and after exposure to agents which elevate the levels of the glucose regulated proteins. *Cancer Res.* **52**:3814–3817.

39. Maines, M. D., and A. Kappas. 1977. Metals as regulators of heme metabolism. *Science* **198**:1215–1221.

40. Sunderman, F. W., Jr. 1995. Metal induction of heme oxygenase. [Review]. *Ann. N.Y. Acad. Sci.* **514**:65–80.

41. Eaton, D. L., N. H. Stacey, K. L. Wong, and C. D. Klaassen. 1980. Dose-response effects of various metal ions on rat liver metallothionein, glutathione, heme oxygenase, and cytochrome P-450. *Toxicol. Appl. Pharmacol.* **55**:393–402.

42. Mitani, K., H. Fujita, Y. Fukuda, A. Kappas, and S. Sassa. 1993. The role of inorganic metals and metalloporphyrins in the induction of haem oxygenase and heat-shock protein 70 in human hepatoma cells. *Biochem. J.* **290**:819–825.

43. Levinson, W., H. Oppermann, J. Jackson. 1980. Transition series metals and sulfhydryl reagents induce the synthesis of four proteins in eukaryotic cells. *Biochem. Biophys. Acta* **606**:170–180.

44. Takeda, K., H. Fujita, and S. Shibahara. 1995. Differential control of the metal-mediated activation of the human heme oxygenase-1 and metallothionein genes. *Biochem. Biophys. Res. Commun.* **207**:160–167.

45. Watowich, S., and R. Morimoto. 1988. Complex regulation of heat shock and glucose responsive genes in human cells. *Mol. Cell. Biol.* **8**:393–405.

46. Taketani, S., H. Sato, T. Yoshinaga, R. Tokunaga, T. Ishii, and S. Bannai. 1990. Induction in mouse peritoneal macrophages of a 34 kDa stress protein and heme oxygenase by sulfhydryl reactive agents. *J. Biochem.* **108**:28–32.

47. Johnston, D., H. Oppermann, J. Jackson, and W. Levinson. 1980. Induction of four proteins in chick embryo cells by sodium arsenite. *J. Biol. Chem.* **255**:6975–6980.

48. Applegate, L. A., P. Luescher, and R. M. Tyrrell. 1991. Induction of heme oxygenase: a general response to oxidant stress in cultured mammalian cells. *Cancer Res.* **51**:974–978.

49. Guzzo, A., C. Karatzios, C. Diorio, and M. S. Dubow. 1994. Metallothionein-II and ferritin H mRNA levels are increased in arsenite exposed HeLa cells. *Biochem. Biophys. Res. Commun.* **205**:590–595.

50. Angel, P., A. Poting, U. Mallick, H. Rahmsdorf, M. Schorpp, and P. Herrlich. 1986. Induction of metallothionein and other mRNA species by carcinogens and tumor promoters in primary human skin fibroblasts. *Mol. Cell Biol.* **6**:1760–1766.

51. Holbrook, N., and A. Fornace. 1991. Response to adversity: molecular control of gene activation following genotoxic stress. *N. Biologist* **3**:825–833.

52. Keyse, S. M., and R. M. Tyrrell. 1990. Induction of the heme oxygenase gene in human skin fibroblasts by hydrogen peroxide and UVA (365nm) radiation: evidence for the involvement of hydroxyl radical. *Carcinogenesis* **11**:787–791.

53. Jornot, L., M. Mirault, and A. Junod. 1991. Differential expression of hsp70 stress proteins in human endothelial cells exposed to heat shock and hydrogen peroxide. *Am. J. Respir. Cell. Mol. Biol.* **5**:265–275.

54. Dalton, T., R. D. Palmiter, and G. K. Andrews. 1994. Transcriptional induction of the mouse metallothionein-1 gene in hydrogen peroxide treated HEPA cells involves a composite major late transcription factor antioxidant response element and metal response promoter elements. *Nucleic Acids Res.* **22**:5016–5023.

55. Bruce, J., B. Price, N. Coleman, and S. Calderwood. 1992. Oxidative injury rapidly activates the heat shock transcription factor but fails to increase levels of heat shock proteins. *Cancer Res.* **53**:12–15.

56. Hiwasa, T., S. Fujimura, and S. Sakiyama. 1982. Tumor promoters increase the synthesis of a 32,000 dalton protein in BALB/c3T3 cells. *Proc. Natl. Acad. Sci. USA.* **79**:1800–1804.

57. Rizzardini, M., M. Terao, F. Falciani, and L. Cantoni. 1993. Cytokine induction of haem oxygenase mRNA in mouse liver. Interleukin 1 transcriptionally activates the haem oxygenase gene. *Biochem. J.* **290**:343–347.

58. Cantoni, L., C. Rossi, M. Rizzardini, M. Gadina, and P. Gezzi. 1991. Interleukin-1 and tumor necrosis factor induce hepatic haem oxygenase. *Biochem. J.* **279**:891–894.

59. Mitani, K., H. Fujita, A. Kappas, and S. Sassa. 1992. Heme oxygenase is a positive acute-phase reactant in human Hep3B hepatoma cells. *Blood* **79**:1255–1259.

60. Fukuda, Y., and S. Sassa. 1993. Effect of interleukin-11 on the levels of mRNAs encoding heme oxygenase and haptoglobin in human HepG2 hepatoma cells. *Biochem. Biophys. Res. Commun.* **193**:297–302.

61. Kutty, R. K., C. N. Nagineni, G. Kutty, J. J. Hooks, G. J. Chader, and B. Wiggert. 1994. Increased expression of heme oxygenase-1 in human retinal pigment epithelial cells by transforming growth factor-β. *J. Cell. Physiol.* **159**:371–378.

62. Gemsa, D., C. H. Woo, H. Fudenberg, and R. Schmid. 1974. Stimulation of heme oxygenase in macrophages and liver by endotoxin. *J. Clin. Invest.* **53**:647–651.

63. Rizzardini, M., M. Carelli, M. R. Cabello Porras, and L. Cantoni. 1994. Mechanisms of endotoxin-induced haem oxygenase mRNA accumulation in mouse liver: synergism by glutathione depletion and protection by *N*-acetylcysteine. *Biochem. J.* **304**:477–483.

64. Snyers, L., and J. Content. 1994. Induction of metallothionein and stomatin by interleukin-6 and glucocorticoids in a human amniotic cell line. *Eur. J. Biochem.* **223**:411–418.

65. Leibbrandt, M. E., and J. Koropatnick. 1994. Activation of human monocytes with lipopolysaccharide induces metallothionein expression and is diminished by zinc. *Toxicol. Appl. Pharmacol.* **124**:72–81.

66. Gomer, C. J., M. Luna, A. Ferrario, and N. Rucker. 1991. Increased transcription and translation of heme oxygenase in chinese hamster fibroblasts following photodynamic stress or photofrin II incubation. *Photochem. Photobiol.* **53**:275–279.

67. Bressoud, D., V. Jomini, and R. Tyrrell. 1992. Dark induction of haem oxygenase messenger mRNA by haematoporphyrin derivative and zinc phthalocyanine; agents for photodynamic therapy. *J. Photochem. Photobiol. B: Biol.* **14**:311–318.

68. Gomer, C. J., A. Ferrario, N. Hayashi, N. Rucker, B. Szirth, and A. L. Murphree. 1988. Molecular, cellular, and tissue responses following photodynamic therapy. *Laser Surg. Med.* **8**:450–463.

69. Xiong, X., K. Arizono, S. H. Garrett, and F. O. Brady. 1992. Induction of zinc metallothionein by calcium ionophore in vivo and in vitro. *FEBS Lett.* **299**:192–196.

70. Lee, A. S. 1987. Coordinated regulation of a set of genes by glucose and calcium ionophores in mammalian cells. *Trends Biochem. Sci.* **12**:20–23.

71. Müller, R. M., H. Taguchi, and S. Shibahara. 1987. Nucleotide sequence and organization of the rat heme oxygenase gene. *J. Biol. Chem.* **262**:6795–6802.

72. Ewing, J. F., and M. D. Maines. 1991. Rapid induction of heme oxygenase-1 mRNA and protein by hyperthermia in rat brain: heme oxygenase-2 is not a heat shock protein. *Proc. Natl. Acad. Sci. USA* **88**:5364–5368.

73. Raju, V. S., and M. D. Maines. 1994. Coordinated expression and mechanism of induction of HSP32 (heme oxygenase-1) mRNA by hyperthermia in rat organs. *Biochim. Biophys. Acta Gene Struct. Expr.* **1217**:273–280.

74. Ewing, J. F., V. S. Raju, and M. D. Maines. 1994. Induction of heart heme oxygenase-1 (HSP32) by hyperthermia: possible role in stress-mediated elevation of cyclic 3′: 5′-guanosine monophosphate. *J. Pharmacol. Exper. Ther.* **271**:408–414.

75. Ewing, J. F., S. N. Haber, and M. D. Maines. 1992. Normal and heat-induced patterns of expression of heme oxygenase-1 (HSP32) in rat brain: hyperthermia causes rapid induction of mRNA and protein. *J. Neurochem.* **58**:1140–1149.

76. Shibahara, S., M. Yoshizawa, H. Suzuki, K. Takeda, K. Meguro, and K. Endo. 1993. Functional analysis of cDNAs for two types of human heme oxygenase and evidence for their separate regulation. *J. Biochem. (Tokyo)* **113**:214–218.

77. Drummond, I., A. Lee, E. Resendez, and R. Steinhardt. 1987. Depletion of intracellular calcium stores by calcium ionophore A23187 induces the genes for glucose regulated proteins in hamster fibroblasts. *J. Biol. Chem.* **262**:12801–12805.

78. Little, E., M. Ramakrishnan, B. Roy, G. Gazit, and A. S. Lee. 1994. The glucose regulated proteins (GRP78 and GRP94): functions, gene regulation, and applications. (Review). *Crit. Rev. Euk. Gene Exp.* **4**:1–18.

79. Kozutsumi, Y., M. Segal, K. Normington, M. Gething, and J. Sambrook. 1988. The presence of malfolded proteins in the endoplasmic reticulum signals the induction of glucose regulated proteins. *Nature* **332**:462–464.

80. Kutty, R. K., M. D. Maines. 1981. Purification and characterization of biliverdin reductase from rat liver. *J. Biol. Chem.* **256**:3956–3962.

81. Hamer, D. 1986. Metallothionein. *Ann. Rev. Biochem.* **55**:913–951.

82. Bauman, J. W., J. M. McKim, Jr., J. Liu, and C. D. Klaassen. 1992. Induction of metallothionein by diethylmaleate. *Toxicol. Appl. Pharmacol.* **114**:188–196.

83. Papathanasiou, M., N. Kerr, J. Robbins, O. McBride, I. Alamo, S. Barret, I. Hickson, and A. Fornace. 1991. Induction by ionizing radiation of the gadd45 gene in cultured human cells: lack of mediation by protein kinase-C. *Mol. Cell. Biol.* **11**:1009–1016.

84. Tyrrell, R. M., S. M. Keyse, E. C. Moraes. 1991. Cellular defense against UVA (320–380 nm) and UVB (290–320 nm) radiations pp. 861–871. *In* E. Riklis, (ed.), *Photobiology, The Science and Its Applications.* Plenum Press, London.

85. Schlesinger, M. J. 1994. How the cell copes with stress and the function of heat shock proteins. [Review]. *Pediatr. Res.* **36**:1–6.

86. Parsell, D. A., J. Taulien, and S. Lindquist. 1993. The role of heat-shock proteins in thermotolerance. [Review]. *Philos. Trans. R. Soc. Lond. Ser. B. Biol. Sci.* **339**:279–286.

87. Fuqua, S. A., S. Oesterreich, S. G. Hilsenbeck, D. D. Von Hoff, J. Eckardt, and C. K. Osborne. 1994. Heat shock proteins and drug resistance. [Review]. *Breast Cancer Res. Treat.* **32**:67–71.

88. Pelham, H. 1986. Speculations on the functions of the major heat shock and glucose regulated proteins. *Cell* **46**:959–964.

89. Fisher, A., A. Ferrario, and C. Gomer. 1993. Adriamycin resistance in chinese hamster fibroblasts following oxidative stress induced by photodynamic therapy. *Photochem. Photobiol.* **58**:581–588.

90. Shen, J., C. Hughes, C. Chao, J. Cai, C. Bartels, T. Gessner, and J. Subjeck. 1987. Coinduction of glucose-regulated proteins and doxorubicin resistance in Chinese hamster ovary cells. *Proc. Natl. Acad. Sci. USA* **84**:3278–3282.

91. Mello-Filho, A., L. Chubatsu, and R. Menghini. 1988. V79 Chinese hamster cells rendered resistant to high cadmium concentration also become resistant to oxidative stress. *Biochem. J.* **256**:475–479.

92. Lee, T. C., and I. C. Ho. 1994. Expression of heme oxygenase in arsenic-resistant human lung adenocarcinoma cells. *Cancer Res.* **54**:1660–1664.

93. Abraham, N. G., Y. Lavrovsky, M. L. Schwartzman, R. A. Stoltz, R. D. Levere, M. E. Gerritsen, S. Shibahara, and A. Kappas. 1995. Transfection of the human heme oxygenase gene into rabbit coronary microvessel endothelial cells: protective effect against heme and hemoglobin toxicity. *Proc. Natl. Acad. Sci. USA* **92**:6798–6802.

94. Abraham, N., S. Mitrione, and R. Levere. 1986. Kinetics and properties of human fetal and adult liver heme oxygenase. *Biochem. Arch.* **2**:253–259.

95. Yoshida, T., and G. Kikuchi. 1979. Purification and properties of heme oxygenase from rat liver microsomes. *J. Biol. Chem.* **254**:4487–4491.

96. Yoshida, T., and G. Kikuchi. 1978. Purification and properties of heme oxygenase from pig spleen microsomes. *J. Biol. Chem.* **253**:4224–4229.

97. Yoshinaga, T., S. Sassa, and A. Kappas. 1982. Purification and properties of bovine spleen heme oxygenase. Amino acid composition and sites of action of inhibitors of heme oxidation. *J. Biol. Chem.* **257**:7778–7785.

98. Beale, S. I., and J. Cornejo. 1984. Enymatic heme oxygenase activity in soluble extracts of the unicellular red alga, *Cyanidium caldarium*. *Arch. Biochem. Biophys.* **235**:371–384.

99. Ishikawa, K., M. Sato, and T. Yoshida. 1991. Expression of rat heme oxygenase in *Escherichia coli* as a catalytically active, full length form that binds to membranes. *Eur. J. Biochem.* **202**:161–165.

100. Shibahara, S., R. M. Müller, H. Taguchi, and T. Yoshida. 1985. Cloning and expression of cDNA for rat heme oxygenase. *Proc. Natl. Acad. Sci. USA* **82**:7865–7869.

101. Maines, M. D., G. M. Trakshel, and R. K. Kutty. 1986. Characterization of two constitutive forms of rat liver microsomal heme oxygenase. *J. Biol. Chem.* **261**:411–419.

102. Braggins, P. E., G. M. Trakshel, R. K. Kutty, and M. D. Maines. 1986. Characterization of two heme oxygenase isoforms in rat spleen: comparison with the hematin-induced and constitutive isoforms of the liver. *Biochem. Biophys. Res. Commun.* **141**:528–533.

103. Trakshel, G. M., R. K. Kutty, and M. D. Maines. 1986. Purification and characterization of the major constitutive form of testicular heme oxygenase. *J. Biol. Chem.* **261**:11131–11137.

104. Trakshel, G., R. Kutty, and M. Maines. 1988. Resolution of rat brain heme oxygenase activity absence of a detectable amount of the inducible form (HO-1). *Arch. Biochem. Biophys.* **260**:732–739.

105. Rotenberg, M. O., and M. D. Maines. 1991. Characterization of a cDNA-encoding rabbit brain heme oxygenase-2 and identification of a conserved domain among mammalian heme oxygenase isozymes: possible heme-binding site? *Arch. Biochem. Biophys.* **290**:336–344.

106. Cruse, I., and M. D. Maines. 1988. Evidence suggesting that the two forms of heme oxygenase are products of different genes. *J. Biol. Chem.* **263**:3348–3353.

107. Trakshel, G. M., J. F. Ewing, and M. D. Maines. 1991. Heterogeneity of heme oxygenase 1 and 2 isoenzymes. *Biochem. J.* **275**:159–164.

108. Kuwano, A., H. Ikeda, K. Takeda, H. Nakai, I. Kondo, and S. Shibahara. 1994. Mapping of the human gene for inducible heme oxygenase to chromosome 22Q12. *Tohoku J. Exp. Med.* **172**:389–392.

109. Kutty, R. K., G. Kutty, I. R. Rodriguez, G. J. Chader, and B. Wiggert. 1994. Chromosomal localization of the human heme oxygenase genes: heme oxygenase-1 (HMOX1) maps to chromosome 22q12 and heme oxygenase-2 (HMOX2) maps to chromosome 16p13.3. *Genomics* **20**:513–516.

110. Tenhunen, R., H. Marver, and R. Schmid. 1968. The enzymatic conversion of heme to bilirubin by microsomal heme oxygenase. *Proc. Natl. Acad. Sci. USA* **61**:748–755.

111. Noguchi, M., T. Yoshida, and G. Kikuchi. 1983. A stoichiometric study of heme degradation catalyzed by the reconstituted heme oxygenase system with special consideration of the production of hydrogen peroxide during the reaction. *J. Biochem.* **93**:1027–1036.

112. Yoshida, T., M. Noguchi, and G. Kikuchi. 1980. Oxygenated form of heme-heme oxygenase complex and requirement for second electron to initiate heme degradation from the oxygenated complex. *J. Biol. Chem.* **255**:4418–4420.

113. Takahashi, S., J. Wang, D. L. Rousseau, K. Ishikawa, T. Yoshida, J. R. Host, and M. Ikeda-Saito. 1994. Heme-heme oxygenase complex: structure of the catalytic site and its implication for oxygen activation. *J. Biol. Chem.* **269**:1010–1014.

114. Sun, J., A. Wilks, P. R. Ortiz de Montellano, and T. M. Loehr. 1993. Resonance Raman and EPR spectroscopic studies on heme-heme oxygenase complexes. *Biochemistry* **32**:14151–14157.

115. Wilks, A., and P. R. Ortiz de Montellano. 1993. Rat liver heme oxygenase. High level expression of a truncated soluble form and nature of the *meso*-hydroxylating species. *J. Biol. Chem.* **268**:22357–22362.

116. Maines, M. D., and A. Kappas. 1975. Cobalt stimulation of heme degradation in the liver. *J. Biol. Chem.* **250**:4171–4177.

117. Drummond, G. S., D. W. Rosenberg, and A. Kappas. 1982. Metal induction of haem oxygenase without concurrent degradation of cytochrome P-450. Protective effects of compound SKF-525A on the heme protein. *Biochem. J.* **202**:59–66.

118. Schmid, R. M., and A. F. McDonagh. 1975. The enzymatic formation of bilirubin. *Ann. N.Y. Acad. Sci.* **244**:533–552.

119. Shibahara, S., T. Yoshida, and G. Kikuchi. 1979. Mechanism of increase of heme oxygenase activity induced by hemin in cultured pig alveolar macrophages. *Arch. Biochem. Biophys.* **197**:607–617.

120. Ishizawa, S., T. Yoshida, and G. Kikuchi. 1983. Induction of heme oxygenase in rat liver. Increase of the specific mRNA by treatment with various chemicals and immunological identity of the enzymes in various tissues as well as the induced enzymes. *J. Biol. Chem.* **258**:4220–4225.

121. Schacter, B., B. Yoda, and L. Israels. 1976. Cyclic oscillations in rat hepatic heme oxygenase and delta-aminolevulinic acid synthase following intravenous heme administration. *Arch. Biochem. Biophys.* **173**:11–17.

122. Bissell, D. M., L. Hammaker, and R. Schmid. 1972. Hemoglobin and erythrocyte catabolism in rat liver: the separate roles of parenchymal and sinusoidal cells. *Blood* **40**:812–822.

123. Pimstone, N. R., R. Tenhunen, P. T. Seitz, H. Marver, and R. Schmid. 1971. The enzymatic degradation of hemoglobin to bile pigments by macrophages. *J. Exp. Med.* **133**:1264–1281.

124. Kurata, S., and M. Matsumoto. 1982. Expression of heme oxygenase and its RNA in mouse liver after injection of heme and splenectomy. *Biochim. Biophys. Acta Gene Struct. Exp.* **1132**:255–258.

125. Srivastava, K. K., E. E. Cable, S. E. Donohue, and H. L. Bonkovsky. 1993. Molecular basis for heme-dependent induction of heme oxygenase in primary cultures of chick embryo hepatocytes. Demonstration of acquired refractoriness to heme. *Eur. J. Biochem.* **213**:909–917.

126. Bissel, D., and L. Hammaker. 1977. Effect of endotoxin on tryptophan pyrrolase and delta-aminolaevulinate synthase, evidence for an endogenous regulatory heme fraction in rat liver. *Biochem. J.* **166**:301–304.

127. Bakken, A., M. Thaler, and R. Schmid. 1972. Metabolic regulation of heme catabolism and bilirubin production, hormonal control of hepatic heme oxygenase activity. *J. Clin. Invest.* **51**:530–536.

128. El Azhary, R., and G. Mannering. 1978. Effects of interferon inducing agents (polyriboinosinic acid-polyribocytidylic acid, tilorone) on hepatic hemoproteins (cytochrome P-450, catalase, tryptophane 2,3, dioxygenase, mitochondrial cytochromes) heme metabolism, and cytochrome P-450 linked monooxygenase systems. *Mol. Pharmacol.* **15**:698–707.

129. Jarvisalo, J., A. H. Gibbs, and F. De Matteis. 1978. Accelerated conversion of heme to bile pigments caused in the liver by carbon disulfide and other sulfer containing chemicals. *Mol. Pharmacol.* **14**:1099–1106.

130. Nutter, L. M., E. E. Sierra, and E. O. Ngo. 1994. Heme oxygenase does not protect human cells against oxidant stress. *J. Lab. Clin. Med.* **123**:506–514.

131. Müller, T., and S. Gebel. 1994. Heme oxygenase expression in Swiss 3T3 cells following exposure to aqueous cigarette smoke fractions. *Carcinogenesis* **15**:67–72.

132. Gemsa, D., C. H. Woo, H. Fudenberg, and R. Schmid. 1973. Erythrocyte catabolism by macrophages in vitro. The effect of hydrocortisone on erythrophagocytosis and on the induction of heme oxygenase. *J. Clin. Invest.* **52**:812–822.

133. Tyrrell, R. M. 1991. UVA (320–380 nm) radiation as an oxidative stress, pp. 57–83. *In* H. Sies (ed.), *Oxidative Stress: Oxidants and Antioxidants.* Academic Press, London.

134. Danpure, H. J., and R. M. Tyrrell. 1976. Oxygen-dependence of near-UV (365nm) lethality and the interaction of near-UV and X-rays in two mammalian cell lines. *Photochem. Photobiol.* **23**:171–177.

135. Keyse, S. M., L. A. Applegate, Y. Tromvoukis, and R. M. Tyrrell. 1990. Oxidant stress leads to transcriptional activation of the human heme oxygenase gene in cultured skin fibroblasts. *Mol. Cell. Biol.* **10**:4967–4969.

136. Basu-Modak, S., R. M. Tyrrell. 1993. Singlet oxygen: a primary effector in the ultraviolet A/near-visible light induction of the human heme oxygenase gene. *Cancer Res.* **53**:4505–4510.

137. Khan, A. U., and M. Kasha. 1994. Singlet molecular oxygen in the Haber Weiss reaction. *Proc. Natl. Acad. Sci. USA* **91**:12365–12367.

138. Griffith, O. W., and A. Meister. 1979. Potent and specific inhibition of glutathione synthesis by buthionine sulfoximine (S-*n*-butyl homocysteine sulfoximine). *J. Biol. Chem.* **254**:7558–7560.

139. Tyrrell, R. M., and M. Pidoux. 1986. Endogenous glutathione protects human skin fibroblasts against the cytotoxic action of UVB, UVA and near-visible radiations. *Photochem. Photobiol.* **44**:561–564.

140. Tyrrell, R. M., and M. Pidoux. 1988. Correlation between endogenous glutathione content and sensitivity of cultured human skin cells to radiation at defined wavelengths in the solar ultraviolet range. *Photochem. Photobiol.* **47**:405–412.

141. Lautier, D., P. Luscher, and R. M. Tyrrell. 1992. Endogenous glutathione levels modulate both constitutive and UVA radiation/hydrogen peroxide inducible expression of the human heme oxygenase gene. *Carcinogenesis* **13**:227–232.

142. Connor, M. J., and L. A. Wheeler. 1987. Depletion of cutaneous glutathione by ultraviolet radiation. *Photochem. Photobiol.* **46**:239–245.

143. Noel, A., R. Tyrrell. 1996. *Photochem. Photobiol.* (submitted).

144. Gomer, C. J., N. Rucker, A. Ferrario, and S. Wong. 1989. Properties and applications of photodynamic therapy. *Radiat. Res.* **120**:1–18.

145. Brunmark, A., and Cadenas E. 1989. Redox and addition chemistry of quinoid compounds and its biological implications. *Free Rad. Biol. Med.* **7**:435–477.

146. Kantengwa, S., and B. S. Polla. 1991. Flavenoids but not protein kinase C inhibitors prevent stress protein synthesis during erythrophagocytosis. *Biochem. Biophys. Res. Commun.* **180**:308–314.

147. Clerget, M., and B. S. Polla. 1990. Erythrophagocytosis induces heat shock protein synthesis by human monocytes-macrophages. *Proc. Natl. Acad. Sci. USA* **87**:1081–1085.

148. Yao, K. S., S. Xanthoudakis, T. Curran, and P. O'Dwyer. 1994. Activation of AP-1 and of a nuclear redox factor Ref-1, in the response of HT29 colon cancer cells to hypoxia. *Mol. Cell. Biol.* **14**:5997–6003.

149. Sakata, K., T. Kwok, K. Laderoute, G. R. Gordon, and R. M. Sutherland. 1991. Hypoxia induced drug resistance: comparison to P-glycoprotein-associated drug resistance. *Br. J. Cancer* **64**:809–814.

150. Kappas, A., and G. S. Drummond. 1984. Control of heme and cytochrome P-450 metabolism by inorganic metals, organometals, and synthetic metalloporphyrins. *Environ. Health Perspect.* **57**:301–306.

151. Maines, M. D., and P. Sinclair. 1977. Cobalt regulation of heme synthesis and degradation in avian embryo liver cell culture. *J. Biol. Chem.* **252**:219–223.

152. Saunders, E. L., M. D. Maines, M. J. Meredith, and M. L. Freeman. 1991. Enhancement of heme oxygenase-1 synthesis by glutathione depletion in chinese hamster ovary cells. *Arch. Biochem. Biophys.* **288**:368–373.

153. Lin, J.H.-C., P. Villalon, P. Martasek, and N. G. Abraham. 1990. Regulation of heme oxygenase gene expression by cobalt in rat liver and kidney. *Eur. J. Biochem.* **192**:577–582.

154. Abraham, N., R. Levere, J. Lin, N. Beru, O. Hermine, and E. Goldwasser. 1991. Coregulation of heme oxygenase and erythropoietin genes. *J. Cell. Biochem.* **47**:43–48.

155. Tomaro, M. L., J. Frydman, and R. B. Frydman. 1991. Heme oxygenase induction by CdCl₂, Co-protoporphyrin IX, phenylhydrazine, and diamide: evidence for oxidative stress involvement. *Arch. Biochem. Biophys.* **286**:610–617.

156. Sardana, M. K., and A. Kappas. 1987. Dual control mechanism for heme oxygenase: tin(IV)-protoporphyrin potently inhibits enzyme activity while markedly increasing content of enzyme protein in liver. *Proc. Natl. Acad. Sci. USA* **84**:2464–2468.

157. Alam, J., J. Cai, and A. Smith. 1994. Isolation and characterization of the mouse heme oxygenase-1 gene. *J. Biol. Chem.* **269**:1001–1009.

158. Alam, J. 1994. Multiple elements within the 5′ distal enhancer of the mouse heme oxygenase-1 gene mediate induction by heavy metals. *J. Biol. Chem.* **269**:25049–25056.

159. Applegate, L. A., A. Noel, G. Vile, E. Frenk, and R. M. Tyrrell. 1995. Two genes contribute to different extents to the heme oxygenase enzyme activity measured in cultured human skin fibroblasts and keratinocytes—implications for protection against oxidant stress. *Photochem. Photobiol.* **61**:285–291.

160. Shelton, K. R., P. M. Egle, and J. M. Todd. 1986. Evidence that glutathione participates in the induction of a stress protein. *Biochem. Biophys. Res. Commun.* **134**:492–498.

161. Freeman, M. L., and M. J. Meredith. 1989. Glutathione conjugation and induction of a 32,000 dalton stress protein. *Biochem. Pharmacol.* **38**:299–304.

162. Koizumi, T., M. Negishi, and A. Ichikawa. 1992. Induction of heme oxygenase by δ-12 prostaglandin J2 in porcine aortic endothelial cells. *Prostaglandins* **43**:121–131.

163. Zollner, H., R. J. Schaur, and H. Esterbauer. 1991. Biological activities of 4-hydroxynonenals, pp. 337–369. *In* H. Sies (ed.), *Oxidative Stress: Oxidants and Antioxidants.* Academic Press, London.

164. Cajone, F., and A. Bernelli-Zazzera. 1988. Oxidative stress induces a subset of heat shock proteins in rat hepatocytes and MH1C1 cells. *Chem. Biol. Interact.* **65**:235–246.

165. Basu Modak, S., P. Luscher, and R. M. Tyrrell. 1996. Lipid Metabolite Involvement In The Activation of The Human Heme Oxygenase-1 Gene, *Free. Rad. Biol. Med.* **20**:887–897.

166. Guy, G. R., J. Cairns, S. B. Ng, and Y. H. Tan. 1993. Inactivation of a redox-sensitive protein phosphatase during the early events of tumor necrosis factor/interleukin-1 signal transduction. *J. Biol. Chem.* **268**:2141–2148.

167. Keyse, S. M., and E. A. Emslie. 1992. Oxidative stress and heat shock induce a human gene encoding a protein-tyrosine phosphatase. *Nature* **359**:644–647.

168. Larsson, R., and P. Cerutti. 1989. Translocation and enhancement of phosphotransferase activity of protein kinase C following exposure in mouse epidermal cells to oxidants. *Cancer Res.* **49**:5627–5632.

169. Abate, C., L. Pate, F. Rauscher, and T. Curran. 1990. Redox regulation of Fos and Jun DNA binding activity *in vitro. Science* **249**:1157–1161.

170. Xanthoudakis, S., and T. Curran. 1992. Identification and characterization of Ref-1, a nuclear protein that facilitates AP-1 DNA binding activity. *EBMO J.* **11**:653–665.

171. Kumar, S., A. Rabson, and C. Gelinas. 1992. The RxxRxRxxC motif conserved in all Rel/kB proteins is essential for the DNA-binding activity and redox regulation of the v-rel oncoprotein. *Mol. Cell. Biol.* **12**:3094–3106.

172. Heinrich, P., J. Castel, and T. Andus. 1990. Interleukin 6 and the acute phase response. *Biochem. J.* **265**:621–636.

173. Kikkawa, U., A. Kishimoto, and Y. Nishizuka. 1989. The Protein Kinase C Family: Heterogeneity and its Implications. *Annu. Rev. Biochem.* **58**:31–44.

174. Hiwasa, T., H. Fujiki, T. Sugimura, and S. Sakiyama. 1983. Increase in the synthesis of a Mr 32,000 protein in BALB/c 3T3 cells treated with tumor promoting indole alkaloids or polyacetates. *Cancer Res.* **43**:5951–5955.

175. Kurata, S. I., and H. Nakajima. 1990. Transcriptional activation of the heme oxygenase gene by TPA in mouse M1 cells during their differentiation to macrophage. *Exp. Cell Res.* **191**:89–94.

176. Muraosa, Y., and S. Shibahara. 1993. Identification of a cis-regulatory element and putative trans-acting factors responsible for 12-O-tetradecanoylphorbol-13-acetate(TPA)-mediated induction of heme oxygenase expression in myelomonocytic cell lines. *Mol. Cell. Biol.* **13**:7881–7891.

177. Shibahara, S., M. Sato, R. M. Müller, and T. Yoshida. 1989. Structural organization of the human heme oxygenase gene and the function of its promoter. *Eur. J. Biochem.* **179**:557–563.

178. Evans, C.-O., J. F. Healey, Y. Greene, and H. L. Bonkovsky. 1991. Cloning, sequencing and expression of cDNA for chick liver haem oxygenase. *Biochem. J.* **273**:659–666.

179. Suzuki, T., M. Sato, K. Ishikawa, and T. Yoshida. 1992. Nucleotide sequence of cDNA for porcine heme oxygenase and its expression in *Escherichia coli*. *Biochem. Int.* **28**:887–893.

180. Okinaga, S., and S. Shibahara. 1993. Identification of a nuclear protein that constitutively recognizes the sequence containing a heat-shock element. Its binding properties and possible function modulating heat-shock induction of the rat heme oxygenase gene. *Eur. J. Biochem.* **212**:167–175.

181. Sato, M., Y. Fukushi, S. Ishizawa, S. Okinaga, R. M. Müller, and S. Shibahara. 1989. Transcriptional control of the rat heme oxygenase gene by a nuclear protein that interacts with adenovirus 2 major late promoter. *J. Biol. Chem.* **264**:10251–10260.

182. Takeda, K., S. Ishizawa, M. Sato, T. Yoshida, and S. Shibahara. 1994. Identification of a cis-acting element that is responsible for cadmium-mediated induction of the human heme oxygenase gene. *J. Biol. Chem.* **269**:22858–22867.

183. Alam, J., and D. Zhining. 1992. Distal AP-1 binding sites mediate basal level enhancement and TPA induction of the mouse heme oxygenase-1 gene. *J. Biol. Chem.* **267**:21894–21900.

184. Alam, J., S. Camhi, and A. Choi. 1995. Identification of a second region upstream of the mouse heme oxygenase-1 gene that functions as a basal level and inducer-dependant transcriptional enhancer. *J. Biol. Chem.* **270**:11977–11984.

185. Sato, M., S. Ishizawa, T. Yoshida, and S. Shibahara. 1990. Interaction of upstream stimulatory factor with the human heme oxygenase gene promoter. *Eur. J. Biochem.* **188**:231–237.

186. Tyrrell, R. M., L. A. Applegate, and Y. Tromvoukis. 1993. The proximal promoter region of the human heme oxygenase gene contains elements involved in stimulation of transcriptional activity by a variety of agents including oxidants. *Carcinogenesis* **14**:761–765.

187. Sawadogo, M., M. Van Dyke, P. Gregor, and R. Roeder. 1995. Multiple forms of the human gene specific transcription factor USF from HeLa cell nuclei. *J. Biol. Chem.* **263**:11985–11993.

188. Nascimento, A. L. T. O., P. Luscher, and R. M. Tyrrell. 1993. Ultraviolet A (320–380 nm) radiation causes an alteration in the binding of a specific protein/protein complex to a short region of the promoter of the human heme oxygenase 1 gene. *Nucleic Acids. Res.* **21**:1103–1109.

189. Waltner, C., A. Tanew, and R. M. Tyrrell. 1994.

190. Solomon, D., B. Amati, and H. Land. 1993. Distinct DNA binding preferences for the c-Myc/Max and Max/Max dimers. *Nucleic Acids Res.* **21**:5372–5376.

191. Lavrovsky, Y., M. L. Schwartzman, and N. G. Abraham. 1993. Novel regulatory sites of the human heme oxygenase-1 promoter region. *Biochem. Biophys. Res. Commun.* **196**:336–341.

192. Lavrovsky, Y., M. L. Schwartzman, R. D. Levere, A. Kappas, and N. G. Abraham. 1994. Identification of binding sites for transcription factors NF-kappaB and AP-2 in the promoter region of the human heme oxygenase 1 gene. *Proc. Natl. Acad. Sci. USA* **91**:5987–5991.

193. Mitani, K., H. Fujita, S. Sassa, and A. Kappas. 1991. A heat inducible nuclear factor that binds to the heat-shock element of the human heme oxygenase gene. *Biochem. J.* **277**:895–897.

194. Schreck, R., P. Rieber, and P. Baeuerle. 1991. Reactive oxygen intermediates as apparently widely used messengers in the activation of the NF-κB transcription factor and HIV-1. *EMBO J.* **10**:2247–2258.

195. Wegenka, U. M., J. Buschmann, C. Lütticken, P. Heinrich, and F. Horn. 1993. Acute-phase response factor, a nuclear factor binding to acute-phase response elements, is rapidly activated by interleukin-6 at the posttranslational level. *Mol. Cell. Biol.* **13**:276–288.

196. Koizumi, T., N. Odani, T. Okuyama, A. Ichikawa, and M. Negishi. 1995. Identification of a cis-regulatory element for delta(12) prostaglandin J(2)-induced expression of the rat heme oxygenase gene. *J. Biol. Chem.* **270**:21779–21784.

197. Camhi, S. L., J. Alam, L. Otterbein, S. L. Sylvester, and A. M. K. Choi. 1995. Induction of heme oxygenase-1 gene expression by lipopolysaccharide is mediated by AP-1 activation. *Am. J. Respir. Cell. Mol. Biol.* **13**:387–398.

198. Basu-Modak, S., L. Richman, and R. Tyrrell. 1996. Hypersensitive site analysis of the human heme oxygenase 1 promoter. Unpublished Observation.

199. Brodersen, R. 1980, Binding of Bilirubin to Albumin. *Crit. Rev. Clin. Lab. Invest.* **11**:305–399.

200. Farrel, G., J. Golan, and R. Schmid. 1980. Efflux of bilirubin into plasma following hepatic degradation of exogenous heme. *Proc. Soc. Exp. Biol. Med.* **163**:504–509.

201. Neuzil, J., and R. Stocker. 1993. Bilirubin attenuates radical-mediated damage to serum albumin. *FEBS Lett.* **331**:281–284.

202. Stocker, R., and Ames B. 1987. Potential role of conjugated bilirubin and copper in the metabolism of lipid peroxides in the bile. *Proc. Natl. Acad. Sci. USA* **84**:8130–8134.

203. Stocker, R., and E. Peterhans. 1989. Antioxidant properties of conjugated bilirubin and biliverdin: biologically relevant scavenging of hypochlorous acid. *Free Rad. Res. Commun.* **6**:57–66.

204. Neuzil, J., and R. Stocker. 1994. Free and albumin-bound bilirubin are efficient co-antioxidants for alpha-tocopherol, inhibiting plasma and low density density lipoprotein lipid peroxidation. *J. Biol. Chem.* **269**:16712–16719.

205. Stocker, R., and E. Peterhans. 1989. Synergistic interaction between vitamin E and the bile pigments bilirubin and biliverdin. *Biochim. Biophys. Acta* **1002**:238–244.

206. Meister, A, and M. E. Anderson. 1983. Glutathione. *Annu. Rev. Biochem.* **52**:711–760.

207. Hershko, C., G. Link, and A. Pinson. 1987. Modification of iron uptake and lipid peroxidation by hypoxia, ascorbic acid, and alpha tocopherol in iron loaded rat myocardial cell cultures. *J. Lab. Clin. Med.* **11**:355–361.

208. Jackson, J., I. Schraufstatter, and P. Hyslop. 1987. Role of oxidants in DNA damage. Hydroxyl radical mediates the synergistic DNA damaging effects of asbestos and cigarette smoke. *J. Clin. Invest.* **80**:1090–1095.

209. Balla, G., G. M. Vercellotti, J. W. Eaton, and H. S. Jacob. 1990. Iron loading of endothelial cells augments oxidant damage. *J. Lab. Clin. Med.* **116**:546–554.

210. Theil, E. 1987. Ferritin: structure, gene regulation, and cellular function in animals, plants, and microorganisms. *Annu. Rev. Biochem.* **56**:289–315.

211. Eisenstein, R. S., and H. Munro. 1990. Translational regulation of ferritin synthesis by iron. *Enzyme* **44**:42–58.

212. Balla, G., H. Jacob, J. Balla, J. Rosenberg, K. Nath, and F. Apple. 1992. Ferritin: a cytoprotective stratagem of endothelium. *J. Biol. Chem.* **267**:18148–18153.

213. Balla, J., H. S. Jacob, G. Balla, K. Nath, J. W. Eaton, and G. M. Vercellotti. 1993. Endothelial-cell heme uptake from heme proteins: induction of sensitization and desensitization to oxidant damage. *Proc. Natl. Acad. Sci. USA* **90**:9285–9289.

214. Cairo, G., L. Tacchini, G. Pogliaghi, E. Anzon, A. Tomasi, and A. Bernelli-Zazzera. 1995. Induction of ferritin synthesis by oxidative stress: transcriptional and post-transcriptional regulation by expansion of the "free" iron pool. *J. Biol. Chem.* **270**:700–703.

215. Yoshinaga, T., S. Sassa, and A. Kappas. 1982. A comparative study of heme degradation by NADPH-cytochrome c reductase alone and by the complete heme oxygenase system. Distinctive aspects of heme degradation by NADPH-cytochrome c reductase. *J. Biol. Chem.* **257**:7794–7802.

216. Kim, E., and A. Sevanian. 1991. Hematin and peroxide catalyzed peroxidation of phospholipid microsomes. *Arch. Biochem. Biophys.* **288**:324–330.

217. Guengerich, P., and T. Macdonald. 1990. Mechanisms of cytochrome P-450 catalysis. *FASEB J.* **4**:2453–2459.

218. Neuzil, J., J. Gebicki, and R. Stocker. 1993. Radical induced chain oxidation of proteins and its inhibition by chain breaking asntioxidants. *Biochem. J.* **293**:601–606.

219. Davies, K., and M. Delsignore. 1986. Protein damage and degradation by oxygen radicals III, modification of secondary and tertiary structure. *J. Biol. Chem.* **262**:9908–9913.

220. Kutty, R. K., R. F. Daniel, D. E. Ryan, W. Levin, and M. D. Maines. 1988. Rat liver cytochrome P-450b, p-420b, p-420c are degraded to biliverdin by heme oxygenase. *Arch. Biochem. Biophys.* **260**:638–644.

221. Yoshinaga, T., S. Sassa, and A. Kappas. 1982. The oxidative degradation of heme c by the microsomal heme oxygenase system. *J. Biol. Chem.* **257**:7803–7807.

222. Yoshida, T., M. Noguchi, and G. Kikuchi. 1982. The step of carbon monoxide liberation in the sequence of heme degradation catalyzed by the reconstituted microsomal heme oxygenase system. *J. Biol. Chem.* **257**:9345–9348.

223. Barinaga, M. 1993. Carbon monoxide: killer to brain messenger in one step. *Science* **259**:309.

224. Stone, J., and M. Marletta. 1994. Soluble guanylate cyclase from bovine lung: activation with nitric oxide and carbon monoxide and spectral characterization of the ferrous and ferric states. *Biochemistry* **33**:5636–5640.

225. Lowenstein, C. J., and S. H. Snyder. 1992. Nitric oxide, a novel biologic messenger. *Cell* **70**:705–707.

226. Moncada, S. E., R. M. Palmer, E. Higgs. 1989. Biosynthesis of Nitric Oxide from L-Arginine a Pathway For the Regulation of Cell Function and Communication. *Biochem. Pharmacol.* **38**:1709–1715.

227. Bredt, D., P. Hwangt, P. Glatt, C. Lowenstein, R. Reed, and S. Snyder. 1991. Cloned and expressed nitric oxide synthase structurally resembles cytochrome P-450 reductase. *Nature* **351**:714–718.

228. Snyder, S. H., and D. S. Bredt. 1992. Biological roles of nitric oxide. *Sci. Am.* **266**:68–77.

229. Weber, C. M., B. C. Eke, and M. D. Maines. 1994. Corticosterone regulates heme oxygenase-2 and NO synthase transcription and protein expression in rat brain. *J. Neurochem.* **63**:953–962.

230. Marks, G. S. 1994. Heme oxygenase—the physiological role of one of its metabolites, carbon monoxide, and interactions with zinc-protoporphyrin, cobalt protoporphyrin and other metalloporphyrins. *Cell. Mol. Biol.* **40**:863–870.

231. Marks, G. S., K. Nakatsu, and J. F. Brien. 1993. Does endogenous zinc protoporphy-rin modulate carbon monoxide formation from heme? Implications for long-term potentiation, memory, and cognitive function. [Review]. *Can. J. Physiol. Pharma-col.* **71**:753–754.

232. Hawkins, R. D., M. Zhuo, and O. Arancio. 1994. Nitric oxide and carbon monoxide as possible retrograde messengers in hippocampal long-term potentiation. *J. Neuro-biol.* **25**:652–665.

233. Dawson, T., and S. Snyder. 1994. Gasses as biological messengers: nitric oxide and carbon monoxide in the brain. *J. Neurosci.* **14**:5147–5159.

234. Stevens, C. F., and Y. Wang. 1993. Reversal of long-term potentiation by inhibitors of haem oxygenase. *Nature* **364**:147–149.

235. Ikegaya, Y., H. Saito, and N. Matsuki. 1994. Involvement of carbon monoxide in long term potentiation in the dentate gyrus of aenesthetized mice. *Jpn. J. Pharma-col.* **64**:225–227.

236. Poss, K. D., M. J. Thomas, A. K. Ebralidze, T. J. Odell, and S. Tonegawa. 1995. Hippocampal long term potentiation is normal in heme oxygenase-2 mutant mice. *Neuron* **15**:867–873.

237. Meffert, M. K., J. E. Haley, E. M. Schuman, H. Schulman, and D. V. Madison. 1994. Inhibition of hippocampal heme oxygenase, nitric oxide synthase, and long-term potentiation by metalloporphyrins. *Neuron* **13**:1225–1233.

238. Greenbaum, N., and A. Kappas. 1991. Comparative photoactivity of tin and zinc porphyrin inhibitors of heme oxygenase: pronounced photolability of the zinc compounds. *Photochem. Photobiol.* **54**:183–192.

239. Rattan, S., and S. Chakder. 1993. Inhibitory effect of CO on internal anal sphincter: heme oxygenase inhibitor inhibits NANC relaxation. *Am. J. Physiol. Gastrointest. Liver Physiol.* **265**:G799–G804.

240. Suematsu, M., S. Kashiwagi, T. Sano, N. Goda, Y. Shinoda, and Y. Ishimura. 1994. Carbon monoxide as an endogenous modulator of hepatic vascular perfusion. *Biochem. Biophys. Res. Commun.* **205**:1333–1337.

16

Induction of Gene Expression by Environmental Oxidants Associated with Inflammation, Fibrogenesis, and Carcinogenesis

Yvonne M. W. Janssen, Cynthia R. Timblin, Christine L. Zanella, L. Albert Jimenez, and Brooke T. Mossman

16.1 Introduction

Several environmental toxicants may cause cell injury and disease through pathways generating active oxygen species (AOS) and/or affecting the redox status of cells.[1,2] Although biochemical pathways of generation of AOS have been characterized for many agents, little is known about how agents affect gene expression in target cells of disease or in cells of the immune system. For many pollutants, such as asbestos and tobacco smoke, oxidant generation occurs by multiple pathways including direct reactions catalyzed by fibers or components of cigarette smoke and indirect mechanisms involving cell interactions and metabolism. Since many organelles may be affected by agents, the regulation of gene expression is complex. In addition, a multiplicity of genes may be activated in specific cell types, which then may act coordinately or in opposing fashions to cause phenotypic changes. Lastly, many oxidant-generating agents cause proliferative effects at low concentrations[3] and cytotoxicity or cell death at high concentrations.[1,2] Thus patterns of gene expression may be governed by dose-response relationships.

Unraveling the complex pathways and patterns of altered gene expression by environmental agents presents a challenge to molecular biologists and toxicologists. However, gaining an understanding of how pathogenic pollutants elicit these events in the target cells of disease is necessary both for preventive and therapeutic approaches to disease. To this end, we have studied the responses of cells of the lung and pleura to the minerals, asbestos, and silica, in an attempt to map the sequence of signaling events elicited by mineral dusts in relationship to patterns of gene expression. Both asbestos and silica are fibrogenic agents in lung, and the processes of asbestosis and silicosis may contribute to the development of lung cancers in smokers.[4] In addition, various asbestos types are associated with induction of the malignancy mesothelioma.[5] The potency of these chemically distinct minerals may be related to their physiochemical nature as well as their insolubility and retention in lung.[6] The ability of various silica polymorphs and asbestos fibers to generate AOS by redox reactions occurring on the mineral surface has been documented by several laboratories.[7-10] A correlation between iron content and mobilization of iron and relative pathogenicity of asbestos types (i.e., crocidolite and amosite) has suggested the importance of iron-driven free radical generation in disease.[11,12] Moreover, fibers and particles can generate AOS after phagocytosis by cells.[13,14]

As discussed in detail in subsequent sections of this chapter, mineral dusts elicit increased expression of a variety genes that are induced by oxidant stresses in other cell types. Many of these genes may be relevant to initiation of disease (i.e., early response genes etc.) whereas others (i.e., genes encoding antioxidant enzymes) may participate in lung defense and repair. In this chapter, we first focus on gene expression as related to cell proliferation, DNA damage, and inflammation, as these are hallmarks of fibrogenesis and carcinogenesis in lung. We then describe genes that may be elicited in cells that may be related to important defense mechanisms such as apoptosis and antioxidant defense. Lastly, we present some data showing the cell signaling cascades important in regulation of gene expression by asbestos and how they may be activated by oxidant-dependent mechanisms.

16.2 Genes Important in Cell Proliferation

A prominent feature of mineral-dust-induced diseases is abnormal and unregu- lated proliferation of target cells of the lung and pleura.[5,15-18] Considerable evidence links asbestos-induced cell proliferation to activation of transcription factors and increased expression of several genes associated with the control of cell replication. Furthermore, AOS generated directly on the asbestos fiber surface or elaborated by recruited inflammatory cells are implicated as mediators of this response. However, it is not clearly understood how asbestos and free radicals interact with cells at the molecular level to cause alterations in cell growth or how these pathways lead to development of disease.

In rodent inhalation models, the development of asbestos-induced fibrosis (asbestosis) is associated with dosage-dependent, increased steady-state mRNA levels of the proliferation-associated genes, c-jun, and the gene encoding ornithine decarboxylase (ODC).[19] ODC, an essential enzyme in the biosynthesis of poly- amines, is required for cellular replication.[20] As discussed in previous chapters, c-jun is a member of a multigene family that is transiently induced in response to a variety of stimuli and encodes the Jun subunit (Jun/Jun homodimers or Jun/ Fos heterodimers) of the transcription factor, activator protein-1 (AP-1).[21] c-jun and members of the c-fos gene family are immediate early response genes involved in the transition of the G1 phase and entry into the S phase of the cell cycle.[21]

Significantly increased steady-state levels of ODC and c-*jun* mRNAs are observed following exposure of rats to crocidolite asbestos, but not after exposure to chrysotile asbestos, highlighting the importance of fiber geometry and composition in changes in gene expression.[22] Moreover, a direct relationship between elevations in c-*jun* and ODC gene expression and pulmonary fibrosis is suggested by the observation that disease occurs in the crocidolite, but not the chrysotile inhalation model.[19]

The molecular events triggered by asbestos that lead to increased expression of proliferation-related genes have been examined in cultures of rodent tracheal epithelial and pleural mesothelial cells, progenitor cell types for bronchogenic carcinoma and malignant mesothelioma, respectively.[5] In pleural mesothelial cells, exposure to crocidolite or chrysotile asbestos causes dose-dependent and persistent increases in the steady-state mRNA levels of c-*jun* and c-*fos* and increased DNA binding of the transcription factor AP-1.[23] In contrast, increased expression of c-*jun*[23] and ODC,[24] but not c-*fos*, is observed in tracheal epithelial cells exposed to asbestos. In tracheal epithelial cells, exposure to asbestos increases AP-1 DNA binding activity and directly activates AP-1-dependent gene transcription.[23,25] Furthermore, tracheal epithelial cells transiently transfected with a plasmid that overexpresses c-*jun* show increased cell proliferation and growth in soft agar, an indicator of cellular transformation.[25] In support of our results, unregulated expression of the photooncogene, c-*jun*, can lead to aberrant proliferation and cellular transformation in a number of different cell types.[21]

In addition to activation of AP-1 activity and increased expression of c-*jun* and c-*fos,* exposure of tracheal epithelial cells to asbestos causes dose-dependent increases in NF-κB DNA-binding activity and NF-κB-dependent gene expression.[26] NF-κB is a transcription factor involved in activation of genes that are involved in cell proliferation and inflammation.[27] The promoter regions of many of these genes (inducible nitric oxide synthase, [iNOS], ODC, and manganese-containing superoxide dismutase [MnSOD]) including c-*myc,* have NF-κB sites in their promoter elements. Tracheal epithelial cells exposed to asbestos show both increased NF-κB-DNA binding complexes and increased expression of c-*myc*.[26] Persistent induction of early response genes such as c-*jun,* c-*fos,* and c-*myc* and other proliferation-related genes, i.e., ODC, may be one mechanism by which asbestos induces chronic cell proliferation, a hallmark of asbestos-induced lung disease.

AOS may be the mediators of transcriptional activation of some, but not other early response genes induced by asbestos. In comparative studies, exposure of tracheal epithelial cells to hydrogen peroxide (H_2O_2) or asbestos causes striking increases in c-*jun,* and ODC mRNA levels.[24,28] However, no increases in c-*fos* mRNA levels are detected in tracheal epithelial cells after exposure to asbestos in contrast to the induction observed following exposure to H_2O_2.[28] Exposure of pleural mesothelial cells to H_2O_2 or the superoxide-generating system, xanthine plus xanthine oxidase, failed to induce expression of c-*jun* or c-*fos* in contrast to the persistent induction of both protooncogenes by asbestos.[29] These observations indicate that the pathways of transcriptional activation of gene expression by asbestos and AOS in tracheal epithelial and pleural mesothelial cells are different.

Activation of AP-1[21] and NF-κB[30-32] and transcriptional activity and expression of a number of genes appears to be regulated by the redox status of the cell. Binding of AP-1[21] and NF-κB[33] to their respective regulatory sequences is abolished by oxidation. This redox regulation is mediated through modification of critical cysteine residues within the proteins. Increases in NF-κB[32] and AP-1[34] DNA-binding activity have been demonstrated in a number of model systems that involve oxidant stress, implicating oxidants as activators of signaling cascades that lead to changes in gene expression.

Studies *in vitro* using pleural mesothelial cells demonstrate that exposure to asbestos causes depletion of cellular glutathione indicating a change in the redox status of the cell.[26,29] Furthermore, asbestos-induced expression of c-*jun* and c-*fos* in these cells is blocked by the glutathione precursor, *N*-acetyl-L-cysteine (NAC) and amplified by pretreatment with buthionine sulfoximine (BSO) which depletes glutathione pools.[29] In tracheal epithelial cells exposed to asbestos, NF-κB DNA-binding and NF-κB-dependent gene expression is also significantly decreased by preexposure to NAC.[26] These results suggest that asbestos-generated oxidants or alterations in the redox status of the cell may contribute to activation of transcription factors and thereby mediate changes in gene expression. Chronic stimulation of proliferation-related genes, i.e., c-*jun*, c-*fos*, c-*myc*, or ODC, also may lead to unregulated cell proliferation and facilitate the fixation of genetic damage into the genome.

16.3 Stress Response Genes

Mammalian cells respond to a diversity of environmental insults, both physical and chemical, by inducing the expression of a group of genes referred to as stress response genes. The induction of these genes and the function of their encoded protein products define a defense mechanism(s) whereby cells detect damaged cellular components and activate appropriate scavenging and repair pathways. Historically, this response was observed in different organisms and cells in response to hyperthermia and was known as the heat shock response. The proteins encoded by these genes are called the heat shock proteins (HSPs). More recently, experimental data has shown that a number of different environmental and physiological agents can elicit this response and hence, it is now more appropriately referred to as the stress response.

The HSPs are ubiquitous, highly conserved proteins that are induced in response to the accumulation of denatured and damaged proteins within the cell. These proteins are grouped into five families based on relative molecular mass and to some extent on their patterns of induction.[35] A diversity of environmental agents including heavy metals, oxidizing agents, sulfhydryl reagents, tobacco smoke, and ultraviolet light as well as physiological conditions, i.e., inflammation, can elicit a stress response.[35-38] During normal cellular metabolism, HSPs function as molecular chaperones to facilitate the correct folding and translocation of proteins.[35,36] In response to stress, HSPs bind denatured and damaged proteins within the cell and prevent their aggregation. Increasing evidence also indicates that some members of the HSP family may also play a role in cell signaling events from subcellular compartments to the nucleus[39] and in cell growth and differentiation.[40,41]

Regulation of HSP synthesis is controlled at the level of gene transcription by an autoregulatory loop that involves the interaction of a constitutively synthesized heat shock factor (HSF) and one member of the HSP family, the HSP70 protein. In the nonstressed cell, HSF is maintained in a non-DNA-binding, monomeric form stabilized by its interaction with HSP70. On accumulation of damaged proteins, the HSP70/HSF complex dissociates, and HSF translocates to the nucleus where a trimeric form of this protein recognizes and binds the heat shock element (HSE) located in the promoter regions of heat shock responsive genes and activates transcription.[42,43] Some agents that induce DNA-binding activity of HSF do not activate transcription indicating a second level of control.[44]

The induction of HSP gene expression by environmental oxidants (i.e., metals, ozone, mineral dusts) has been examined in a number of model systems. In human mesothelial cells, exposure to asbestos causes a dose-dependent increases in mRNA levels for heme oxygenase.[45] The oxidant-generating system, xanthine plus xanthine oxidase, induces similar increases in heme oxygenase mRNA expression.[45] Heme oxygenase is an enzyme that catalyzes the breakdown of heme to biliverdin. In subsequent enzymatic reactions, biliverdin is converted to bilirubin, thereby reducing heme pools that can provide iron for Fenton reactions.[46] Thus, induction of heme oxygenase indicates a response to oxidative stress.[47] Increased levels of heme oxygenase mRNA are also observed in rat pleural mesothelial cells exposed to asbestos.[29] The induction of heme oxygenase is consistent with the proposal that generation of oxidants is one mechanism of cell injury by asbestos fibers.

Studies in our laboratory employing rat lung epithelial cells have shown that expression of heme oxygenase is also induced by exposure of cells to cadmium.[1] Cadmium is a sulfhydryl reactive molecule that can interact with proteins and modify cellular glutathione levels promoting an environment of oxidative stress.[48] Sodium arsenite, another modulator of glutathione pools, also increases the expression of heme oxygenase mRNA.[47] Exposure to cadmium has also been linked to the induction of several other heat shock genes. Expression of the genes encoding HSP60 and HSP70 is induced following exposure of cells to cadmium, although the kinetics of induction are different for the two genes.[49] HSP60 functions as a molecular chaperone in protein translocation and assembly of a number of mitochondrial enzyme complexes.[35] HSP70 also functions in protein chaperoning and in regulation of HSP gene transcription as described above. Regulation of heat shock gene transcription by cadmium requires dissociation of a negative transcriptional regulatory factor from the promoter region and binding of HSF to the heat shock element.[50] In addition, cadmium-induced transcription of *hsp*70 is modulated by cellular glutathione levels, implicating a role for free radicals in upregulation of this gene.[48]

Changes in HSP expression are also observed in animal models of oxidant-induced lung toxicity and disease. Dose-dependent increases in HSP72 protein is observed in rat lungs following intravenous administration of sodium arsenite.[51] Increased synthesis of heme oxygenase also is observed in the lungs of animals exposed to zinc oxide (ZnO), but not to ozone.[52] ZnO and ozone are components of metal fumes generated in some occupational settings and are implicated in the development of welding-related respiratory disease. Ozone, while not a free radical itself, can interact with different cellular molecules to produce free radicals and generate oxidant stress.[53]

Although induction of heat shock proteins in response to hyperthermia is well documented, the characterization of the pattern(s) of expression of these proteins in specific target organs and/or cells induced by different environmental stress agents is just beginning. In response to hyperthermic treatment of cells, the induction of all HSPs has been shown to be coordinately regulated. However, it remains to be determined whether this is a general phenomenon of HSP synthesis or if there are agent- and/or cell-type-specific components to the induction of this stress response. Differential patterns of stress protein expression in pulmonary target cells of disease, as well as other tissue types, may provide insight into the molecular mechanisms triggered by environmental oxidants within cells that lead to adaptation and maintenance of normal function, or to cytotoxicity and the development of the disease state.

The induction of the stress response is not the only response of cells to cellular damage caused by environmental oxidants. Many physical and chemical agents that cause proteotoxicity and elicit a stress response also damage DNA (genotoxic stress) and lead to the induction of a different set of genes.[54] The cellular response to genotoxic stress includes the transcriptional activation of genes that function to arrest cell growth presumably to provide time for repair of DNA damage. In addition, many environmental oxidants stimulate the transcriptional activation of genes encoding antioxidants.[55] These proteins function to scavenge and eliminate excessive AOS and thus, comprise another line of cellular defense. The cellular response to oxidant-induced damage is not an isolated event, but involves many different cellular compartments and components.

16.4 Genes Important in the Inflammatory Response

A variety of reactive environmental pollutants damage the lung through mechanisms that involve an inflammatory response. Pulmonary inflammation can be triggered by different mechanisms that involve direct activation of phagocytic cells or chronic irritation of pulmonary target cells.[56] For instance, inhaled particulate material such as asbestos or silica is removed from the lung by phagocytic cells that become activated following contact.[4,10,14,57,58] Sustained activation of macrophages can result in a prolonged release of AOS and proinflammatory cytokines that can recruit remote phagocytic cells to the site of insult and can cause chronic damage.[56] Following inhalation, asbestos fibers, due to their fibrous structure are not successfully phagocytosed, which causes sustained activation of the cell with a prolonged release of inflammatory mediators that include, tumor necrosis factor (TNF),[57] interleukin-1 (IL-1),[57,58] and AOS.[10,13,14] Other particulate matter, including silica or titanium dioxide can evoke similar responses. Due to the insoluble nature of these particulates, cell death occurs with the subsequent renewed release of these pathogenic materials into the lung causing a continuous inflammatory stimulus.[56,58]

Another pathway of activation of the inflammatory cascade that occurs after exposure to reactive environmental pollutants involves acute damage of pulmonary target cells of disease.[59,60] On injury, these cells can release cytokines to recruit inflammatory cells to the site of cell damage. Prolonged irritation may therefore cause a persistent inflammatory response that contributes to pulmonary disease associated with exposure to reactive intermediates.[59,60,61]

The elicitation of an inflammatory response in lung requires networking of various cytokines and chemokines that act on specific target cells,[62] activation of transcription factors, and increases in expression of inflammatory genes.[63] During the inflammatory process, a number of proteins that are produced are controlled at the transcriptional level. As mentioned above, exposure to asbestos fibers or silica particulates causes release of TNF and IL-1 from macrophages.[58,64] These increases are accompanied by elevations in mRNA expression of these cytokines.[65] Both TNF and IL-1 are potent inducers of the transcription factor, NF-κB. NF-κB in its inactive form is sequestered in the cytoplasm by the inhibitor protein IκB. During the process of activation of NF-κB, IκB is phosphorylated and degraded though the proteasome pathway, and the active transcription factor translocates to the nucleus to activate transcription of NF-κB-dependent genes.[66]

Recent studies in our laboratory have demonstrated that asbestos fibers cause activation of NF-κB in a variety of pulmonary target cells.[26,67] Evidence for an oxidant-sensitive step in activation of the transcription factor by asbestos stems from studies that demonstrate an amelioration of NF-κB induction in cells that were pretreated with the glutathione precursor and antioxidant, NAC.[26] Furthermore, nonfibrous and nonpathogenic particulates do not cause activation of NF-κB, suggesting a direct correlation between the toxicity of particulates and the activation of this transcription factor.[67] To further determine the role of NF-κB in lung, we measured p65, the active component of NF-κB in lungs of rats exposed to chrysotile asbestos via inhalation. Increases in p65 protein levels, as evidenced by enhanced immunofluorescence occurred in mesothelium and bronchial epithelium compartments and in the parenchyma.[67] These findings are indicative of activation of NF-κB *in vivo*, in response to inhaled pathogenic particulates.

Activation of NF-κB has traditionally been observed in models employing inflammatory mediators where inhibition by antioxidant compounds also is observed.[68] In addition, a number of models that cause oxidative stress such as ultraviolet radiation, ionizing radiation, and chemical generating systems of AOS also result in activation of NF-κB.[69,70] In concert, these data suggest that oxidative stress may be a critical denominator in the activation of NF-κB.

Genes that are transcriptionally controlled by NF-κB have a diversity of functions. However, a number of genes play a role in the inflammatory process and include, TNF,[71] iNOS,[72] interleukin-6 (IL-6),[73] interleukin-8 (IL-8),[74] monocyte chemotactic protein -1 (MCP-1),[75] and macrophage inflammatory protein-2 (MIP-2).[76] Some of these gene products are chemotactic factors that direct recruitment of phagocytic cells to the site of injury whereas others are proinflammatory cytokines involved in a number of other processes necessary to induce inflammation.

Inhalation or intratracheal instillation of the inflammatory and fibrogenic minerals, asbestos, silica, or ultrafine titanium dioxide into rodent lung causes increases in mRNA levels of cytokine induced neutrophil chemotactic factor (CINC),[77] MIP,[262] and TNF[78] in lung tissue. Furthermore, increased release of IL-8 from bronchial epithelial[79] or mesothelial cells[80] occurs in response to asbestos and may contribute to the development of asbestos-associated disease.[81] Lavaged cells from rodents exposed to asbestos or from workers with asbestotic lung disease release increased levels of many cytokines regulated by NF-κB.[82] These findings indicate that activation of NF-κB-dependent pathways occur after exposure to asbestos and related pathogenic minerals.

In summary, a number of inflammatory mediators are elaborated by pathogenic minerals that cause oxidative stress. The induction of an inflammatory response requires networking of various cytokines and chemokines, as depicted in Figure 16.1. Current studies also suggest that an NF-κB-dependent pathway may be contributing to the inflammatory response that is observed in these models of pulmonary disease.

16.5 Antioxidant Defenses

A variety of environmental pollutants (asbestos, silica, oxidant gases, etc.) albeit chemically and physically distinct have a common feature, i.e., the ability to cause oxidative stress. Asbestos fibers, many of which contain iron have the intrinsic ability to generate the highly reactive hydroxyl radical through Fenton-like chemical reactions catalyzed on the fiber surface.[7,8] Similarly, silica particles, especially when freshly crushed to generate silanol groups, display this ability.[9,10]

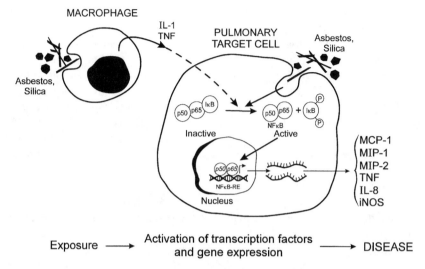

Figure 16.1. Cytokine networking and activation of NF-κB in pulmonary target cells exposed to asbestos or silica. TNF and IL-1, released from macrophages on contact of these particulates, can interact with pulmonary cells to activate NF-κB. Alternatively, particulates may interact with target cells directly to activate this transcription factor. Activation of NF-κB and additional transcription factors up-regulate expression of a variety of genes important in inflammation. MCP-1, MIP-1, MIP-2, TNF, IL-8, and iNOS are genes under control of NF-κB, and known to be induced by mineral dusts. Prolonged activation of the inflammatory cascade may contribute to pulmonary disease observed after exposure to asbestos or silica. IL-1, interleukin-1; TNF, tumor necrosis factor; NF-κB, nuclear factor-kappa B; NF-κB-RE, nuclear factor kappa B response element; MCP-1, monocyte chemotactic protein-1; MIP-1, macrophage inflammatory protein-1; MIP-2, macrophage inflammatory protein-2; IL-8, interleukin-8; iNOS, inducible nitric oxide synthase.

In addition, insoluble particulates such as asbestos and silica are phagocytosed, causing an oxidative burst of phagocytic cells. Activation of nitric oxide synthase in these cells cause release of nitric oxide, another potent free radical species.[83]

Oxidant gases such as ozone, nitrogen dioxide, sulfur dioxide, and cigarette smoke contain a variety of free radical species with oxygen, nitrogen, sulfur, or carbon centers.[55] In the lung, a variety of intermediates are formed with a different reactivity. It is therefore unclear which free radical species are involved in the development of disease associated with exposure to these reactive gases.

The lung has developed an intricate antioxidant defense mechanism to counteract the deleterious effects of oxidants. This system is localized in various compartments of the cell, or in the extracellular milieu.[2,55] In addition to traditional antioxidant enzymes, heme oxygenase may have an antioxidant function due to the removal of heme groups that may be a source of OH· radicals formed in

Fenton reactions.[46,47] As mentioned before, heme oxygenase mRNA levels are up-regulated by a variety of agents that cause oxidative stress or deplete glutathione levels.[46,47] Therefore, heme oxygenase may be a sensitive molecular marker to detect oxidative stress in cells of the lung. Metallothionein is a cysteine-rich molecule that binds and detoxifies reactive metals, but also serves as an antioxidant.[84]

Whether lung disease develops after exposure to reactive pollutants is dependent on the balance between oxidant stress and antioxidant defenses. In conditions of extensive or prolonged oxidant stress, antioxidant defenses are overwhelmed and disease ensues. During mild oxidant stress, the lung is able to mount an oxidant stress response that can protect against subsequent AOS, in a phenomenon referred to as adaptation.[2]

Manganese-containing superoxide dismutase (MnSOD), compartmentalized in mitochondria, is the major antioxidant enzyme that appears to be induced at the gene and protein level in a variety of models that involve pulmonary oxidative stress. Studies in our laboratory have demonstrated that inhalation of fibrogenic particulates such as asbestos, silica, and ultrafine minerals cause increases in mRNA levels of MnSOD in rat lung that are accompanied by increases in MnSOD protein levels and enzyme activity.[85,86] The extent of induction of MnSOD mRNA levels correlates directly with the inflammatory response elicited by these minerals. These observations led to the hypothesis that MnSOD is a biomarker for oxidant stress triggered by inflammatory and fibrogenic minerals.[86] Ultrastructural studies to localize MnSOD in rat lung revealed that it occurred predominantly in mitochondria of type II pneumocytes.[87] Furthermore, MnSOD levels were increased in this cell type in rat lung after inhalation of fibrogenic concentration of asbestos or silica.[87] This suggests an antioxidant defense response of the type II cell that is insufficient to protect against the development of asbestos or silica-induced pulmonary disease. This finding may represent a condition where the oxidant stress response is insufficient to protect against development of disease.

Other laboratories have also demonstrated increases in mRNA or protein levels of MnSOD in lungs exposed to a variety of oxidants that include hyperoxia[88] and ozone.[89] Examination of other antioxidant enzymes showed no or modest increases at the level of gene expression after exposure to oxidants, mineral dusts or chemical generating system of AOS in vitro.[90] For example, our laboratory has demonstrated minor increases in gene expression of copper-zinc-containing superoxide dismutase (CuZnSOD), glutathione peroxidase, or catalase, in response to inhalation of asbestos, silica, or titanium dioxide, in particular when compared to the increases of MnSOD mRNA or protein levels.[85,86] These observations may indicate that the mitochondria is a primary target of AOS by oxidant stresses. Alternatively, the control of other antioxidant enzymes in lung may not be at the level of gene expression.

Studies in target cells of pulmonary disease have also demonstrated oxidant stress responses that are characterized by increases in MnSOD levels or gene expression of heme oxygenase. For instance, normal human pleural mesothelial cells exposed to asbestos or a xanthine plus xanthine oxidase generating system of AOS exhibit increased expression of MnSOD and heme oxygenase that are accompanied by increases in MnSOD protein levels.[45] Concurrent examination of normal adult lung fibroblasts revealed only a minor stress response to asbestos, indicated by lack of increases in MnSOD mRNA or protein levels and only slight increases in expression of heme oxygenase.[45] Human adult lung fibroblasts are more resistant to the toxic effects of asbestos in comparison to mesothelial cells. Thus, the relative lack of an oxidant stress response appears to correlate with a higher resistance to asbestos. To determine a protective role for antioxidant enzymes in conditions of oxidative stress, a number of approaches have been undertaken to enhance cellular levels of antioxidants. Liposome-entrapped or polyethylene-glycol-conjugated antioxidant enzymes, when administered to rats have been shown to protect against lung damage associated with exposure to asbestos[16] or hyperoxia.[91] Furthermore, our laboratory and others utilized transfection techniques to increase MnSOD in hamster tracheal epithelial and other cell types.[92-96] MnSOD ameliorated the cell damage induced by asbestos and other stresses such as hyperoxia, paraquat, etc.[92-96] Furthermore, transgenic mice with increased levels of antioxidant enzymes are also protected against oxidative stress.[97] Conversely, mice lacking extracellular superoxide dismutase exhibit an increased sensitivity to oxidants.[98]

In summary, after exposure to reactive environmental pollutants such as pathogenic minerals or oxidant gases, the lung can mount an antioxidant response to limit oxidant-induced damage. MnSOD appears to be the major antioxidant enzyme that is up-regulated in a variety of models, whereas the other antioxidant enzymes remain largely unaffected. However, the extent of increases in MnSOD that occur in these models is insufficient to prevent the development of disease. Exogenous administration of antioxidant enzymes or overexpression of these factors are approaches that have proven to be successful in the amelioration of lung disease caused by asbestos and other oxidants in experimental models.

16.6 Gene Expression in Apoptosis

Apoptosis, a form of programmed cell death, is hallmarked by DNA fragmentation, chromatin condensation, and membrane blebbing. This form of cell death has been implicated in such cellular events as thymic selection, cell-mediated cytotoxicity, and activation-induced cell death.[99] Many chemical or physical agents induce apoptosis by elaboration of AOS.[100,101] Under oxidative conditions, cell survival appears to be determined by a balance between AOS and antioxidants. A direct link between oxidative stress and programmed cell death has been shown at low doses of H_2O_2 in a variety of cell lines while at high doses, this oxidant induces necrosis.[102] Nitric oxide, another important oxidant, also initiates apoptosis in macrophages and monocytes.[103] Furthermore, the exposure of cells to buthionine sulfoximine, an agent that reduces intracellular glutathione leads to a greater sensitivity to oxidant-mediated apoptosis. Thus, the diminution of cellular scavengers or detoxifiers of AOS can also induce oxidative stress[104] and prime cells to undergo apoptosis.

Apoptosis is associated with the induction of AOS in some cell types. For example, TNF, a cytokine that stimulates production of intracellular AOS, can initiate apoptosis. Addition of thioredoxin, a thiol reductant and free radical scavenger, or NAC prevents TNF-induced cell death.[105,106] Silica, presumably by elaboration of AOS, induces apoptosis in alveolar macrophages.[107] Moreover a recent study with crocidolite asbestos also has demonstrated a marked increase in the percentage of apoptotic cells in asbestos-treated confluent monolayers of rat pleural mesothelial cells at concentrations known to induce the protooncogenes, c-fos and c-jun (Berubé, Quinlan, Fung, Magae, Vacek, Taatjes, Mossman, 1996). Since asbestos is thought to mediate its deleterious effects by either generating or inducing the formation of AOS, these data further corroborate the notion that oxidative stress can mediate apoptosis in a wide array of cell types.

Apoptosis is an active process requiring the induction of specific genes. Withdrawal of growth factors from density-arrested murine Balb/c-3T3 cells causes rapid cell death and the upregulation of c-Myc, c-Fos, c-Jun, cdc2, and phosphorylation of the retinoblastoma protein.[108] In addition, the presence of a proliferation-specific nuclear antigen (PCNA) is also detected. These data indicate that apoptosis induced by serum deprivation results in the expression of genes induced in the early G1 phase and typically present at the commencement of the cell cycle. Bcl-2, a protein that forms either homodimers or heterodimers with Bax (Bcl-2-associated X protein) is intimately involved in the regulation of apoptosis. This protein localizes to the endoplasmic reticulum, mitochondria, and nuclear membranes, sites of intracellular production of AOS.[109,110] Bax and Bcl-2 are found associated with one another prior to the transduction of an apoptotic stimulus and dictate whether a cell will undergo proliferation or cell death. Excess of Bcl-2 results in the formation of Bcl-2 homodimers and inhibition of cell death. In contrast, elevated levels of Bax leads to Bax homodimerization and the promotion of apoptosis. The expression of Bcl-2 has been shown to rescue serum-deprived fibroblasts from c-Myc-mediated apoptosis.[111] Furthermore, Bcl-2 protects cells from programmed cell death induced by lethal concentrations of H_2O_2, t-butyl hydroperoxide, and menadione.[110,112] Although Bcl-2 does not inhibit the oxidative burst elicited by menadione, it does appear to function by interfering with the intermediate stage between peroxide generation and lipid membrane peroxidation.[113]

The discovery of the "death genes," ced3 and ced4, in the development of the nematode Caenorhabditis elegans has led to the speculation that activation-induced apoptosis may also result in induction of genes strictly associated with apoptosis. In the case of AOS-induced cell death, this could potentially occur through a transcription factor upregulated by oxidative stress, such as NF-κB. The transcription factor, AP-1, has recently been classified as an antioxidant responsive factor[114] that can negatively regulate *c-myc*,[115] which is involved in the initiation of activation-induced apoptosis in T-cell hybridomas.[116]

Genes that are involved in the induction of apoptosis induced by environmental oxidants such as asbestos remain to be discovered. Possible molecular pathways of apoptosis induced by mineral dusts and oxidants are presented in Figure 16.2. Knowledge of function of these genes may provide a better understanding of

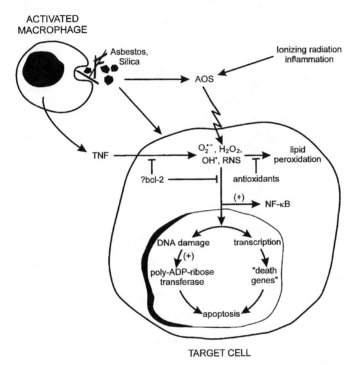

Figure 16.2. Proposed schema for apoptotic molecular pathways initiated by oxidative stress. Mineral dusts, oxidants, and ionizing radiation elicit the formation of AOS either directly or indirectly. As described in the text, AOS are generated directly from mineral dusts through redox reactions occurring on the surface of fibers or particles, or after phagocytosis by inflammatory cells. Activated macrophages also release TNF which can increase endogenous levels of AOS in target cells. Moreover, macrophages can also elevate intracellular levels of antioxidants and GSH through the release of thiol precursors. The protein product of *bcl-2* plays a role in cellular defense against apoptosis by decreasing levels of AOS in some cell types. Oxidative stress may also activate transcription factors such as NF-κB, which may be important in regulation of inflammatory genes such as TNF, which are associated with apoptosis. AOS-induced DNA damage also may activate poly-ADP-ribose transferase leading to severe decreases in reducing potential and ATP stores and ultimately to apoptosis. RNS, reactive nitrogen species.

mechanisms by which these agents damage pulmonary target cells and cells of the immune system.

16.7 Signal Transduction Pathways

Inflammatory mediators either released from cells of the immune system or elaborated by target cells of disease include AOS; cytokines such as interleukins, interferons, and TNF; and eicosanoids, i.e., prostaglandins, thromboxane, leukotrienes, etc. These agents may exacerbate dust-induced effects on gene expression. There is growing evidence that mineral fibers directly and indirectly through oxidants and/or cytokines induce signal transduction pathways regulated by protein kinases. These different signal transduction pathways may affect gene expression critical to the production of inflammatory mediators by alveolar macrophages and other cell types and phenotypic changes involved in cell injury and/or defense.

One pathway of cell signaling elicited by oxidant stress to lung may involve elaboration of eicosanoids. In this pathway, stimuli activate phospholipase A_2 or C (PLA_2, PLC), which then liberate arachidonic acid (AA) from membrane phospholipids. AA can be converted by the lipoxygenase pathway to leukotrienes or by the cyclooxygenase pathway to prostaglandins or thromboxane. The initial activation of phospholipases is controlled by the phosphorylation status of membrane bound proteins (e.g., receptor tyrosine kinases [RTK]).

Ultraviolet B (UVB) is another oxidant stress that depletes antioxidant stores and leads to increased prostaglandin E_2 (PGE_2) synthesis.[117] Pretreatment of epidermal cells with the tyrosine kinase inhibitors, tyrophostin-23 or genistein, blocks the UVB-stimulated PGE_2 synthesis as well as the tyrosine phosphorylation of the epidermal growth factor (EGF) receptor. The glutathione precursor and antioxidant, NAC, decreases both UVB oxidant-induced EGF receptor phosphorylation and PGE_2 synthesis, suggesting a coupling of events.

Mineral dusts also induce alveolar macrophages to release AOS and eicosanoids from AA in the synthesis of prostaglandins.[118] The antioxidant, α-tocopherol, reduces the dust-induced synthesis of PGE_2 and thromboxane A_2 in vitro and in vivo. Since calcination of silica, which removes exposed $OH^{\cdot}$ groups, inhibits eicosanoid production, metabolism of AA by silica may be mediated by oxidant stress much like UVB-induced AA metabolism in epidermal cells.[118]

Mineral dusts may also cause increases in gene expression of cytokines important in the inflammatory response. Differential screening of mouse macrophage-subtracted cDNA libraries indicates that silica causes a five-fold increase in TNF mRNA expression.[119] This is accompanied by an eight-fold increase in TNF secretion, indicating that TNF mRNA is a good biomarker for the activation state of the macrophage.[119] The ability of amosite asbestos to stimulate the release of TNF from rat alveolar macrophages in vitro is substantially increased with opsonization of the fibers by immunoglobulin G (IgG).[120] Since these elevations could be inhibited with the protein kinase C (PKC) inhibitor, staurosporine, kinase signaling may be critical to the production of cytokines following exposure to mineral dusts.

Increased activity of PKC is also observed after exposure of hamster tracheal epithelial cells to asbestos[121] and may be critical to c-*fos* and c-*jun* induction in rat pleural mesothelial cells by asbestos (Fung et al., in preparation). In the latter studies, down-modulation of PKC and pretreatment of cells with the PKC inhibitor, calphostin, blocks increased steady-state mRNA levels of c-*fos* and c-*jun* normally observed after exposure of cells to asbestos.

The mitogen-activated protein kinases (MAPK) both phosphorylate and activate $cPLA_2$[122] and are also activated by AA in epithelial cells via a PKC-dependent mechanism.[123] A requirement for MAPK activity in interferon-stimulated gene expression has recently been demonstrated,[124] indicating the importance of this signaling pathway in regulation of cytokine production.

The MAPK cascade is a major pathway through which signals proceed to generate intracellular responses and activate nuclear transcription factors.[125] There are at least three MAPK subtypes: (1) the extracellular-signal-regulated kinases (ERKs); (2) the c-Jun NH_2-terminal kinases/stress activated protein kinases (JNKs/SAPKs)[126]; and (3) the mammalian HOG1 homologue p38.[127] The various subfamilies of mammalian MAPKs can be activated simultaneously by distinct, parallel cascades in response to the same stimulus.[128] Conversely, individual MAPK subtypes may be selectively activated.[128,129]

In mammalian cells, the ERKs are the best characterized members of the MAPK family. In mitogen-stimulated cells, a series of phosphorylation events is initiated usually from a cell surface receptor (e.g., growth factor receptors and integrins). Activation can involve a c-Ras-mediated membrane translocation and phosphorylation of c-Raf,[130] a serine kinase upstream of ERKs, and subsequent phosphorylation of the downstream kinase, MAPK/ERK kinase (MEK). MEK is a dual-specificity kinase that activates the ERKs by phosphorylating them at threonine and tyrosines within a TEY site. Phosphorylated ERKs then translocate to the nucleus to phosphorylate substrates including the ternary complex factor (TCF)/Elk-1.[131–133] TCF/Elk-1 is phosphorylated by ERKs *in vitro* on sites essential for c-*fos* transactivation *in vivo*[132,133] and elevated AP-1 activity via c-*fos* induction. TCF binds with the serum response factor (SRF) to the serum response element (SRE) to mediate increases in gene expression.[134] Figure 16.3 shows these various pathways as well as a possible model of MAPK activation by asbestos fibers based on preliminary data from our laboratory.

The ERKs can be rapidly phosphorylated and activated in response to a number of external factors that promote growth and differentiation. These stimulators include such factors as ionizing radiation and H_2O_2, which activate MAPK by the production of AOS[135] and do not interact with cell surface elements in a

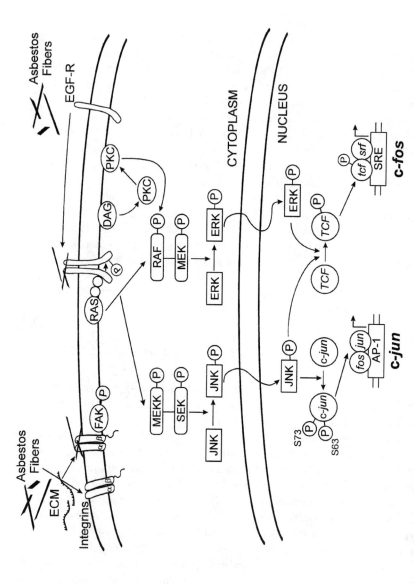

Figure 16.3. Multiple mechanisms of cell signaling by asbestos. Work in our laboratory indicates that asbestos fibers interact with the EGF receptor (EGF-R), thus initiating the ERK cascade. Present studies are focused on the JNK/p38 cascade, which may be important in mediating increases in c-*jun* expression by asbestos. In addition, asbestos fibers may directly interact with components of extracellular matrix (ECM) to induce increased expression of integrins and phosphorylation of p125 focal adhesion kinase (FAK).

specific manner. Signals triggered by specific ligand interactions at the cell surface can also activate MAPK. These events include oligomerization of RTK[136] or engagement of integrins.[137]

Work in our laboratory has shown that MAPKs are activated after exposure of rat pleural mesothelial cells to asbestos at concentrations inducing increased mRNA levels of early response protooncogenes (Zanella, Posada, Tritton, and Mossman, 1996). Both crocidolite and chrysotile asbestos as well as the mesotheliomagenic fiber, erionite, phosphorylate and activate the MAPK ERK subfamily in contrast to a number of noncarcinogenic fibers and particulates. Additionally, crocidolite asbestos induces the phosphorylation of c-Raf at multiple sites.

There may be multiple mechanisms by which mineral dusts activate ERKs. Several observations also suggest that oxidant stress has a role in the stimulation of MAPK in cells of the respiratory tract. For example, both H_2O_2 and menadione cause phosphorylation of ERKS in rat pleural mesothelial cells. Exposure to asbestos depletes total glutathione levels in these cell types,[29] and pretreatment of cells with NAC before exposure to asbestos abrogates both asbestos-induced activation of ERKs (Jimenez, Zanella, Fung, Janssen, Vacek, Charland, Goldberg, and Mossman, 1997) and protooncogene expression.[29]

In addition to pathways mediated by AOS, asbestos fibers may physically activate signaling pathways from the cell surface (Fig. 16.3). We have shown that the ERK response to asbestos is inhibited by suramin, an inhibitor of growth factor receptor interactions (Zanella, Posada, Tritton, and Mossman, 1996). Moreover, a specific and potent tyrphostin inhibitor of the EGF receptor (EGF-R) was able to block asbestos-induced ERK activation. *In vitro* kinase assays of the EGF-R showed that crocidolite asbestos had effects on the EGF-R similar to those induced by EGF itself.

Dimerization of RTKs is necessary for activation of their intrinsic protein kinase activity and autophosphorylation. Autophosphorylation of sites within the cytoplasmic catalytic domain is crucial for control of kinase activity and interaction with downstream signaling molecules.[136] In addition to Ras, activated RTK can also interact with and activate PLC inducing the release of diacylglycerol (DAG) and inositol phosphate (IP_3) from membrane phospholipids. These signaling pathways may be important in cell proliferation by tumor promoters and extracellular stimuli. Tracheal epithelial cells exposed to crocidolite asbestos demonstrate significant elevations in DAG and IP_3 levels.[138] These events precede crocidolite asbestos-induced proliferation and indirectly support a role for activation of RTK by asbestos.

The JNK/SAPK pathway is parallel to the ERK pathway and contains homologous elements from the c-Raf level to the ERK level (Fig. 16.3). Active JNKs phosphorylate the amino terminus of c-*jun* at serine residues, which transcriptionally activates c-*jun*.[126,139] While the ERK family is activated by various peptides and phorbol esters, the JNK/SAPK family is more strongly activated by various stresses such as UV light,[126,139] TNF,[140,141] and heat shock.[142] The p38 MAPK pathway is also activated by proinflammatory cytokines (TNF, IL-1) and environmental stress (UVC).[143] Furthermore, the ability of TNF to stimulate AP-1 activity has been shown to occur through the prolonged activation of JNK.[140] We are presently examining whether asbestos fibers activate these pathways either alone or coordinately with other cytokines.

Summary

The cellular mechanisms essential to gene expression by inhaled mineral dusts and other oxidant stresses are complex. Signals generated extracellularly, at the cell surface, or by oxidants as "second messengers" can proceed through various pathways to produce a change in the phosphorylation status of transcription factors. This may then either activate or inactivate transcription factors to produce change in gene expression and phenotypic ramifications (i.e., proliferation, differentiation, cell death, apoptosis). The balance between various pathways, which may be cell-type-specific, may govern gene expression and the endpoints of cell response. For example, increased expression of c-*fos,* c-*jun,* and c-*myc* may be related to both cell proliferation and apoptosis.[144] Moreover, the activation of a particular signaling cascade may be dependent on both the stimulus and cell type.

Many agents induce multiple genes that can be related to cell proliferation, inflammation, adaptation and/or defense, and cell death. Although little is known about the signaling mechanisms involved in activation of many of these genes by oxidants, it is clear that many disparate genes have NF-κB or AP-1 sites in their promoter regions which may function coordinately or competitively in gene expression. Further studies are needed to define the regulation of gene expression by inflammatory and pathogenic agents in lung to enable preventive and therapeutic strategies to pulmonary disease.

Acknowledgments

This research was supported by Grants ES06499 and ES07038 from the National Institute of Environmental Health Sciences and Grant HL39469 from the National Heart, Lung, and Blood Institute. Y.M.W.J. is a fellow of the Parker B. Francis Foundation for Pulmonary Research. C.R.T. is a recipient of a NIOSH SERCA Award.

References

1. Timblin, C. R., Y. M. W. Janssen, and B. T. Mossman. In press. Free radical-mediated alterations of gene expression by xenobiotics. *In* K. Wallace (ed.), *Free Radical/Toxicology.* Raven Press.

2. Janssen, Y. M. W., P. J. A. Borm, B. Van Houten, and B. T. Mossman. 1993. Cell and tissue responses to oxidative damage. *Lab. Invest.* **69**:261–274.

3. Remacle, J., M. Raie, O. Tousaint, P. Renard, and G. Rao. 1995. Low levels of reactive oxygen species as modulators of cell function. *Mutat. Res.* **316**:103–122.

4. Mossman, B. T., L. A. Jimenez, K. BeruBe, T. Quinlan, and Y. M. W. Janssen. 1995. Possible mechanisms of crystalline silica-induced lung disease. *Appl. Occupat. Environ. Hygiene* **10**:1115–1117.

5. Mossman, B. T., and J. B. L. Gee. 1989. Medical progress. Asbestos-associated diseases. *N. Engl. J. Med.* **320**:1721–1730.

6. Guthrie, G., Jr., and B. T. Mossman (eds.). 1993. *Health Effects of Mineral Dusts.* Reviews in Mineralogy, Vol. 28. Mineralogical Society of America, Washington, D.C.

7. Weitzman, S. A., and P. Graceffa. 1984. Asbestos catalyzes hydroxyl and superoxide radical generation from hydrogen peroxide. *Arch. Biochem. Biophys.* **228**:373–376.

8. Gulumian, M., and J. A. Van Wyk. 1987. Hydroxyl radical production in the presence of fibers by a Fenton-type reaction. *Chem. Biol. Interact.* **62**:89–97.

9. Vallyathan, V., X. Shi, N. S. Dalal, W. Irr, and V. Castranova. 1988. Potential role of reactive silicon-oxygen radicals in acute silica-induced lung injury. *Am. Rev. Respir. Dis.* **138**:1213–1219.

10. Vallyathan, V., J. F. Mega, X. Shi, and N. S. Dalal. 1992. Enhanced generation of free radicals from phagocytes induced by mineral dusts. *Am. J. Respir. Cell Mol. Biol.* **6**:404–413.

11. Lund, L. G., and A. E. Aust. 1991. Iron-catalyzed reactions may be responsible for the biochemical and biological effects of asbestos. *BioFactors* **3**:83–89.

12. Mossman, B. T. 1993. Editorial: Mechanisms of asbestos carcinogenesis and toxicity: the amphibole hypothesis revisited. *Br. J. Indust. Med.* **50**:673–676.

13. Goodglick, L., and A. Kane. 1986. Role of reactive oxygen metabolites in crocidolite asbestos toxicity to mouse macrophages. *Cancer Res.* **46**:5558–5566.

14. Hansen, K., and B. T. Mossman. 1987. Generation of superoxide (O_2^-) from alveolar macrophages exposed to asbestiform and nonfibrous particles. *Cancer Res.* **47**:1681–1686.

15. Mossman, B. T., J. Bignon, M. Corn, A. Seaton, and J. B. L. Gee. 1990. Asbestos: scientific developments and implications for public policy. *Science* **247**:294–301.

16. Mossman, B. T., J. P. Marsh, A. Sesko, S. Hill, M. A. Shatos, J. Doherty, J. Petruska, K. B. Adler, D. Hemenway, R. Mickey, P. Vacek, and E. Kagan. 1990. Inhibition of lung injury, inflammation and interstitial pulmonary fibrosis by polyethylene glycol-conjugated catalase in a rapid inhalation model of asbestosis. *Am. Rev. Respir. Dis.* **141**:1266–1271.

17. Ames, B., and L. Gold, 1990. Mitogenesis increases mutagenesis. *Science* **249**:970–971.

18. Preston-Martin, S., M. C. Pike, R. K. Ross, P. A. Jones, B. E. Henderson. 1990. Increased cell division as a cause of human cancer. *Cancer Res.* **50**:415–421.

19. Quinlan, T. R., J. P. Marsh, Y. M. W. Janssen, K. O. Leslie, D. Hemenway, P. Vacek, and B. T. Mossman. 1994. Dose-responsive increases in pulmonary fibrosis after inhalation of asbestos. *Am. J. Respir. Crit. Care Med.* **150**:200–206.

20. Pegg, A. E. 1988. Polyamine metabolism and its importance in neoplastic growth and as a target for chemotherapy. *Cancer Res.* **48**:759–774.

21. Angel, P., and M. Karin. 1991. The role of Jun, Fos and the AP-1 complex in cell proliferation and transformation. *Biochim. Biophys. Acta* **1072**:129–157.

22. Quinlan, T., K. BeruBe, J. Marsh, Y. Janssen, P. Taishi, K. Leslie, D. Hemenway, P. O'Shaughnessy, P. Vacek, and B. Mossman, 1995. Patterns of inflammation, cell proliferation and related gene expression in lung after inhalation of chrysotile asbestos. *Am. J. Pathol.* **147**:728–739.

23. Heintz, N. H., Y. M. W. Janssen, and B. T. Mossman, 1993. Persistent induction of c-*fos* and c-*jun* protooncogene expression by asbestos. *Proc. Natl. Acad. Sci. USA* **90**:3299–3304.

24. Marsh, J. P., and B. T. Mossman. 1991. Role of asbestos and active oxygen species in activation and expression of ornithine decarboxylase in hamster tracheal epithelial cells. *Cancer Res.* **51**:167–173.

25. Timblin, C., Y. M. M. Janssen, and B. T. Mossman. 1995. Transcriptional activation of the proto-oncogene, c-*jun,* by asbestos and H_2O_2 is directly related to increased proliferation and transformation of tracheal epithelial cells. *Cancer Res.* **55**:2723–2726.

26. Janssen, Y., A. Barchowsky, M. Treadwell, K. Driscoll, and B. Mossman. 1995. Asbestos induces nuclear factor κB (NF-κB) DNA-binding activity and NF-κB-dependent gene expression in tracheal epithelial cells. *Proc. Natl. Acad. Sci. USA* **92**: 8458–4862.

27. Liou, H. C., and D. Baltimore. 1993. Regulation of the NF-κB/rel transcription factor and IκB inhibitor system. *Curr. Opin. Cell Biol.* **5**:477–487.

28. Janssen, Y., N. Heintz, J. Marsh, P. Borm, and B. Mossman. 1994. Induction of c-*fos* and c-*jun* protooncongenes in target cells of the lung and pleura by carcinogenic fibers. *Am. J. Respir. Cell. Mol. Biol.* **11**:522–530.

29. Janssen, Y., N. Heintz, and B. Mossman, 1995. Induction of c-*fos* and c-*jun* proto-oncogenes by asbestos is ameliorated by N-acetyl-L-cysteine in mesothelial cells. *Cancer Res.* **55**:2085–2089

30. Hayashi, T., Y. Ueno, T. Okamoto. 1993. Oxidoreductive regulation of nuclear factor κB. Involvement of a cellular reducing catalyst thioredoxin. *J. Biol. Chem.* **268**:11380–11388.

31. Galter, D., M. I. H. M. Sabine, W. Dröge. 1994. Distinct effects of glutathione disulphide on the nuclear transcription factors κB and the activator protein-1. *Eur. J. Biochem.* **221**:639–648.

32. Meyer, M., W. H. Caselmann, V. Schlüter, R. Schreck, P. H. Hofschneider, and P. A. Baeuerle. 1992. Hepatitis B virus transactivator MHBs[t]: activation of NF-κB, selective inhibition by antioxidants and integral membrane localization. *EMBO J.* **11**:2991–3001.

33. Staal, F. J. T., M. Roederer, L. A. Herzenberg, L. A. Herzenberg. 1990. Intracellular thiols regulate activation of nuclear factor kappa B and transcription of human immunodeficiency virus. *Proc. Natl. Acad. Sci. USA* **87**:9943–9947.

34. Devary, Y., R. A. Gottlieb, T. Smeal, M. Karin. 1992. The mammalian ultraviolet response is triggered by activation of Src tyrosine kinases. Cell 71:1081–1091.

35. Nover, L. (ed.) *Heat Shock Response.* 1991. CRC Press, Boca Raton, FL.

36. Welch, W. J. 1992. Mammalian stress response: cell physiology, structure/function of stress proteins, and implications for medicine and disease. *Physiol. Rev.* **72**:1063–1081.

37. Jacquier-Sarlin, M. R., K. Fuller, A. T. Dinh-Xuan, M. J. Richards, and B. S. Polla. 1994. Protective effects of hsp70 in inflammation. *Experientia* **50**:1031–1038.

38. Jäättelä, M., and D. Wissing. 1992. Emerging role of heat shock proteins in biology and medicine. *Ann. Med.* **24**:249–258.

39. Little, E., and A. S. Lee. 1995. Generation of a mammalian cell line deficient in glucose-regulated protein stress induction through targeted ribozyme driven by a stress-inducible promoter. *J. Biol. Chem.* **270**:9526–9534.

40. Milarski, K. L., and R. Morimoto. 1986. Expression of human HSP70 during the synthetic phase of the cell cycle. *Proc. Natl. Acad. Sci. USA* **83**:9517–9521.

41. Sistonen, L., K. D. Sarge, K. Abravaya, B. Phillips, and R. I. Morimoto. 1992. Activation of heat shock factor 2 (HSF2) during hemin-induced differentiation of human erythroleukemia cells. *Mol. Cell Biol.* **12**:4104–4111.

42. Morimoto, R. I. 1993. Cells in stress: transcriptional activation of heat shock genes. *Science* **259**:1409–1410.

43. Morimoto, R. I., K. D. Sarge, and K. Abrayaya. 1992. Transcriptional regulation of heat shock genes. *J. Biol. Chem.* **267**:21987–21990.

44. Bruce, J. L., B. D. Price, C. N. Coleman, and S. K. Calderwood. 1993. Oxidative injury rapidly activates the heat shock transcription factor but fails to increase levels of heat shock proteins. *Cancer Res.* **53**:12–15.

45. Janssen, Y., J. Marsh, and M. Absher, 1994. Oxidant stress responses in human pleural mesothelial cells exposed to asbestos. *Am. J. Respir. Crit. Care Med.* **149**:795–802.

46. Keyse, S. M., and R. M. Tyrell. 1989. Heme oxygenase is the major 32-kDa stress protein induced in human skin fibroblasts by UVA radiation, hydrogen peroxide, and sodium arsenite. *Proc. Natl. Acad. Sci. USA* **86**:99–103.

47. Applegate, L. A., P. Luscher, and R. M. Tyrrell. 1991. Induction of heme oxygenase: a general response to oxidant stress in cultured mammalian cells. *Cancer Res.* **51**:974–978.

48. Abe, T., T. Konishi, T. Katoh, H. Hirano, K. Matsukuma, M. Kashimura, and K. Higashi. 1994. Induction of heat shock 70 mRNA by cadmium is mediated by glutathione suppressive and non-suppressive triggers. *Biochim. Biophys. Acta* **1201**:29–36.

49. Hiranuma, K., K. Hirata, T. Abe, T. Hirano, K. Matsuno, H. Hirano, K. Suzuki. 1993. Induction of mitochondrial chaperonin, hsp60, by cadmium in human hepatoma cells. *Biochem. Biophys. Res. Commun.* **194**:531–536.

50. Liu, R. Y., P. M. Corry, and Y. J. Lee. 1995. Potential involvement of a constitutive heat shock element binding factor in the regulation of chemical stress-induced hsp70 gene expression. *Mol. Cell. Biochem.* **144**:27–34.

51. Ribeiro, S. P., J. Villar, G. P. Downey, J. D. Edelson, and A. S. Slutsky. 1994. Sodium arsenite induces heat shock protein-72 kilodalton expression in the lungs and protects rats against sepsis. *Crit. Care Med.* **22**:922–929.

52. Cosma, G., H. Fulton, T. DeFeo, and T. Gordon. 1992. Rat lung metallothionein and heme oxygenase gene expression following ozone and zinc oxide exposure. *Toxicol. App. Pharmacol.* **117**:75–80.

53. Pryor, W. A. 1994. Mechanisms of radical formation from reactions of ozone with target molecules in the lung. *Free Rad. Biol. Med.* **17**:451–465.

54. Fornace Jr., A. J. 1992. Mammalian genes induced by radiation; Activation of genes associated with growth control. *Annu. Rev. Genet.* **26**:507–526.

55. Quinlan, T., S. Spivak, and B. T. Mossman. 1994. Regulation of antioxidant enzymes in lung after oxidant injury. *Environ. Health Persp.* **102**:79–87.

56. Sibille, Y., and H. Y. Reynolds. 1990. Macrophages and polymorphonuclear neutrophils in lung defense and injury. *Am. Rev. Respir. Dis.* **141**:471–501.

57. Dubois, C. M., E. Bissonnette, and M. Rola-Pleszczynski. 1989. Asbestos fibers and silica particles stimulate rat alveolar macrophages to release tumor necrosis factor. *Am. Rev. Respir. Dis.* **139**:1257–1264.

58. Driscoll, K. E., J. K. Maurer, J. Higgins, and J. Poyner. 1995. Alveolar macrophage cytokine and growth factor production in a rat model of crocidolite-induced pulmonary inflammation and fibrosis. *J. Toxicol. Environ. Health* **46**:155–169.

59. Crapo, J., F. J. Miller, B. T. Mossman, W. A. Pryor, and J. P. Kiley. 1992. Environmental lung disease. *Am. Rev. Respir. Dis.* **145**:1506–1512.

60. Gaston, B., J. M. Drazen, J. Loscalzo, and J. S. Stamler. 1994. The biology of nitrogen dioxides in the airways. *Am. J. Respir. Crit. Care Med.* **149**:538–551.

61. Devalia, J. L., A. M. Campbell, R. J. Sapsford, C. Rusznak, D. Quint, P. Godard, J. Bousquet, and R. J. Davies. 1993. Effect of nitrogen dioxide on synthesis of inflammatory cytokines by human bronchial epithelial cells *in vitro. Am. J. Respir. Cell Mol. Biol.* **9**:271–278.

62. Driscoll, K. E., J. K. Maurer, D. Hassenbein, J. Carter, Y. M. W. Janssen, B. T. Mossman, M. Osier, and G. Oberdorster. 1994. Contribution of macrophage-derived cytokines and cytokine networks to mineral dusts-induced lung inflammation. pp. 176–189. *In* U. Mohr (ed.), *Toxic and Carcinogenic Effects of Solid Particles in the Respiratory Tract.* ILSI Press, Washington, D.C.

63. Ohmori, Y., and T. A. Hamilton. 1994. Regulation of macrophage gene expression by T-cell derived lymphokines. *Pharmacol. Ther.* **63**:235–264.

64. Driscoll, K. E., R. C. Lindenschmidt, J. M. Maurer, J. M. Higgins, and G. Ridder. 1990. Pulmonary response to silica or titanium dioxide: inflammatory cells, alveolar

macrophage-derived cytokines and histopathology. *Am. J. Respir. Cell Mol. Biol.* **2**:381–390.

65. Zhang, Y., T. C. Lee, B. Guillemin, M. C. Yu, and W. N. Rom. 1993. Enhanced IL-1 beta and tumor necrosis factor-alpha release and messenger RNA expression in macrophages from idiopathic pulmonary fibrosis or after asbestos exposure. *J. Immunol.* **150**:4188–4196.

66. Read, M. A., M. Z. Whitley, A. J. Williams, and T. Collins. 1994. NF-κB and IκBα: an inducible regulatory system in endothelial activation. *J. Exper. Med.* **179**:503–512.

67. Janssen, Y. M. W., K. E. Driscoll, B. Howard, T. R. Quinlan, M. Treadwell, A. Barchowsky, and B. T. Mossman. 1996. Asbestos increases in NF-κB DNA binding activity and translocation of p65 protein in rat lung epithelial and pleural mesothelial cells. Submitted.

68. Beg, A. A., and A. S. Baldwin, Jr. 1994. Activation of multiple-NF-κB/Rel DNA-binding complexes by tumor necrosis factor. *Oncogene* **9**:1487–1492.

69. Anderson, M. T., F. J. T. Staal, C. Gitler, L. A. Herzenberg, and L. A. Herzenberg. 1994. Separation of oxidant-initiated and redox-regulated steps in the NF-κB signal transduction pathway. *Proc. Natl. Acad. Sci. USA* **91**:11527–11531.

70. Mohan, N., and M. L. Meltz. 1994. Induction of nuclear factor κB after low-dose ionizing radiation involves a reactive oxygen intermediate signalling pathway. *Radia. Res.* **140**:97–104.

71. Collart, M. A., P. Baeuerle, and P. Vassalli. 1990. Regulation of tumor necrosis factor alpha transcription in macrophages: involvement of four κB like motifs and of constitutive and inducible forms of NF-κB. *Mol. Cell. Biol.* **10**:1498–1506.

72. Xie, Q.-W., Y. Kashiwabara, and C. Nathan. 1994. Role of transcription factor NF-κB/Rel in induction of nitric oxide synthase. *J. Biol. Chem.* **269**:4705–4708.

73. Matsusaka, T., K. Fujikawa, Y. Nishio, N. Mukaida, K. Matsushima, T. Kishimoto, and S. Akira. 1993. Transcription factors NF-IL6 and NF-κB synergistically activate transcription of the inflammatory cytokines, interleukin 6 and interleukin 8. *Proc. Natl. Acad. Sci. USA* **90**:10193–10197.

74. Kunsch, C., and C. A. Rosen. 1993. NF-κB subunit-specific regulation of the interleukin-8 promoter. *Mol. Cell. Biol.* **13**:6137–6146.

75. Ueda, A., K. Okuda, S. Ohno, A. Shirai, T. Igarashi, K. Matsunaga, J. Fukushima, S. Kawamoto, Y. Ishigatsubo, and T. Okubo. 1994. NF-kappa B and Sp1 regulate transcription of the human monocyte chemoattractant protein-1 gene. *J. Immunol.* **153**:2052–2063.

76. Widmer, U., K. R. Manogue, A. Cerami, and B. Sherry. 1993. Genomic cloning and promoter analysis of macrophage inflammatory protein (MIP)-2, MIP-1α, and MIP-1β, members of the chemokine superfamily of proinflammatory cytokines. *J. Immunol.* **150**:4996–5012.

77. Driscoll, K. E., D. G. Hassenbein, J. M. Carter, S. L. Kunkel, T. R. Quinlan, and B. T. Mossman. 1995. TNF-alpha and increased chemokine expression in rat lung after particle exposure. *Toxicol. Lett.* **82/83**:483–489.

78. Driscoll, K. E., J. Strzelecki, D. Hassenbein, Y. M. W. Janssen, J. Marsh, G. Oberdorster, and B. T. Mossman. 1994. Tumour necrosis factor (TNF): evidence for the role of TNF in increased expression of manganese superoxide dismutase after inhalation of mineral dusts. *Ann. Occupat. Hygiene* **38**:375–382.

79. Rosenthal, G. J., D. R. Germolec, M. E. Blazka, E. Corsini, P. Simeonova, P. Pollock, L. Y. Kong, J. Kwon, and M. I. Luster. 1994. Asbestos stimulates IL-8 production from human lung epithelial cells. *J. Immunol.* **153**:3237–3244.

80. Griffith, D. E., E. J. Miller, L. D. Gray, S. Idell, and A. R. Johnson. 1994. Interleukin-1-mediated release of interleukin-8 by asbestos-stimulated human pleural mesothelial cells. *Am. J. Respir. Cell Mol. Biol.* **10**:245–252.

81. Boylan, A. M., C. Ruegg, K. J. Kim, C. A. Hebert, J. M. Hoeffel, R. Pytela, D. Sheppard, I. M. Goldstein, and V. C. Broaddus. 1992. Evidence of a role for mesothelial cell-derived interleukin 8 in the pathogenesis of asbestos-induced pleurisy in rabbits. *J. Clin. Invest.* **89**:1257–1267.

82. Zhang, Y., T. C. Lee, B. Guillemin, M. C. Yu, and W. N. Rom. 1993. Enhanced IL-1 beta and tumor necrosis factor alpha release and messenger RNA expression in macrophages from idiopathic pulmonary fibrosis or after asbestos exposure. *J. Immunol.* **150**:4188–4196.

83. Thomas, G., T. Ando, K. Verma, and E. Kagan. 1994. Asbestos fibers and interferon-gamma upregulate nitric oxide production in rat alveolar macrophages. *Am. J. Respir. Cell Mol. Biol.* **11**:707–715.

84. Bauman, J. W., J. Liu, Y. P. Liu, and C. D. Klaassen. 1991. Increase in metallothionein produced by chemicals that induce oxidative stress. *Toxicol. Appl. Pharmacol.* **110**:347–354.

85. Janssen, Y. M. W., J. P. Marsh, M. P. Absher, D. Hemenway, P. M. Vacek, K. O. Leslie, P. J. A. Borm, and B. T. Mossman. 1992. Expression of antioxidant enzymes in rat lungs after inhalation of asbestos or silica. *J. Biol. Chem.* **267**:10625–10630.

86. Janssen, Y. M. W., J. P. Marsh, K. E. Driscoll, P. J. A. Borm, G. Oberdorster, and B. T. Mossman. 1994. Increased expression of manganese-containing superoxide dismutase in rat lungs after inhalation of inflammatory and fibrogenic minerals. *Free Rad. Biol. Med.* **16**:315–322.

87. Holley, J. A., Y. M. W. Janssen, B. T. Mossman, and D. J. Taatjes. 1992. Increased manganese superoxide dismutase protein in type II epithelial cells of rat lungs after inhalation of crocidolite asbestos or cristobalite silica. *Am. J. Pathol.* **141**:475–485.

88. Clerch, L. B., and D. Massaro. 1993. Tolerance of rats to hyperoxia. Lung antioxidant enzyme gene expression. *J. Clin. Invest.* **91**:499–508.

89. Rahman, I., L. Clerich, and D. Massaro. 1991. Rat lung antioxidant enzyme induction by ozone. *Am. J. Physiol.* **260**:412–418.

90. Shull, S., N. H. Heintz, M. Periasamy, M. Manohar, Y. M. W. Janssen, J. P. Marsh, and B. T. Mossman. 1991. Differential regulation of antioxidant enzymes in response to oxidants. *J. Biol. Chem.* **266**:24398–24403.

91. Turrens, J. F., J. D. Crapo, and B. A. Freeman. 1984. Protection against oxygen toxicity by intravenous injection of liposome-entrapped catalase and superoxide dismutase. *J. Clin. Invest.* **73**:87–95.

92. Mossman, B., P. Surinrut, B. Brinton, J. Marsh, N. Heintz, B. Lindau-Shepard, and J. Shaffer. 1996. Transfection of a manganese-containing superoxide dismutase gene into hamster tracheal epithelial cells ameliorates asbestos-mediated cytotoxicity. *Free Rad. Biol. Med.* **21**:125–131.

93. Lindau-Shepard, B., J. B. Shaffer, and P. J. Del Vecchio. 1994. Overexpression of manganous superoxide dismutase (MnSOD) in pulmonary endothelial cells confers resistance to hyperoxia. *J. Cell. Physiol.* **161**:237–242.

94. Hirose, K., D. L. Longo, J. J. Oppenheim, and K. Matsushima. 1993. Overexpression of mitochondrial manganese superoxide dismutase promotes the survival of tumor cells exposed to interleukin-1, tumor necrosis factor, selected anticancer drugs, and ionizing radiation. *FASEB J.* **7**:361–368.

95. St. Clair, D. K., T. D. Oberley, and Y.-S. Ho. 1991. Overproduction of human Mn-superoxide dismutase modulates paraquat-mediated toxicity in mammalian cells. *FEBS Lett.* **293**:199–203.

96. Warner, B., R. Papes, M. Heile, D. Spitz, and J. Wispe. 1993. Expression of human MnSOD in Chinese hamster ovary cells confers protection from oxidant injury. *Am. J. Physiol.* **264**(*Lung Cell. Mol. Physiol.* **8**):L598–L605.

97. Wispe, J. R., B. B. Warner, J. C. Clark, C. R. Dey, J. Neumann, S. W. Glasson, J. D. Crapo, L. C. Chang, and J. A. Whitsett. 1992. Human Mn superoxide dismutase in pulmonary epithelial cells of transgenic mice confers protection from oxygen injury. *J. Biol. Chem.* **267**:23937–23941.

98. Carlsson, L. M., J. Jonsson, T. Edlund, and S. L. Marklund. 1995. Mice lacking extracellular superoxide dismutase are more sensitive to hyperoxia. *Proc. Natl. Acad. Sci. USA* **92**:6264–6268.

99. Buttke, T. M., and P. A. Sandstrom. 1994. Oxidative stress as a mediator of apoptosis. *Immunol. Today* **15**:7–11.

100. Halliwell, B., and J. M. C. Gutteridge. 1990. Role of free radicals and catalytic metal ions in human disease: an overview. *Methods Enzymol.* **186**:1–85.

101. Benchekroun, M. N., P. Pouquier, B. Schott, and J. Robert. 1993. Doxorubicin-induced lipid peroxidation and glutathione peroxidase activity in tumor cell lines selected for resistance to doxorubicin. *Eur. J. Biochem.* **211**:141–146.

102. Lennon, S. V., S. J. Martin, and T. G. Cotter. 1991. Dose-dependent induction of apoptosis in human tumour cell lines by widely diverging stimuli. *Cell Prolif.* **24**:203–214.

103. Albina, J. E., S. Cui, R. B. Mateo, and J. S. Reichner. 1993. Nitric oxide-mediated apoptosis in murine peritoneal macrophages. *J. Immunol.* **150**:5080–5085.

104. Zhong, L.-T., T. Sarafian, D. J. Kane, A. C. Charles, S. P. Mah, R. H. Edwards, and D. E. Bredesen. 1993. *bcl-2* inhibits death of central neural cells induced by multiple agents. *Proc. Natl. Acad. Sci. USA* **90**:4533–4537.

105. Matsuda, M., H. Masutani, H. Nakamura, J. Miyajima, A. Yamauchi, S. Yonehara, A. Uchida, K. Irimajiri, A. Horiuchi, and J. Yodoi. 1991. Protective activity of adult T cell leukemia-derived factor (ADF) against tumor necrosis factor-dependent cytotoxicity on U937 cells. *J. Immunol.* **147**:3837–3841.

106. Chang, D. J., G. M. Ringold, and R. A. Heller. 1992. Cell killing and induction of manganous superoxide dismutase by tumor necrosis factor-α is mediated by lipoxygenase metabolites of arachidonic acid. *Biochem. Biophys. Res. Commun.* **188**:538–546.

107. Sarih, M., V. Souvannavong, S. C. Brown, and A. Adam. 1993. Silica induces apoptosis in macrophages and the release of interleukin-1α and interleukin-1β. *J. Leukocyte Biol.* **54**:407–413.

108. Pandey, S., and E. Wang. 1995. Cells en route to apoptosis are characterized by the upregulation of c-*fos*, c-*myc*, c-*jun*, *cdc2,* and RB phosphorylation, resembling events of early cell-cycle traverse. *J. Cell. Biochem.* **58**:135–150.

109. Jacobson, M. D., J. F. Burne, M. P. King, T. Miyashita, J. C. Reed, and M. C. Raff. 1993. Bcl-2 blocks apoptosis in cells lacking mitochondrial DNA. *Nature* **361**:365–369.

110. Hockenbery, D. M., Z. N. Oltvai, X. Yin, C. L. Milliman, and S. J. Korsmeyer. 1993. Bcl-2 functions in an antioxidant pathway to prevent apoptosis. *Cell* **75**:241–251.

111. Evan, G. I., A. H. Wylie, C. S. Gilbert, T. D. Littlewood, H. Land, M. Brooks, C. M. Waters, L. Z. Penn, and D. C. Hancock. 1992. Induction of apoptosis in fibroblasts by c-myc protein. *Cell* **69**:119–128.

112. Kane, D. J., T. A. Sarafian, R. Anton, H. Hahn, E. B. Gralla, J. S. Valentine, T. Ord, and D. E. Bredesent. 1993. Bcl-2 Inhibition of neural death: decreased generation of reactive oxygen species. *Science* **262**:1274–1277.

113. Korsmeyer, S. I. 1995. Regulators of cell death. *Trends Genet.* **11**:101–105.

114. Meyer, M., R. Schreck, and P. A. Baeuerle. 1993. H_2O_2 and antioxidants have opposite effects on activation of NF-κB and AP-1 in intact cells: AP-1 as secondary antioxidant-responsive factor. *EMBO J.* **12**:2005–2015.

115. Schrier, P. I., and L. T. C. Peltenburg. 1993. Relationship between *myc* oncogene activation and MHC class I expression. *Adv. Cancer Res.* **60**:181–246.

116. Shi, Y., J. M. Glynn, L. J. Guilbert, T. G. Cotter, R. P. Bissonnette, and D. R. Green. 1992. Role for c-myc in activation-induced apoptotic cell death in T cell hybridomas. *Science* **257**:212–214.

117. Miller, C. C., P. Hale, and A. P. Pentland. 1994. Ultraviolet B injury increases prostaglandin synthesis through a tyrosine kinase-dependent pathway. *J. Biol. Chem.* **269**:3529–3533.

118. Demers, L. M., and D. C. Kuhn. 1994. Influence of mineral dusts on metabolism of arachidonic acid by alveolar macrophage. *Environ. Health Perspect.* **102**:(S10)97–100.

119. Segade, F., E. Claudio, K. Wrobel, S. Ramos, and P. S. Lazo. 1995. Isolation of nine gene sequences induced by silica in murine macrophages. *J. Immunol.* **154**:2384–2392.

120. Donaldson, K., X. Y. Li, S. Dogra, B. G. Miller, and G. M. Brown. 1992. Asbestos-stimulated tumour necrosis factor release from alveolar macrophages depends on fibre length and opsonization. *J. Pathol.* **168**:243–248.

121. Perderiset, M., J. P. Marsh, and B. T. Mossman. 1991. Activation of protein kinase C by crocidolite asbestos in hamster tracheal epithelial cells. *Carcinogenesis* **12**:1499–1502.

122. Lin, L. L., M. Wartmann, A. Y. Lin, J. L. Knopf, A. Seth, and R. J. Davis. 1993. cPLA$_2$ is phosphorylated and activated by MAP kinase. *Cell* **72**:269–278.

123. Hii, C. S. T., A. Ferrante, Y. S. Edwards, Z. H. Huang, P. J. Hartfield, D. A. Rathjen, A. Poulos, and A. W. Murray. 1995. Activation of mitogen-activated protein kinase by arachidonic acid in rat liver epithelial WB cells by a protein kinase C-dependent mechanism. *J. Biol. Chem.* **270**:4201–4204.

124. David, M., E. Petricon III, C. Benjamin, R. Pine, M. J. Weber, and A. C. Larner. 1995. Requirement for MAP kinase (ERK2) activity in interferon α- and interferon β-stimulated gene expression through STAT proteins. *Science* **269**:1721–1723.

125. Karin, M. 1995. The regulation of AP-1 activity by mitogen-activated protein kinases. *J. Biol. Chem.* **270**:16483–16486.

126. Hibi, M., A. Lin, T. Smeal, A. Minden, and M. Karin. 1993. Identification of an oncoprotein-and UV-responsive protein kinase that binds and protentiates the c-Jun activation domain. *Genes Dev.* **7**:2135–2148.

127. Han, J., J. D. Lee, L. Bibbs, and R. J. Ulevitch. 1994. A MAP kinase targeted by endotoxin and hyperosmolarity in mammalian cells. *Science* **265**:808–811.

128. Cano, E., and L. C. Mahadevan. 1995. Parallel signal processing among mammalian MAPKs. *Trends Biochem. Sci.* **20**:117–122.

129. Dérijard, B., J. Raingeaud, T. Barrett, I. H. Wu, J. Han, R. J. Ulevitch, and R. J. Davis. 1995. Independent human MAP kinase signal transduction pathways defined by MEK and MKK isoforms. *Science* **267**:682–685.

130. Stokoe, D., S. G. Macdonald, K. Cadwallader, M. Symons, and J. F. Hancock. 1994. Activation of Raf as a result of recruitment to the plasma membrane. *Science* **264**:1463–1466.

131. Gille, H., A. D. Sharrocks, and P. E. Shaw. 1992. Phosphorylation of transcription factor p62^{TCF} by MAP kinase stimulates ternary complex formation at c-*fos* promoter. *Nature* **358**:414–421.

132. Marais, R., J. Wynne, and R. Treisman. 1993. The SRF accessory protein Elk-1 contains a growth factor-regulated transcriptional activation domain. *Cell* **73**:381–393.

133. Zinck, R., R. A. Hipskind, V. Pingoud, and A. Nordheim. 1993. C-*fos* transcriptional activation and repression correlate temporally with the phosphorylation status of TCF. *EMBO J.* **12**:2377–2387.

134. Whitmarsh, A. J., P. Shore, A. D. Sharrocks, and R. J. Davis. 1995. Integration of MAP kinase signal transduction pathways at the serum response element. *Science* **269**:403–407.

135. Stevenson, M. A., S. S. Pollock, C. N. Coleman, and S. K. Calderwood. 1994. X-irradiation, phorbal esters, and H_2O_2 stimulate mitogen-activated protein kinase activity in NIH-3T3 cells through the formation of reactive oxygen intermediates. *Cancer Res.* **54**:12–15.

136. Schlessinger, J., and A. Ullrich. 1992. Growth factor signalling by receptor tyrosine kinases. *Neuron* **9**:383–391.

137. Chen, Q., M. S. Kinch, T. H. Lin, K. Burridge, and R. L. Juliano. 1994. Integrin-mediated cell adhesion activates mitogen-activated protein kinases. *J. Biol. Chem.* **269**:26602–26605.

138. Sesko, A., M. Cabot, and B. T. Mossman. 1990. Hydrolysis of inositol phospholipids precedes cellular proliferation in asbestos-stimulated tracheobronchial epithelial cells. *Proc. Natl. Acad. Sci. USA* **87**:7385–7389.

139. Dérijard, B., M. Hibi, I. H. Wu, T. Barrett, B. Su, T. Deng, M. Karin, and R. J. Davis. 1994. JNK-1: A protein kinase stimulated by UV light and Ha-Ras that binds and phosphorylates the c-Jun activation domain. *Cell* **76**:1025–1037.

140. Westwick, J. K., C. Weitzel, A. Minden, M. Karin, and D. A. Brenner. 1994. Tumor necrosis factor α stimulates AP-1 activity through prolonged activation of the c-Jun kinase. *J. Biol. Chem.* **269**:26396–26401.

141. Sluss, H. K., T. Barrett, B. Dérijard, and R. J. Davis. 1994. Signal transduction by tumor necrosis factor mediated by JNK protein kinases. *Mol. Cell. Biol.* **14**:8376–8384.

142. Rouse, J., P. Cohen, S. Trigon, M. Morange, A. Alonso-Liamazares, D. Zamanillo, T. Hunt, and A. R. Nebreda. 1994. A novel kinase cascade triggered by stress and heat shock that stimulates MAPKAP kinase-2 and phosphorylation of the small heat shock proteins. *Cell* **78**:1027–1037.

143. Raingeaud, J., S. Gupta, J. S. Rogers, M. Dickens, J. Han, R. J. Ulevitch, and R. J. Davis. 1995. Pro-inflammatory cytokines and environmental stress cause p38 mitogen-activated protein kinase activation by dual phosphorylation on tyrosine and threonine. *J. Biol. Chem.* **270**:7420–7426.

144. Xia, Z., M. Dickens, J. Raingeaud, R. J. Davis, and M. E. Greenberg. 1995. Opposing effects of ERK and JNK-p38 MAP kinases on apoptosis. *Science* **270**:1326–1331.

17

Regulation of Glutathione Peroxidases

*Leopold Flohé, Edgar Wingender, and
Regina Brigelius-Flohé*

17.1 Introduction

The glutathione peroxidases (Gpx) belong to a superfamily of phylogenetically related proteins of diverse functions.[1] The members of the superfamily containing a selenocysteine residue in their catalytic centers are highly efficient peroxidases reacting with a variety of hydroperoxides at rate constants of greater than 10^6 $M^{-1} s^{-1}$.[1] A cysteine residue in homologous position is catalytically less effective by about three orders of magnitude.[2,3] Such GPx-like proteins, although still potential redox catalysts, can not be rated as peroxidases, if this term is to designate enzymes that must remove peroxides rapidly from a biological environment.

The real selenium-containing glutathione peroxidases have been extensively studied in mammals, where at least four distinct types coexist[1]: (1) the "classical" cytosolic form (cGPx), a homotetrameric enzyme with a broad specificity for hydroperoxide substrates, but inactive on hydroperoxides of complex lipids (for review see Ref. 4); (2) the extracellular or plasma GPx (pGPx), also tetrameric and apparently exhibiting an even broader specificity[5-7]; (3) the phospholipid hydroperoxide GPx (PHGPx), which, apart from low-molecular-weight hydroperoxides, also acts on hydroperoxides of complex lipids[8] and cholesterol[9]; and (4) a gastrointestinal form giGPx, which is functionally not yet fully characterized.[10] More recently glutathione peroxidases were also detected in the plathelminth worm *Schistosoma mansoni*,[11,12] in the phytophlagellate *Chlamydomonas reinhardtii*[13] and in the plant *Aloe vera*.[14] Most commonly, their biological role is seen in the defense against oxidative stress, but unusual tissue distribution, hormone-dependent expression, and activity changes during differentiation, as is, e.g., observed with PHGPx, suggest more specific functions for some members of this family.

Here we summarize data on the regulation of GPx expression and function in higher animals. This compilation will by no means reveal an identical pattern of regulation for the different types of GPx, rather, particular regulatory phenomena might draw the attention to the biological context of the individual enzyme.

17.2 Regulation by Selenium

It might appear trivial that selenoproteins can not be formed in the presence of selenium deficiency. However, the various selenoproteins respond to selenium deprivation in a characteristic pattern. Under selenium restriction, the selenoproteins incorporate the trace element according to a fixed hierarchy, which is tissue-specific and possibly reflects the relative importance of the individual proteins. The tissue specificity of selenoprotein expression is obviously achieved by transcriptional regulation (see below), but it is modulated by selenium availability at the translational level. To clarify the selenium effects, the most complex pathway of selenoprotein biosynthesis must be briefly reviewed[15-18] (Fig. 17.1). In both, prokaryotes and eukaryotes selenocysteine is encoded by the termination codon TGA. In both cases, a characteristic stem-loop structure of the mRNA is required for recoding FUGA as the selenocysteine codon. In bacteria this stem-loop is located immediately downstream the UGA, i.e., within the coding region; in eukaryotes the stem-loop system, called SECIS for *se*lenocysteine *i*nsertion *s*equence), is found in the 3' flanking region. This secondary mRNA structure is recognized by a translation factor, called SELB, which in bacteria has been characterized as being homologous to elongation factor Tu,[19] yet remains unidentified in eukaryotes. SELB "codes from a distance" by binding selenocysteyl tRNA (SELC), which carries an anticodon for UGA, and directing the selenocysteine-loaded tRNA to the UGA codon of the ribosome-bound mRNA (Fig. 17.1B). The specific tRNA (tRNA[(Ser)Sec]) is first loaded with serine and then transformed into selenocysteyl-tRNA[(Ser)Sec] by a selenocysteyl-tRNA synthetase (SELA), which requires selenophosphate as substrate. Selenophosphate, in turn, is made available by a synthetase, called SELD, from ATP and inorganic selenide (Fig. 17.1A).

Most of the steps of the cascade being common to the synthesis of all selenoproteins, the reasons for differential selenium incorporation must be sought in the structure of the genes or the pertinent mRNAs. Theoretically, three possibilities can be considered: (1) selenium-responsive elements in selenoprotein genes might differentially enhance transcription in response to selenium, (2) the mRNA stability might be affected by selenium, or (3) the SECIS elements might have different affinities to the SELB translation factor (still hypothetical in animals) and thereby allow for a translation efficiency that is characteristic for the individual selenoprotein.

So far, the search for selenium-responsive elements regulating transcription has remained unsuccessful. Moreover, no difference in transcription rates of the cGPx gene between selenium-deprived and selenium-adequate tissue could be

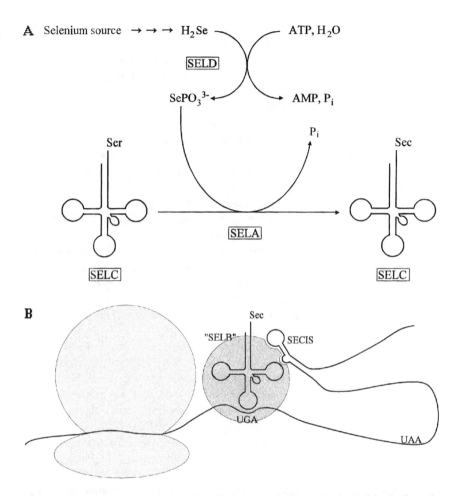

Figure 17.1. Schematic representation of selenoprotein biosynthesis. A. Metabolic pathway leading to selenocysteyl-tRNA[(Ser)Sec]. B. Presumed translation mechanism of UGA as codon for selenocysteine in eukaryotes. For further explanations see text.

found in a variety of nuclear run-on experiments[20-24]. This evidence for lacking selenium-dependent transcriptional regulation of the cGPx gene should, however, not be uncritically generalized.

As a rule, the mRNA levels of selenoproteins decrease on long-term selenium deprivation *in vivo*,[25-27] and the cGPx mRNA drops most dramatically[22,23,26,28-30] despite, as mentioned above, unchanged transcription rate. A similar instability of the cGPx mRNA could also be observed in Hep3B cells[20] and MCF-7 cells[31] grown in selenium-deficient media. A notable exception in this respect is the mRNA of PHGPx, which was reported to remain unaffected by long-term sele-nium deprivation.[26] These observations imply that the mRNAs of the various selenoproteins are stabilized by selenium, although to a different extent, and this differential mRNA stabilization is believed to be of outstanding importance for the maintenance of the hierarchy of selenoproteins. For instance, the mRNA for selenoprotein P (and 5'-deiodinase) decreases more slowly than the mRNA of cGPx in the liver of rats kept on a selenium-deficient diet (19 versus 3% of control after 14.5 weeks), and correspondingly the selenoprotein P levels were not depleted to the same extent (4.3%) as those of cGPx (0.8%).[26] Also, selenopro-tein P returns to normal values more rapidly than did cGPx on refeeding sele-nium.[32] The mRNA levels of pGPx were reported to be slightly less decreased in selenium-deficient cell lines than those of cGPx.[33] An even more pronounced difference was seen between PHGPx mRNA (unchanged) and cGPx mRNA in liver (6%) and heart (12%) of selenium-deprived rats, but not, however, in testis (unchanged for both despite selenium deprivation); pertinent cGPx activities were 1% in liver and 6% in heart, whereas PHGPx activity did not fall to less than 40% of normal.[30] The exceptional stability of PHGPx mRNA may also explain the fast recovery of PHGPx activities (plateau reached within 10 h) in selenium-deficient rat basophilic leukemia cells on reexposure to selenium, where cGPx activities did not recover but after 3 days.[34]

Differential mRNA stability can, however, not explain the entire selenium regulation. As already mentioned, cGPx mRNA and PHGPx mRNA remained unaffected in the testes of selenium-deficient rats, yet the pertinent enzyme activities were significantly reduced.[30] Also a 30-fold decrease of GPx activity associated with unchanged mRNA levels has been reported for the human myeloid cell line HL-60 cultured in selenium-deficient media.[35] Most importantly, when-ever mRNA and pertinent enzyme activities were measured simultaneously, the fall in activity was substantially larger. This implies that the presence of a particular mRNA, there can only not be a bottle neck in selenoprotein synthesis, but the availability of selenocysteyl-tRNA must be equally important and some-times appears to be the only limiting factor.

In this respect, the tRNA[(Ser)Sec] itself and its loading with selenocysteine may be limiting. Transfection of cells with the tRNA[(Ser)Sec] gene similarly increased selenoprotein (5'-deiodinase) biosynthesis as did selenium supplementation of the medium.[36] Note also that the chemical nature of the selenium compound offered is of crucial importance for its appropriate utilization for selenoprotein synthesis and such utilization does not parallel "bioavailability" in terms of tissue uptake in vivo. Selenomethionine, for instance, up to 100 nM proved to be unable to induce any significant biosynthesis of cGPx and PHGPx in a variety of cultured cells, whereas sodium selenite already at 50 nM guaranteed optimum expression of these enzymes.[37] This demonstrates that the ease of metabolization of any selenium compound to selenide as a source for selenophosphate and, in con-sequence, selenocysteyl-tRNA[(Ser)Sec] determines its utility for selenoprotein synthesis. Finally, in a tissue-specific manner, selenium affects the editing of selenocysteyl-tRNA$^{(Ser)Sec}$.[38,39] In response to selenium supplementation, the 5-methyl-carboxymethyluridine at the wobble position in the anticodon of tRNA$^{(Ser)-Sec}$ is O-methylated at the 2' position of its ribose.[39] The functional relevance of this effect, however, remains to be determined. In short, while mRNA instability is a major determinant for selenoprotein biosynthesis in long-term selenium deficiency, factors affecting the level of selenocysteyl-tRNA$^{(Ser)Sec}$ (and possibly its appropriate editing) are equally important and probably more relevant in short-term deprivation and repletion studies.

Returning to the problem of how a differential biosynthesis of selenoproteins at limiting supply could be achieved, we must also consider the inequivalence of their different SECIS elements. The relevance of a variable SECIS efficiency is stressed by the observation that the appropriate translation of the UGA codon as selenocysteine, in general, is not a particularly efficient process, but is accompanied by a considerable production of truncated proteins.[36] Based on consensus analysis and mutational analysis[18,40] a stem-loop system qualifies as SECIS by a strictly conserved triple A in the loop plus some conserved nucleotides forming bulges at a defined distance from the loop. Further determinants of SECIS efficiency are the thermodynamic stability of the stem, its shape, and its distance from the coding region.[40] Interestingly, the distance of the stem-loop from the translated region appears to become almost irrelevant, once a certain length is exceeded[18] which implies that only a minimum flexibility of the 3' flanking region must be guaranteed to allow a bending-back of the SELB-SECIS-selenocysteyl-tRNA$^{(Ser)Sec}$-complex to the UGA codon on the ribosome-bound mRNA (see Fig. 17.1B). Unfortunately, only a limited number of different SECIS elements have been directly compared for relative efficiencies. Fusing the SECIS elements of 5' deiodinase, selenoprotein P, and cGPx to the coding region of the deiodinase and monitoring the expression thereof yielded the SECIS efficiency ranking: selenoprotein P > 5'-deiodinase > cGPx[18]. Another system measuring suppression of UGA, placed between two indicator genes followed by various SECIS elements, revealed that the PHGPx SECIS is more efficient than that of the 5' deiodinase and selenoprotein P.[40] The tentative ranking derived from these experiments, PHGPx > selenoprotein P > 5' deiodinase > cGPx, strikingly parallels the resistence of these proteins to selenium depletion and also against degradation of the pertinent mRNAs on long-term selenium deprivation.

The hierarchy of selenoprotein biosynthesis thus appears to be determined by two different phenomena, the differential SECIS efficiency and the differential mRNA stability. In view of the parallelism of the events, one is, however, tempted to speculate about a common underlying physical phenomenon. The formation of the ternary complex of selenocysteyl-tRNA$^{(Ser)Sec}$, SELB, and SECIS could be envisaged to occur sequentially. If selenocysteyl-tRNA$^{(Ser)Sec}$ binding to SELB were a prerequisite for binding of SELB to the SECIS element, the bioavailability of selenium would in principle determine the degree of complex formation, yet the formation of the ternary complex (tRNA-SELB-SECIS) would still depend on the affinity of the binary complex (tRNA-SELB) to a particular SECIS. If now the complex were more resistent against degradation than free mRNA, selenium via selenocysteyl-tRNA[(Ser)Sec] binding to SELB would stabilize selenoprotein mRNA according to SECIS efficiency.

In summary, the regulation of selenoprotein expression by selenium is complex. In general, selenoprotein synthesis depends on adequate availability of selenocysteine-loaded tRNA[(Ser)Sec]. Among the selenoperoxidases, PHGPx is most resistent, cGPx is most affected by limited selenium supply, pGPx ranks in between,[26] and giGPx has not been analyzed in this respect. The most relevant findings explaining the preferential biosynthesis of certain selenoproteins at limited selenium supply are selenium-dependent differential mRNA stabilities and different SECIS efficiencies.

17.3 Regulation by Redox Status

Elevated oxygen pressure alone or in combination with endotoxin leads to increased cGPx activity in the lungs of rabbits[41] and rats,[42] respectively. A corresponding induction of cGPx by high oxygen tension is also observed *in vitro*, e.g., in human myocytes[43,44] and human umbilical vein endothelial cells[45] and the increase of enzyme activity was reported to be associated with an increased transcription rate of the cGPx gene.[44,45] A recent analysis indeed revealed the presence of two oxygen responsive elements, referred to a ORE 1 and ORE 2, in the 5' flanking region of the human cGPx gene.[46] By means of reporter gene constructs, footprinting and gel mobility shift assays the oxygen-responsive elements could be localized upstream from the transcription starting at positions −1232 to −1213 and −282 to −275.[46] The proteins binding to the OREs have, however, not yet been characterized and correspondingly also the effector ultimately triggering transcriptional activation has remained unknown until now.

In this context, it may be revealing that cGPx is also induced by a variety of oxidative stress conditions at normoxic conditions such as ischemia reperfusion,[47,48] and exposure to redox cyclers,[49,50] endotoxin,[42] or interleukin-1.[50] These conditions have long been known to be associated with massive superoxide and, in consequence, hydrogen peroxide formation,[51–53] and thus any of the reaction products thereof usually referred to as reactive oxygen species (ROS) could be discussed as the proximal inducer of cGPx. Circumstantial evidence, however, favors the assumption that the major substrate of cGPx, H_2O_2, might be the inducing agent. In patients with trisomy 21, Cu/Zn superoxide dismutase is substantially elevated due to gene dose effects, but cGPx is similarly elevated[54] despite unchanged dosage of the cGPx gene, which is located on chromosome 3.[55–57] A marked increase of cGPx is also observed in transfected murine fibroblasts overexpressing Cu/Zn superoxide dismutase[58] and inhibition of superoxide dismutase by *N, N'*-diethyl-dithiocarbamate in umbilical vein endothelial cells decreased cGPx mRNA levels.[59] These observations imply that cGPx transcription may be enhanced whenever the superoxide/hydrogen peroxide is shifted toward the peroxide and vice versa.

Certainly, however, any generalization with respect to ROS-dependent induction of antioxidant enzymes should be viewed with caution, since many observations suggest a differential regulation of the cellular antioxidant devices depending on the site of ROS production and the triggering mechanism. Exogenous H_2O_2, for instance, strongly induces catalase in tracheobronchial epithelia, but only slightly GPx and Mn superoxide dismutase (MnSOD), whereas ROS generation by the xanthine/xanthine oxidase system selectively induces MnSOD.[60] Tumor necrosis factor (TNF) is believed to preferentially induce mitochondrial superoxide formation and leads to a selective increase of the mitochondrial MnSOD, leaving cGPx levels usually unaffected.[61–63] Such different responses to ROS generation may in part be explained by the emerging evidence that a large variety of nuclear transcription factors is subject to redox regulation.

It is well documented that both TNF and ROS may lead to the activation of the nuclear transcription factor NF-κB thereby inducing its translocation and finally allowing transcriptional activation of several genes.[64] However, the genomic sequences of the GPx genes known so far do not contain any recognizable NF-κB-responsive elements. This signaling pathway is thus not likely directly involved in the transcriptional activation of GPx genes.

Among several stress factors, oxidizing agents as well as hypoxia have been demonstrated to induce DNA binding by heat shock factor 1 (HSF-1).[65,66] HSF-1 was first detected as the responsible transcription factor that activates the gene for cytoprotective protein HSP70 in yeast, *Drosophila*, mouse, and human cells in response to heat and other stress conditions. Since overexpressed cGPx inhibits hsp70 induction, activation of this gene appears to involve ROS,[67] which may suggest a common regulator for both hsp70 and GPx genes. HSF-1 binds to a two- to threefold repeat of GAA trinucleotides in alternative orientation with two spacing nucleotides, thus following the (idealized) pattern GAANNTTCNNG AA.[68] Several similar elements can be identified in some of the known GPx genes. For example, arrays of GAA/TTC trinucleotides can be found around −1618/−1576[46] of the human cGPx transcription start site, in a region demonstrated to positively affect the basal transcription rate,[56] and also within the first intron of the androgen-dependent, epididymis-specific GPx (+729/+755)[69] (Fig. 17.2), but not in any of the pGPx and PHGPx genes.

Another transcription factor that is involved in redox regulation of gene expression is activator protein-1 (AP-1). Although assigned to hypoxic rather than oxygen-mediated induction processes, it may be considered in this context as well, since a factor mediating positive responses in general may also be involved in repression events (e.g., by inhibiting binding of a more potent and/or more specific activator, by activating expression of a repressor gene, etc.). The main

```
cGPx:              TTCaaGGAatTTCttaacTTCttgcTTCtTTCtaGAAagaGAA
GPxh:              GAAtaGAAc.TTCttgtcTTCcatGAA
human hsp70:       GTCggGAAtaTTCcaggggtTTC
```

Figure 17.2. DNA sequences of GPx genes (cGPx, human cytosolic Gpx; GPxh, murine androgen-dependent, eipdidymis-specific GPx) and of human heat shock protein 70 gene (hsp70) resembling the GAANNTTCNNGAA motif characteristically binding stress factors such as heat shock factors (HSFs).

constituents of AP-1 heterodimers, c-Jun and c-Fos, and both inactivated by a regulated redox mechanism that involves Cys-269 (c-Jun) and Cys-154 (c-Fos), respectively.[70,71] Moreover, the c-Fos is up-regulated by antioxidants,[72] whereas the c-Jun is induced by hydrogen peroxide.[73] Instead, the cognate JunB is up-regulated during hypoxia.[74] Putative AP-1 sites are also present in many superoxide dismutase (SOD) genes, and perfect AP-1 consensus matches are found in intron 1 of the human and murine cGPx genes, both of them exhibiting a purine-rich 3'-flanking sequence, but neither in pGPx nor in PHGPx genes. Antioxidant responsive elements (AREs), at least some of them comprising AP-1(-like) binding sequences, have been defined in several target genes such as the genes for glutathione S-transferase P and Ya subunit and NAD(P)H:quinone oxidoreductase[75,76] and tyrosine hydroxylase.[74] Interestingly, the basic pattern of their AREs exhibit some similarity to the two sequences found in mediate oxygen response of the cGPx gene (Fig. 17.3).

It can be seen that the AREs reveal significant similarity with the NF-E2-type of AP-1 site extension. NF-E2 is a heterodimer of a Maf component and of a

```
                    TGASTCA                            AP-1 consensus
                    * * * *    *
TGGCATTgctaatggTGACAAAGCAacttt                         ARE of RGSTYa
  GCAGTCAcag     TGACTCAGCAgaatcTGAGCCT                 ARE of HNQO1
              *   * * * * *
    gagggctccgttttTTCTCAGAggttcctga                     ORE2 (inverse)
            *  *  * * * *  * *          *              (identities between OREs)
              TCATCCTCAAAGAAAGTGTA                      ORE1

                    TYCTCARAG                           ORE consensus
                    * * * * * * *
               KGSNKNCTCAG                              c-Maf consensus
               RTGASTCAGCR                              NF-E2 consensus
```

Figure 17.3. Oxygen-responsive elements (OREs) of the human cGPx gene and their similarity with antioxidant responsive elements (AREs) of the genes for rat glutathione S-transferase Ya subunit (RGSTYa) and human NAD(P)H:quinone oxidoreductase (HNQO1). These elements are aligned with the consensus binding patterns for the transcription factors AP-1, c-Maf, and the erythroid-enriched factor NF-E2. Sequences shown in capital letters have previously been suggested as significant. Notice the similarity between AREs and AP-1 binding sites.[75,76] Consensus sequences are given in the 15-letter IUPAC code (K: G or T; M: A or C; S: C or G; W: A or T; R: A or G; Y: C or T).

p45 component, both belonging to the bZIP family. While the members of the Maf family are ubiquitously present, the NF-E2 p45 protein exhibits an expression pattern that overlaps with that of cGPx in that it is found in erythroid, liver, lung, spleen, and, at low levels, in the testis; however, it is not present in the kidney.[77,78] Since the bZIP family of transcription factors comprises numerous members, it could be easily envisaged that specific combinations are involved in transcriptional activation of GPx genes in a variety of tissues.

The region of the tyrosine hydroxylase gene that governs O_2 response additionally comprises a binding site for the transcription factor HIF-1 (hypoxia-inducible factor 1).[74] This factor belongs to the bHLH-PAS family[79] and was previously detected to bind to and activate the human erythropoietin gene.[80] It reveals similarity with the aryl hydrocarbon (dioxin) receptor (AhR), which mediates genetic responses to toxic aromatic compounds through ACGTGCK-containing sequence elements. Several similar elements can be found in GPx genes, one of them in the known transcription regulating region of human cGPx (at $-472/-465$, where it is part of a larger imperfect palindrome), but this type of element has not yet been sufficiently characterized to be used for predictive analyses.

In summary, there is a clear evidence for a redox-regulation of cGPx genes from tissue culture experiments and from the identification of OREs at the DNA level. Further, many putative regulatory elements such as AP-1, AP-1-like, HSF-1, and NF-E2-type sites found in the cGPx genes suggest the possibility of fine-tuned response to various events summarized as oxidative stress, although the biological significance of these putative regulatory elements awaits experimental verification. For any of the other GPx types, experimental evidence for redox regulation is lacking and their gene structures, as far as available, do not reveal cGPx-type OREs, nor any other unequivocal consensus sequences suggestive of redox regulation.

17.4 Tissue-Specific Expression

17.4.1 Source and Fate of Extracellular GPx (pGPx)

The extracellular glutathione peroxidase (pGPx), first described by Takahashi et al.,[5] was purified and characterized from human blood plasma. Low levels of this enzyme have also been found in other extracellular fluids such as milk[81] and bile[82] but only the activity in plasma and milk has been confirmed as authentic pGPx by precipitation with specific antisera.

The source of pGPx in plasma remained unclear until recently. Originally, liver was proposed to synthesize and secrete pGPx, since also a human hepatoblastoma line, HepG2, was shown to be able to secrete pGPx into the cell culture medium,[33] whereas myeloid or endothelial cells were not. However, mRNA could not be detected in normal human liver.[6] It has, however, become clear that the kidney is the main source of pGPx in rats,[83] mice,[84] and men,[85] as demonstrated by Northern hybridization and immunoblotting. In situ hybridization localized pGPx transcripts to proximal tubules and parietal epithelial cells of Bowman's capsule in humans[85] and to proximal tubules in mice.[84] When compared to pertinent kidney tissue, much smaller amounts of pGPx mRNA were found in rat lung[83]; in human heart, lung, breast, and liver[86]; and in all murine tissues investigated with the exception of liver.[84] Which of these tissues secrete pGPx into the blood plasma is unknown. The bronchioalveolar lavage fluid may be the destination of lung pGPx, whereas the pGPx produced in bovine ocular ciliary epithelium appears to be secreted into the aqueous humor, a fluid that inter alia protects the avascular ocular tissues from oxidative stress.[87] pGPx produced in human placenta is apparently secreted into the maternal circulation but, surprisingly, not into the fetal circulation.[88]

17.4.2 Recent Advances in Tissue Specificity of cGPx

Since cGPx has been known for almost 40 years,[4,89,90] its tissue distribution has been amply studied in many species (for review see Ref. 4) but is nevertheless not really understood. Under adequate selenium supply, all cells express at least some cGPx. Particularly high levels are generally found in tissues with a high rate of peroxide production, such as erythrocytes, liver, kidney, or lung.[4] In rats, the lowest cGPx levels were measured in testis.[30,91]

An exceptional tissue distribution of cGPx was observed in guinea pigs. Despite normal selenium concentrations cGPx, they were extremely low or absent in liver, kidney, heart, lung, brain, and testis, whereas in erythrocytes cGPx activities were similar to those of mice and rats.[92] The reason for this low cGPx activity was attributed to a poor transcription rate by means of nuclear run-on assays. The lack of GPx activity in most guinea pig tissues is obviously compensated for by a two to three times higher activity of catalase, mainly localized in the cytosol.[93] This metabolic uniqueness not only lends further support to the assumption that the guinea pig is misclassified as a usual rodent,[94,95] it also may offer an excellent opportunity to study the mutual impact of expression of peroxide metabolizing enzymes.

The molecular basis responsible for high-level expression of cGPx in specific tissues was found to be up-regulation of mRNA. Detailed analysis of the murine cGPx gene revealed that its promoter region only marginally contributes to tissue-specific expression in erythroid cells; 1.7 kb of the upstream region gave rise to only 1.5 to 2-fold higher efficiency in erythroid cells than in an epithelial cell line.[96] In contrast, an enhancer element was identified in the 3'-flanking region, which contains a cluster of four binding sites for transcription factors of the GATA family and two sites for CACC/GT factors.[96] This enhancer mediates an up to 11-fold higher transcription in the erythroid than in the epithelial cell line, depending on the intact assembly of all factor binding sites in the proper orientation. Factors binding to GATA and CACC/GT sites (GATA-½ and EKLF) determine the cell-specific expression of several erythroid proteins, e.g., the α- and β-globin or the haem biosynthetic enzyme porphobilinogen deaminase.[97] In addition to be involved in the cGPx expression in erythroid cells, they may also play a role in the transcription of this gene in other tissues since GATA-2 has also been found in the kidney and (embryonic) liver, and EKLF is present in the testis and, at lower levels, in the spleen,[98,99] while its cognate LKLF is enriched in lung cells.[100] Moreover, the above-mentioned AP-1-related factor NF-E2 has originally been discovered in the erythroid context (Ref. 101, and the references therein).

Another factor that matches with the tissue-specific expression of cGPx, particularly with the high expression in liver and kidney, is the hepatocyte nuclear factor 1 (HNF-1). Most suggestive is a cluster of three perfectly matching HNF half-sites at −1839/−1708, all of them also revealing imperfect palindromic complementations, in a region of known transcriptional importance of the human cGPx gene (−1851/−1590).[56] However, the biological significance of putative HNF-1 sites is difficult to predict because of their highly degenerate recognition sequence, and the mechanisms that govern the tissue-specific expression of cGPx in nonerythroid cells has not yet been analyzed experimentally.

Phenomenological studies in developing animals revealed that cGPx contents in rat lungs increased after birth, especially when exposed to high oxygen tensions.[42] This was first interpreted as adaption to the higher oxygen exposure of lungs after birth.[102] An increase in cGPx mRNA was also observed in mouse liver, spleen, and kidney,[102] but could not be confirmed in another study, where cGPx already rose 2 weeks before delivery.[103] The time course of cGPx mRNA in fetal and neonatal tissues thus apparently do not respond in a logical way to oxygen exposure and the developmental phases of expression and the related fluctuations are not easily interpreted.

Developmental changes in cGPx expression have also been observed in rat intestine.[104] Two weeks after birth, cGPx protein became undetectable in duodenum, and disappeared in ileum after weaning. Unfortunately, the selenium status of the animals was not reported and the specificity of the antibodies applied must also be questioned. In any event, the data demonstrate that either cGPx or giGPx in the intestine is subject to pronounced changes during postnatal development, which could be due to a genetic program, the alimentary status or, more likely, to both.

17.4.3 Unusual Pattern of PHGPx Tissue Distribution

One of the peculiarities of PHGPx, when compared to cGPx, is its unusual tissue distribution. Its activity is generally lower in all organs with the exception of testis. Rat testis, for example, has the highest PHGPx activity of all organs tested[30,105]; it is 15-fold more active there than cGPx is in liver and 25-fold more active in heart.[30] Surprisingly, PHGPx activity in rat testis appears only after puberty (day 25 after birth) and reaches a plateau around day 60.[106] The expression of PHGPx in testis was shown to be under control of hormones (see below) and only marginally decreased in selenium deficiency.[30] A remarkable stability of PHGPx activity was also observed in the thyroid under selenium deficiency.[107] In other organs, such as heart, liver, kidney, and lung, PHGPx decreased when selenium became limited, but the decrease never reached values as low as generally found for cGPx.[30,91,107] The preferential channeling of selenium into PHGPx in endocrine tissue and reproductive organs may indicate that PHGPx has not only antioxidant functions but rather plays a role, e.g., in sexual maturation and differentiation.

Obviously, also a post-translational regulation of PHGPx activity occurs in a tissue-specific way.[105] In most tissue, immunologic estimation of PHGPx content and PHGPx activities do not match. The discrepancies are particularly pronounced in muscle and liver, where no detectable or low activities despite high antigene contents were reported.[105] The chemical nature of the post-translational modification modulating PHGPx activity is still unclear.[108] But a phosphorylation at Tyr 95 of pig heart PHGPx has been suggested[109] and significant discrepancies between the molecular mass calculated from the amino acid sequence and determined by laser desorption mass spectroscopy have also been reported.[110]

The cellular distribution of PHGPx has not been studied extensively. In human kidney, PHGPx antigene was found to be localized in podocytes and parietal epithelial cells of the glomeruli, in the cytoplasm of tubular cells, and in the epithelial cells of Bowman's capsule.[111] Whether this distribution reflects a special need for peroxide removal in the respective tissues remains to be demonstrated. It may, in fact, be doubted, since the mesangial cells, the site of oxygen radical formation in kidney,[112] did not contain significant amounts of PHGPx.[111]

The subcellular distribution of PHGPx is equally puzzling. When measured immunologically, PHGPx content in almost all rat tissues is found to be higher in membrane fractions than in the soluble fraction. Again, activity measurements do not match. Based on enzyme activity, PHGPx was undetectable in the membrane fractions of 13 rat tissues investigated with the exception of testis, heart, liver, and lung, and only in testis membranes was the PHGPx activity significantly higher than in the cytosol.[105]

In general, PHGPx has been found in the cytosol, in mitochondria, and associated with nuclear membranes, but not with the outer cell membrane, which is consistent with the lack of glycosylation.[108] In rat brain mitochondria, PHGPx was found to be associated with the inner membrane and there, particularly with the contact sites,[113] whereas cGPx prevailed in the mitochondrial matrix space, as was also reported for rat liver mitochondria.[114] Similarly, in rat testis mitochondria, PHGPx is localized in the contact sites of mitochondrial membranes.[115] Interestingly, the subcellular localization of PHGPx changes during sperm maturation. In spermatogonia, it is seen evenly distributed in the cytosol and gradually becomes associated with membranes during differentiation until it is almost exclusively linked to mitochondria in mature spermatozoa in an immunologically detectable, but is almost inactive in form.[106]

Certainly, the post-translational modifications of PHGPx tyrosine phosphorylation,[109] partial N-terminal blockade,[109,110] and the discrepancy between calculated and measured molecular weight[110] may also be implicated for the variable subcellular distribution. More recently, however, also an alternative mechanism was proposed. In the genomic DNA of PHGPx, two in-frame ATG codons, which could both be used as translation starts, were detected.[110,116] We[110] and others[116] have reasoned that the second ATG, referred to as ATG[141,117] is used as translation start in pig liver or pig blastocysts, resulting in a 170–amino-acid protein. However, Pushpa-Rekha et al.[117] described that in the reticulocyte system a full-length cDNA derived from testis directed the synthesis of a 197-amino-acid protein starting from ATG[61]. The N-terminal extension thereby obtained was discussed as a potential mitochondrial targeting signal.[117]

17.4.4 GiGPx Restricted to the Gastrointestinal Tract?

The fourth member of the GPx family, the giGPx, was found as a cDNA sequence present in cDNA libraries of bovine liver and HepG2 cells probed with a sequence of bovine cGPx cDNA.[10] The detected sequence turned out to be almost identical to the one obtained by Akasada et al.[118] from a rabbit liver cDNA library. The new sequence was *in vitro* translated and the resulting protein used for preliminary functional characterization and antibody production. By immunhistochemistry giGPx was then found in the human and rodent gastrointestinal tract and in human liver, but not in human heart, kidney, and breast, nor in rat liver, skeletal muscle, heart, or kidney. The specificity for the hydroperoxide substrate of giGPx was reported to be similar to that of cGPx. A preferential reaction with organic hydroperoxides was, however, also observed. It was therefore concluded that giGPx may protect the organism from dietary organic hydroperoxides.[10] Some newly established human breast tumor cell lines, however, expressed giGPx either as the only GPx or together with cGPx and PHGPx.[119] It thus appear premature to speculate about a specific physiological role of this new glutathione peroxidase in the gastrointestinal tract.

17.5 Regulation by Hormones

Surprisingly, several members of the GPx superfamily are obviously regulated by sexual hormones, and the pattern of hormonal control, in some cases at least, suggests a specific role of such enzymes in sexual maturation rather than unspecific protection against oxidative stress.

cGPx is appreciably higher in female than in male rat liver,[120] and a comparable sex difference is also observed with the mitochondrial fraction.[121] Also, liver cGPx increases in oophorectomized female rats when treated with estrogens. In male rats liver, GPx activity increases on castration and decreases following testosterone application.[120,122,123]

Neither the physiological relevance nor the mechanism of the prevalence of cGPx in females has been seriously investigated so far. Two potential estrogen-regulatory elements (ERE) can be found in the upstream sequences of the cGPx gene close to the TATA box (−54/−40), the other (−1682/−1669) within the known transcription regulating region at −1851 to −1590.[56] However, their significance has not yet been proven, and it can even be ruled out that cGPx expression strictly depends on estrogen, since cGPx levels and estrogen receptor density were reported to be inversely correlated in a variety of human breast tumor cells.[119]

In contrast to cGPx, PHGPx appears to be highest in male rat gonads, but is built only after puberty and remains practically zero on hypophysectomy, unless gonadal development is initiated by gonadotropin treatment.[106] Within the testicular tissue, PHGPx is particularly high in spermatogonia, where it is localized in the cytosol. During sperm maturation it migrates to the nuclear membrane and almost disappears in mature spermatozoa.[106] Interestingly, PHGPx ion the plathelminth worm Schistosoma mansoni also appears to be prevalent in sexual organs, in this case, however, in the female vitelline cells.[12] Epididymis of various mammalian species contain selenium-free GPx homologs (GPxh), which are closely related to pGPx.[1] These GPx-like proteins are apparently secreted from the epididymal tissue and bind to the head of spermatozoa. Although their biological role is still obscure, their formation has been shown to depend on androgens.[124] An unusual steroid hormone regulatory element (glucocorticoid regulatory element [GRE]) has been shown to govern androgen induction of the GPxh gene[69]; it comprises two properly spaced half-sites, one of them perfectly matching the pattern of the conserved GRE half-site TGTYCT, but in an everted instead of an inverted relative orientation. No similar element is obvious in any of the other GPx genes, not even in the PHGPx gene, which is most suspected to be androgen regulated. Whether one of the other half-site-matching sites could possess androgen-regulatory properties remains to be elaborated. It therefore may be questioned whether the induction of PHGPx by gonadotropins is mediated by steroid hormones. Instead, chorionic gonadotropin could act directly via cAMP-responsive elements (CRE), since resembling the canonical CRE motif (TGACGTCA) are present in the PHGPx gene.[110]

cGPx also appears to be controlled by additional hormones. The enzyme is known to rise steeply in the lung and to a lesser degree in other tissues during the perinatal phase, which is consistent with the concept of necessary adaptation to life in high oxygen tension after delivery.[102] Thyreotropin-releasing hormone (TRH) administered prenatally largely depressed the adaptive rise of cGPx (and other antioxidant enzymes) on increased oxygen tension and this depressed adaption was associated with a correspondingly low mRNA.[125,126] Dexamethasone enhanced the inhibitory effect of TRH.[125,126] These regulatory phenomenons are not easily explained at the molecular level.

There is little information available about the signal transduction pathways used by TRH or the genomic signals it may aim at. It is known that TRH causes enhanced calcium concentration in the lactotrophs of the pituitary gland, which then stimulates phosphorylation of transcription factor Pit-1/GHF-1 to bind to, and stimulate transcription of the prolactin gene.[127,128] The same factor governs transcription of the thyroid-stimulating hormone beta (TSH-β) gene in response to TRH, which has been suggested to be mediated by protein kinase C (PKC).[129] The TRH response is associated with phosphorylation of a threonine residue within the DNA-binding domain of Pit-1 that appears to shift the sequence specificity of this transcription factor from AWWTATNCAT to AWWAATN CAT.[129,130] Although Pit-1 is a pituitary-specific factor, it is a representative of a large family of related transcription factors (the Pit-₁ Oct-₂ and Unc-86 homologous factors (POU) domain class), and thus similar mechanisms may conceivably work in other cell types. However, sequence motifs matching the pattern are not found in cGPx genes, but in the murine GPxh gene only. On binding to the glucocorticoid receptor (GR), glucocorticoids may act directly on negative glucocorticoid regulatory elements (nGRE) to repress gene activity. nGREs exhibit a slightly different sequence pattern than "normal" (positive) GREs.[131] However, convincing matches of the nGRE pattern can not be identified in any of the known GPx gene sequences. Also, no perfect "classical" GRE can be seen unless one allows for considerable mismatches: The numerous TGTYCT 3′ half-sites are not even partially complemented by the less well conserved 5′ half (GG-TACA). One striking exception is a cluster of three potential GREs within the first intron on the human pGPx gene (+985/+1187), which exhibit 10/12, 9/12, and 9/12 matches to the consensus pattern GGTACAnnnTGTYCT. The nuclear receptors for progesterone, androgens, and mineralocorticoids, however, exhibit the same DNA-binding specificity as the glucocorticoid receptor. Therefore, the putative GRE cluster of the human pGPx gene, which is in fact expressed in the placenta, could physiologically act as a progesteron-responsive element (PRE), and definitely the cGPx gene structures do not provide any obvious key to easily interpret the dexamethasone effect. Most likely, both, TRH and dexamethasone act on cGPx genes via AP-1. As mentioned above, TRH triggers the calcium/PKC signal transduction pathway and thus could negatively affect the AP-1 elements found in the GPx genes by an AP-1-like component. In view of the dexamethasone effect on cGPx, it is tempting to apply this model to GPx gene regulation, since there are several examples known where AP-1 and corticosteroids either synergistically or antagonistically act in concert.[132,133]

As is obvious from these considerations, the genes of the various GPx types and related proteins contain quite a number of elements suggestive of hormonal control, but it will require a detailed analysis to work out which ones are relevant to the hormonal effects observed *in vivo*. Available data, however, already reveal that the individual members of the GPx family are under distinct hormonal control suggesting specific roles in sex-specific or stress-modulated antioxidant defense, or peculiar peroxidative processes in gonadal functions.

17.6 Summary

Within the GPx family complex, regulatory phenomena are observed most of which are neither plausible physiologically nor completely understood at the molecular level. Emerging evidence, however, indicates that the regulatory pattern is characteristic for each individual member of the family, thus suggesting distinct physiologic roles.

The biosynthesis of all GPx types, as expected from their selenoprotein nature, is regulated by selenium. Selenium determines the availability of selenocysteine-loaded t-RNA$^{(Ser)Sec}$ and the stability of mRNA encoding selenoproteins. The dependence of mRNA stability on selenium and the SECIS efficiencies are characteristic for each selenoprotein and determine into which particular protein limiting amounts of selenium are directed, probably according to physiological demand. In this respect, cGPx appears to be the most, PHGPx the least dispensible type of GPx.

Induction by various conditions subsummized as oxidative stress is documented for cGPx, and H_2O_2 can be rated as a likely proximal inductor by circumstantial evidence. Also, oxygen-responsive elements were identified in the cGPx gene. These findings and the tissue distribution of cGPx are in line with the assumption that cGPx is a major antioxidant defence device. Although antioxidant roles in certain compartments are equally likely for pGPx and giGPx, experimental evidence for oxidant-dependent induction of these enzymes appears lacking and neither oxygen-responsive elements nor other consensus sequences suggestive of redox regulation have so far been clearly identified in the pertinent genes. Neither the tissue distribution nor the gene structure appears to classify PHGPx as an antioxidant enzyme.

Regulation by hormones has been documented for cGPx and PHGPx. cCPx levels *in vivo* are up-regulated by estrogen and down-regulated by testosterone and, under specific conditions, also depressed by TRH and corticosteroids. The physiological implications thereof as well as the molecular mechanism remains to be established. A selenium-free homolog of GPx in rat epididymis of unknown function responds to androgens via an unusual GRE in the pertinent gene. Most striking is the strict dependence of rat testis PHGPx on gonadotropin *in vivo*. This effect, the abundance in testis, and shifts in subcellular distribution during cell differentiation point to a specific role of PHGPx in gonadal maturation or function.

References

1. Ursini, F., M. Maiorino, R. Brigelius-Flohé, K. D. Aumann, A. Roveri, D. Schomburg, and L. Flohé. 1995. Diversity of glutathione peroxidases. *Methods. Enzymol.* **252**:38–53.

2. Rocher, C., J.-L. Lalanne, and J. Chaudière. 1992. Purification and properties of a recombinant sulfur analog of murine selenium-glutathione peroxidase. *Eur. J. Biochem.* **205**:955–960.

3. Maiorino, M., K. D. Aumann, R. Brigelius-Flohé, D. Doria, J. van den Heuvel, J. McCarthy, A. Roveri, F. Ursini, and L. Flohé. 1995. Probing the presumed catalytic triad of selenium-containing peroxidases by mutational analysis of phospholipid hydroperoxide glutathione peroxidase (PHGPx). *Biol. Chem. Hoppe-Seyler* **376**:651–660.

4. Flohé, L. 1989. The selenoprotein glutathione peroxidase, pp. 643–731. *In* D. Dolphin, R. Poulson, O. Avramovic (eds.), *Glutathione: Chemical, Biochemical and Medical Aspects—Part A.* John Wiley & Sons, New York.

5. Takahashi, K., N. Avissar, J. Whitin, and H. Cohen. 1987. Purification and characterization of human plasma glutathione peroxidase: a selenoglycoprotein distinct from the known cellular enzyme. *Arch. Biochem. Biophys.* **256**:677–686.

6. Takahashi, K., M. Akasaka, Y. Yamamoto, C. Kobayashi, J. Mizoguchi, and J. Koyama. 1990. Primary structure of human plasma glutathione peroxidase deduced from cDNA sequences. *J. Biochem.* **108**:145–148.

7. Yamamoto, Y., and K. Takahashi. 1993. Glutathione peroxidase isolated from plasma reduces phospholipid hydroperoxides. *Arch. Biochem. Biophys.* **305**:541–545.

8. Ursini, F., M. Maiorino, M. Valente, L. Ferri, and C. Gregolin. 1982. Purification from pig liver of a protein which protects liposomes and biomembranes from peroxidative degradation and exhibits glutathione peroxidase activity on phosphatidylcholine. *Biochim. Biophys. Acta* **710**:197–211.

9. Thomas, J. P., M. Maiorino, F. Ursini, and A. W. Girotti. 1990. Protective action of phospholipid hydroperoxide glutathione peroxidase against membrane-damaging lipid peroxidation. *J. Biol. Chem.* **265**:454–461.

10. Chu, F.-F., J. H. Doroshow, and R. S. Esworthy. 1993. Expression, characterization, and tissue distribution of a new cellular selenium-dependent glutathione peroxidase, GSH-Px-GI. *J. Biol. Chem.* **268**:2571–2576.

11. Roche, C., J. L. Liu, T. LePresle, A. Capron, and R. J. Pierce. 1996. Tissue localisation and stage-specific expression of the phospholipid hydroperoxide glutathione peroxidase of *Schistosoma mansoni. Mol. Biochem. Parasitol.* **75**, 187–195.

12. Maiorino, M., C. Roche, M. Kieβ, K. Koenig, D. Gawlik, M. Matthes, E. Naldini, R. Pierce, and L. Flohé. 1996. A selenium-containing phospholipid hydroperoxide glutathione peroxidase from *Schistosoma mansoni. Eur. J. Biochem.*, **238**, 838–844.

13. Shigeoka, S., T. Takeda, and T. Hanaoka. 1991. Characterization and immunological properties of selenium-containing glutathione peroxidase induced by selenite in *Chlamydomonas reinhardtii*. *Biochem. J.* **275**:623–627.

14. Sabeh, F., T. Wright, and S. J. Norton. 1993. Purification and characterization of a glutathione peroxidase from the *Aloe vera* plant. *Enzyme Protein* **47**:92–98.

15. Stadtman, T. C. 1990. Selenium biochemistry. *Annu. Rev. Biochem.* **59**:111–127.

16. Böck, A., K. Forchhammer, J. Heider, and C. Baron. 1991. Selenoprotein synthesis: an expansion of the genetic code. *Trends Biochem. Sci.* **16**:463–467.

17. Heider, J., C. Baron, and A. Böck. 1992. Coding from a distance: dissection of the mRNA determinants required for the incorporation of selenocysteine into protein. *EMBO J.* **11**:3759–3766.

18. Berry, M., L. Banu, J. W. Harney, and P. R. Larsen. 1993. Functional characterization of the eukaryotic SECIS elements which direct selenocysteine insertion at UGA codons. *EMBO J.* **12**:3315–3322.

19. Forchhammer, K., W. Leinfelder, and A. Böck. 1989. Identification of a novel translation factor necessary for the incorporation of selenocysteine into protein. *Nature* **342**:453–456.

20. Baker, R. D., S. S. Baker, K. LaRosa, C. Whitney, and P. E. Newburger. 1993. Selenium regulation of glutathione peroxidase in human hepatoma cell line Hep3B. *Arch. Biochem. Biophys.* **304**:53–57.

21. Chang, M., and C. C. Reddy. 1991. Active transcription of the selenium-dependent glutathione peroxidase gene in selenium-deficient rats. *Biochem. Biophys. Res. Commun.* **181**:1431–1436.

22. Christensen, M. J., and K. W. Burgener. 1992. Dietary selenium stabilizes glutathione peroxidase mRNA in rat liver. *J. Nutr.* **122**:1620–1626.

23. Toyoda, H., S. Himeno, and N. Imura. 1990. Regulation of glutathione peroxidase mRNA level by dietary selenium manipulation. *Biochim. Biophys. Acta* **1049**:213–215.

24. Christensen, M. J., P. M. Cammack, and C. D. Wray. 1995. Tissue specificity of selenoprotein gene expression in rats. *Nutr. Biochem.* **6**:367–372.

25. Burk, R. F. 1991. Molecular biology of selenium with implications for its metabolism. *FASEB J.* **5**:2274–2279.

26. Hill, K. E., P. R. Lyons, and R. F. Burk. 1992. Differential regulation of rat liver selenoprotein mRNAs in selenium deficiency. *Biochem. Biophys. Res. Commun.* **185**:260–263.

27. Burk, R. F., and K. E. Hill. 1993. Regulation of selenoproteins. *Annu. Rev. Nutr.* **13**:65–81.

28. Toyoda, H., S. Himeno, and N. Imura. 1989. The regulation of glutathione peroxidase gene expression relevant to species difference and the effects of dietary selenium manipulation. *Biochim. Biophys. Acta* **1008**:301–308.

29. Saedi, M. S., C. G. Smith, J. Frampton, I. Chambers, P. R. Harrison, and R. A. Sunde. 1988. Effect of selenium status on mRNA levels for glutathione peroxidase in rat liver. *Biochem. Biophys. Res. Commun.* **153**:855–861.

30. Lei, X. G., J. K. Evenson, K. M. Thompson, and R. A. Sunde. 1995. Glutathione peroxidase and phospholipid hydroperoxide glutathione peroxidase are differentially regulated in rats by dietary selenium. *J. Nutr.* **125**:1438–1446.

31. Chu, F. F., R. S. Esworthy, S. Akman, and J. H. Doroshow. 1990. Modulation of glutathione peroxidase expression by selenium: effect on human MCF-7 breast cancer cell transfectants expressing a cellular glutathione peroxidase cDNA and doxorubicin-resistant MCF-7 cells. *Nucleic Acids Res.* **18**:1531–1539.

32. Burk, R. F., K. E. Hill, R. Read, and T. Bellew. 1991. Response of rat selenoprotein P to selenium administration and fate of its selenium. *Am. J. Physiol.* **261**:E26–E30.

33. Avissar, N., E. A. Kerl, S.S. Baker, and H. J. Cohen. 1994. Extracellular glutathione peroxidase mRNA and protein in human cell lines. *Arch. Biochem. Biophys.* **309**:239–246.

34. Weitzel, F., and A. Wendel. 1993. Selenoenzymes regulate the activity of leukocyte 5-lipoxygenase via the peroxide tone. *J. Biol. Chem.* **268**:6288–6292,

35. Chada, S., C. Whitney, and P. E. Newburger. 1989. Post-transcriptional regulation of glutathione peroxidase gene expression by selenium in the HL-60 human myeloid cell line. *Blood* **74**:2435–2541.

36. Berry, M. J., J. W. Harney, T. Ohama, and D. L. Hatfield. 1994. Selenocysteine insertion or termination: factors affecting UGA codon fate and complementary anticodon: codon mutations. *Nucleic Acids Res.* **22**:3753–3759.

37. Brigelius-Flohé, R., K. Lötzer, S. Maurer, M. Schultz, and M. Leist. 1996. Utilization of selenium from different chemical entities for selenoprotein biosynthesis by mammalian cell lines. *BioFactors.* **5**:125–137.

38. Diamond, A. M., Y. Montero-Puerner, B. J. Lee, and D. Hatfield. 1990. Selenocysteine inserting tRNAs are likely generated by tRNA editing. *Nucleic Acids Res.* **18**:6727.

39. Diamond, A. M., I. S. Choi, P. F. Crain, T. Hashizume, St. C. Pomerantz, R. Cruz, C. J. Steer, K. E. Hill, R. F. Burk, J. A. McCloskey, and D. L. Hatfield. 1993. Dietary selenium affects methylation of the wobble nucleoside in the anticodon of selenocysteine tRNA. *J. Biol. Chem.* **268**:14215–14223.

40. Kolmus, H., L. Flohé, and J. E. G. McCarthy. 1996. Analysis of eukaryotic mRNA structure directing cotranslational incorporation of selenocysteine. *Nucleic Acids Res.* **24**:1195–1201.

41. Sosenko, I. R., Y. Chen, L. T. Price, and L. Frank. 1995. Failure of premature rabbits to increase lung antioxidant enzyme activities after hyperoxic exposure: antioxidant enzyme gene expression and pharmacologic intervention with endotoxin and dexamethasone. *Pediatr. Res.* **37**:469–475.

42. Clerch, L. B., and D. Massaro. 1993. Tolerance of rats to hyperoxia. Lung antioxidant enzyme gene expression. *J. Clin. Invest.* **91**:499–508.

43. Li, R. K., D. A. Mickle, R. D. Weisel, L. C. Tumiati, G. Jackowski, T. W. Wu, and W. G. Williams. 1989. Effect of oxygen tension on the antioxidant enzyme activities of tetralogy of Fallot ventricular myocytes. *J. Mol. Cell. Cardiol.* **21**:567–575.

44. Cowan, D. B., R. D. Weisel, W. G. Williams, and D. A. Mickle. 1992. The regulation of glutathione peroxidase gene expression by oxygen tension in cultured human cardiomyocytes. *J. Mol. Cell. Cardiol.* **24**:423–433.

45. Jornot, L., and A. F. Junod. 1995. Differential regulation of glutathione peroxidase by selenomethionine and hyperoxia in endothelial cells. *Biochem. J.* **306**:581–587.

46. Cowan, D. B., R. D. Weisel, W. G. Williams, and D. A. G. Mickle. 1993. Identification of oxygen responsive elements in the 5′-flanking region of the human glutathione peroxidase gene. *J. Biol. Chem.* **268**:26904–26910.

47. Ter Horst, G. J., S. Knollema, B. Stuiver, H. Hom, S. Yoshimura, M. H. Ruiters, and J. Korf. 1994. Differential gluthatione peroxidase mRNA up-regulations in rat forebrain areas after transient hypoxia-ischemia. *Ann. N. Y. Acad. Sci.* **738**:329–333.

48. Das, D. K., R. M. Engelman, and Y. Kimura. 1993. Molecular adaptation of cellular defences following preconditioning of the heart by repeated ischemia. *Cardiovasc. Res.* **27**:578–584.

49. Krall, J., M. J. Speranza, and R. E. Lynch. 1991. Paraquat-resistant HeLa cells: increased cellular content of glutathione peroxidase. *Arch. Biochem. Biophys.* **286**:311–315.

50. Niwa, Y., O. Iizawa, K. Ishimoto, H. Akamatsu, and T. Kanoh. 1993. Age-dependent basal level and induction capacity of copper-zinc and manganese superoxide dismutase and other scavenging enzyme activities in leukocytes from young and elderly adults. *Am. J. Pathol.* **143**:312–320.

51. Flohé, L., and H. Giertz. 1987. Endotoxins, arachidonic acid, and superoxide formation. *Rev. Infect. Dis.* **9**:S553–S561.

52. Flohé, L., R. Beckmann, H. Giertz, and G. Loschen. 1985. Oxygen-centered free radicals as mediators of inflammation, pp. 403–435. *In* H. Sies (ed.), *Oxidative Stress.* Academic Press, London.

53. Meier, B., H. H. Radeke, S. Selle, M. Younes, H. Sies, K. Resch, and G. G. Habermehl. 1989. Human fibroblasts release reactive oxygen species in response to interleukin-1 or tumour necrosis factor-α. *Biochem. J.* **263**:539–545.

54. Michelson, A. M., K. Puget, and P. Durosay. 1977. Clinical aspects of the dosage of erythrocuprein, pp. 467–499. *In* A. M. Michelson, J. M. McCord, and J. Fridovich (eds.), *Superoxide and Superoxide Dismutases.* Academic Press, London, New York, San Francisco.

55. McBride, O. W., A. Mitchell, B. Jae Lee, G. Mullenbach, and D. Hatfield. 1988. Gene for selenium-dependent glutathione peroxidase maps to human chromosomes 3, 21 and X. *BioFactors* **1**:285–292.

56. Moscow, J. A., C. S. Morrow, R. He, G. T. Mullenbach, and K. H. Cowan. 1992. Structure and function of the 5′-flanking sequence of the human cytosolic selenium-dependent glutathione peroxidase gene (hgpx1). *J. Biol. Chem.* **267**:5949–5958.

57. Johannsmann, R., B. Hellkuhl, and K.-H. Grzeschik. 1981. Regional mapping of human chromosome 3. Assignment of a glutathione peroxidase-1 gene to 3p13 leads to 3q12. *Hum. Genet.* **56**:361–363.

58. Kelner, M. J., and R. Bagnell. 1990. Alternation of endogenous glutathione peroxidase, manganese superoxide dismutase, and glutathione transferase activity in cells transfected with a copper-zinc superoxide dismutase expression vector. Explanation for variations in paraquat resistance. *J. Biol. Chem.* **265**:10872–10875.

59. Maitre, B., L. Jornot, and A. F. Junod. 1993. Effects of inhibition of catalase and superoxide dismutase activity on antioxidant enzyme mRNA levels. *Am. J. Physiol.* **265**:L636–L643.

60. Shull, S., N. H. Heintz, M. Perisamy, M. Manohar, Y. M. Janssen, J. P. Marsh, and B. T. Mossman. 1991. Differential regulation of antioxidant enzymes in response to oxidants. *J. Biol. Chem.* **266**:24398–24403.

61. Del Maestro, R. F., M. Lopez-Torres, W. B. McDonald, E. C. Stroude, and I. S. Vaithilingam. 1992. The effect of tumor necrosis factor-alpha on human malignant glial cells. *J. Neurosurg.* **76**:652–659.

62. Kizaki, M., A. Sakashita, A. Karmakar, C. W. Lin, and H. P. Koeffler. 1993. Regulation of manganese superoxide dismutase and other antioxidant genes in normal and leukemic hematopoietic cells and their relationship to cytotoxicity by tumor necrosis factor. *Blood* **82**:1142–1150.

63. Sato, M., M. Sasaki, and H. Hojo. 1995. Antioxidative roles of metallothionein and manganese superoxide dismutase induced by tumor necrosis factor-alpha and interleukin-6. *Arch. Biochem. Biophys.* **316**:738–744.

64. Schreck, R., P. Rieber, and P. A. Baeuerle. 1991. Reactive oxygen intermediates as apparently widely used messengers in the activation of the NF-kappaB transcription factor and HIV-1. *EMBO J.* **10**:2247–2258.

65. Benjamin, I. J., B. Kröger, and R. S. Williams. 1990. Activation of the heat shock transcription factor by hypoxia in mammalian cells. *Proc. Natl. Acad. Sci. USA* **87**:6263–6267.

66. Benjamin, I. J., S. Horie, M. L. Greenberg, R. J. Alpern, and R. S. Williams. 1992. Induction of stress proteins in cultured myogenic cells. *J. Clin. Invest.* **89**:1585–1689.

67. Mirochnitchenko, O., U. Palnitkar, M. Philbert, and M. Inouye. 1995. Thermosensitive phenotype of transgenic mice overproducing human glutathione peroxidase. *Proc. Natl. Acad. Sci. USA* **92**:8120–8124.

68. Bienz, M., and H. R. B. Pelham. 1986. Heat shock regulatory elements function as an inducible enhancer in the Xenopus *hsp 70* gene and when linked to a heterologous promoter. *Cell* **45**:753–760.

69. Ghyselinck, N. B., I. Dufaure, J. J. Lareyre, N. Rigaudière, M. G. Mattéi, and J. P. Dufaure. 1993. Structural organization and regulation of the gene for the androgen-dependent glutathione peroxidase-like protein specific to the mouse epididymis. *Mol. Endocrinol.* **7**:258–272.

70. Abate, C., L. Patel, F. J. Rauscher III, and T. Curran. 1990. Redox regulation of Fos and Jun DNA-binding activity in vitro. *Science* **249**:1157–1161.

71. Xanthoudakis, S., and T. Curran. 1992. Identification and characterization of Ref-1, a nuclear protein that facilitates AP-1 DNA-binding activity. *EMBO J.* **11**:653–665.

72. Meyer, M., R. Schreck, and P. A. Baeuerle. 1993. H_2O_2 and antioxidants have opposite effects on activation of NF-κB and AP-1 in intact cells: AP-1 as secondary antioxidant-responsive factor. *EMBO J.* **12**:2005–2015.

73. Stein, B., P. Angel, H. van Dam, H. Ponta, P. Herrlich, A. van der Eb, and H. J. Rahmsdorf. 1992. Ultraviolet-radiation induced c-*jun* gene transcription: two AP-1 like binding sites mediate the response. *Photochem. Photobiol.* **55**:409–415.

74. Norris, M. L., and D. E. Millhorn. 1995. Hypoxia-induced protein binding to O_2-responsive sequences on the tyrosine hydroxylase gene. *J. Biol. Chem.* **270**:23774–23779.

75. Rushmore, T. H., M. R. Morton, and C. B. Pickett. 1991. The antioxidant responsive element. *J. Biol. Chem.* **266**:11632–11639.

76. Xie, T., M. Belinsky, Y. Xu, and A. K. Jaiswal. 1995. ARE- and TRE-mediated regulation of gene expression. *J. Biol. Chem.* **270**:6894–6900.

77. Chan, J. Y., X. L. Han, and Y. W. Kan. 1993. Isolation of cDNA encoding the human NF-E2 protein. *Proc. Natl. Acad. Sci. USA* **90**:11366–11370.

78. Ney, P. A., N. C. Andrews, S. M. Jane, B. Safer, M. E. Purucker, S. Weremowicz, C. C. Morton, S. C. Goff, S. H. Orkin, and A. W. Nienhuis. 1993. Purification of the human NF-E2 complex: cDNA cloning of the hematopoietic cell-specific subunit and evidence for an associated partner. *Mol. Cell. Biol.* **13**:5604–5612.

79. Wang, G. L., B. H. Jiang, E. A. Rue, and G. L. Semenza. 1995. Hypoxia-inducible factor 1 is a basic-helix-loop-helix-PAS heterodimer regulated by cellular O_2 tension. *Proc. Natl. Acad. Sci. USA* **91**:5510–5514.

80. Wang, G. L., and G. L. Semenza. 1993. Characterization of hypoxia-inducible factor 1 and regulation of DNA binding activity by hypoxia. *J. Biol. Chem.* **268**:21513–21518.

81. Avissar, N., J. R. Slemmon, I. S. Palmer, and H. J. Cohen. 1991. Partial sequence of huma plasma glutathione peroxidase and immunologic identification of milk glutathione peroxidase as the plasma enzyme. *J. Nutr.* **121**:1243–1249.

82. Esworthy, R. S., F.-F. Chu, R. J. Paxton, S. Akman, and J. H. Doroshow. 1991. Characterization and partial amino acid sequence of human plasma glutathione peroxidase. *Arch. Biochem. Biophys.* **286**:330–336.

83. Yoshimura, S., K. Watanabe, H. Suemizu, T. Onozawa, J. Mizoguchi, K. Tsuda, H. Hatta, and T. Moriuchi. 1991. Tissue specific expression of the plasma glutathione peroxidase gene in rat kidney. *J. Biochem.* **109**:918–923.

84. Maser, R. L., B. S. Magenheimer, and J. P. Calvet. 1994. Mouse plasma glutathione peroxidase. *J. Biol. Chem.* **269**:27066–27073.

85. Avissar, N., D. B. Ornt, Y. Yagil, S. Horowitz, R. H. Watkins, E. A. Kerl, K. Takahashi, I. S. Palmer, and H. J. Cohen. 1994. Human kidney proximal tubules are the main source of plasma glutathione peroxidase. *Am. J. Physiol.* **266**:C367–C375.

86. Chu, F. F., R. S. Esworthy, J. H. Doroshow, K. Doan, and X.-F. Liu. 1992. Expression of plasma glutathione peroxidase in human liver in addition to kidney, heart, lung, and breast in humans and rodents. *Blood* **79**:3233–3238.

87. Martin-Alonso, J.-M, S. Ghosh, and M. Coca-Prados. 1993. Cloning of the bovine plasma selenium-dependent glutathione peroxidase (GP) cDNA from the ocular ciliary epithelium: expression of the plasma and cellular forms within the mammalian eye. *J. Biochem.* **114**:284–291.

88. Avissar, N., C. Eisenmann, J. G. Breen, S. Horowitz, R. K. Miller, and H. J. Cohen. 1994. Human placenta makes extracellular glutathione peroxidase and secretes it into maternal circulation. *Am. J. Physiol.* **267**:E68–E76.

89. Mills, G. C. 1957. Hemoglobin catabolism. I. Glutathione peroxidase, an erythrocyte enzyme which protects hemoglobin from oxidative breakdown. *J. Biol. Chem.* **229**:189–197.

90. Flohé, L. 1971. Die Glutathionperoxidase: énzymologie und biologische Aspekte. *Klin. Wochenschr.* **49**:669–683.

91. Weitzel, F., F. Ursini, and A. Wendel. 1990. Phospholipid hydroperoxide glutathione peroxidase in various mouse organs during selenium deficiency and repletion. *Biochim. Biophys. Acta* **1036**:88–94.

92. Himeno, S., A. Takekawa, H. Toyoda, and N. Imura. 1993. Tissue-specific expression of glutathione peroxidase gene in guinea pigs. *Biochim. Biophys. Acta* **1173**:283–288.

93. Himeno, S., A. Takekawa, and N. Imura. 1993. Species difference in hydroperoxide-scavenging enzymes with special reference to glutathione peroxidase in guinea-pigs. *Comp. Biochem. Physiol.* **104B**:27–31.

94. Graur, D., W. A. Hide, and W.-H. Li. 1991. Is the guinea-pig a rodent? *Nature* **351**:649–652.

95. Wolf, B., K. Reinecke, K.-D. Aumann, R. Brigelius-Flohé, and L. Flohé. 1993. Taxonomical classification of the guinea pig based on its Cu/Zn superoxide dismutase sequence. *Biol. Chem. Hoppe-Seyler* **374**:641–649.

96. O'Prey, J., S. Ramsay, I. Chambers, and P. R. Harrison. 1993. Transcriptional up-regulation of the mouse cytosolic glutathione peroxidase gene in erythroid cells is due to a tissue-specific 3′ enhancer containing functionally important CACC/GT motifs and binding sites for GATA and Ets transcription factors. *Mol. Cell. Biol.* **13**:6290–6303.

97. Harrison, P. R., M. Plumb, J. Frampton, B. Thiele, K. MacLeod, J. Chester, I. Chambers, J. Fleming, J. O'Prey, M. Walker, H. Wainwright, S. Lowe, and S. Janetzki. 1990. Regulation of erythroid-specific gene expression. *Biomed. Biochim. Acta* **49**:5–10.

98. Crossley, M., A. P. Tsang, J. J. Bieker, and S. H. Orkin. 1994. regulation of the erythroid Krüppel-like factor (EKLF) gene promoter by the erythroid transcription factor GATA-1. *J. Biol. Chem.* **269**:15440–15444.

99. Feng, W. C., C. M. Southwood, and J. J. Bicker. 1994. Analyses of b-thalassemia mutant DNA interactions with erythroid Krüppel-like factor (EKLF), an erythroid cell-specific transcription factor. *J. Biol. Chem.* **269**:1493–1500.

100. Anderson, K. P., C. B. Kern, S. C. Crable, and J. B. Lingrel. 1995. Isolation of a gene encoding a functional zinc finger protein homologous to erythroid Krüppel-like factor: identification of a new multigene family. *Mol. Cell. Biol.* **15**:5957–5965.

101. Andrews, N. C., H. Erdjument-Bromage, M. B. Davidson, P. Tempst, and S. H. Orkin. 1993. Erythroid transcription factor NF-E2 is a haematopoietic-specific basic-leucine zipper protein. *Nature* **362**:722–728.

102. De Haan, J. B., M. J. Tymms, F. Cristiano, and I Kola. 1994. Expression of copper/ zinc superoxide dismutase and glutathione peroxidase in organs of developing mouse embryos, fetuses, and neonates. *Pediatr. Res.* **35**:188–196.

103. Chen, Y., and L. Frank. 1993. Differential gene expression of antioxidant enzymes in the perinatal rat lung. *Pediatr. Res.* **34**:27–31.

104. Tauchi, K., Y. Tsutsumi, H. Tsukamoto, H. Hasegawa, S. Yoshimura, and K. Watanabe. 1991. Glutathione peroxidase and glutathione S-transferase, class α, in rat intestine. *Acta Pathol. Jpn.* **41**:573–580.

105. Roveri, A., M. Maiorino, and F. Ursini. 1994. Enzymatic and immunological measurement of soluble and membrane-bound phospholipid-hydroperoxide glutathione peroxidase. *Methods Enzymol.* **233**:203–212.

106. Roveri, A., A. Casasco, M. Maiorino, P. Dalan, A. Calligaro, and F. Ursini. 1992. Phospholipid hydroperoxide glutathione peroxidase of rat testis. *J. Biol. Chem.* **267**:6142–6146.

107. Bermano, G., F. Nicol, J. A. Dyer, R. A. Sunde, G. J. Beckett, J. R. Arthur, and J. E. Hesketh. 1995. Tissue-specific regulation of selenoenzyme gene expression during selenium deficiency in rats. *Biochem. J.* **311**:425–430.

108. Roveri, A., M. Maiorino, C. Nisii, and F. Ursini. 1994. Purification and characterization of phospholipid hydroperoxide glutathione peroxidase from rat testis mitochondrial membranes. *Biochim. Biophys. Acta* **1208**:211–221.

109. Schuckelt, R., R. Brigelius-Flohé, M. Maiorino, A. Roveri, J. Reumkens, W. Straβburger, F. Ursini, B. Wolf, and L. Flohé. 1991. Phospholipid hydroperoxide glutathione peroxidase is a selenoenzyme distinct from the classical glutathione peroxidase as evident from cDNA and amino acid sequencing. *Free Rad. Res. Commun.* **14**:343–361.

110. R. Brigelius-Flohé, K. D. Aumann, H. Blöcker, G. Gross, M. Kiess, K.-D. Klöppel, M. Maiorino, A. Roveri, R. Schuckelt, F. Ursini, E. Wingender, and L. Flohé. 1994. Phospholipid-hydroperoxide glutathione peroxidase. Genomic DNA, cDNA, and deduced amino acid sequence. *J. Biol. Chem.* **269**:7342–7348.

111. Conz, P. A., P. A. Bevilacqua, G. La Greca, D. Danieli, M. P. Rodighiero, L. Cavarretta, M. Maiorino, A. Roveri, and F. Ursini. 1993. Phospholipid hydroperoxide glutathione peroxidase in the normal human kidney: a possible role in protecting cell membranes. *Exp. Nephrol.* **1**:376–378.

112. Radeke, H. H., B. Meier, N. Topley, J. Flöge, G. G. Habermehl, and K. Resch. 1990. Interleukin 1-α and tumor necrosis factor-α induce oxygen radical production in mesangial cells. *Kidney Int.* **37**:767–775.

113. Panfili, E., G. Sandri, and L. Ernster. 1991. Distribution of glutathione peroxidases and glutathione reductase in rat brain mitochondria. *FEBS Lett.* **290**:35–37.

114. Flohé, L., and W. Schlegel. 1971. Glutathione-peroxidase IV. Intrazelluläre Verteilung des Glultathion-Peroxidase-Systems in der Rattenleber. *Hoppe-Seyler's Z. Physiol. Chem.* **352**:1401–1410.

115. Godeas, C., G. Sandri, and E. Panfili. 1994. Distribution of phospholipid hydroperoxide glutathione peroxidase (PHGPx) in rat testis mitochondria. *Biochim. Biophys. Acta* **1191**:147–150.

116. Sunde, R. A., J. A. Dyer, T. V. Moran, J. K. Evenson, and M. Sugimoto. 1993. Phospholipid hydroperoxide glutathione peroxidase: full-length pig blastocyst cDNA sequence and regulation by selenium status. *Biochem. Biophys. Res. Commun.* **193**:905–911.

117. Pushpa-Rekha, T. R., A. L. Burdsall, L. M. Oleksa, G. M. Chisolm, and D. M. Driscoll. 1995. Rat phospholipid-hydroperoxide glutathione peroxidase. cDNA cloning and identification of multiple transcription and translation start sites. *J. Biol. Chem.* **270**:26993–26999.

118. Akasada, M., J. Mizoguch, S. Yoshimura, and K. Watanabe. 1989. Nucleotide sequence for rabbit glutathione peroxidase. *Nucleic Acids Res.* **17**:2136.

119. Esworthy, R. S., M. A. Baker, and F.-F. Chu. 1995. Expression of selenium-dependent glutathione peroxidase in human breast tumor cell lines. *Cancer Res.* **55**:957–962.

120. Pinto, R. E., and W. Bartley. 1969. The nature of the sex-linked differences in glutathione peroxidase activity and aerobic oxidation of glutathione in male and female rat liver. *Biochem. J.* **115**:449–456.

121. Flohé, L., and R. Zimmermann. 1970. The role of GSH peroxidase in protecting the membrane of rat liver mitochondria. *Biochim. Biophys. Acta* **233**:210–213.

122. Capel, I. D., and A. E. Smallwood. 1983. Sex differences in the glutathione peroxidase activity of various tissues of the rat. *Res. Commun. Chem. Pathol. Pharmacol.* **40**:367–378.

123. Igarashi, T., T. Satoh, K. Iwashita, S. Ono, K. Ueno, and H. Kitagawa. 1985. Sex difference in the subunit composition of hepatic glutathione S-transferase in rats. *J. Biochem. Tokyo* **98**:117–123.

124. Rigaudiere, N., N. B. Ghyselinck, J. Faure, and J. P. Dufaure. 1992. Regulation of the epididymal glutathione peroxidase-like protein in the mouse: dependence upon androgens and testicular factors. *Mol. Cell Endocrinol.* **89**:67–77.

125. Chen, Y., P. L. Whitney, and L. Frank. 1993. Negative regulation of antioxidant enzyme gene expression in the developing fetal rat lung by prenatal hormonal treatments. *Pediatr. Res.* **33**:171–176.

126. Chen, Y., I. R. Sosenko, and L. Frank. 1995. Positive regulation of pulmonary antioxidant enzyme gene expression by prenatal thyrotropin releasing hormone plus dexamethasone treatment in premature rats exposed to hyperoxia. *Pediatr. Res.* **37**: 611–616.

127. Hoggard, N., J. R. E. Davis, M. Berwaer, P. Monget, B. Peers, A Belayew, and J. A. Martial. 1991. Pit-1 binding sequences permit calcium regulation of human prolactin gene expression. *Mol. Endocrinol.* **5**:1748–1754.

128. Yan, G. Z., and C. Bancroft. 1991. Mediation by calcium of thyrotropin-releasing hormone action on the prolactin promoter via transcription factor Pit-1. *Mol. Endocrinol.* **5**:1488–1497.

129. Steinfelder, H. J., S. Radovick, and F. E. Wondisford. 1992. Hormonal regulation of the thyrotropin b-subunit gene by phosphorylation of the pituitary-specific transcription factor PIT-1. *Proc. Natl. Acad. Sci. USA* **89**:5942–5945.

130. Kapiloff, M. S., Y. Farkash, M. Wegner, and M. G. Rosenfeld. 1991. Variable effects of phosphorylation of Pit-1 dictated by the DNA response elements. *Science* **253**:786–789.

131. Beato, M. 1989. Gene regulation by steroid hormones. *Cell* **56**:335–344.

132. Zhang, X. K., M. J. Dong, and J. F. Chiu. 1991. Regulation of a-fetoprotein gene expression by antagonism between AP-1 and the glucocorticoid receptor at their overlapping binding site. *J. Biol. Chem.* **266**:8248–8254.

133. Beato, M., P. Herrlich, and G. Schütz. 1995. Steroid hormone action: many players in search of a plot. *Cell* **83**:851–857.

18

Gene Expression of DT-Diaphorase in Cancer Cells

Venugopal Radjendirane, Pius Joseph, and Anil K. Jaiswal

18.1 DT-Diaphorases [*N*AD(P)H:*Q*uinone *O*xidoreductases (NQOs)]

NQOs are flavoproteins that catalyze the obligatory two-electron reductive metabolism and detoxification of quinones and its derivatives.[1,2] Several reports indicate the presence of two or more isozymic forms of NQOs in the rat, mouse, and human liver.[1,2] In humans, genetic evidence indicates that the different forms of NQOs are encoded by four gene loci.[3] Two of these loci (designated as NQO$_1$ and NQO$_2$) have been cloned and sequenced.[4-7] NQO$_1$ activity has also been cloned from the rat and mouse liver.[8-11] Among various NQOs, NQO$_1$ is the best-studied enzyme. NQO$_1$ is a cytosolic protein, occurs as a dimer *in vivo*, is made up of two homomonomeric subunits with individual molecular weights of 32 kDa, requires NADH or NADPH as a cofactor for the enzymatic activity, and is strongly inhibited by low concentrations of dicoumarol.[12-15] Both the human and rat cDNA encode for a protein of 274 residues.[4] The NQO$_1$ protein is shown to metabolize 2,6-dichlorophenolindophenol, menadione, vitamin K, benzo(α)-pyrene-3,6-quinone, 2,6-dimethylbenzoquinone, methylene blue, p-benzoquinone, 1,4-naphthoquinone, 2-methyl-1,4-benzoquinone, and several other quinones with high affinity.[12,13,16] The second isoenzymic form of the NQO (designated as NQO$_2$) was recently identified by cDNA cloning from human liver.[5] The human NQO$_2$ cDNA and protein are 54 and 49% similar to human liver cytosolic NQO$_1$ cDNA and protein, respectively.[5] The initiation codon (ATG) of the NQO$_2$ cDNA aligns perfectly with the initiation codons of human and rat NQO$_1$ cDNAs.[5,17] However, the NQO$_2$-encoded protein is 43 amino acids shorter at the carboxy terminal end as compared to that of NQO$_1$ and failed to catalyze high-affinity reduction of 2,6-dichlorophenolindophenol and menadione. Interestingly, both proteins were equally active in catalyzing the nitroreduction of antitumor drug CB10-200.[7]

18.2 NQOs and Chemoprevention

Quinones (e.g., benzo(α)pyrene quinones, naphthoquinones, benzoquinones, etc.) are highly abundant in nature and human exposure to them is extensive. Quinones are found in all burnt organic materials, including automobile exhaust, cigarette smoke, and urban air particulates.[18] They are also found naturally in many foods we eat and compounds containing the quinoid nucleus are widely employed as antitumor agents. Quinones are reactive molecules and readily undergo one- or two-electron reductive metabolism, leading to its activation and/or detoxification. One-electron reductive activation of quinones by enzymes that include NADPH-cytochrome P-450 reductase (P-450 reductase) (EC 1.6.2.4), NADH-b5 oxidore-ductase (EC 1.6.2.2), and NADPH-ubiquinone oxidoreductase (EC 1.6.5.3) generates unstable semiquinones that undergo redox cycling in the presence of molecular oxygen leading to the formation of highly reactive oxygen species causing lipid peroxidation, membrane and DNA damage, oxidative stress, cytotoxicity, mutagenicity, and possibly carcinogenicity (Fig. 18.1).[17,18] On the other hand, the cytosolic NQO_1 competes with the one-electron reducing enzymes and catalyzes two-electron reductive metabolic conversion of the quinones to hydroquinones leading to its detoxification by conjugation with UDP-glucuronic acid and gluta-thione (Fig. 18.1).[1,2,16,18–24] The two-electron reduction of quinones by NQO_1 generally does not lead to the formation of semiquinones and highly reactive oxygen species, resulting in the prevention of quinone-induced redox cycling, oxidative stress, DNA and membrane damage, cytotoxicity, mutagenicity, and carcinogenicity. As a protective agent, NQO_1 activity has been shown to prevent the formation of highly mutagenic quinone metabolites[25]; to detoxify benzo(α)pyr-ene-3,6-quinone via glucuronide formation[26]; and to reduce chromium (VI), thereby reverting its mutagenicity.[27] Recently, we reported that cDNA-derived cytochrome P-450 1A1 (CYP1A1) and NADPH:cytochrome P-450 reductase (P-450 reductase) metabolically activated benzo(α)pyrene to generate eight DNA adducts (Fig. 18.2E, DNA adducts 1 to 8).[16] Interestingly, coexpression of cDNA-derived NQO_1 with CYP1A1 and P-450 reductase specifically prevented the formation of two of the P450 reductase activated benzo(α)pyrene quinone-DNA adducts (adducts 5 and 6 in Fig. 18.2E; compare Figs. 18.2E and 18.2F).[16] More recently, we have found that binding of P-450 reductase activated benzo(α)pyrene quinones to the DNA induced mutagenesis, which was prevented in the presence of NQO_1.[28] In summary, NQO_1 competed with P-450 reductase and prevented the generation of benzo(α)pyrene quinones induced DNA adducts, and benzo(α)-pyrene quinone mutagenesis (Fig. 18.2).[16]

The induction of NQO_1 plays an important role in reducing the neoplastic transformation of cells due to exposure to mutagens and carcinogens.[24,29,30] As early as 1929, a variety of chemicals including polycyclic aromatic hydrocarbons, metal ions, and 4-nitroquinoline N-oxide blocked chemical induction of tumors in rat liver and mammary glands.[31] Today, we know that the capacity of many

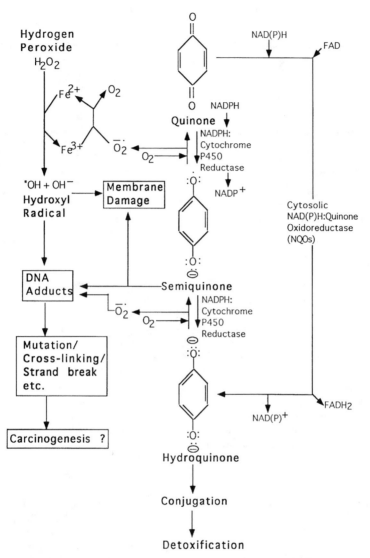

Figure 18.1. Metabolic activation and detoxification of quinones. One-electron reduction of quinones by NADPH:P-450 reductase, cytochrome b_5 reductase, xanthine oxidoreductase, and ubiquinone oxidoreductase results in the formation of mutagenic semiquinones. The semiquinones in the presence of molecular oxygen enter in the redox cycling to generate reactive oxygen species which stimulate oxidative stress and DNA/membrane damage. On the other hand, two-electron reduction of similar quinones catalyzed by cytosolic NQO_1 and related enzymes result in the formation of hydroquinones, which are much more stable than the semiquinones and are removed by conjugation reactions. The NQO_1 pathway does not generate reactive oxygen species, the cytotoxicity of which is well known. Therefore, NQO_1 and related enzymes protect the cells against redox cycling and oxidative stress.

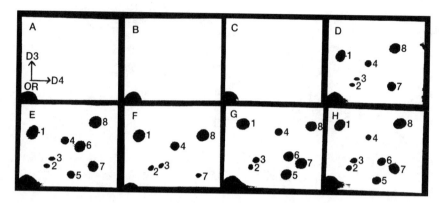

Figure 18.2. Cytosolic NQO_1 specifically prevents the formation of benzo(α)pyrene quinones-DNA adducts. The monkey kidney (COS1) cells (deficient in CYP1A1 and NQO_1) were transfected with pMT2 (expression vector) alone, pMT2-CYP1A1, pMT2-P-450 reductase, and pMT2-NQO_1 individually and in combinations. The extracts from transfected cells containing intact nuclei were incubated with benzo(α)pyrene and DNA adducts were analyzed by ^{32}P-postlabeling method by the procedure, as described in Ref. 16. The ^{32}P-labeled DNA adducts were separated by multidirectional Polyethylene Imide (PEI)-cellulose thin-layer chromatography (TLC). The adduct numbering is based on Fig. 2E, which contains the maximum number of DNA adducts. (A) Nontransfected COS1cells; (B) COS1 cells + P-450 reductase; (C) COS1 cells + NQO_1; (D) COS1 cells + CYP1A1; (E) COS1 cells + CYP1A1 + P-450 reductase; (F) COS1 cells + CYP1A1 + P-450 reductase + NQO_1; (G) COS1 cells + CYP1A1 + P-450 reductase + NQO_1 + 10-μM dicoumarol; and (H) COS1 cells + CYP1A1 + P-450 reductase + NQO_1(3' to 5'). Results with nontransfected and COS1 cells transfected with pMT2 (vector control) did not reveal any adduct formation. Therefore, results are shown with non-transfected COS1 cells as control. The analysis of results indicated that benzo(α)pyrene-DNA adducts were not observed in 18.2A to 18.2C. This was because benzo(α)pyrene is a procarcinogen and requires its activation by CYP1A1 to produce DNA binding metabolites. The COS1cells deficient in CYP1A1 failed to metabolize benzo(α)pyrene and thus failed to generate benzo(α)pyrene-DNA adducts. The metabolism of benzo(α)pyrene by cDNA derived CYP1A1 produced six distinct adducts as observed in 18.2D. Inclusion of P-450 reductase with CYP1A1 increased the number of adducts to eight, as seen in 18.2E. The coexpression of NQO_1 with CYP1A1 and P-450 reductase eliminated adducts 5 and 6 (18.2F). The disappearance of adducts 5 and 6 was specific to the presence of NQO_1, as suggested by appearance of adducts 5 and 6 in presence of dicoumarol, an inhibitor of NQO_1 (18.2G) and transfection with antisense NQO_1 (18.2H). Further experiments conclusively identified that adducts 5 and 6 generated due to binding of benzo(α)pyrene semiquinones to the deoxy guanosine residues of the DNA.[16]

different chemicals of diverse structure to block carcinogenesis correlates with their capacity to induce NQO_1[24,29,30]. The induction of the NQO_1 by "Sulforaphane" from Saga broccoli has been shown to block the formation of mammary tumors in Sprague-Dawley rats treated with a single dose of 9,10 dimethyl-1,2-benzan-thracene.[32] It has been reported that the induction of NQO_1 is a part of the activation of defensive mechanisms within the cells on exposure to the xenobiotics, drugs and carcinogens.[33,34] These defensive mechanisms also include induction of other detoxifying enzymes that[33,34] catalyze the metabolic detoxification of xenobiotics, drugs, and carcinogens.[30,33,34] The other detoxifying enzymes include glutathione S-transferases (GSTs), which conjugate hydrophobic electrophiles and reactive oxygen species with glutathione (GSH)[35–37]; UDP-glucuronosyl transferases (UDP-GT), which catalyze the conjugation of glucuronic acid with xenobiotics and drugs for their excretion[38,39]; epoxide hydrolase (EH), which inactivates epoxides[40]; γ-glutamylcysteine synthetase (γ-GCS), which plays a key role in the regulation of glutathione metabolism[41] and so on. In other words, the exposure of cells to xenobiotics, drugs, and carcinogens results in the coordinated increases in the expression of several defensive genes that produce enzymes for the metabolic detoxification of the agents and provide protection to the cells against DNA and membrane damage, cytotoxicity and neoplastic transformation. The role of NQO_2 in chemoprevention has been suggested.[7] However, it remains to be determined in future.

18.3 NQOs and Bioreductive Chemotherapy

Interestingly, the exclusive role of NQO_1 in cellular protection is questionable. In fact, the activation by NQO_1 had been suggested in case of quinones possessing reactive groups such as 3-bromomethyl menadione.[42] The possible contribution to the metabolic activation of important mutagens and carcinogens, such as heterocyclic amines and components of cigarette smoke tar, further supports an alternate role for NQO_1 in these instances.[43] It was reported that depending on the chemistry of a particular compound, the hydroquinones generated undergo further reactions to generate reactive oxygen species or reactive alkylating species that could damage the DNA and membranes leading to cell death.[44] This property of NQO_1 is important for bioreductive activation of certain antitumor drugs in tumors, which generally contain much higher levels of NQO_1 than do normal tissues of the same origin.[1,2] NQO_1 has been shown to catalyze the bioreductive activation of several antitumor drugs that include alkylating aziridinyl benzoquinones such as diaziquinone (AZQ), mitomycin C (MMC), and indoloquinone EO9, as well as the dinitrophenyl aziridine CB 1954, the benzotriazine-di-N-oxide, SR 4233, and nitroimidazoles and is considered one of the most important factors in bioreductive chemotherapy.[1,17,45,46] Most studies on the role of NQO_1 in bioreductive chemotherapy were done for mitomycin C, a prototypical alkylating antitumor agent that is effective against tumors of the breast, bladder, colon, and the head and neck.[47] Bioreductive metabolism leading to the formation of electrophilic metabolites capable of binding with DNA to cause its damage is the basic cause of cytotoxicity and antitumor activity of mitomycin C.[48,49] However, the actual involvement of the NQO_1 enzyme in the activation of mitomycin C remains controversial.[45] Although, as mentioned earlier, there are a number of model systems where aerobic resistance to mitomycin C correlates with a loss of NQO_1 activity, a lack of correlation was reported in the "Oxford" panel of 15 cell lines.[50] In addition, mitomycin C was reported to inhibit the kidney NQO_1 enzyme.[51] The role of NQO_1 in the activation of mitomycin C under hypoxic conditions is also questionable.[45] This is clearly evident by the observation of the high sensitivity of human colon carcinoma (BE) cells that lack NQO_1 to mitomycin C and indoloquinone EO9 under hypoxia.[52] It is expected that other enzymes and factors, not yet known, play important role in the bioactivation of mitomycin C and its derivatives. The distribution of these enzymes and factors along with NQO_1 may determine the amount of activation of mitomycin C in a given cell type. Therefore, the role of additional factors or enzymes, not yet known, should be investigated to completely understand the extent of involvement of NQO_1 in the bioactivation of mitomycin C. One such factor that potentiates the mitomycin C activation is lower pH.[53,54] The quinone methide of mitomycin C acts as an electrophile at neutral and higher pH and alkylates the enzyme, resulting in loss of the catalytic activity. Under acidic conditions (e.g., pH 5 to 6), the quinone methide is protonated, giving rise to 2,7-diaminomitosene as a detectable reaction product and preserve the catalytic activity. In addition to pH, we have identified a unique cytosolic activity that bioactivates mitomycin C with much higher affinity than NQO_1 within the cells.[23] We believe that NQO_1 by itself is a detoxifying enzyme.[16] However, NQO_1 in combination with an unknown cytosolic factor catalyzes the activation of antitumor drug mitomycin C.[23]

18.4 Expression and Regulation of Genes Encoding NQO_1 and NQO_2

18.4.1 Constitutive and Tissue-Specific Expression

NQO_1 is ubiquitous throughout the eukaryotes.[12,55] In mammals, it is present in many organs but is most abundant in the liver.[12] In man, however, NQO_1 transcripts and activity are significantly higher in many extrahepatic tissues (Fig. 18.3).[7,56] The expression of the NQO_1 gene was higher in human kidney, skeletal muscle, lung, and liver followed by heart and placenta and lower in brain and pancreas (Fig. 18.3).[7] The pattern of NQO_2 gene expression was very similar to that of NQO_1 in various human tissues with the exception that NQO_2 gene expression was absent in human placenta (Fig. 18.3).[7]

18.4.2 Expression in Normal and Tumor Tissues

The information on the expression and regulation of NQO genes are limited to the NQO_1 gene only. Higher NQO_1 levels have been reported in several cell types including non-small-cell lung cancer, colon cancer, breast cancer, ovarian cancer, glioma, and melanoma.[50,57-60] Among the human tumor materials taken directly from the patients, higher NQO_1 levels have been observed in liver, lung, colon, and breast tumors (Fig. 18.4).[61-64] However, down-regulation of the NQO_1

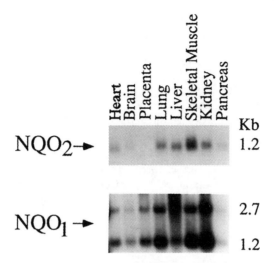

Figure 18.3. Tissue-specific expression of NQO_1 and NQO_2. Northern blot containing 2 µg of poly(A)$^+$RNA from various human tissues was hybridized to the ^{32}P-nick translated human NQO_2 cDNA. The blot after exposure was stripped and reprobed with human NQO_1 cDNA.

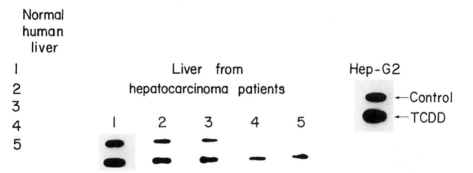

Figure 18.4. High levels of expression of NAD(P)H:quinone oxidoreductase (NQO_1) gene in human hepatoblastoma (Hep-G2) cells and liver tumors and in tissues surrounding the tumors in patients with hepatocarcinoma compared to the livers of normal human individuals. The RNA was isolated from DMSO (control) and 2,3,7,8-tetra-chlorodibenzo-*p*-dioxin (TCDD)-treated Hep-G2 cells, livers of patients with hepatocarcinoma and normal human individuals and analyzed by slot blotting (each slot containing 15 µg of respective RNA) and hybridization to human NQO_1 cDNA. In case of patients with hepatocarcinomas, the RNA was isolated from the center of the tumors and tissue surrounding the tumors. In the lanes of liver from patients with hepatocarcinoma the upper and lower slots indicate the NQO_1 cDNA hybridizable RNA from center of the tumor (top panel) and 6 cm from the center (lower panel), respectively.

gene expression was reported in the stomach and kidney tumors as compared to the normal tissues.[64]

18.4.3 Induction of NQO₁ Gene Expression

The transcription of the NQO_1 gene is activated in response to exposure to a variety of chemicals and other agents (Fig. 18.5). These include polycyclic and planar aromatic compounds, phenolic antioxidants, Michael reaction acceptors, isothiocynates, hydrogen peroxide, mercaptans, heavy metal salts, phorbol esters, anthracyclins, ionizing radiations, and hypoxia.[2] Some of the inducers that have frequently been used by various investigators include β-napthoflavone (β-NF)[8,65–69]; tert-butylhydroquinone[65,68–70] (t-BHQ); 2(3)-tert-butyl-4-hydroxy-ani-sole (BHA)[66,69]; 12-O-tetradecanoylphrobol-13-acetate (TPA)[68,71]; and hydrogen peroxide.[68,71] Many of the inducers of the NQO_1 gene have also been shown to increase the transcription of NQO_2, GST Ya, GST P, and γ-GCS genes.[33,37,72] It

(1). Polycyclic and Planar Aromatic Compounds
β-Naphthoflavone
3-Methylcholenthrene

(2). Phenolic Antioxidants
t-butyl hydroquinone (t-BHQ)
3-(2)-tert-butyl-4-hydroxyanisole (BHA)
3,5-di-tert-butyl-4-hydroxytoluene (BHT)

(3). Michael Reaction Acceptors
1-Nitro-1-cyclohexane

(4). Isothiocyanates:
Sulforaphane

(5). Peroxides:
Hydrogen peroxide

(6). Mercaptans
2-Mercaptoethanol
Dithiothreitol

(7). Heavy Metal Salt

(8). Phorbol Esters
TPA

(9). Anthracyclins
Mitomycin C
Adriamycin

(10). Ionizing Radiations

(11). Hypoxia

Figure 18.5.   Inducers of NQO₁ gene expression.

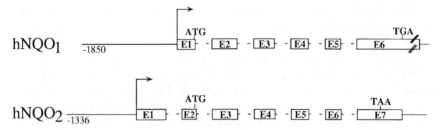

Figure 18.6. Human NQO₁ and NQO₂ genes structure. The human NQO₁ (hNQO₁) and human NQO₂ (hNQO₂) gene structures are shown. The NQO₁ gene contained six exons as compared to the NQO₂ gene, which contained seven exons. The second exon of the NQO₂ gene corresponded to the first exon of the NQO₁ gene and so on. The sizes and sequences of exons 3 to 5 were highly conserved between the two NQO genes.

is believed that a set of detoxifying enzyme genes are coordinately induced on exposure of cells to the xenobiotics, antioxidants, drugs, and carcinogens.[33,34,37] The coordinated increase in various detoxifying enzyme genes may be required to provide protection to the cells against adverse effects of exposure to chemicals as discussed earlier. These results raised important questions regarding a single or different mechanism(s) of induction of NQO₁ and other detoxifying enzyme genes by chemicals of diverse structures and agents such as ionizing radiation and hypoxia. The common feature among various inducers was their capacity to generate electrophiles and/or reactive oxygen species. Either the electrophiles or the reactive oxygen species or both may be the driving force for the activation of NQO₁ and other detoxifying enzyme genes and is a subject of great future interest.

18.4.4 Regulation of NQO Genes Expression and Induction

The human and rat NQO₁ and human NQO₂ genes with their 5′ and 3′ flanking regions have been cloned and sequenced (Fig. 18.6).[6,7,11] The exon-intron composition of human and rat NQO₁ genes are highly conserved. The exon-intron structure of human NQO₁ gene was also conserved in human NQO₂ gene. However, the NQO₂ gene contained an additional noncoding first exon. The proteins encoded by NQO₁ and NQO₂ were highly homologous except that NQO₂ protein was 43 amino acids shorter than NQO₁ at its C terminus.

18.4.4.1 hARE-Mediated Expression and Induction by Xenobiotics and Antioxidants

18.4.4.1.1 hARE

Deletion mutagenesis in the human NQO₁ gene promoter identified 24 base pairs of the human antioxidant response element (hARE) that was essentially required for high constitutive expression of the NQO₁ gene in hepatoblastoma cells as compared to normal (primary hepatocytes) of the same origin (Fig. 18.7).[6,66,68,73] It may be noteworthy that 90% or more of the constitutive expression of the NQO_1 gene in various mammalian carcinoma cells is mediated by hARE. A similar kind of element has also been reported in the upstream region of human NQO_2 gene[7]; rat NQO_1 gene[71]; rat GST Ya subunit gene[37,74-77]; mouse GST Ya gene[78,79]; rat GST P gene,[80,81] which mediated the constitutive and inducible expression of respective genes in cancer cells. The alignment of the nucleotide sequences of various genes AREs is shown in Figure 18.8. The human NQO_1 gene ARE contained two activator protein-1 (AP-1)/AP-1-like elements in reverse orientation at the interval of three nucleotides followed by a 'GC' box (Fig. 18.8).[66-68,73] Mutations in the 5' AP-1-like element, 3' AP-1 element, and GC box of the hARE revealed that 3' AP-1 is the most important for constitutive expression and β-NF induction of the NQO_1 gene expression.[73] The 5' AP-1-like element within the hARE was required for β-NF response and the 'GC' box was required for optimal expression and induction of the NQO_1 gene. These mutational studies clearly established that hARE is a unique element even though it contains AP-1 and AP-1-like elements.[73] This conclusion was also supported by the fact that human collagenase gene TRE (a single AP-1) element was found nonresponsive to β-NF.[73] Similar to the hARE, rat NQO_1, human NQO_2, and rat GST P genes, AREs also contain two AP-1-like elements arranged in reverse orientation at the interval of three nucleotides followed by a 'GC' box. However, the two AP-1-like elements were found arranged as direct repeats at the spacing of eight nucleotides followed by the 'GC' box in the rat and mouse GST Ya genes. The recently identified ARE in human γ-GCS gene contained two AP-1-like elements facing each other. It may be noteworthy that the minimum ARE sequence required

Figure 18.7. Human NQO_1 gene promoter: hARE-mediated constitutive and inducible expression. (18.7A) The 1850 base pairs of human NQO_1 gene promoter and the nucleotide positions and sequences of the various *cis*-elements. AP-2, Binding site for transcription factor AP-2; XRE, xenobiotic response element; hARE, human antioxidant response element; CAT, CCAAT binding site; Alu, human repetitive sequence. (18.7B). The promoterless pBLCAT3 and thymidime kinase basal promoter containing pBLCAT2 and CAT vectors containing either different lengths of NQO_1 gene promoter in kilobase pairs, internal deletion (represented by Δ) and enhancer elements were cotransfected with RSV-β-galactosidase plasmid in human hepatocytes and Hep-G2 cells and analyzed for β-galactosidase normalized CAT activities. Values represent pmoles of CAT activities per minute per milligram protein. (18.7C). Hep-G2 cells were cotransfected with RSV-β-galactosidase plasmid and pBLCAT2, hARE-tk-CAT and mutant hARE-tk-CAT in separate experiments. The transfected cells were treated with 50 μM concentration of either β-NF or *t*-BHQ and analyzed for β-galactosidase normalized CAT activities. Values represent picomoles of CAT activities per minute per milligram protein. Mutant hARE contained mutations in the 3' perfect activator protein-1 (AP-1) element, which is known to result in the loss of constitutive and induced expression.[66]

A. Human NQO1 Gene Promoter

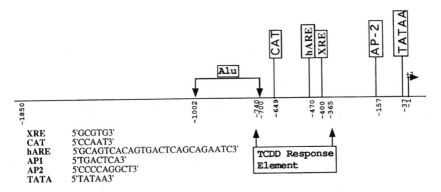

XRE	5'GCGTG3'
CAT	5'CCAAT3'
hARE	5'GCAGTCACAGTGACTCAGCAGAATC3'
AP1	5'TGACTCA3'
AP2	5'CCCCAGGCT3'
TATA	5'TATAA3'

TCDD Response Element

B. Constitutive Expression in primary liver and hepatocarcinoma cells

	Human hepatcytes	Human Hep-G2
pBLCAT3 (CAT Vector Alone)	1.0±0.2	1.7±0.2
pNQO1CAT1.85	3.4±1.0	53.5±4.1
pNQO1CAT1.85(Δ-0.587-0.379)	1.5±0.3	4.9±0.5
pBLCAT2 (CAT Vector/NO Enhancer)	2.4±1.0	3.9±1.1
pNQO1tkCAT-0.587-0.379	5.6±0.6	48.3±3.9
pNQO1hARE-tk-CAT	3.6±0.2	56.3±4.7
pNQO1 mutant hARE-tk-CAT	1.8±0.3	2.3±0.7

C. Inducible Expression in Hep-G2 cells

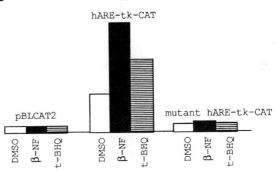

Figure 18.7

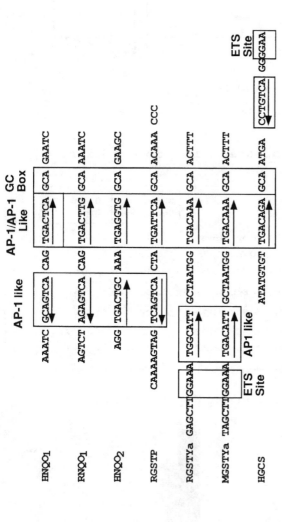

HNQO1,   Human NAD(P)H:Quinone Oxidoreductase Gene ARE
RNQO1,   Rat NAD(P)H:Quinone Oxidoreductase Gene ARE
HNQO2,   Human NAD(P)H:Quinone Oxidoreductase Gene ARE
RGSTP,   Rat Glutathione S-transferase P Subunit Gene ARE
RGSTYa, Rat Glutathione S-transferase Ya Subunit Gene ARE
MGSTYa, Mouse Glutathione S-transferase Ya Subunit Gene ARE
HGCS,   Human γ-Glutamylcysteine Synthetase Heavy Subunit Gene ARE

Figure 18.8. Alignment of the antioxidant response elements (AREs) from seven genes. The AP-1 and AP-1-like elements and their orientations are indicated. The 3′ AP-1 site in the human NQO_1 gene ARE is a perfect consensus sequence and has been separated from AP-1-like (imperfect consensus) elements in all other genes. The 'GCA' box and ETS binding sites are also shown in boxes. The rat NQO_1 and GST Ya gene AREs have been characterized to contain only one AP-1-like element and 'GC' box (5′puGTGACNNNGC3′). In these cases, additional sequences at 5′ and 3′ ends were included for alignment purposes.

452

for the expression and induction of rat GST-Ya and rat NQO$_1$ genes have been determined as 5'puGTGACNNNGC3', which contains only one AP-1-like element.[37] Fine mutational analysis in the rat GST-Ya gene ARE showed that six base pairs 'TGAC***GC' are essential for induction, and four out of these six base pairs (TGAC) are also required for basal expression.[76] This is in contrast to the reports on rat GST-P gene and mouse GST-Ya subunit gene AREs, which require the presence of both the AP-1-like elements for their proper function.[69,79] Recently, it is reported that replacement of two AP-1-like elements in the mouse GST Ya gene ARE with perfect AP-1 or TPA response elements (TREs) resulted in the loss of induction by antioxidants and other inducers.[82] The GST-P gene ARE, containing two Ap-1-like elements, exhibited a strong transcriptional enhancing activity in F9 embryonal carcinoma and HeLa cells.[81] In summary, the AREs in various genes usually contain two AP-1/AP-1-like elements arranged in varying orientations separated by either three or eight nucleotides. The orientations and spacing between the two AP-1 elements could be important in determining the levels of basal and induced expression and is of great future interest. It will be interesting to determine if these differences could lead to significant changes in the mechanisms of induction mediated by various AREs.

18.4.4.1.2 hARE-Nuclear Proteins Interaction

Nuclear proteins have been shown to bind to the AREs from various genes and mediate constitutive and induced expression of respective genes.* However, the complete identification of the nuclear proteins that bind to the AREs from various genes and regulate the expression and induction of the detoxifying enzyme genes is still under investigation.

Band shift assays were used to demonstrate the binding of nuclear protein complexes to the hARE and rARE from different mammalian carcinoma cells (Figs. 18.9A and 18.9B)[66–68,71,73] The supershift assays revealed that Jun-D, Jun-B, and c-Fos proteins from mouse hepatoma (Hepa-1) cells bind to the hARE (Fig. 18.9C).[66–68, 73] The binding of Jun and Fos proteins (AP-1 complex) to the hARE was also reported by using *in vitro* translated proteins in gel mobility shift assays.[71,73] However, in a similar experiment, the *in vitro* translated Jun and Fos proteins (AP-1 complex) did not bind to the rARE.[71] This difference in the binding of Jun and Fos proteins to the hARE and rARE was attributed to the presence of a perfect AP-1 (Jun and Fos binding) site in the hARE but not in the rARE (Fig. 18.8).[71,86] This makes sense because it was found that binding of nuclear protein complexes at the hARE was specifically competed with cold (non-radioactive) hARE but poorly competed by rARE (Fig. 18.9B, left panel). The hARE and rARE both bind to nuclear protein complexes of similar mobilities in various mammalian cells (Fig. 18.9B, left panel). However, the hARE- and rARE-com-

---

*References 66–68, 70, 71, 73, 76, 77, 79, 81, 83–85.

A. hARE and rARE nucleotide sequence:

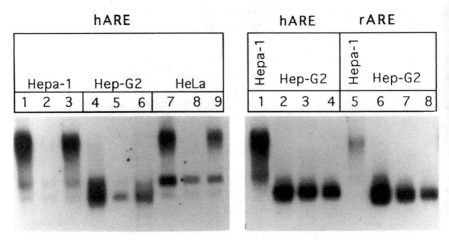

hARE: | GCAGTCA | CAG | TGACTCA | GCA | GAATC

rARE: AGTCT AGAGTCA CAG TGACTTG GCA AAATC

B. hARE-nuclear proteins interaction- Band Shift:

C. hARE-nuclear proteins interaction- Supershift:

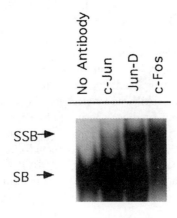

Figure 18.9.

plexes were much smaller in Hep-G2 cells as compared to those of the Hepa-1 cells (Fig. 18.9B, left and right panels). The reasons for the difference in mobility of hARE- and rARE-complexes in Hep-G2 as compared to the Hepa-1 and HeLa cells remain unknown. Because of binding of Jun and Fos proteins to the hARE, questions were raised regarding the role of Jun and Fos proteins in the hARE-mediated regulation of constitutive expression and induction of human NQO_1 gene expression. Questions were also raised concerning whether the ARE-mediated regulation of human and rat NQO_1 genes was different, which appeared unlikely, not only because of highly conserved sequences, but also because both of them regulated the respective orthologous genes expression in a similar fashion.[86] Therefore, in an unpublished report, we studied the regulation of hARE-mediated chloramphenicol acetyl transferase (CAT) gene expression in Jun and Fos overexpressed Hep-G2 cells (Fig. 18.10). The results indicated that all the possible combinations of Jun (c-Jun, Jun-B, Jun-D) and c-Fos failed to up-regulate the hARE-mediated CAT gene expression in Jun and Fos overexpressed Hep-G2 cells. Interestingly, simultaneous overexpression of Jun and Fos repressed the hARE-mediated CAT gene expression. Further experiments clearly indicated that it was the c-Fos and not the Jun proteins, which repressed the hARE-mediated CAT gene expression (Fig. 18.10). These results indicated that the *in vitro* binding of the Jun and Fos to the hARE was due to the presence of a perfect AP-1 binding site and that Jun and Fos proteins are not involved in the up-regulation of hARE-mediated human NQO_1 gene expression and induction. Identification of the nuclear proteins responsible for the hARE and rARE-mediated regulation of NQO_1 gene expression and induction remains unknown. However, we believe that unknown but Jun and Fos related proteins are involved in the *in vivo* regulation

---

Figure 18.9. Band and supershift assays: hARE-nuclear protein interaction. (18.9A) Nucleotide sequences of hARE and rARE used for band and supershift assays. (18.9B) Human and rat NQO_1 gene AREs (hARE and rARE) were end labeled with γP^{32}-ATP and incubated with nuclear extracts from mouse hepatoma (Hepa-1), human hepatoblastoma (Hep-G2), and human cervical carcinoma (HeLa) cells. The incubation mixtures were separated on 5% nondenaturing polyacrylamide gel, dried under vacuum and autoradiographed. Here 100 ng of cold (nonradioactive) hARE and rARE (rARE) were used as competitors. Left panel: lanes 1, 4, and 7, hARE + nuclear extract; lanes 2, 5, and 8, competition with cold hARE; lanes 3, 6, and 9, competition with cold rARE. Right panel: lane 1, hARE + Hepa-1 nuclear extract; lane 2, hARE + Hep-G2 nuclear extract; lanes 3 and 4, competetion with 50 and 100 ng of cold hARE, respectively; lane 5, rARE + Hepa-1 nuclear extract; lane 6, rARE + Hep-G2 nuclear extract; lanes 7 and 8, competetion with 50 and 100 ng of cold rARE. Only shifted bands are observed. (18.9C) The nuclear extracts from Hepa-1 cells were incubated with preimmune serum (no antibodies) and antibodies for c-Jun, Jun-D, and c-Fos before performing band shift with end-labeled hARE. SB, shifted band; SSB, supershifted band.

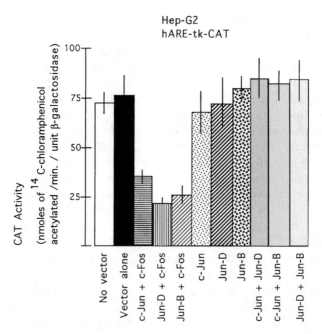

Figure 18.10. Effect of overexpression of Jun and Fos on hARE-mediated CAT gene expression. Hep-G2 cells were cotransfected with the control plasmid RSV-β-galactosidase, reporter plasmid hARE-tk-CAT, and expression plasmid Incx alone or Incx containing Jun (c-Jun, Jun-D, Jun-B) and Fos in combinations as shown. The transfected cells were analyzed for the β-galactosidase normalized CAT activities.

of hARE-mediated constitutive expression and induction of NQO₁ and NQO₂ genes.

Multiple factors including c-Jun, c-Fos, and novel transcription factors have been shown to bind to the ARE enhancer element of the GST-P gene promoter.[83] Interestingly, these novel factors were shown to be present not only in the differentiated HeLa cells but also in the nondifferentiated F9 cells, which lack the expression of c-Jun and c-Fos but do express Jun-D and Jun-B.[83] The ARE mediated CAT activity in F9 cells were thought to be because of novel transcription factors and not because of Jun-Fos heterodimer (AP-1 complex). More recently, Jun and Fos (AP-1 complex) proteins were shown to bind to the GST P gene ARE (GPE I) and modulate the expression of rat GST P gene.[87] To the contrary, Suzuki et al.[84] reported lack of correlated expression between the GST P and c-Jun and c-Fos in rat tissues and preneoplastic hepatic foci.

Friling et al.[79] demonstrated that *in vitro* translated c-Jun and c-Fos (AP-1 complex) bind to the mouse GST-Ya gene ARE. In addition, they have shown activation of CAT gene expression on cotransfection of GST Ya gene ARE-CAT with c-Jun and c-Fos into mouse embryonic F9 cells. In contrast to the work of

Friling et al.,[79] Nguyen and Pickett[85] used photochemical cross-linking techniques to demonstrate that a heterodimer with subunit molecular mass of approximately 28,000 and 45,000 Da bind to the ARE.[85] The same authors showed that *in vitro* translated c-Jun and c-Fos do not bind to the rat GST gene ARE and that TRE (a single AP-1 element) from human collagenase gene did not compete for binding of the nuclear proteins to the ARE.[77] These observations led them to conclude that transacting factors that bind to the ARE is not a Jun-Fos heterodimer. Recently, Yoshioka et al.[70] have reported that c-Jun protein along with Fra1 (non-AP-1 complex) bind to the mouse GST Ya gene ARE and mediate its induction by *t*-BHQ.

In summary, Jun, Fos, Fra, and unknown proteins bind to the AREs of various genes in *in vitro* band shift assays. However, significant differences were observed in the relative binding of Jun and Fos proteins with various AREs. These differences may presumably be because of several reasons that include, the presence of perfect AP-1 (human NQO_1 gene ARE) and imperfect AP-1 (rat NQO_1, mouse GST Ya, rat GST Ya, rat GST P, and γ-GCS genes AREs) elements; spacing and orientation of the AP-1/AP-1-like elements within various gene AREs; differences in the procedures of making nuclear extracts and the gel shift assays. Further studies are required to answer several interesting questions that include complete identification of the nuclear proteins in the hARE-nuclear protein complexes; identification of the nuclear proteins that bind to the ARE within the cells and mediate *in vivo* regulation of expression and induction of detoxifying enzyme genes and if similar or different nuclear proteins bind to the various genes AREs and mediate the expression and induction of the detoxifying enzymes genes.

The methylation interference and protection assays have suggested that the transcription factor(s) bind in the major groove and involve contact with the CpG dinucleotide and the G residue within the 'TGAC' tetramer on the coding strand of the rat GST-Ya gene ARE.[85]

18.4.4.1.3 Mechanism of Signal Transduction

The studies also raise important questions regarding the mechanism of signal transduction from the xenobiotics and antioxidants to the ARE resulting in increased transcription of the NQO_1, NQO_2, and other detoxifying enzyme genes. These include the mechanism of signal transduction from the antioxidants and xenobiotics to the nuclear proteins that bind to the ARE and regulates the transcription and induction of the various detoxifying enzyme genes; similar or different mechanisms with various classes of structurally different compounds; and differences/similarities in the mechanisms among the AREs from various genes. Xenobiotics and drugs, antioxidants, hydrogen peroxide, UV light, and heavy metals, all induce the expression of the detoxifying enzyme genes. These are structurally different, but have one characteristic in common that they are capable of generating electrophiles and/or reactive oxygen species (Fig. 18.11).[88,89] The reactive

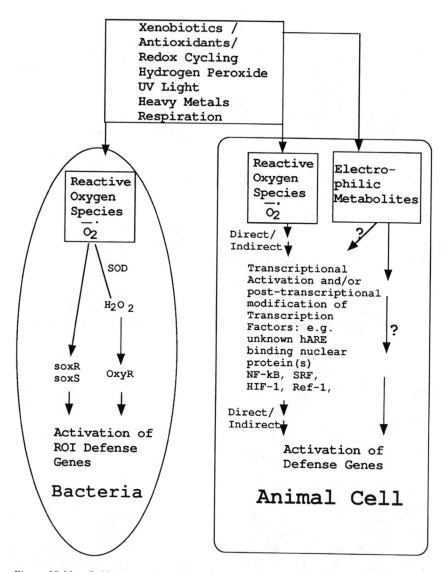

Figure 18.11. Oxidative regulon in bacterial and animal cells. SOD, superoxide dismutase; Sox RS and Oxy R are bacterial transcription factors that activate more than a dozen genes encoding oxidative stress reducing (defensive) proteins. Unknown hARE binding nuclear proteins, NF-kB, AP-1 (Jun and Fos), SRF (serum response factor), and HIF-1 (hypoxia factor) are mammalian transcription factors that regulate the expression of several genes in immune response and antagonistic signals.

oxygen has been implicated in the induction of a number of transcription factors in bacteria, yeast, and mammalian cells (Fig. 18.11).[90] These include, Oxy R and Sox RS, which activate more than a dozen defense genes in the bacteria (Fig. 18.11)[91–93]; yAP-1 and yAP-2 in the yeast, which regulate the expression of oxidative stress response genes superoxide dismutase, glutathione reductase, and glucose 6-phosphate dehydrogenase,[94,95] and Mac-1, which activates catalase gene expression[96]; and NF-κB and AP-1 in higher eukaryotic cells which regulate the expression of several genes in immune response and antagonistic signals.[97] Two other eukaryotic transcription factors, serum response factor (SRF) and hypoxia response factor HIF-1 have also been suggested to be altered by reactive oxygen.[98] Recently, it has been reported that the quinone-mediated generation of hydroxyl radicals induces the ARE mediated expression of the mouse glutathione S-transferase gene.[99] This correlation remains unknown in the case of AREs from other genes including NQO_1 and NQO_2 genes. However, hydrogen peroxide (H_2O_2) has been shown to induce the ARE mediated expression of rat GST Ya, rat NQO_1, and human NQO_1 genes[68,71,76] and may be an important intermediate in the induction of these genes by the antioxidants and xenobiotics. The role of electrophiles, if any, in the ARE mediated expression and induction of detoxifying enzyme genes remains unknown.

Alterations in the transcription and expression of Jun, Fos and Fra proteins in response to the antioxidants and xenobiotics have been reported.[70,100] It has been shown that the intracellular glutathione levels and the induced expression of Jun and Fos proteins mediate the expression and induction of mouse GST-Ya gene.[101] In contrast, Yoshioka et al.[70] reported that the c-Fos expression is not induced by *t*-BHQ and that the increased transcription and expression of Jun and Fra proteins mediate the mouse GST Ya gene expression. The involvement of protein kinase C in the induction mechanism mediated by the mouse GST-Ya subunit gene ARE was ruled out.[79] The involvement of protein kinase C and tyrosine kinases in the transcriptional activation of the hARE mediated NQO_1 gene expression by the bifunctional and monofunctional inducers had also been ruled out.[68] However, it is our opinion that the involvement of protein kinase C, tyrosine kinases, and other kinases should be further studied after the complete identification of the proteins that mediate the ARE regulated detoxifying genes expression. Li and Jaiswal[68] have also reported the β-NF induced sulfhydryl modification of the nuclear proteins that bind to the ARE and increase the expression of the NQO_1 gene. This is interesting because nuclear protein Ref-1, a major redox protein in HeLa cells regulate the AP-1 (c-Jun and c-Fos) proteins binding at the AP-1 site by reducing the cysteine in the DNA binding domain of Jun and Fos proteins.[102–104] The cysteine of c-Jun and c-Fos proteins that is regulated by Ref-1 is highly conserved in the DNA binding domains of other Jun- and Fos-related proteins including Jun-D, Jun-B, Fos-B, Fra 1, and Fra2. It may be possible that unknown nuclear proteins that regulate hARE-mediated *in vivo* expression and induction of NQO_1 gene also contain cysteine in their DNA binding domains and regulated by Ref-1. However, it is not clear if Ref-1 is activated by the antioxidants and xenobiotics or if it plays a role in the ARE mediated detoxifying enzyme genes expression. In summary, the xenobiotics and antioxidants undergo metabolism to generate metabolites (electrophiles) and reactive oxygen species that may increase the transcription/expression of the nuclear proteins and/or catalyze direct/indirect modification of the nuclear proteins that bind to the ARE (Fig. 18.11). The increased expression and/or modification of the nuclear proteins result in the increased binding of these proteins to the ARE and subsequently the increased transcription of NQO_1 and other detoxifying enzyme genes. The indirect modification of the proteins may or may not involve the redox factor Ref-1 and remains to be determined.

18.4.4.2 XRE-Mediated Expression and Induction

Deletion mutagenesis in the human NQO_1 gene promoter also revealed that TCDD induction of the human NQO_1 gene expression was mediated by a DNA segment between the region −740 to −365.[6] This fragment of the NQO_1 gene promoter contained a single copy of the hARE and one copy of the xenobiotic response element (XRE) (Fig. 18.7). The XRE was originally identified in the 5′ flanking region of the cytochrome P-450 1A1 (CYP1A1) gene and is known to mediate the Ah-receptor-TCDD-mediated induction of the CYP1A1 gene expression.[105] Recently, it has been reported that TCDD induction of the rat NQO_1 gene expression is mediated by XREs and does not involve AREs.[65] However, several reports[106,107] suggest otherwise. In these studies, TCDD is suggested to induce Jun and Fos, which may or may not be effective in the induction of the NQO_1 and GST genes by ARE, and remains to be investigated. There is clearly a need to determine the mechanism of TCDD induction of NQO_1 gene expression. These studies are significant because they should reveal information on alternate (ARE/oxidative stress mediated) pathways of mechanism of the TCDD induction of detoxifying enzyme genes including NQO_1 and NQO_2 genes.

18.4.4.3 AP-2-Mediated Expression

In addition to the ARE and XRE, AP-2-mediated expression and cAMP-induced expression of the NQO_1 gene has also been reported.[108] The AP-2-mediated expression of CAT gene and the binding of nuclear proteins to the AP-2 element were observed in HeLa (AP-2-positive) cells but not in Hep-G2 (AP-2-deficient) cells, indicating the involvement of AP-2 transcription factor in the regulation of NQO_1 gene expression. The involvement of AP-2 in the NQO_1 gene regulation was confirmed by the upregulation of AP-2-mediated CAT gene expression in AP-2-overexpressed Hep-G2 cells (Fig. 18.12). The nucleotide sequence analysis of human NQO_2 gene promoter did not reveal the presence of a AP-2 binding site.[7]

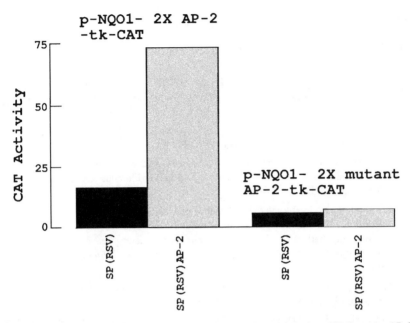

Figure 18.12. AP-2-mediated regulation of NQO₁ gene expression. NQO₁ gene AP-2-tk-CAT and mutant AP-2-tk-CAT were cotransfected with the expression plasmid SP(RSV) alone or the expression vector containing human AP-2 cDNA SP(RSV)AP-2 in separate experiments. RSV-β-galactosidase plasmid was included in each case as the control of transfection efficiency. The transfected cells were analyzed for β-galactosidase and AP-2-mediated CAT activities. CAT activities are expressed as pmoles of ¹⁴C-chloramphenicol acetylated per minute per unit of β-galactosidase activity.

18.4.4.4 *Other* cis- *and* trans-*Elements*

Additional *cis*-elements including a unique basal element (Jaiswal, 1996) and a NF-κB binding site in the upstream region of the NQO₁ gene have also been identified.[109] The unique basal element, although much weaker than hARE, does contribute to the basal expression of the NQO₁ gene expression. At this time, however, it is not clear if NF-κB binding site plays any role in the regulation of NQO₁ gene expression. Additional elements that include SP1 element was also identified in the 5′ flanking region of the human NQO₂ gene by sequence comparison.[7] However, the role of transcription factor SP1 in the regulation of NQO₂ gene expression remains unknown.

Acknowledgments

This work was supported by NIH grants GM 47466 and ES 07943 and Tobacco Council grant 3176A. We thank our colleagues for helpful discussions.

References

1. Riley, R. J., and P. Workman. 1992. DT-diaphorase and cancer chemotherapy. *Biochem. Pharmacol.* **43**:1657–1669.

2. Joseph, P., T. Xie, Y. Xu, and A. K. Jaiswal. 1994. NAD(P)H:quinone oxidoreductase₁ (DT diaphorase): expression, regulation, and role in cancer. *Oncol. Res.* **6**:525–532.

3. Edwards, Y. H., J. Potter, and D. A. Hopkinson. 1980. Human FAD-dependent NAD(P)H diaphorase. *Biochem. J.* **187**:429–436.

4. Jaiswal, A. K., O. W. McBride, M. Adesnik, and D. W. Nebert. 1988. Human dioxin-inducible cytosolic NAD(P)H:Quinone oxidoreductase. *J. Biol. Chem.* **263**:13572–13578.

5. Jaiswal, A. K., P. Burnett, M. Adesnik, and O. W. McBride. 1990. Nucleotide and deduced amino acid sequence of a human cDNA (NQO₂) corresponding to a second member of the NAD(P)H:quinone oxidoreductase gene family. Extensive polymorphism at the NQO₂ gene locus on chromosome 6. *Biochemistry* **29**:1899–1906.

6. Jaiswal, A. K. 1991. Human NAD(P)H:quinone oxidoreductase gene structure and induction by dioxin. *Biochemistry* **30**:10647–10653.

7. Jaiswal, A. K., 1994. Human NAD(P)H:quinone oxidoreductase₂: gene structure, activity, and tissue-specific expression. *J. Biol. Chem.* **269**:14502–14508.

8. Williams, J. B., A. Y. H. Lu, R. G. Cameron, and C. B. Pickett. 1986. Rat liver NAD(P)H:quinone reductase. *J. Biol. Chem.* **261**:5524–5528.

9. Robertson, J. A., H. C. Chen, and D. W. Nebert. 1986. NAD(P)H:menadione oxidoreductase: novel purification of enzyme. cDNA and complete amino acid sequence and gene regulation. *J. Biol. Chem.* **261**:15794–15799.

10. Vasiliou, V., M. J. Theurer, A. Puga, S. F. Reuter, and D. W. Nebert. 1994. Mouse dioxin-inducible NAD(P)H:menadione oxidoreductase: NM01 cDNA sequence and genetic differences in mRNA(−) levels. *Pharmacogenetics* **4**:341–348.

11. Bayney, R. M., M. R. Morton, L. V. Favreau, and C. B. Pickett. 1989. Rat liver NAD(P)H:quinone reductase: regulation of quinone reductase gene expression by planar aromatic compounds and determination of the exon structure of the quinone reductase structural gene. *J. Biol. Chem.* **264**:21793–21797.

12. Ernster, L. 1967. DT-diaphorase. *Methods Enzymol.* **10**:309–317.

13. Amzel, L. M., S. H. Bryant, H. J. Prochaska, and P. Talalay. 1986. Preliminary crystallographic X-ray data for NAD(P)H:quinone reductase from mouse liver. *J. Biol. Chem.* **261**:1379.

14. Shaw, P. M., A. Reiss, M. Adesnik, D. W. Nebert, J. Schembri, and A. K. Jaiswal. 1991. The human dioxin-inducible NAD(P)H:quinone oxidoreductase cDNA-encoded protein expressed in COS-1 cells is identical to diaphorase 4. *Eur. J. Biochem.* **195**:171–176.

15. Li, R., M. A. Bianchet, P. Talalay, and L. M. Amzel. 1995. The three-dimensional structure of NAD(P)H:quinone reductase, a flavoprotein involved in cancer chemo-

protection and chemotherapy: mechanism of the two-electron reduction. *Proc. Natl. Acad. Sci USA* **92**:8846–8850.

16. Joseph, P., and A. K. Jaiswal. 1994. NAD(P)H:quinone oxidoreductase₁ (DT diaphorase) specifically prevents the formation of benzo(α)pyrene quinone-DNA adducts generated by cytochrome P4501A1 and P450 reductase. *Proc. Natl. Acad. Sci. USA* **91**:8413–8417.

17. Belinsky, M., and A. K. Jaiswal. 1993. NAD(P)H:quinone oxidoreductase₁ (DT diaphorase) expression in normal and tumor tissues. *Cancer Metast. Rev.* **12**:103–117.

18. Monks, T. J., R. P. Hanzlik, G. M. Cohen, D. Ross and D. G. Graham. 1992. Quinone chemistry and toxicity. *Toxicol. Appl. Pharmacol.* **112**:2–16.

19. Lind, C., P. Hochstein, and L. Ernster. 1982. DT-diaphorase: properties, reaction mechanism, metabolic function. A progress report. *In* T. E. King, H. S. Mason, and M. Morrison (eds.), *Oxidases and Related Redox Systems*. Pergamon Press, Oxford. 321–347.

20. Thor, H., M. T. Smith, P. Hartzell, G. Bellomo, S. Jewell, and S. Orrenius. 1982. The metabolism of menadione (2-methyl-1,4-naphthoquinone) by isolated hepatocytes. *J. Biol. Chem.* **257**:12419–12425.

21. Ernster, L. 1987. DT-diaphorase: a historical review. *Chem. Scripta* **27A**:1–13.

22. Prochska, H. J., P. Talalay, and H. Sies. 1987. Direct protective effect of NAD(P)H: quinone reductase against menadione-induced chemiluminescence of postmitochondrial fractions of mouse liver. *J. Biol. Chem.* **262**:1931–1934.

23. Joseph, P., Y. Xu, and A. K. Jaiswal. 1995. Non-enzymatic and enzymatic activation of mitomycin C: identification of a unique cytosolic activity. *Int. J. Cancer* **65**:263–271.

24. Talalay, P. 1989. Mechanisms of induction of enzymes that protect against chemical carcinogenesis. *Adv. Enzyme Reg.* **28**:149–159.

25. Chesis, P. L., D. E. Levin, M. T. Smith, L. Ernster, and B. N. Ames. 1984. Mutagenicity of quinones: pathways of metabolic activation and detoxification. *Proc. Natl. Acad. Sci. USA* **81**:1696–1700.

26. Lind, C. 1985. Formation of benzo(a)pyrene-3,6-quinol mono- and diglucuronides in rat liver microsomes. *Arch. Biochem. Biophy.* **280**:226–235.

27. De Flora, A. 1985. Prominent role of DT-diaphorase as a cellular mechanism reducing chromium (VI) and reverting its mutagenicity. *Cancer Res.* **45**:3188–3196.

28. Joseph, P., and A. K. Jaiswal. 1996. Mutagenicity of benzo(α)pyrene quinones: role of specific reductases. *Abstract SOT*.

29. Talalay, P., M. J. De Long, and H. J. Prochaska. 1989. Molecular mechanisms in protection against carcinogenesis, pp. 197–182. *In* J. G. Cory and A. Szentivanyi (eds.), *Cancer Biology*, Plenum, New York. 197–216.

30. Talalay, P., J. W. Fahey, W. D. Holtzclaw, T. Prestera and Y. Zhang. 1995. Chemoprotection against cancer by phase 2 enzyme induction. *Toxicol. Lett.* **82–83**:173–179.

31. Wattenberg, L. W. 1972. Inhibition of carcinogenic and toxic effects of polycyclic aromatic hydrocarbons by phenolic antioxidants and ethoxyquin. *J Natl. Cancer Inst.* **48**:1425–1430.

32. Zhang, Y., T. W. Kensler, C. Cho, G. H. Posner, and P. Talalay. 1994. Anticarcinogenic activities of sulforaphane and structurally related synthetic norbornyl isothiocyanates. *Proc. Natl. Acad. Sci. USA* **91**:3147–3150.

33. Jaiswal, A. K. 1994. Antioxidant response element. *Biochem. Pharmacol.* **48**:439–444.

34. Nebert, D. W. 1994. Drug-metabolizing enzymes in ligand-modulated transcription. *Biochem. Pharmacol.* **47**:25–37.

35. Pickett, C. B., and A. Y. H. Lu. 1989. Glutathione S-transferases: gene structure, regulation and biological function. *Annu. Rev. Biochem.* **58**:743–764.

36. Tsuchida, S., and K. Sato. 1992. Glutathione transferases and cancer. *Crit. Rev. Biochem. Mol. Biol.* **27**:337–384.

37. Rushmore, T. H., and C. B. Pickett. 1993. Glutathione S-transferases, structure, regulation and therapeutic implications. *J. Biol. Chem.* **268**:11475–11478.

38. Tephly, T., and B. Burchell. 1990. UDP-glucuronosyl transferases: a family of detoxifying enzymes. *Trends Pharmacol.* **11**:276–279.

39. Kashfi, K., E. K. Yang, J. R. Chowdhury, N. R. Chowdhary and A. J. Dannenberg. 1994. Regulation of uridine diphosphate glucuronosyltransferase expression by phenolic antioxidants. *Cancer Res.* **54**:5856–5859.

40. Oesch, F., I. Gath, T. Igarashi, H. R. Glatt, and H. Thomas. 1991. Epoxide hydolases, pp. 447–461. *In* E. Arinc, J. B. Schenkman, and E. Hodgson (eds.), *Molecular Aspects of Monooxygenases and Bioactivation of Toxic Compounds.* Plenum Press, New York.

41. Meister, A. and M. E. Anderson. Glutathione. 1983. *Annu. Rev. Biochem.* **52**:711–760.

42. Talcott, R. E., M. Rosenblum, and V. A. Levin. 1983. Possible role of DT-diaphorase in the bioactivation of anti-tumor quinones. *Biochem. Biophys. Res. Commun.* **11**:346–351.

43. De Flora, S., C. Bernnicelli, A. Camoirano, D. Serra, and P. Hochstein. 1988. Influence of DT diaphorase on the mutagenicity of organic and inorganic compounds. *Carcinogenesis* **9**:611–617.

44. Cadenas, E. 1995. Antioxidant and pro-oxidant functions of DT diaphorase in quinone metabolism. *Biochem. Pharmacol.* **49**:127–140.

45. Workman, P. 1994. Enzyme-directed bioreductive drug development revisited: a commentary on recent progress and future prospects with emphasis on quinone anticancer agents and quinone metabolizing enzymes, particularly DT-diaphorase. *Oncol. Res.* **6**:461–475.

46. Ross, D., H. Beall, R. D. Traver, D. Siegel, R. M. Phillips, and N. W. Gibson. 1994. Bioactivation of quinones by DT-diaphorase, molecular, biochemical, and chemical studies. *Oncol. Res.* **6**:493–500.

47. Verweij, J., and H. H. Pinedo. 1990. Mitomycin C, pp. 63–73. *In* H. M. Pinedo, B. A. Cabner, and D. L. Longo (eds.), *Cancer Chemotherapy and Biological Modifyers.* Elsevier Science Publishers, Amsterdam.

48. Iyer, V. N., and W. Szybalski. 1963. A molecular mechanism of mitomycin C action: linking of complementary DNA strands. *Proc. Natl. Acad. Sci. USA* **50**:355–362.

49. Iyer, V. N., and Szybalski W. 1964. Mitomycins and porfiromycin: chemical mechanism of activation and cross-linking of DNA. *Science* **145**:55–58.

50. Robertson, N., I. J. Stratford, S. Houlbrook, J. Carmichael, and G. E. Adams. 1992. The sensitivity of human tumor cells to quinone bioreductive drugs: what role for DT-diaphorase? *Biochem. Pharmacol.* **44**:409–412.

51. Schlager, J. J., and G. Powis. 1988. Mitomycin C is not metabolized by but is an inhibitor of human kidney NAD(P)H:(quinone acceptor) oxidoreductase, *Cancer Chemother. Pharmacol.* **22**:126–130.

52. Plumb, J. A., and P. Workman. 1994. Unusually marked hypoxic sensitization to indoloquinone EO9 and mitomycin C in a human colon-tumour cell line that lacks DT diaphorase activity. *Int. J. Cancer* **56**:134–139.

53. Ross, D., D. Siegel, H. Beall, A. S. Prakash, T. Mulcahy, and N. Gibson. 1993. DT diaphorase in activation and detoxification of quinones. *Cancer Metast. Rev.* **12**:83–101.

54. Siegel, D., H. Beall, C. Senekowitsch, M. Kasai, H. Arai, N. W. Gibson, and D. Ross. 1992. Bioreductive activation of mitomycin C by DT-diaphorase. *Biochemistry* **31**:7879–7885.

55. Ernster, L., and F. Navazio. 1958. Soluble diaphorases in animal tissues. *Acta Chem. Scand.* **12**:595, 1958.

56. Martin, L. F., S. D. Patrick, and R. Wallin. 1987. DT-diaphorase in morbidly obese patients. *Cancer Lett.* **36**:341–347.

57. Philips, R. M., P. B. Hulbert, M. C. Bibbey, N. R. Sleigh, and J. A. Double. 1992. *In vitro* activity of the novel indoloquinone EO9 and the influence of pH on cytotoxicity. *Br. J. Cancer* **65**:359–364.

58. Smitskamp-Wilms, E., G. J. Peters, H. M. Pinedo, J. VanArk-Otte, and G. Giaccone. 1994. Chemosensitivity to the indoloquinone EO9 is correct with DT-diaphorase activity and its gene expression. *Biochem. Pharmacol.* **47**:1325–1332.

59. Plumb, J. A., M. Gerritsen, and P. Workman. 1994. DT-diaphorase protects cells from the hypoxic cytotoxicity of indoloquinone EO9. *Br. J. Cancer* **70**:1136–1143.

60. Robertson, N., A. Haigh, G. E. Adams, and I. J. Stratford. 1994. Factors affecting the sensitivity to EO9 in rodent and human tumor cells *in vitro*: DT-diaphorase and hypoxia. *Eur. J. Cancer* **30A**:1013–1019.

61. Koudstaal, J., B. Makkins, and S. H. Overdiep. 1975. Enzyme histochemical pattern in human tumors-II. Oxidoreductases in carcinoma of colon and breast. *Eur. J. Cancer* **11**:111–115.

62. Schor, N. A., and C. J. Cornelisse. 1983. Biochemical and quantitative histochemical study of reduced pyridine nucleotide dehydrogenation by human colon carcinomas. *Cancer Res.* **43**:4850–4855.

63. Schlager, J. J., and G. Powis. 1990. Cytosolic NAD(P)H:(quinone-acceptor) oxido-reductase in human normal and tumor tissue: effects of cigarette smoking and alcohol. *Int. J. Cancer* **45**:403–409.

64. Cresteil, T., and A. K. Jaiswal. 1991. High levels of expression of the NAD(P)H: quinone oxidoreductase (NQO₁) gene in tumor cells compared to normal cells of the same origin. *Biochem. Pharmacol.* **42**:1021–1027.

65. Favreau, L. V., and C. B. Pickett. 1991. Transcriptional regulation of the rat NAD(P)H:quinone reductase gene. Identification of regulatory elements controlling basal level expression and inducible expression by planar aromatic compounds and phenolic antioxidants. *J. Biol. Chem.* **266**:4556–4561.

66. Li, Y., and A. K. Jaiswal. 1992. Regulation of human NAD(P)H:quinone oxidore-ductase gene: role of AP1 binding site contained within human antioxidant response element. *J. Biol. Chem.* **267**:15097–15104.

67. Li, Y., and A. K. Jaiswal. 1992. Identification of Jun-B as third member in human antioxidant response element-nuclear protein complex. *Biochem. Biophys. Res. Commun.* **188**:992–996.

68. Li, Y., and A. K. Jaiswal. 1994. Human antioxidant response element mediated regulation of type 1 NAD(P)H:quinone oxidoreductase gene expression: effect of sulfhydryl modifying agents. *Eur. J. Biochem.* **226**:31–39.

69. Prestera, T., W. D. Holtzclaw, Y. Zhang, and P. Talalay. 1993. Chemical and molecular regulation of enzymes that detoxify carcinogens. *Proc. Natl. Acad. Sci. USA* **90**:2965–2969.

70. Yoshioka, K., T. Deng, M. Cavigelli, and M. Karin. 1995. Antitumor promotion by phenolic antioxidants: inhibition of AP-1 activity through induction of Fra expression. *Proc. Natl. Acad. Sci. USA* **92**:4972–4976.

71. Favreau, L. V., and C. B. Pickett. 1993. Transcriptional regulation of the rat NAD(P)H:quinone reductase gene. Characterization of a DNA-protein interaction at the antioxidant responsive element and induction by 12-O-tetradecanoylphorbol 13-acetate. *J. Biol. Chem.* **268**:19875 –19881.

72. Mulcahy, R. T., and J. J. Gipp. 1995. Identification of a putative antioxidant response element in the 5′ flanking region of the human γ-glutamylcysteine synthetase heavy subunit gene. *Biochem. Biophys. Res. Communun.* **209**:227–233.

73. Xie, T., M. Belinsky, Y. Xu, and A. K. Jaiswal. 1995. ARE- and TRE-mediated regulation of gene expression: response to xenobiotics and antioxidants. *J. Biol. Chem.* **270**:6894–6900.

74. Rushmore, T. H., R. G. King, K. E. Paulson, and C. B. Pickett. 1990. Regulation of glutathione S- transferase Ya subunit gene expression: identification of a unique xenobiotic-responsive element controlling inducible expression by planar aromatic compounds. *Proc. Natl. Acad. Sci. USA* **87**:3826–3830.

75. Rushmore, T. H., and C. B. Pickett. 1990. Transcriptional regulation of the rat glutathione S-transferase Ya subunit gene: characterization of a xenobiotic-respon-sive element controlling inducible expression by phenolic antioxidants. *J. Biol. Chem.* **265**:14648–14653.

76. Rushmore, T. H., M. R. Morton, and C. B. Pickett. 1991. The antioxidant responsive element. Activation by oxidative stress and identification of the DNA consensus sequence required for functional activity. *J. Biol. Chem.* **266**:11632–11639.

77. Nguyen, T., T. H. Rushmore, and C. B. Pickett. 1994. Transcriptional regulation of a rat liver glutathione S-transferase Ya subunit gene. *J. Biol. Chem.* **269**:13656–13662.

78. Friling, R. S., A. Bensimon, Y. Tichauer, and V. Daniel. 1990. Xenobiotic-inducible expression of murine glutathione S-transferase Ya subunit gene is controlled by an electrophile-responsive element. *Proc. Natl. Acad. Sci. USA* **87**:6258–6262.

79. Friling, R. S., S. Bergelson, and V. Daniel. 1992. Two adjacent AP-1-like binding sites form the electrophile responsive element of the murine glutathione S-transferase Ya subunit gene. *Proc. Natl. Acad. Sci.* **89**:668–672.

80. Sakai, M., A. Okuda, and M. Muramatsu. 1988. Multiple regulatory elements and phorbol 12-O-tetradecanoate 13-acetate responsiveness of the rat placental glutathione transferase gene. *Proc. Natl. Acad. Sci. USA* **85**:9456–9460.

81. Okuda, A., M. Imagawa, Y. Maeda, M. Sakai, and M. Muramatsu. 1990. Functional co-operativity between two TPA responsive elements in undifferentiated F9 embryonic stem cells. *EMBO J.* **9**:1131–1135.

82. Prestera, T., and P. Talalay. 1995. Electrophile and antioxidant regulation of enzyme that detoxify carcinogens. *Proc. Natl. Acad. Sci. USA* **92**:8965–8969.

83. Diccianni, M. B., M. Imagawa, and M. Muramatsu. 1992. The dyad palindromic glutathione transferase P enhancer binds multiple factors including AP1. *Nucleic Acids Res.* **20**:5153–5158.

84. Suzuki, S., K. Satoh, H. Nakano, I. Hatayama, K. Sato, and S. Tsuchida. 1995. Lack of correlated expression between the glutathione S-transferase P-form and the oncogene products c-Jun and c-Fos in rat tissues and preneoplastic hepatic foci. *Carcinogenesis* **16**:567–571.

85. Nguyen, T., and C. B. Pickett. 1992. Regulation of rat glutathione-S-transferase Ya subunit gene expression: DNA-protein interaction at the antioxidant response element. *J. Biol. Chem.* **267**:13535–13539.

86. Jaiswal, A. K. In press. Antioxidant responses and regulation of gene expression. *In* D. Massaro (ed.). *Oxygen, Gene Expression.* Marcel Dekker, New York.

87. Oridate, N., S. Nishi, Y. Inuyama, and M. Sakai. 1994. Jun and Fos related gene products bind to and modulate the GPE I, a strong enhancer element of the rat glutathione transferase P gene. *Biochem. Biophys. Acta* **1219**:499–504.

88. Halliwell, B., and J. M. C. Gutteridge. 1989. *Free Radicals in Biology and Medicine.* Clarendon Press, Oxford.

89. De Long, M. J., A. B. Santamaria, and P. Talalay. 1987. Role of cytochrome P_1-450 in the induction of NAD(P)H:quinone reductase in a murine hepatoma cell line and its mutants. *Carcinogenesis* **8**:1549–1553.

90. Pahl, H. L., and P. A. Baeuerle. 1994. Oxygen and the control of gene expression. *Bioessays* **16**:497–502.

91. Demple, B., and C. F. Amabile-Cuevas. 1991. Redox redux: the control of oxidative stress responses. *Cell* **67**:837–839.

92. Li, Z., and B. Demple. 1994. Sox S, an activator of superoxide stress genes in *Escherichia coli*: purification and interaction with DNA. *J. Biol. Chem.* **269**:18371–18377.

93. Toledano, M. B., I. Kullik, F. Trinh, P. T. Baird, T. D. Schneider, and G. Storz. 1994. Redox-dependent shift of OxyR-DNA contacts along an extended DNA-binding site: a mechanism for differential promoter selection. *Cell* **78**:887–909.

94. Zitomer, R. S., and C. V. Lowry. 1992. Regulation of gene expression by oxygen in *Saccharomyces cerevisiae*. *Microbiol. Rev.* **56**:1–11.

95. Kuge, S., and N. Jones. 1994. YAP1 dependent activation of TRX2 is essential for the response of *Saccharomyces cerevisiae* to oxidative stress by hydroperoxides. *EMBO J.* **13**:655–664.

96. Jungmann, H., H. A. Reins, J. Lee, A. Romeo, R. Hassett, D. Kosman, and S. Jentsch. 1993. Mac-1, a nuclear regulatory protein related to Cu-dependent transcription factors is involved ion Cu/Fe utilization and stress resistance in yeast. *EMBO J.* **12**:5051–5056.

97. Schreck, R., and P. A. Baeuerle. 1994. Assessing oxygen radicals as mediators in activation of inducible eukaryotic transcription factor NF-κB. *Methods Enzymol.* **234**:151–163.

98. Meyer, M., R. Schreck, and P. Baeuerle. 1993. H₂O₂ and antioxidants have opposite effects on activation of NF-kB and AP-1 in intact cells: AP-1 as secondary antioxidant-responsive factor. *EMBO J.* **12**:2005–2015.

99. Pinkus, R., L. M. Weiner, and V. Daniel. 1995. Role of quinone-mediated generation of hydroxyl radicals in the induction of glutathione S-transferase gene expression. *Biochemistry* **34**:81–88.

100. Bergelson, S., R. Pinkus, and V. Daniel. 1994. Induction of AP1 (Fos/Jun) by chemical agents mediates activation of glutathione S-transferase and quinone reductase gene expression. *Oncogene* **9**:565–571.

101. Bergelson, S., R. Pinkus, and V. Daniel. 1994. Intracellular glutathione levels regulate Fos/Jun induction and activation of glutathione S-transferase gene expression. *Cancer Res.* **54**:36–40.

102. Abate, C., L. Patel, F. J. Rauscher III, and T. Curran. 1990. Redox regulation of Fos and Jun DNA-binding activity *in vitro*. *Science* **249**:1157–1161.

103. Xanthoudakis, S., and T. Curran. 1992. Identification and characterization of Ref-1, a nuclear protein that facilitates AP-1 DNA binding activity. *EMBO J.* **11**:653–665.

104. Xanthoudakis, S., G. Miao, F. Wang, Y. E. Pan, and T. Curran. 1992. Redox activation of Fos-Jun DNA binding activity is mediated by a DNA repair enzyme. *EMBO J.* **11**:3323–3335.

105. Whitlock, J. P. 1993. Mechanistic aspects of dioxin action. *Chem. Res. Toxicol.* **6**:754–763.

106. Rao, G. N., B. Lasségue, K. K. Griendling, R. W. Alexander, and B. C. Berk. 1993. Hydrogen peroxide-induced c-fos expression is mediated by arachidonic acid release: role of protein kinase C. *Nucleic Acids Res.* **21**:1259–1263.

107. Puga, A., D. W. Nebert, and F. Carrier. 1992. Dioxin induces expression of c-fos and c-jun proto-oncogenes and a large increase in transcription factor AP1. *DNA Cell Biol.* **11**:269–281.

108. Xie, T., and A. K. Jaiswal. 1996. AP-2-mediated regulation of human NAD(P)H: quinone oxidoreductase$_1$ (NQO$_1$) gene expression. *Biochem. Pharmacol.* **51**:771–778.

109. Yao, K. S., and P. J. Odwyer. 1995. Involvement of NF-kB in the induction of NAD(P)H:quinone oxidoreductase (DT diaphorase) by hypoxia, oltipraz and mitomycin C. *Biochem. Pharmacol.* **49**:275–282.

Index